To Jay

with best wishes

[signature]

Apr. 21, 99

at Beltsville, MD.

Agriculture in China
1949-2030

T.C. Tso
Francis Tuan
Miklos Faust
Editors

IDEALS
5010 Sunnyside Ave, Suite 301
Beltsville, MD 20705, USA

1998

I

Trade names are used in this publication solely for the purpose of providing specific information. The mention of trade names, proprietary products , or specific equipment does not constitute guarantee or warranty by IDEALS Inc., and does not imply approval or the exclusion of other products that may be suitable.

Copyright 1998 by IDEALS Inc.
Library of Congress Catalog Card No: 98-072208
ISBN No. 1-891998-00-5

IDEALS Inc.
5010 Sunnyside Avenue, Suite 301
Beltsville, MD 20705, USA

Manufactured in the United States of America

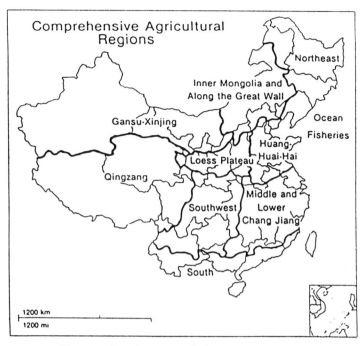

The agricultural regions of China.

UNITS, MEASUREMENTS AND EQUIVALENTS

Officially, and in principle, China uses the metric system of measurements. Such has been adopted throughout this volume. In Chinese practice, however, traditional units, now defined in terms of metric units, are also used. The most distinctive and widespread usage is the mu or mou, a measurement of area which consists of 667 square meters or one-fifteenth of a hectare or one sixth of an acre. Yields per unit area are often expressed as jim per mu. A table of comparative conversion units for areas and weights for the Chinese, Metric and U.S. systems follows:

CHINESE	METRIC	U.S. (ENGLISH)
1 mou (mu)	0.0667 hectare	0.1647 acre
15 mou (mu)	1.0 hectare	2.4711 acres
1 jin (catty)	0.5 kilogram	1.1023 pounds
1 dan (100 jin)	50.0 kilograms	110.23 pounds
1 dun (ton)	1,000.00 kilograms (1 metric ton)	2,204.6 pounds
1 jin/mu	7.5 kilograms/hectare	6.93 pounds/acre

Chronology of Events and Policy Changes Affecting Agriculture

1949   Founding of People's Republic of China.

1952   Land reform completed.

1958   "Great Leap Forward" Campaign: "To accelerate industrialization and agricultural productivity – 20 years in 1 day." Commune system established.

1959-61   Economic crisis and famine, "three bad years" resulting from Great Leap Forward.

1966-76   The Cultural Revolution.

1966   Campaign against "Four Old Traditions"; higher education system, including agriculture, was shut down completely.

1968   Urban youth were sent to countryside "to learn from the peasants."

1969   Intellectuals were sent to countryside for "political education."

1977   Urban youth were allowed to return to cities; colleges and universities were reopened.

1978   Deng Xiaoping launched reforms.

1978-79   Rural reforms began, transforming central planning to household decision-making system. Production responsibility system stimulated farmers' incentive.

1982-83   Commune system abolished.

1991-93   Change of grain policy . Fertilizer subsidy plan to farmers also revised.

1996   Official declaration of promoting agricultural development through science and technology revolution.

Prologue

Nothing more can be attempted than to

establish the

beginning and the direction of an infinitely

long road.

The pretension of any systematic and

definitive completeness

would be, at least, a self illusion. Perfection

can here be

obtained by the individual student only in the

subjective sense

that he communicates everything he has been

able to see.

George Simmel, From Carlos Castenda's
The teachings of Don Juan: A Yaqui way of
knowledge.

Table of Contents

Part IV Toward 2030

Science and technology

V

VII

SUMMARY OF RECOMMENDATIONS

China will play an important role in the world economy in the next century. The Chinese economy can be strong only when Chinese agriculture is strong. The rapid growth of the Chinese economy would not be able to continue without sustained agricultural growth based on continuous development of modern science and technology. This is so, despite the fact that government actions in the 1980's greatly promoted agricultural growth.

Based on the official statistics on population growth in China, it is expected that the population of China will reach a peak of 1.6 billion at Year 2030. Assuming China maintains the present grain consumption level of 400 kg per capita, including food. feed, and industrial use it would need 640 million tons in the year 2030 to meet the total demand. China may not need to produce all this grain; however, China should develop its agriculture to the full capacity in accordance to its natural resources and ecological conditions to support the well-being of its people.

There are many factors which affect China's agriculture, its grain production and ability to achieve agricultural self-sufficiency. These factors include increasing population, decreasing arable land, deterioration of natural resources and environment, lack of adequate investment for science and technology, infrastructure, slow in implementation of policy changes, and most important, the need for dynamic leadership at the highest agricultural level. We are approaching the new century in only 700 days, and agricultural improvement takes time. China must act now before it is too late. The world's agricultural scientific community is ready to cooperate with China for the development of its agriculture.

The following recommendations and suggestions reflect our deep affection for China, and with great concern for the world's food security. This recommendation represents the collective efforts of forty-eight world-renowned authors with, an accumulation of more than 2000 man-years of experience in research, teaching, extension, administration, international development and leadership. They include a Nobel Laureate, a World Food Prize Laureate, university presidents, deans, professors, presidents of international and national learned academic societies. We jointly authored this monograph entitled "Agriculture in China: 1949-2030," with the purpose of examining how China can approach agricultural self-sufficiency. All the authors contributed their best wisdom toward the objective of developing sustained modern agriculture in harmony with the natural environment. Some views may differ from current official Chinese policies and guidelines. We do believe this is the best approach toward the year 2030. Our collective thoughts as presented in this summary, are of great value toward China's agricultural development, and of the world's stability.

We, the authors respect China's concerns regarding agriculture and place development of agricultural science and technology at national top priority. We fully support such a decision, and identify, more precisely, what approaches must be taken to attain this goal. Following are our key recommendations:

Leadership:
Dynamic agricultural leadership is essential to make the best use of human and financial resources. The officers responsible for agriculture, whether in the central government or in the provinces, should be selected based on (1) true understanding and knowledge of agricultural science and technology; (2) the ability to initiate and promote international agricultural cooperation and exchange; and (3) leadership and personality which commands the respect of all segments of the agricultural arena. We strongly recommend that such leadership must understand the whole spectrum of agriculture, and enjoy the respect and cooperation of the national and international agricultural community, and should never be a bureaucratic administrator.

Institutional Change:
Agriculture is a natural and physical science as well as an art. The institutions of Agricultural research, teaching and extension is comprised of three interdependent parts that make up one whole unit. Chinese agricultural institutes are, unfortunately, separated from comprehensive universities following the former Soviet system. Agricultural research, education, and extension systems are independent from each other, inefficient, lack of communication, isolated from each other and from farmers. We recommend that agricultural teaching, research, and extension be united into one system complementary to each other, and to serve the farmers. We further recommend that agricultural college be a part of a comprehensive university, and closely linked to the local community.

Funding:

The Chinese population continues to increase and arable land continues to decrease. In the near future, there will be no more arable land to be reclaimed, and grain increase will be dependent totally on yield increases, by 30-40 percent. Such a large increase in yield will depend on the development of science and technology, traditional and modern science, such as biotechnology. Our estimate is that such an effort would need an increase of at least fivefold funding from the government over the next 10 years. Agricultural research is the duty of the government and should be supported by the government. In recent years, despite the call that agriculture is the nation's top priority, the input of government funding for agricultural-related development, including agriculture, forestry, animal husbandry, fishery, water, rural development and side-business are in fact decreasing every year in relation to the agricultural gross domestic product (GDP). In 1997, it was less than 0.39 percent, and agriculture itself only 0.19 percent of the total agricultural GDP. We strongly recommend that agricultural research alone, including teaching, research, and extension be increased to the level of at least one to two percent of the agricultural GDP.

Policy Issues:

We recommend that tillers should own their land, and the marketing system should be totally relaxed and free to stimulate economic growth and farmers' incentive. Sustainability of land can only be maintained and enhanced when farmers have ownership of their land.

Production Issues:

The success or failure of agricultural production involves numerous variables. Human interference is, however, of the most important and manageable. We, therefore recommend the following: Guarantee the abundant supply, availability and affordability of agricultural input materials, such as seeds (certified improved seeds, including hybrids), fertilizer (various formulations, quantity, timely availability and affordability); plastic films (made from biodegradable materials), strengthening the mechanism of harvesting, drying, storage, transportation, and processing systems, including trading and marketing. In the animal husbandry area, improve the feed conversion ratio and improve mass production, processing technology.

Water
Water is the key to all life, and the water supply in China is insufficient and unevenly distributed. The Northwest area of China occupies 42 percent of the total area, but with only 8 percent of the population. Building a new Northwest China is a must, and it needs water. We strongly recommend that in addition to increasing water-usage efficiency, new water resources must be found. The current plan of moving water from South to North is insufficient. New concepts must be considered and action taken now. For example, water from Qinghai-Tibet sources should be seriously considered. Use those water sources to generate electricity, and use the same electricity to move water to develop the Northwest region. Particularly, we recommend the diversion of water from the Yarlung Zangbo River from Tibet to the Northwest region. Studies show that it is practical and feasible.

Environment:
Reduce environmental pollution is the government policy. For example, China is the world leader in using coal. The pollution effect on health and on the economy is obvious. Acid rain greatly reduces crop production. Other factors, such as overuse of fertilizer, pesticides, and water causes erosion; also pollutes land, water and air. Industrial wastes also are major pollutants. We recommend that this trend of pollution must be reduced and stopped with much stronger determination and much greater effort from all levels of government, and act as soon as possible.

Town and Village Enterprises:
Development of town and village enterprises (TVE's) has to be supported by public and private resources, with special emphasis on Western and Central China. We recommend strengthening its management and technology to ensure efficiency, product quality and marketability, to add benefit to consumers, to TVE shareholders, and to whole society. Avoid TVE-induced pollution and occupancy of arable land. These will greatly increase the export market, create rural wealth and improve the quality of life.

Agricultural Sustainability:
Sustainability is an overused term in China but with little understanding in practice. We must truly recognize the significance and meaning of agricultural sustainability. We strongly recommend that never strive for "high yield, high quality, high efficiency" at the expense of natural resources and damaging the environment. Current methods involve over fertilization, excessive use of pesticides, waste of water, intensive use of

Land which will result in long-term destruction of natural resources. We inherited this earth in good condition, and we should leave this earth in a better condition to generations to come.

Information and Statistics:

<u>We strongly recommend</u> that China should establish a sound, credible, and timely information and statistical system and open up the internet system. All data must be collected by scientific sampling methods instead of by comprehensive reporting. This is essential for national credibility, for scientific communication and exchange, for project planning, and most importantly, for long-term projection and policy decisions.

Editors' foreword

During the past two decades, the People's Republic of China (PRC) became a major power in the world economic community, but the Chinese agricultural sector remained weak and may not be able to support the overall development. With a large share of the world population and small share of natural resources, there is much concern whether China will exhaust the world's grain supply in the next century, particularly when China will have strong purchasing power. There are different projections available as well as many variables. The objective of this monograph is to examine China's resources, agricultural past, current status, major concerns, and to make recommendations for its agricultural future.

The general assumptions for the future in agriculture maybe summarized in the following key points:

• There will be less land, more people, and a need to increase production per unit area. Therefore, agricultural related policies and production practices must be reexamined and improved including land policy, marketing systems, containment of pollutants, protection of the ecosystem, increase biocontrol, intensify recycling plant and animal residues, and better utilization of agricultural products and of soil microbial biomass.

• Recognize the fact that the progress of technology and advancement of scientific knowledge probably will not be as ready and rapid in the future as it was in the past 30 years. The major breakthrough will depend upon advancement of basic science and technology such as the increase of photosynthetic efficiency, broad adaptation of nitrogen fixation by plants, improve production and utilization efficiency, stress resistance through biotechnology, and rapid transfer of science and technology into practice.

• Soil and water conservation alone is not enough. We must improve soil quality, increase water-use efficiency and search for new water resources.

This monograph is a collective contribution by scholars and experts who communicate all what they have been able to see, read, and reason. The authors do not pretend that their views are perfect, as they are limited by current understanding of science and technology, as well as unforeseen human interventions. Each author is responsible for his or her own expressions. As scholars, they express to their best opinions concerning the correct path in achieving food security while preserving natural resources. It is our belief that this communication is not only important for China, but for the world as well.

During the period between 1949 and 1978, the agriculture sector in China showed very little improvement, nor was there any significant change in the living standard in the rural area. Since 1979, China has made considerable progress in the agriculture sector, mainly due to changes in structure and policy, such as abolishing the commune system and reverting farm management decisions to private households, reinstating material incentives, and permitting partial return of free markets and private enterprises. As a result, agricultural production and farm income soared. Livelihood in both rural and urban areas in China became the envy of other centrally-planned economy countries. However, progress made in the agriculture sector is considered significant only when it is measured against the low pre-1978 standard and the minimal base of 1949. In fact, a number of fundamental problems stemmed from the past negligence and misdirections still prominently exist.

More importantly, if China is to achieve its stated goal of moving toward an advanced stage of economic development and targets set for Year 2010, more basic changes in policy and structure must be instituted. Looking toward Year 2030, when the Chinese population is expected to peak at 1.6 billion and grain demand will be of 640 million tons, such changes are urgent and necessary.

Until these existing and emerging problems are properly addressed, they will act and/or interact as serious bottlenecks to progress and create restrictive constraints in China's effort to maintain the current growth in agriculture, reach its goals set for the Year 2010 and beyond, and remain a nearly insurmountable barrier to feed its huge population.

Attempting to present a succinct discussion to cover all the problems that have a direct or indirect bearing on China's future development in agriculture is not an easy task. In China, because of its sheer size in land area, almost one-fourth of the world's population, and a large portion of semi-literate populace, a problem, that is simple in a small country, becomes much more involved and complicated. In particular, recent strong economic growth led to industrialization and urbanization, resulting in negligence in agricultural development makes the whole situation even more complex.

The monograph is organized into one orientation section and four parts. Orientation paper provides an overview of major issues and the four parts are as follows:

Part I:
Describing the physical settings, which include natural resources, population

and human intervention in providing a general background.

Part II.

Dealing with the past agricultural performance from 1949 to 1992. During those 43 years, the total population more than doubled (from 542 million to 1172 million), arable land per capita reduced more than half (from 0.18 ha to 0.08 ha), and grain production per capita increased 80 percent (from 208 kg to 378 kg). Remarkable progress was made beginning in 1979, due mainly to introduced reforms and reversals from previous structure and policy.. The years between 1993 and 1996, were a staggering period in Chinese agriculture, particularly concerning grain output amidst rapid economic growth at the end of the eighth, and the beginning of the ninth Five-Year -Plan. During these years, there was an awakening of national concern for the agricultural future, investment and policy adjustment. Expectations and strategies for food security in the next century ,and the significant contributions through China-World Bank partnership are also discussed.

Part III:

Examining the major concerns in agricultural development, including ecological, economic, political, and social concerns which are closely related to and are limiting factors of agricultural development.

Part IV:

Toward 2030. This parts includes various independent subjects and specific areas. Each author provides practical comments and precise recommendations. Those discussions cover all aspects of Chinese agriculture with one simple objective: how to achieve food security toward Year 2030. Subjects include policy, science and technology, resource management, agricultural production, and postharvest management. In total, this part provides collective recommendations to China, as well to the world for food security, and for peace and stability.

With a very few exceptions, all numbers taken from Chinese sources are official statistics from the State Statistical Bureau (SSB). The readers should be reminded that differences in statistical data among Chinese, United States, and the Food and Agriculture Organization (FAO) may exist, primarily due to differences in definition of terms (such as the definition used for grain) and not necessarily due to calculation errors. Methods of data collection may also contribute to certain discrepancies.

T.C. Tso , Francis Tuan and Miklos Faust

XV

CONTRIBUTORS

Jill S. Auburn
Universityof California. Davis. CA, U.S.A.

Bai Ji-Xun
State Bureau of Foreign Experts
Beijing Friendship Hotel, Beijing 100873,China

Norman Borlaug, Nobel Price Lauerate
CIMMYT, Apdo Postal 6-641 Lesboa 27, 06600 Mexico D.F., Mexico and
Texas A&M University, Collage Station TX, U.S.A.

Colin Carter
University of California - Davis, California, U.S.A

Chen Chuanyou
Commission for Integrated Survey of Natural Resources, Beijing, 100101,
China

Hsien-Jer Chen
Hangzhou Want Food Ltd.
Donghu Road, Linping, Yuhang, Hangzhou
311100 China

S. H. Chen,
Department of Agronomy, Kansas State University, Manhattan, KS, U.S.A.

H.H. Cheng
Department of Soil, Water and Climate, College of Agricultural, Food and
Environmental, Sciences, University of Minnesota, St. Paul, Minnesota
55108-6028, U.S.A.

Cheng Xu
Ministry of Agriculture, Beijing 100026, .China

W.F. Chiu
China Agricultural University,College of Plant Sciences and Technology
Beijing, 1000094, China

Dai Lun-kai
Institute of Botany, Chinese Academy of Sciences. Beijing, China

Shenggen Fan,
International Food Policy Research Institute,
Washington, D.C. 20036-3006, U.S.A.

Miklos Faust
USDA, Agricultural Research Service, Beltsville Agricultural Research
Center, Beltsville, MD 20705-2350, U.S.A.

Shu Geng
University of California, Davis, CA

Joseph Goldberg
The World Bank, 1818 H Street, N.W., Washington, D.C. 20433, U.S.A.

Gong Yan-Ming
State Bureau of Foreign Experts, Beijing , China

Yin-yen Guo
University of California - Davis, California, U.S.A

He Kang, World Price Lauerate Former Minister, Ministry of Agriculture
Building #3, Forestry Survey Bureau Residence, He Ping Li Dong Jie, Beijing
100013,China

Charles E. Hess
University of California, Davis, CA, U.S.A.

Mei-Ling Hsu
Department of Geography, University of Minnesota, Minneapolis, Minnesota
55455, U.S.A.

Jikun Huang,
Center for Chinese Agricultural Policy, Institute of Agricultural Economics.
Chinese Academy of Agricultural Sciences, Beijing 100081, China

C. Huang Purdue University, West Lafayette, Indana. U.S.A.

Jiang Jian-Ping,
Macro Agriculture Research Department.Chinese Academy of Agricultural
Science Beijing 100081 China

Hiroyushi Kawashima
National Institute of Agro-environmental Sciences, Japan

Shain-Dow Kung .
Hong Kong University of Science and technology, Clearwater Bay, Kowloon, Hong Kong

Klaus Lampe Former Director general IRRI
Karl-Bieber Hoehe 29 D-60437 Frankfurt M. Germany

Uma Lele
The World Bank, 1818 H Street, N.W., Washington, D.C. 20433, U.S.A.

George H. Liang,
Department of Agronomy, Kansas State University Manhattan, KS 66506-5501 U.S.A.

Justin Yifu Lin
Peking University, Beijing and Hong Kong University of Science and Technology, Hong Kong

Shouying Liu
Development Research Center (DRC)
of State Council of China. Beijing, China

Y. C. Luo,
Department of Agronomy, Kansas State University
Manhattan, KS 66506

Ma Junru, Former Director General
State Bureau of Foreign Experts Beijing, Beijing 100873, China

Ma Ming
Commission for Integrated Survey of Natural Resources, Beijing, 100101, China

Paul C. Ma
965 Erica Drive, Sunnyvale, California 94086

Mei Fangquan,
Macro Agriculture Research development
Chinese Academy of Agricultural Sciences, Beijing 100081, China

Albert Nyberg
Rural and Social Development Operations Division, The World Bank, 1818 H Street, N.W., Washington, D.C. 20433, U.S.A.

I. C. Pan
Tansui Research Institute for Animal Diseases, Tansui, Taiwan, China

David Pimentel College of Agriculture and Life Sciences Cornell University, Ithaca, NY 14853,U.S.A.

Samuel Pohoryles
Agrindus International Ltd., 13 Kdoshey Kahir Holon 58309, Israel

Shi Yuan-Chun
President Emeritus, China Agricultural University, Yuan-Ming-Yuan Road Beijing 100094, China

James Simpson
Ryukoku University, Japan

Song Jian,
State Science and Technology Commission, Beijing 100862 P.R. China

Samuel S.M. Sun
The Chinese University of Hong Kong

William Tai
University of Maryland, College Park, MD. U.S.A.

Clair Terrill
USDA, Agricultural Research Service, Beltsville, Maryland 20705, U.S.A.

T. C. Tso,
IDEALS, Incorporated
5010 Sunnyside Avenue, Suite 301, Beltsville, MD 20705, U.S.A.

Francis Tuan,
USDA Economic Research Service. Washington D.C. 20036, U.S.A.

Alexander von der Osten
CGIAR - The World Bank,J Building, 18th Street, N.W.
Washington, D.C. 20433, U.S.A.

Chien-Yi Wang
USDA, ARS, Horticultural Crops Quality Laboratory,Beltsville, MD 20705, U.S.A.

Wang Hal-Yang
State Bureau of Foreign Experts, Beijing 100873, China

Anning Wei
The World Bank, 1818 H Street, NW, Washington D.C.20433, U.S.A.

Wen Dazhong
Institute of Applied Ecology, The Chinese Academy of Sciences, Shenyang 110015 China

Ray Wu
Section of Biochemistry, Molecular and Cell Biology, Cornell University, Ithaca, NY 14853 U.S.A.

Jui-Sen Yang
Institute of Marine Biology, National , Taiwan Ocean University, Hsinchi, Taiwan, China

T.L. Yuan, Professor Emeritus
University of Florida, 1020 Northwest 60th street, Gainsville, Florida 32605 U.S.A.

Howard Zhang
Beltsville Agricultural Research Center, Beltsville, MD 20705, U.S.A.

Zhou Yi-xing
Agricultural Regional Planning Center, Wuhan, Hubei 430071, China

Min Zhu
Bank of China, Beijing, China

SPONSORS

IDEALS acknowledges with thanks the generous financial support of the following and several anonymous organizations which made this publication possible.

- Agricultural Research Service, U.S. Department of Agriculture, Washington, DC.

- Rockefeller Foundation, New York, NY

- Pioneer Hy-bred International, Inc. Johnston, Iowa.

- Philip Morris, International Inc. Rye Brook, NY

ORIENTATION

Orientation

Norman Borlaug
Nobel Price Laureate
CIMMYT and Texas A&M University

Assessing the level of food production needed to meet the future needs of a small and stable country such as New Zealand is fraught with immense uncertainty. Assessing the future level of the food production needed for China, the world's most populous country, is even more uncertain. China is a vast, sprawling and fast-changing nation containing about a sixth of all humanity; judging how much food it will need and how much food it will produce in the years 2020 or 2030 is extraordinarily difficult.

Nonetheless, by confronting the facts from China itself and by applying the knowledge from agricultural developments taking place in other parts of the world, it is possible to grasp some central features of the task China faces in raising enough food to satisfy its large and growing population over the coming decades. It is possible also to derive from that a measure of informed opinion about its chances of achieving a well-fed populace.

My interest in China and its agricultural problems goes back to my initial visit in 1974 as a member of the first American delegation permitted to tour farming regions and agricultural research facilities. I was there again in 1977 and several times during the 1980s and 1990s. I have therefore seen the tremendous improvements that have taken place since the collapse of the Cultural Revolution in the watershed year of 1978. Below are summarized some of my conclusions about the present and future situation regarding the food supply.

First and foremost, the expected rise of China's population from today's 1.2 billion to the expected 1.6 billion in 2030 is not unprecedented. When I first began work in India in 1963 the population there was *450* million and today it has risen by more than half a billion souls, and they are eating as well as, or better than, those of thirty-five years ago.

With the even greater knowledge of agricultural science today, China should be able to improve on India's accomplishment. Indeed, China has already achieved amazing advances in food production. As of now, it is the world's number one producer of wheat, rice, potatoes and sweet potatoes and the world's number two producer of maize. Nevertheless, to meet the burgeoning demands of its vast and increasing population China must do much more. Indeed, its experts indicate that it must achieve a production of 400 kilograms of grain per capita per year.

That amount is needed just to produce the basic foods, including the feeds needed for expanding meat production, primarily from pork but also from poultry. Such a level of production, say the experts, will provide at least 90 percent of the grain the nation needs and the remainder can be bought on the international market. I think that this target of 400 kilos of basic grains per capita per year can be achieved. However, it will require that various technological and policy changes be implemented and carried through with diligence and efficiency. Authors in this monograph have pointed out many issues of great importance to their respective explanations, I wish to single out the following for special attention.

1. Balancing Soil Fertility.

Balancing the soil fertility is the number one need if China is to meet its goal. This is made more urgent because even now the formerly rising yields of some grains seem to be leveling off. This is happening even with the greater use of fertilizer, and some observers have taken this fall off in yield increase as an inexorable trend that cannot be reversed. But that gloomy scenario is unlikely. I think that the cause is inadequate farming practices and poor soil management.

China is the largest producer and consumer of nitrogen in the world but it is the other fertilizer nutrients that are increasingly deficient these days. In particular there is a shortage of potash in China's soils and, I suspect, there is a deficiency of minor elements such a copper, cobalt, zinc and molybdenum on some soils as well. In addition, the adequacy of phosphate in many of the soils is questionable. These mineral ingredients are probably the primary culprits in the diminishing yield increases, and a concentrated research effort devoted to restoring their balance in the soils will likely return the yield-growth curve to normal.

It is also necessary to improve the efficiency of organic fertilizer use. I've seen from personal experience that the quality of the organic fertilizer is often very variable from place to place and year to year. The composts used in north China, for instance, are typically much less decomposed than that those used in south China where the warmer climate induces higher levels of microbial activity. Using poorly decomposed compost robs crops of nutrients. The situation is worsened by the fact that farmers in the northern provinces plant as early as they can get onto the land in the spring. The temperatures are then very low and that coupled with the poorly decayed organic material means that the plants get not enough nitrogen in the first month or two. Seedlings that should be well tillered and ten centimeters tall are less than three centimeters tall and the primordia tissues eventually will form the head of grain-are stunted and incapable of ever producing a full yield at harvest time.

As with most of the needed agricultural research, this work must be done with field scientists fully involved. Tests must be run in farmers' fields rather than in experiment stations. They must measure the crop responses to variable amounts of the different nutrients and be correlated with chemical soil analysis so that they guide real-life recommendations for the better use of all fertilizers.

2. Increasing Research Funding and strengthening the education and extension system.

In the last decade China's research effort in the agricultural sciences has been greatly expanded and the number of scientists vastly increased but the funds for carrying out the day-to-day research has not been lifted commensurately. Indeed, because of inflation and the greater number of scientists the operating capital for conducting actual experiments has gone down in real terms. In other words, China's agricultural laboratories have been improved but because budgets for operations have not kept up they cannot function up to their potential. Even in the face of looming food deficits beautiful modern equipment is being under employed and the level of results and the ultimate benefit to the nation curtailed thereby.

Correcting these deficiencies is a matter of policy not science. Too often the guiding philosophy seems to be that the researchers must generate their own funds for their projects. I don't think they can do this in a country with a centralized government and with no recent heritage of private entrepreneurship. Making such a policy succeed is difficult enough in a country like United States; for China at the present time it is beyond reason and the research budgets need to be readjusted to provide adequate funding for the work expected.

Research, education and extension system must be strengthened as one coordinated unit to be functional. Education should not limited to train scientists or technicians but also to train farmers. Efficient transfer of research findings to production level depends on adequate extension system managed by scientists and not through administrative actions.

3. Utilizing Quality-Protein Maize

One of the big opportunities still to be exploited in China is the newly developing form of maize that has an almost perfect protein for supplementing the diets of people and pigs. Both humans and hogs share the same basic digestive system that requires that certain amino acids be included in the diet. Because quality-protein maize (QPM) is the only cereal containing these essential nutrients it has outstanding merit for future use.

In particular, the new maize has tremendous potential for helping China produce more pork and poultry meat. However it is not being extensively exploited. Research on QPM began in China in the mid-1980s and Professor Li Jingxiong led a very dynamic program to produce QPM hybrids for use in the northern provinces. He produced varieties of a type with soft kernels and was well along in producing the somewhat more practical hard-kernel type when he was felled by a stroke. His project now needs to be aggressively revived. One way to do this quickly is to add substantial support to the current low-level collaboration between Sam Vassal of the international center known as CIMMYT and Chinese scientists in both north and south China.

Although QPM can help boost human nutrition its main immediate value for China is likely to be in pork production. Pigs need the essential amino acids lysine and tryptophan that are lacking in virtually all cereals but that QPM uniquely provides. Presently, China has no balanced rations for pigs and farmers take at least six months to produce a 70kg hog. By producing QPM on a large scale China can achieve much better efficiency of the use of feed without requiring great increases in use of soybean meal (which is more expensive and is in chronic short supply).This has to be given high priority. Breeding QPM will help China achieve its targets in meat production. QPM is good also for poultry and also for direct human consumption. The Chinese people already consume maize and employing more QPM-both directly and via animals-will boost their nutritional status without any extra cost or effort.

4. Removing Bottlenecks in Transportation

The open access between the producers and consumers is a vital necessity, but there are still difficulties in getting food from the farms to the consumers in China. This is especially a problem with perishable foods such as potatoes, vegetables and fruit that can be lost to rot and decay if they're not marketed in time. This open access issue is something that needs a higher order of priority in national policy. Farm-to-market roads and other transportation links that help food move to the cities efficiently are critical to the feeding of the future. Moreover, if China is to import more basic grains to supplement its domestic production it is essential that deepwater ports, unloading facilities and ground transport to the interior be improved. If this is not done tremendous food losses and increased costs will result.

5.   Raising Potato Yields

China is the world's largest producer of potatoes. Out of 1995 global production of 285 million tons, China produces 45 million tons, U.S. 20 million tons, and India 18 million tons. Nonetheless, China's average per-hectare potato yield falls far below that of. India and United States. Thus, there is vast potential for boosting potato production to far greater levels than those of today .

My belief is that the real cause of the trouble is virus-infected seed tubers. China's growers are planting potatoes that are already doomed to low production because the plants are burdened with disease from the start. Actually, it is fairly easy to control or eliminate the viral infection but it demands good organization to produce and distribute healthy seed tubers to the farmers in time for planting each year. This is a task that requires no delay for research experimentation and is well within the capacity of the government or private enterprise to fulfill immediately.

Viruses are not the only problem with the potato. Late blight also suppresses yield and can sometimes destroy the crop entirely, but potato varieties resistant to this fungus are available from the International Potato Center (CIP) in Peru as well as from Mexico.

6. Securing adequate water supply

Water is one of the most important factor limiting crop production. China has inadequate water distribution and insufficient total water supply. Measures must be taken to provide sufficient water for agriculture, industry and human needs in quantity and quality. It is essential to increase water use efficiency by reducing waste, eliminate pollution, plus seeking new water resources to garantee that future demand of water can be met.

In the current cultivated land, irrigation area can be expanded to as much as 10 million hectares. Other areas, that potentially can be irrigated, should be developed and reclaimed especially in North China and Northeast China where irrigation rate is very low. Irrigation systems, irrigation facilities and irrigation technics should be coordinated in total as one package so as to maintain water availability without waste or contamination. Current situation that certain areas waste water by over-irrigation and other areas have no water supply must be corrected.

In addition, the vast Northwest of China, which consist 40 % of total China land area is not well developed primarily due to shortage of water.

New water resources must be found to provide water supply and irrigation system for this area. With adequate water supply and irrigation system a new Nortwest can be a great agricultural base for China.

7.Obtaining Reliable Statistics.

It is absolutely essential that the statistical services be greatly improved. At present there are wide discrepancies (estimated to be between 25 and 40 percent) regarding just how much agricultural land is actually under cultivation. Unless such basic aspects of the overall agricultural endeavor are properly accounted for it is impossible to make good policy. Of course, complications arise when there is multiple cropping (more than one crop in the same plot of land each year). However, with good statistical services this age-old farming practice, common in certain parts of China, can be reliably accounted for.

In summary, then, I support the general conclusions of the documents in this publication. The feeding of China in the coming decades is achievable. At least 90 percent of the food can be produced within the country and China's expanding exports of industrial goods will give it the financial capability to make up the rest through imports. It is not likely that imports on this scale will cause international economic disruptions.

However, it must be said that for any country to increase in population by half a billion people in hardly more than thirty years is to put tremendous stress on the whole farming and agricultural-research enterprise. Should any single crop-wheat, rice, maize, potatoes or sweet potatoes-fail in any year, the consequences will be a catastrophe of unimaginable proportions. Such a calamity will affect the stability of societies around the globe through the collapse of trade, monstrous migrations of desperately hungry peoples, and political unrest that brings violence on a scale unprecedented in peacetime.

This looming possibility of global turmoil hangs on the challenge of keeping China's food production up and its population growth down. Foreign critics who chastise China for one or another perceived failing towards its people should always keep clearly in mind that the consequences of chaos in that country include this threat of global economic and social devastation.

Finally, I'd like to note my opinion that any revision of China's domestic policies must not reduce the incentives for farmers to produce as much as they can. Indeed, every incentive for increasing the food supply should be diligently fostered where possible. Some 70 percent of China's people live in rural areas and the gap between rural incomes and urban industrial incomes cannot continue to widen without affecting food production and at the same time spawning social, economic and political stresses. All in all, there must be real incentives to stimulate farmers to grow more food and to give the members of the largest sector of the Chinese populace - the rural peoples - satisfaction and success in their lives.

Institute of International Development and Education in Agriculture and Life Sciences, Inc.

PART I

THE PHYSICAL SETTING

CHAPTER 1.1

Considerations involving land, climate and population of China.

T.C.Tso
IDEALS Inc.

Location and Territory

The People's Republic of China (PRC), located in the eastern part of the continent of Eurasia, facing the Pacific Ocean with 18,000 km coastline, is the third largest country in the world, after Russia and Canada. Its land mass of 9.6 million km^2 lies between 73°E and 135°E and stretches from 53°52'N to 18°20'N (southern tip of Hainan Island) with a group of tiny coral reefs called Nansha Islands located as far south as 5°N in the South China Sea. It extends more than 5,200 km from the Pamirs in the west to the confluence of the Heilongjiang River and the Wusuli River in the east, and stretches more than 5,500 km from the border town of Mohe in the north to the Nansha Islands in the south.

Topography

A great part of China is occupied by mountains and plateaus. About two-thirds of the country is higher than 1,000 m above the sea level, and only 16 percent is lower than 500 m. Topographically, China can be divided into three levels of elevation like three gigantic steps. Each step occupies approximately one-third of the country. The highest step is the Qinghai-Tibetan Plateau in the southwest, with an average elevation of more than 4,000 m. The plateau is rimmed by mountain ranges with peaks rising to more than 6,000 m.

The second step, between 1,000 to 2,000 m, consists of plateaus and basins. They are the Junggar Basin and the Tarim Basin in Xinjiang, the Mongolian Plateau in Neimonggol, the Loess Plateau in Shanxi and northern Shaanxi, the Sichuan Basin in Sichuan and the Yun-Gui Plateau in Yunnan and Guizhou. Rising above the general level of the second step are mountain ranges with an average elevation of 3,000-4,000 m above the sea level. Among them the most geographically important one is the Qinling Range, which trends west-east across the central part of the country and functions as a divide between the dry, wheat-growing lands to the north and the humid, rice-growing lands to the south.

The lowest step comprises the territory east of the line running from the eastern edge of the Mongolian Plateau to the Eastern edge of the Yun-Gui Plateau. The land for the most part drops to less than 500 m above the sea level and includes China's most extensive alluvial plains, such as the Northeast China Plain, the North China Plain and the plains in the middle and lower reaches of the Changjiang (Yangtze) River. The region east of the Northeast China Plain and that south of the plains along the Changjiang River consist of hills and low mountains with an elevation of less than 500-1000 m above the sea level.

Soil

The broad area and the wide topography differences of China covers various types of soil. In the East, as affected by east-south slope seasonal wind, soils distributed from south gradually toward the north by advancing systematic zones, from laterite (latosol), lateritic red earth, red soil (red earth or Krasnozem), dark brown earth, and finally bleaching podzolic soil.

Next, the northwest inland arid area, there is little effect by seasonal wind, the distribution of soil can be divided from east to west as chestnut soil (castanozem), calcic brown soil, Sierozem, and desert soil. Further down, from the Loess Plateau toward northeast to the west edge of Da-Hinggan-Ling, there are zones of Cinnamon soil (drab soil), Heilu soil, Chernozen, grey cinnamanic soil, chestnut soil, again grey cinnamanic soil, and finally grey black soil.

China has many mountains of various heights which affect water and temperature differences, that in turn affect soil distribution.

Climate

China's climate is dominated by monsoons. In the winter, cold and dry continental air masses originated in Siberia spread steadily southward over China as the winter monsoon. In South China, the confrontation of the winter continental air masses with the maritime air masses form the tropical Pacific may bring some winter rainfall, whereas north China's winter is extremely dry.

In the summer, warm maritime air masses originated on the tropical ocean invade northward as the summer rain-bearing monsoon. The duration of the summer monsoon varies from 6 months in the south to 3 months in the north. For the most part of China, the rainfall is mostly concentrated in the summer monsoon period. Annual rainfall decreases gradually from 2,000 mm along the southeast coast to 1,000-1,500 mm in central China, 500-700 mm in the North China Plain and 400-600 mm in the Northeast China Plain. The great part of the northwest has less than 200 mm. Temperature also varies considerably between summer and winter and from north to south. In summer, with the exception of the Qinhai-Tibetan Plateau, it is hot all over China. The July mean temperature is above 20°C throughout the country. It is higher than 28°C south of the Qinling Range, between 24°C and 28°C in the North China Plain, and between 20°C and 24°C in the northeast and northwest.

In the winter, temperature decreases rapidly from south to north. The January mean temperature is above 0°C south of the Qinling Ranges and rises to more than 10°C along the southeast coast. It is generally between 0°C and -10°C in the North China Plain and the Loess Plateau, and between -10°C and -24°C in the northeast and northwest with the lowest of -30°C north of 50°N.

Since China is under strong influence of the high pressure system over Siberia in winter, the intrusion of cold air masses into south China is quite frequent and often causes sharp fall of temperature. Sometimes the temperature may fall below 0°C under the invading cold air masses and causes tremendous damages to subtropical crops.

Land Use Pattern

Land and Water Resources

China's land resources can be classified into four broad categories: (1) lands suitable for agriculture; (2) lands suitable for animal husbandry; (3) lands suitable for forestry; and (4) lands unsuitable for all the three primary industries which includes deserts in the northwest, high cold

Qinghai-Tibetan Plateau, high mountains above the snow-line and all the naked precipitous rocky mountains. This category accounts for 25 percent of the country's land mass. Land suitable for animal husbandry accounts for 37.1 percent and land suitable for agriculture accounts for only 10.2 percent.

China has a limited supply of underground water. The surface water resources are comparatively plentiful, amounting to 2,720 million M^3, but 80 percent of which is distributed in the south [including the reaches of the Changjiang (Yangtze) River and all other rivers in the south], and only 20 percent is in the north [including the reaches of the Huang (Yellow) River and all other rivers in the northeast and northwest]. Since 64 percent of the country's arable land is distributed in the north, it is obvious that the distribution of surface resources is unfavorable for the agricultural development. In order to improve the water supply for agriculture in the north, projects of diversion of water from the south to the north have long been under consideration and there are some limited action. However, the total amount of water, in view of future demand, is insufficient.

Natural Calamities

Floods and droughts are the two most damaging natural calamities in China, either in terms of the area inflicted or in terms of economic losses. China's floods and droughts are mainly caused by high intensity of rainfall, high variability of annual rainfall and untimely advance and retreat of the rain-bearing summer monsoon. In recent years, reduced capacity to retain water from deforestation in the interior and to absorb water from shrinking lake areas (Dongting Lake and Buoyang Lake) has further exacerbated the severity of flooding in the middle and lower reaches of Changjiang (Yangtze) River. Agricultural production has suffered frequently from the recurring floods and droughts in the past. And it is still rare to see a year without some region in China inflicted by floods or droughts. All the water conservation projects undertaken in recent decades can only alleviate the damages to some extent at the best. Soil erosion is a natural process. Under the interference of human's practice of irrational land use it can bring tremendous damages to the land resources. Soil erosion is especially serious in the hilly lands south of the Changjiang (Yangtze) River, in the Loess Plateau and in the northern border zones of the semi-arid regions in the northwest. In the first two regions, it is water erosion that destroys large tracks of land. In the semi-arid regions it is wind erosion in action. Wind erosion accelerated by overgrazing and mismanagement of cultivated land has resulted in desertification over large areas. The desertified land increased from 137,000 km^2 in the 1950's to 170,000 km^2 annually in the 1990's and it still keeps increasing.

Land Utilization

The arable land as reported by the State Statistical Bureau (SSB), of 95.65 million hectares is about 10 percent of the total area. Irrigated area is slightly over 50 percent, the rest is rainfed land. Other potential area that could be developed from current wasteland amounts to 35.35 million hectares, mostly located north of 35° N, and concentrated in northeast Inner Mongolia and the northwest region. It should be noted here that SSB data has frequently been questioned as being much lower than the actual area. Forest area has 131 million hectares. Total coverage rate is 13.63 percent, but very unevenly distributed. Most forest is in the northeast and southwest. The coverage rate is only 11 percent in densely populated area, and only 1.1 percent in the wide northwest region.Grassland has 400 million hectares, of which 313.3 million hectares are being utilized. Pasture resources occupy 68 percent of China's total hilly grass area. Overgrazing, poor management, and desertification are the major problems. Potential development is limited due to limited water resources. Only those in the hilly area in the East, South and Southeast, which has a combined total of 53 million hectares may be developed as animal husbandry production base. Water area includes fresh water and shallow shoals and tidal wetland.

Human Interferences and Interventions

Human interventions of land in China can be viewed from two different directions. One is population distribution induced uneven demand, or heavy burden to the land. The other is to protect the natural resources to prolong sustainability and bring the land back into a "healthy" condition to improve productivity.

Population Distribution

China's population is very unevenly distributed, which can be best illustrated by using a very important geographical demarcation line running from Heihe, a northern border town in Heilongjiang Province, to

Ruli, a southern border town in Yunnan Province. The region west of the line consists mainly of the Qinhai-Tibetan Plateau and the extensive arid and semi-arid areas in the northwest. It accounts for 57 percent of the country's territory, but only less than 10 percent of the country's population. For most part of the region, the average population density is no more than ten persons per km^2. In certain areas in west and north of Tibet the population is very sparse, less than one person per km^2. Population density reaches 100 persons per km^2 only in places where irrigation is available or around a few urban centers (such as Urumqi).

More than 90 percent of the population is distributed in the region east of the demarcation line, which accounts for only 43 percent of China's land mass. This area includes almost all of China's subtropical humid areas and temperate humid and sub-humid areas. Within this region, population density varies from place to place. The most densely populated areas are on the North China Plain, the Sichuan Basin and the plains in the middle and lower reaches of the Changjiang (Yangtze) and Zhujiang River Deltas. Population density reaches as high as 600 per km^2 (on the Chengdu Plain); whereas the density in the Northeast China Plain, the Loess Plateau the Yun Gui Plateau and the hilly land south of Changjiang (Yangtze) River is mostly less than 200 persons per km^2.

To achieve a better balance of population distribution, China must seek ways to develop the vast west region. This would release the high burden of the east region, particularly, it would give relief of overtaxed natural resources, of water and soil, and also for reduction of pollution.

Sustainable Agriculture and Resource Conservation
Field cultivation is the major activity for agricultural production. The multiple cropping index, due to high demand for agricultural products, is continually increasing in China. Intensive use of land, plus little or no rotation, absence of green manure, high application of chemical fertilizer, pesticides, etc., resulted in lower land productivity and more pollution and erosion.

All these practices need to be corrected soon. It must be realized that not all low or medium productive land can be improved by human activities through higher inputs. Not all "wasteland" can be reclaimed. Forest and grassland must be preserved to prevent soil and water erosion. Desertification, salination must be stopped by all means. When water is polluted, not only people have no water to drink, there will also be no industrial development.

Water is short worldwide, especially in China. Soon water resources will be exhausted if no effective action is taken now. Soil erosion, according to available data, destroys 150 km^2 of arable land per year. Soil loss into the Huang (Yellow) River from the Loess Plateau is 1.6 billion tons annually. The practice of making new land through filling lakes had already affected micro-climate in addition to reducing water surface and water supply. Heavy use of coal as fuel also make "acid rain" a serious problem in China. For example, the acidity of rain in south China is pH 4-5, but in north China is pH 5-6. All these problems require intensive and immediate human intervention. Human activities induced environmental and resources destruction, human intervention can also prevent or restore ecological balance. This is particularly important for China, as China is overpopulated and her resources are limited.

PART II

AGRICULTURE FROM 1949-1996

AND FUTURE OUTLOOK

CHAPTER 2.1

AGRICULTURE in 1949 - 1992

T.C. Tso
IDEALS Inc.

Chinese agriculture has five thousand years experience. Attempts to reform agriculture in the past were few and short lived. In the 7th century Empress Wu Zetian, in the 11th century Prime Minister Wang Anshi of Song Dynasty tried unsuccessfully reform agriculture. These should be compared with land reform in Taiwan in the early 1950's that was a resounding success. During the period of 1949-1977, China's agriculture sector experienced drastic changes, disturbances, great improvements and set backs. In June 1950, the "PRC Land Reform Act" declared that land reform should be completed by the end of 1952,with the exception of some minority areas in Xinjiang and Tibet.

As a result, 300 million tenant farmers were given 750 million mu, or 50 million hectares, of land, and other production assets. They were relieved from the obligation of paying rent to private landlords, (35 million tons of grain annually). This reform eliminated an age-old feudal structure, in which the rich landlords, representing 10 percent of rural population owned 70-80 percent of farm land. This reform act at the beginning greatly stimulated farmers' incentive, and fulfilled their productive capacity. The grain yield in 1951 and 1952 reached a historical high of 163.9 million tons, or 9.3 percent higher than the 1936 record of 150 million tons.

After the 1958 "Great Leap Forward" campaign, three years of severe crop shortfall and famine followed. Grain production hit its lowest point. In 1960, the Gross Value of Agricultural Product (GVAO) decreased 26 percent from 1958 level. It took another five years to return to the size of 1958 output. During the 27 years between 1949 and 1977, conditions for agricultural production improved considerably. Machine power increased, so did the irrigated areas, along with the number of rural hydroelectric power stations, and electric generating capacity. Rural electric consumption increased from 0.05 to 22.2 billion kw/h. Chemical fertilizer usage increased from780 thousand tons to 6.48 million tons or an annual increase of 8.7 percent.

However, the progress had not been translated into high productivity or a better living standard in general. One reason was due to continuous political movements and another reason was that the basic condition was very backward in Chinese agriculture and the population increased tremendously during that period.

The period of 1978-1989 saw significant development of agriculture in China, including all subsectors of crops, forestry, animal husbandry, and fishery. Annual growth in gross value averaged 6.64 percent. Grain production in 1984 and 1987 reached a record 407 million tons. Average grain yield per hectare increased from 169 kg to 242 kg, or 43 percent. The gross value of other agriculture subsectors also progressed with rapid paces. The progress during the period from 1978-1989 may be divided into two stages. The first 5-year period maintained a high speed increase with rise in gross value at an annual growth rate of 17.3 percent and for the first time in history, grain production surpassed the 400 million tons barrier. For the next five years, the speed of development began to slow and fluctuated greatly. Average annual growth rate in agriculture was only 1.04 percent.

Chinese agricultural development had made another advance in 1989 to 1992. The GVAO in 1990 increased 7.6 percent over 1989 and 1992 increased 6.4 percent over 1991. Total grain production reached 407.6 million t, a return to the 1984 record level. During the following three years, total grain output stood firmly above the 430 million t level. Other subsectors of crops, forestry, animal husbandry, and fishery kept up with similar or faster strides.

At the end of 1978, the Chinese Communist Party held its Third Plenary Meeting of the Eleventh Central Committee. This meeting acknowledged the mistakes of the "Cultural Revolution," and called for a total review of the national economic system. It particularly pointed out that modernization of the nation's economy should be the key policy. In agricultural sector, the Party document of "Certain Decisions Related to Agricultural Development" clearly stated the rural economic policy.

- Agriculture was the very foundation of the people's economy.

- Agriculture, forestry, animal husbandry, sideline, and fishery should have a balanced development, with grain as the most important key element.

- Farmers' incentive and material benefit was fully recognized, and be rewarded by one's labor, not by sharing return equally.

- The People's Commune was to be maintained with three levels (State, collective, and individual) ownership, and the brigade as the basic accounting unit.

- There would be no change to commune members' private plots,

- Family sideline, and rural free cooperative markets, as these are necessary supplements to the socialistic economy.

- Investment in agriculture would be increased by the State, and grain procurement quota for government would be lowered.

- Prices of agricultural input, such as machinery, chemical fertilizer, and pesticides would also be lowered.

With success of the above policy, the State implemented further reform in agriculture with the "Household Production Responsibility System" to eventually replace the commune system. The system started with production contracts to households and evolved to more advance forms of production contract to farm groups and contract for specialized production, all with joint accounting for returns. The transformation from commune to the responsibility system was very slow, stemming from concerns whether the change would be permanent. In March 1980, only 28 percent of the commune brigades adopted "responsibility system." The Party Central Committee had to give assurance that the new system was compatible with socialism and would be maintained.

The Party document of January 1982, "Records on National Rural Work Conference" emphasized the need to produce according to local conditions. As a result, "responsibility system" in rural area reached 92.3 percent nationwide by November 1982. A survey of 10,938 farmer families in January 1988 determined the farmers' reasons for making planting decisions. Planting decisions by 66.1 percent of farmers was based on best self interest, 25 percent on State quota requirement, and 4.8 percent on general village plan. In addition, some villages collectively arranged machinery use, plowing, over 98.6 percent of all rural villages and 97.7 percent of all farming households. Under the responsibility system, a rural household enters into a contract with the township government, with specific terms on theirrigation and drainage, plant protection, and supply of seeds to serve individual households. By he end of 1992, "responsibility system," encompassed over 98.6 percent of all rural villages and 97,7 percent of farming households. Under the responsibility system a rural household enters into contract with the township government, with specific terms on plot size to be farmed, length of contract, and types and quantities (quota) of farm products to be sold to the government at a fixed price.

As long as contract obligations are fulfilled, the household is free to make own farm management decisions of what and how much to produce, to sell and to whom, and is responsible for its own profit and loss. The government may purchase additional amount of farm products above the contracted quantity at a negotiated price, which is usually above the fixed price. In addition, the household is allowed total ownership of production materials and facilities.

In the early 1950's, all commodities were under government control, with Government buying and selling these commodities at fixed price , and little regard to market supply and demand. As a result, farmers did not produce what the market demanded. Without synchronization of market supply and demand and with only a minimum price paid to farmers for agricultural products, this system reduced incentives, and low production led to even tighter controls. The situation was reversed under the "responsibility system." In 1989, the State raised procurement prices for 18 major agricultural items, including grains, cottons, oilseeds, hogs, etc., and offered a 50 percent premium for quantity purchased above contracted quota. At the same time, the State allowed open market and prices for these products. Actually, relaxation of State control began in 1985 on certain medicinal herbs, and was extended further to fresh-water-fishery products, and finally, with a few exemptions, most meat, eggs, fish, and vegetables were free from control. Government also encouraged organizations of rural cooperatives, with farmer ownership and management participation. By the end of 1992, there were 79,188 free markets nationwide, facilitating activities for producers, consumers, and traders. Trade volume reached 353 billion Yuan, of which grain and oil accounted for 21.3 billion Yuan. This trend of free markets further induced other business, such as storage, farmer-industry-commerce cooperative enterprise, farmers cooperative, specialty markets, transportation, etc. Historically, the only business for Chinese farmers was to grow crops. Beginning in 1979, a total reform and readjustment took place under the State plan. The basic policies were:

- to maintain grain production as key but to increase economic crops production;

- to maintain crop production as key but to increase forestry, fishery, and sideline activity; and

- to maintain rural agricultural production as key, but to promote secondary rural enterprises which included production of construction materials, construction, service, and transportation.

Fast rural economic growth in the recent 10 years can be reflected by rural savings, which experienced an impressive two-digit annual increase. Comparing with 1979, total rural deposit in the Agricultural Bank of China (ABC) and rural credit cooperatives (RCC) jumped from 30.35 billion Yuan to 516.2 billion Yuan in 1991, or an annual increase of 26.6 percent. Of which saving deposit was 379.4 billion Yuan, at an annual increase 35.73 percent. In 1992, fixed savings deposits were 347.7 billion Yuan, of which savings deposits by farmers were 286.7 billion or 82.44 percent of total deposit.

Agriculture Sector

During the 25 years before 1978, little had changed in the composition of agricultural production, of which, crops accounted for about three-fourth of gross value of agricultural product (GVAO), with forestry, animal husbandry, fishery and sideline earning the remaining one-fourth. But since 1978, the composition has undergone a vast change. The proportion of crops dropped drastically while that of the other four subsectors rose, particularly with marked upswing for animal husbandry.

Comparing 1992 with 1952, ratio of crops in GVAO was lowered from

14

73.5 to 55.5 percent, down by 18.1 percent. At the same time, animal husbandry increased from 11.2 to 27.1 percent, up by 15.8 percent, forestry from 1.6 to 4.7 percent, and fishery from 1.3 to 6.8 percent.

This notable change is the result of Government policies to promote:

• balanced development of the subsectors in agriculture and vertical integration of agriculture, industry and commerce.

• the rational development of agricultural resources, and

• striving for a high yield and an efficient agriculture sector.

Grain Production

According to traditional Chinese definition, there was no distinction between grains for food or feed. Grain crops consist of cereal crops, soybeans and tubers (conversion rate: 5 kg tubers = 1 kg grains); of which, rice, wheat and corn are the three major cereal crops, accounting for 42, 23, and 22 percent of total grain production (TGP) respectively, and the sum of the three crops' production accounts for 87 percent of TGP.

Grains has always been regarded as the main stay of agriculture sector in China. In 1949-92, TGP rose from 113.2 to 442.7 million tons, an increase of 391 percent, which notably exceeded the population growth rate of 216 percent for the same period. Also in this period, per capita grain production rose from 208 to 380 kg, a net increase of 171.1 kg. Though the figure is not high , but it has played an important role in feeding the large population and maintaining stable development of China's economy.

With adjustment and improvement of agricultural composition during 1978-92, the ratio of grain sown area to total sown acreage had gone down from 80.3 to 74.2 percent, a drop of 6.1 percent; and the actual area had decreased from 120.6 to 110.6 million hectares, a 10.0 million hectares or 8.3 percent drop. But because of grain yield per hectare had increased from 2,535 kg to 4,004 kg, an increase of 57.9 percent, which had rendered TGP to increase from 304.8 to 442.7 million tons, or an increase of 45.2 percent.

Rice. Rice is mainly grown in 1) the single cropping belt of Central China, an area that is south of Qinling Mountain and Huai River, north of Nanling Mountain and east of Chengdu, and 2) the double cropping belt in the South located south of Nanling Mountains. These two major rice belts produce over 80 percent of total rice production. In addition, rice is also grown in: 1) single crop belt in North China, 2) early ripen and single crop belt in Northeast, 3) dry-area rice variety belt in Northwest and 4) high-altitude rice variety belt on high plateaus of Southwest.

In 1992, there were 9 major rice producing provinces, each produced over 10 million tons; namely, Hunan, Sichuan, Hubei, Jiangsu, Guangdong, Jiangxi, Zhejiang, Anhui, and Guangxi. Altogether they produced 79.9 percent of a total production of 186.2 million tons.

Wheat. Wheat production regions in China are: 1) northern winter wheat area, and 2) southern winter and Nortwest spring wheat area. In 1992, there were 6 major wheat producing provinces, each produced over 5 million tons; namely, Shandong, Henan, Jiangsu, Hebei, Sichuan, and Anhui. Altogether they produced 66.8 percent of a total production of 101.6 million tons.

Corn. Corn production is spread over a wide range of areas, including northern spring-sown area, Huang-Huai-Hai River Delta inter-cropping summer-sown area, and southwest hilly area. In 1992, there were 7 major producing provinces, each produced over 5 million tons of corn; namely, Jilin, Shandong, Heilungjiang, Liaolin, Hebei, Henan and Sichuan. Altogether they produced 69.7 percent of a total production of 95.4 million tons.

Economic crops in China are defined to include cotton, oilseed crops, sugar crops, coarse fiber crops, tobacco, mulberry for silk worms, tea, and fruits. In 1992, areas sown for economic crops numbered 24.3 million hectares, or about 16.3 percent of total sown areas and an increase of 6.7 percent over 1978.

Production of other crops

Cotton. Planting of cotton mainly is centered in five major areas; namely, Yellow River Basin area, Yangtze River Basin area, northwest interior area, northern early ripening area, and southern area. In recent years, growing area for cotton has had notable shifts with Yellow River Basin area now as the leading producer. In 1992, there were 6 major cotton producing provinces, each produced over 300 thousand tons of cotton; namely, Shandong, Xinjiang, Henan, Hubei, Jiangsu, and Hebei. Altogether they produced 76.5 percent of a total production of 4.5 million tons. During the 1978-92 period, cotton production has had rapid expansion, with total cotton output increased 108 percent. But production in 1992 dropped 20.5 percent from the year before, and a 28 percent decrease from 6.3 million tons in 1984.

Oilseed Crops. Major oilseed crops are rapeseeds and peanuts, accounting jointly for 82.9 percent of the total output. Prominent producing provinces are Shandong, Sichuan, Anhui, Henan, Jiangsu, Hubei, Hunan. Since 1978, there were quite brisk expansion, with 1992 total production reaching 16.4 million tons or a 215 percent increase over 1978. During the same period, processing capacity also experienced similar growth so to keep up with rising edible-oil demand in the nation.
Sugar Crops. Sugar crops in China virtually mean sugarcane and sugar beets, with the former grown in Guangdong, Guangxi, and Yunnan, and the latter in Heilungjiang, Nei   Monggol, and Jilin. In 1992, China produced 73.0 million tons of sugarcane and 15.1 million tons of sugar beets, a 246 and 458 percent increase respectively over 1978. With domestic production plus imports, per capita sugar consumption was 5.42 kg in 1992.

Fruits. Because of China's vast land areas and wide varieties of topography, so the same can be said about it's fruits. Since 1978, growth of fruit output has been quite notable, with 1992 total production at 24.4 million tons, a 271 percent increase over 1978, or 1,930 percent over 1949. Of the total, the output had 6.6 million tons of apple, 5.2 million t of citrus, 28.5 million t of pears, 2.5 million t of bananas, and 1.1 million t of grapes, or 26.9, 21.1, 11.7, 10.0, and 4.6 percent of total fruit output respectively. Major fruit producing provinces are Guangdong, Shandong, Hebei, Liaoning, Guangxi, and Sichuan. Mainly due to efforts to increasing production, opened to free marketing earlier than other farm products, and continuous improvement in storage and transporting, it has reached the realization of year-round fruit supply, and greatly transformed from the vast shortage in the old days. In 1992, per capita output was 20.8 kg, a 270 percent increase over 1978.

One of the major problems limiting fruit industry development in China is that all kinds of fruits are widely distributed in small acreage plots over a broad area. In other words, not concentrated in desired locations for efficient management, production, processing, storage, transportation and marketing.

Vegetables. China has had long tradition to consume a large quantity and wide variety of vegetables. Following an early opening of markets and decontrol of prices, vegetable production had leaped forward, with planted areas increasing to 7.0 million hectares or 88 percent over 1978, and per capita consumption to over 120 kg in 1992, mostly around populated areas. Instead of open-air cultivation, using the technique of growing vegetables under large, protective, plastic sheet covering shelters caused a large scale increase in production. Currently year-round vegetable supply is common in large and medium size urban areas with ever-increasing varieties, despite of weak marketing links, poor storage capacity, transportation, packaging and processing.

Composition of Livestock Production

Livestock subsector had always been considered as a supplementary farm endeavor in China's traditional agriculture, with low meat production and consumption levels. Large animals were utilized for field supplementing humans. Raising hogs was, in a limited sense, was considered for meat consumption, and for manure production. This was reflected in Government agricultural policies before 1978, pushing hard to raise hogs for manure reflected in a high year-end inventory number of hogs rather than pork output. Thus for many years, hogs was in first place, followed by draft animal in livestock composition. Only after 1978, livestock animal composition began to improve. In 1992, year-end hog inventory reached 384.2 million head, an increase of 570 and 275 percent over 1949 and 1978 respectively; for cattle, year-end inventory was 107.8 million head, or 150 and 52 percent over 1949 and 1978; of the cattle, over 90 percent were draft yellow cattle and water buffaloes with the remaining as dairy cows and beef cattle. Beside hogs and cattle, the others were 10.0 million head of horses and 11.0 million head of donkeys.

Following large increase in production of wool textile goods, numbers of sheep and goats also increased rapidly, reaching 109.7 and 97.6 million head respectively, or a 320 and 510 percent increase over 1949.

As a result of rapid rise in per capita income and consumer demand for more animal protein products in food diet, livestock production responded. In 1992, total meat production was 34.1 million tons; of which pork was 26.4 million tons (77 %), poultry 4.5 million tons (13.3 %), beef 1.8 million tons (5.3%), lamb and mutton 1.3 million tons (3.7%), rabbit meat 185 thousand tons (0.5%). The above numbers depict the obvious decline of pork in its ratio to total meat output, the reverse for poultry, beef and lamb and mutton, and large change in livestock product composition.

By its rate of growth, pork output in 1985-92 increased 59 percent, poultry 180 percent, beef 290 percent, and lamb and mutton 110 percent, a reflection of consumption pattern changes from single product of pork to multiple-choices of meat.The change in composition was also much enhanced by very rapid growth in poultry industry. China became the world leading producer of poultry eggs in 1992 with a production of 10.2 million tons, an increase of 90.7 percent over 1985.

Dairy product production also made notable progress, with total output reaching 5.63 million tons in 1992, an increase of 94.9 percent over 1985. But due to slow growth in companion processing industry and low consumption in rural areas, average per capita dairy consumption in China is still very low, mainly centered in large and medium size urban areas.

Wool production in 1992 was 255 thousand tons, an increase of 33.7 percent over 1985; of which sheep wool was 238 thousand tons. Of the sheep wool, 66 percent or 158 thousand tons was fine and semi-fine wool.

Regionally, almost every province produces all types of livestock products. But only a small number of provinces produce over 1 million tons of pork production each; they are Sichuan, Hunan, Shandong, Jiangsu, Guangdong, Hubei, Hebei, Jiangxi. For other livestock products, beef production is centered in Henan, Shandong, Anhui; poultry and eggs in Shandong, Jiangsu, Henan and Hebei; sheep wool in Nei Monggol, Xinjiang, Qinghai, Shandong, and Gansu.

Fishery

Fishery was a subsector that was allowed to have open market and prices soon after 1978. With market forces and other companion programs, fishery enjoyed an unprecedented development. In 1992, China became the leading producer of fish in the world, with total production reaching 15.6 million tons and an increase of 3,360 percent over 1949, 234 percent over 1978, and 15.3 percent over a year before. On the demand side, per capita consumption of fishery products amounted to 7.3 kg per person per year, an increase of 108 percent over 1978 and a change from past vast shortage. Salt water fishery production accounted for 9.3 million tons (60 %), with fresh water production 6.2 million tons (40 %) of the total output. Fish farming developed rapidly, producing 7.8 million tons while natural production (or fish catching) produced 7.8 million tons, with fish farming output increasing from 1978's from 26 percent of the total to 1992 to about 50 percent of the total fish output.

Considering product types, fish accounted for 11.2 million tons or 71.7 percent of total production, shrimp and crabs for 1.4 million tons or 9 percent production, mussels for 2.2 million tons or 13.8 percent production, sea weeds for 568 thousand tons or 3.6 percent of production, and other products for 297 thousand tons or 1.9 percent of production.

Considering regional distribution, interior provinces had rapid development in fresh water fish farming, but production is still heavily centered along the coastal provinces. In 1992, there were 6 major fish producing provinces, each produced over 1 million tons of output; namely, Guangdong, Shandong, Zhejiang, Fujian, Jiangsu. and Liaoning. Altogether they produced 70 percent of total fishery output.

Forestry

Despite Government's continuous emphasis on the importance of forestry development and implementation of "Forestry Law", its efforts has consistently failed to effectively counter the chronic shortage of fuels, rampart poverty in interior areas and age old ignorance of and negligence of trees by the populace, all of which had contributed to serious reduction of forest acreage and denuding of countryside. Currently, Chinese statistics reported that forest covered area reached 131 million hectares with a forest coverage rate of 13.6 percent, an increase of 12.2 million or 1.27 percent since the Second Survey of National Forest Resources in 1977-81. Also the Survey reported timber reserve from living and standing forests as 10.9 billion cubic meter, an increase of 202 million.

Major problems facing China's forestry sector are:

• Per capita forest area is only 0.11 hectare, and per capita living and standing timber reserve is only 9.3 m^3, very low in world ranking.

• The 20 million hectares of natural forest owned by the State Forestry Bureau, through 40 years of cutting, is now reduced to an alarming condition from illegal and indiscriminate cutting.

• The current 2 million forest employees and workers lack modern

• operating and managerial knowledge and skill to conduct reforms of structure and management.

Aiming to solve these problems, the Government introduced policies as follows:

• The Government should actively enforce the "Forestry Law", expand educational program for forest planting and protection, and implement policies and programs relating to forestry.

• The Government should implement reforms in forestry management with improved production responsibility system; meanwhile, shift away from old, narrow concept of Forestry is only for wood producing." to comprehensive forestry management, resource renewal and utilization.

• In implementing policies, the Government should emphasize overall attention to a combined concept of economic efficiency, welfare of the society and ecological welfare, so to maintain balanced relations between forest renewal and timber cutting by strict control of cutting not to exceed the natural growth of forests. The Government should continue the development and tree planting programs started in 1981, and further implement: 1) the green countryside movement in Northeast, North China, Northwest, all coastal provinces, Yangtze River middle and lower reaches, and 2) construction of forest ecological zones in areas around Taixing Mountains, Beijing and Tianjin.

• The policy should establish and expand nature protection regions by increasing the number from the current 420 to 500 and from 44 million hectares to 50 million hectares

• The policy should: 1) enforce the practice of reverting land with 25 degrees of slope, which are currently under farming cultivation, back to forestation and/or livestock grazing, 2) expand forest acreage, and 3) expand programs in prevention of water and soil erosion.

Agricultural research

Historically, support from Government to science and technology, or percentage of gross expenditure, was maintained around 4 or 5 percent since the mid 1960s, and it has a slightly decreasing trend since 1978. The ratios of total S&T funding to total expenditure in past years are as fellows: 1953 - 0.3%; 1966 - 4.60/o, 1978 - 4 8%, 1988 - 4.4% and in 1991 - 4.1%. However, the funding for S&T in agriculture is only a fraction of the total S&T funding, about 10.1% in 1991. Or four4enth of one percent of gross expenditure.

Through efforts of past 4 decades, China had made marked advancement in establishing agricultural research network. In 1991, there were 1,142 argricultural research institutions above the prefecture level; of which, 61 were institutes in the Chinese Academy of Agricultural Sciences (CAAS), under the jurisdiction of Chinese Ministry of Agriculture; 457at provincial level, and 624 at prefectural level. Total employees were numbered at 129.6 thousand, with an annual budget (including basic conrtruction) of 1.63 billion yuan; of which, appropriation from the Government accounted for 8.59 billion yuan. There were 16,538 research projects.

Since 1980, China had introduced a series of advanced technology with notable progress in such fields as hybrid rice, hybrid corn, crop management through economic modeling, crop cultivation with plastic sheet covering, rice pest research and forecasts, selective breeding of Holstein dairy cattle and lean pork type hogs and -try, research and prevention of livestock disease such as cattle plague, swine fever, cattle pneumonia, prawn farming and indoor cultivation.

Agricultural extension

Through the past years, China has established an agricultural extension. In 1991, various Government agricultural agencies had set up 22,000 extension entities, with extension agents numbered 920 thousand, with 10.22 million hectares for extension uees.

Administratively, the whole agricultural extension structure is under the jurisdiction of the Ministry of Agriculture. At central level, there is National Agricultural Extension General Station in the Ministry of Agriculture. At provincial level and in each Provincial Bureau of Agriculture, there is a General Station to serve as central level's arm in carrying out central policies and programs. The same goes down to prefecture, county, township levels.

In 1991, 68.7 percent of all counties had extension centers (numbered 11.4 million), 82 percent of all townships had extension station (totaled 459 thousand) and 66 percent of all villages had either extension service group or full time extension agents. Among various extension groups and at various agency levels, there were frequent confusion and overlapping in activities and responsibilities for years, until 1992. In order to strengthen extension work, the State Council had enacted in 1993 the "Law of Agricultural Extension in the PRC," to be implemented by six agriculturally related Ministerial agencies, including the Ministry of Agriculture.

CHAPTER 2.2

China's Agricultural Growth During 1993 - 1996

Jiang Jianping[1]
Chinese Academy of Agricultural Sciences (CAAS)

Review and Comments on Agricultural Development

During the Eighth Five -Year Plan (1986 - 1990), Chinese agriculture grew at 4.2 percent each year . During that time, China was in the process of transition from a traditional planned system into one that market oriented, accordingly, new conflicts and problems emerged, and while industrial development grew at a rapid pace, agriculture development stagnated. Comparative advantages in agriculture began to diminish and farmer incentives lessened, leading to less cultivated land. Cultivated land fell dramatically because it was diverted to non - agricultural uses: establishing industrial development districts. As a result, cultivated land for grain deceased 1152 thousand hectares and 2906 thousand hectares in 1991 and 1992 respectively, and total grain production (TGP) decreased 10.95 million tons in 1991 and 3.58 million tons in 1992. In 1993, the decline of rice production in southern China caused grain prices to soar, which aggravated overall inflation. Retail price increased 5.4 percent in 1992 and 21.7 percent in 1994. Such inflation harmed not only for development of agriculture and the national economy but also was detrimental for social security.

The Central Committee of the Communist Party in China (CCCP) and the State Council (SC) took effective measures to address above problems. Firstly, the CCCP held two conferences on agriculture, one in October 1993 and another in March 1994, at these meetings, all leaders from different levels of responsibility were asked to pay more attention to agriculture, rural areas and farmers, and attempt to speed up agricultural and rural development. It is rare for central government to take decisive action and measurement in such short times since establishment of the People's Republic of China. Next, a

---

[1] Editors' note: certain terms used by this author may be different from other writers in this monograph

comprehensive set of measures were formed to promote agricultural development: acreage planted to grain, cotton and vegetables were stabilized, price ceilings were imposed on input price, procurement systems were reformed and grain anti-risk founds were established.

Table 2.2.1. Gross Value of Agricultural Output and Proportion

Item	1992	1993	1994	1995	1996	1996 Over 1992 %
GVAO	898.9	969.3	1,052.6	1,167.1	1,267.1	42.1
% of Total	100.0	100.0	100.0	100.0	100.0	
Crops	54.7	59.3	56.4	54.9	54.1	-0.6*
Forestry	4.9	4.9	4.9	4.7	4.9	0.0*
Animal Husbandry	27.0	27.7	29.8	30.8	34.1	7.1*
Fishery	7.4	8.1	8.9	10.6	10.9	3.5*

Source: Chinese Statistic Yearbook,1992, 1993 and1994. Abstract of Nationwide Agricultural Statistics,1995 and1996. * growth rate. calculated by the author. GVEO numbers are at 1990 constant price.

The provincial governor responsibility for grain production called "rice bag program" and increases in quota price in 1994 and 1996 took great effect in growth of grain.The above measures helped reverse the stagnation of agricultural growth, enhanced development in rural areas and stabilized prices. The gross value of agricultural output reached 1276.9 billion Yuan (1990 constant price) in 1996, which was increased by 42.1 percent as comparing with 1992.

Cultivated land and total grain production were 112.548 hectares and 490 million tons(a record grain harvest) in 1996, increased by 1988 thousand hectares (Table 2.2.2) and 47.34 million tons or 10.7 percent as comparing with 1992. Other major agricultural products grew rapidly also (Table 2.2.3). Due to the good harvests in 1995 and 1996, the supply of agricultural products increased. Further more, government used moderately restrictive fiscal and monetary policies, commodity retail price growth fell from 21.7 percent in 1994 to 14.8 percent in 1995 and 6.1 percent in 1996. This "soft landing" policy was successful in helping the transition into a more market oriented system, and has been recognized in the world.

Table-2.2.2   Crop Sown Area

Item	1992	1993	1994	1995	1996	1996 over 1992 (%)**
Sown Area˝1000 ha.	149,007	147,741	148,241	149,879	152,445	3,438
Grains	110,560	110,509	109,544	110,061	112,548	1,988
Cereals	92,520	88,912	87,534	89,310	92,208	312
Rice	32,090	30,355	30,171	30,745	31,406	684
Wheat	30,496	30,235	28,981	28,860	29,611	885
Corn	21,044	20,694	21,152	22,776	24,498	3,454
Soybeans	8,983	12,377	12,736	11,232		
Potato	9,057	9,220	9,270	9,519	9,796	739
Cash Crops	24,275	20,689	21,537	22,468	21,581	2,694
Cotton	6,835	4,985	5,528	5,422	4,722	2,113
Oil Crops	11,489	11,142	12,081	13,101	12,555	1,066
Peanuts	2,976	3,379	3,776	3,809	3,616	640
Rapeseeds	5,976	5,300	5,783	6,907	6,734	758
Sugar Crops	1,906	1,687	1,755	1,820	1,846	60
Sugarcane	1,246	1,088	1,057	1,125	1,189	57
Sugar Beets	660	599	708	695	657	3
Fiber	434	420	372	376	349	85
Tobacco	2,093	2,089	1,490	1,470	1,853	240
Vegetable and melon	7,030	9,262	10,042	10,616	11,693	4,663
Other Crops	7,142	7,356	7,117	6,731	6,623	519
Cropping Index	156.0	154.9	156.0	157.8	159.7	3.7*

Source: Chinese Statistic Yearbook ,1996. Abstract of Nationwide Agricultural Statistics,1996.
* represents percentage of growth rate.** The figure is calculated by the author.

In reviewing the above, I think the following comments can be made: (1). We should have never neglected agriculture, rural area and farmers in China, which has 1.2 billion population and 900 million farmers, otherwise, it may lead to disastrous situation and can negatively impact economic development and social security. It was just as Mr. Deng Xiaoping suggested that "if there is setback in agriculture, it will take at least three to five years for recovery", "if there is a problem in economy, it is possible to occur in agriculture". We must pay more attention to agriculture to avoid such things happening.

(2). While transforming from a centrally planned to a market oriented system, it is necessary to support and protect agriculture because labor productivity and comparative advantages in agriculture are lower than other sectors, and market and competition do not favor agriculture. It is important to plan the overall economic environment so that agriculture can benefit and can coordinate with industrial development. It is essential to narrow price gap (called scissors differential) between agricultural and industrial products for protecting farmers' benefit in farming and to increase investment in agricultural R&D.

(3). Given the importance of grain in our economy, grain production should never be ignored when adjusted sector structure in rural area. In China, grain is a merit commodity because it is the basis of subsistence for 1.2 billion population and also it influences overall stabilized prices and smooth transformation of the system. Further more, grain production not only connects with farmers' life and benefits but also impacts on development of animal husbandry and aquaculture, and sector adjustment in rural area. We should remember and take warning from the recession of grain production in the early 1990's.

Characteristics in Agriculture and Rural Development

Grain production broke through three years of stagnation, and made the highest record in history.

After three years of decline, total grain production begun to increase in 1995, and reached 490 million tons in 1996: the largest harvest in China's recorded history. It took less than four years to realize the minimum target of the Ninth Five - Year Plan.

There were four characteristics important in changes of grain production in China:

Table-2.2.3 Major Agricultural Products

Item	1992	1993	1994	1995	1996	1996 Over 1992
Grains (mil t)	442.66	456.49	445.10	466.62	490.00	10.7
Rice	186.22	177.51	175.93	185.22	195.10	4.8
Wheat	101.59	106.39	99.30	102.21	110.57	8.8
Corn	95.38	102.70	99.28	111.99	127.47	33.6
Soybean	10.30	15.31	16.00	13.50	13.22	28.3
Potato	28.44	31.81	30.25	32.63	35.36	24.3
Cotton(1000t)	4,508	3,739	4,341	4,768	4,203	-6.8
Oilseeds (1000t)	16,412	18,039	19,896	22,503	22,106	34.7
Peanut	5,953	8,421	9,68 2	10,235	10,139	70.3
Rapeseed	7,653	6,939	7,492	9,777	9,201	20.2
Fiber(1000 t)	938	960	747	897	795	-15.2
Sugar Crops(mil t)	88.08	76.24	73.45	79.40	83.60	-5.1
Sugarcane	73.01	64.19	60.93	65.42	66.88	-8.4
Sugarbeet	15.07	12.05	12.53	13.98	16.73	11.0
Tobacco (1000 t)	3,499	3,451	2,238	2,314	3,234	-7.6
Flue-cured	3,119	3,036	1,940	2,072	2,946	-5.5
Veg. & melon (1000t)					333.11	
Vegetable			166.02	257.23	303.79	
Melon					29.32	
Fruits(mil t)	24.40	30.11	35.00	42.15	46.53	90.7
Apples	6.56	9.07	11.13	14.01	17.05	159.5
Citrus	5.16	6.56	6.80	8.22	8.46	64.0
Pears	2.85	3.22	4.04	4.94	5.81	103.9
Bananas	2.45	2.70	2.90	3.13	2.54	3.7
Tea(1000 t)	560	600	588	589	593	5.9

Source: Chinese Statistics Yearbook,1996. Abstract of Nationwide Agricultural Statistics,1995 and 1996.

(1). The proportion of rice, wheat and corn in total grain production rose from 86.6 percent in 1992 to 88.8 percent in 1996, among which corn production surpassed wheat since 1995 and became the second major grain crop in China. This change was caused by rapid growth in feed demand.

(2). After a long period of stagnation, soybean production made the highest record, 16 million tons in 1994. Although it fell in the next two years because of the price and other reasons, soybean production still reached 13.22 million tons or increase of 28.3 percent over that in 1992: the highest growth rate in the major grain crops.

(3). Grain yield per hectare grew at a high rate, increasing 12.7 percent during 1992 - 1996, or an increase of 137.8 kg annually. Corn production grew fastest at 14.8 percent or 167.5 kg annually. This was due mostly to adoption of hybrid seed and advanced cultivating technologies.

(4). Quality and profit of grain were emphasized on the basis of increase in grain production. Breeding and extension of qualified varieties were promoted since "Development in High Yield, High Quality and High Efficiency Agriculture " was issued in September, 1992. To solve the problem of inferior longgrained nonglutinous rice, some advanced varieties represented by "Zhong You Zao 5" had been bred recently and were in extension, it was predicted that sown area of long - grained nonglutinous rice with high quality would increase from 1 million hectares present to 3.33 million hectares by 2000 and the share would increase from 12 percent to more than 40 percent of total sown area of early rice. At the same time, wheat and corn with high quality will also develop.

Among cash crops, oil crop, fruit and vegetable production increased rapidly, tea production increased slightly, while cotton, sugar, fiber and tobacco production decreased

Demand for cash crops will increase with improvement of living standard and change in food structure, particularly demand for fruit, vegetables and oil increases significantly. During 1992 - 1996, output of fruit grew 90.7 percent, or 5.53 million tons annually; oil crops increased 34.7 percent (peanut production increased 70.3 percent); vegetables grew from 166 million tons to 304 million tons or 83 percent during the years 1994 to 1996. Total output of peanut, oil seeds, apple, pear, tomato and yellow melon ranked among the top in the world.

However, output of cotton, sugar crops, fiber crops and tobacco fell. Output of cotton fluctuated between 4.2 and 4.76 million tons after reaching the record 5.675 million tons in 1991, Cotton production fell (even to a lowest point since 1987) primarily because of bad weather, unfavorite cotton prices, procurement system and problems associated with the transforming economic system.

Table-2.2.4 Major Crops Yield (KG/Ha.)

Item	1992	1993	1994	1995	1996	1996 over 1992 %
Cereals	4,342	4,557	4,500	4,659	4,893	12.7
Rice	5,803	5,854	5,831	6,024	6,212	7.0
Wheat	3,331	3,518	4,426	3,541	3,374	12.1
Corn	4,533	4,962	4,693	4,917	5,203	14.8
Cotton	660	750	785	879	890	34.8
Peanut	2,000	2,491	2,564	2,586	2,804	40.2
Rapeseed	1,281	1,309	1,295	1,415	1,366	6.6
Fibers	2,160	2,449	2,020	2,534	2,483	15.0
Sugar cane	46,217	59,012	57,670	58,134	56,232	21.7
Suggar beet	22,832	20,120	17,935	20,129	25,470	11.6
Tobacco	1,672	1,652	1,502	1,573	1,745	4.4
Vegetable			18,610	27,034	28,959	
Melon			19,410		24,369	

Source: Chinese Statistics Yearbook,1996. Chinese Agricultural Statistics Information, 1993 and 1994

When cotton harvests were good, there were problems with farmers selling their cotton, and they could seldom received cash directly and immediately (called Da Bai Tiao --- blank receipt). Moreover, input price soared. Consequently, rents from growing cotton fell. For example, the quota price of cotton was 420 Yuan per 50 kg in1994, while profit of cotton was only 2550 Yuan per hectare in major cotton province in northern China; whereas profit of wheat and corn was above 3000 Yuan per hectare. The quota price of cotton increased

64.8 percent in 1994 and 29.6 percent or 770 Yuan per 50 kg in 1995, however, the problems of low profit in cotton production still existed, while cotton prices in China are near and even greater than that in the international market. Price fluctuations in the international market and a large amount of imports had side effects on domestic cotton production. Under these conditions, the government controlled cotton businesses, cotton markets and prices in order that the demand for cotton fibers in factories was met. Because of its special importance, in a long run, cotton needs to adjust price, to regulate market, to strengthen store system and to improve government macro control. Meanwhile, advanced technologies for growing cotton should be developed to root out the problems of comparative disadvantages in cotton farming.

Animal husbandry grew at a high rate and the supply of animal products improved significantly.

Share of animal husbandry in total agricultural value increased from 27.0 percent to 34.1 percent during 1992 - 1996 (Table 2.2.1), it resulted from continuously high growth rate in animal husbandry and sector adjustment in agriculture.

There were main four characteristics in the development of animal husbandry:

(1). Output of animal products rose significantly and continuously. Total output of meats, eggs, dairy products and sheep's wool were 59.15 million tons, 19.54 million tons, 7.36 million tons and 0.333 million tons respectively in 1996, an increase of 72,4 percent, 91.6 percent, 30.5 percent and 30.0 percent respectively over 1992 (Table 2.2.5).

(2). Meat structure modified further, proportion of pork fell gradually, other meats accounted for an increased composition. Pork output was 40.38 million tons, accounting for 68.3 percent of total meat output in 1996, a decrease of 8.5 percent over 1992 level. Poultry production was 10.75 million tons, accounting for 18.2 percent of total meats in 1996, an increase of 5 percent over that in 1992. Beef and mutton increased from 8.9 percent to 12.4 percent during the same period. It begun to change meats production dominated by pork.

Table 2.2.5 Production of Animal Husbandry

Item	1992	1993	1994	1995	1996	1996 over 1992 %
Output of Animal Products						
Slaughter(1000 hd)						
Hogs	351,697	378,243	421,032	480,491	526,634	49.7
of Inventory (%)	95.1	98.4	107.1	115.9	119.2	24.1*
Beef Cattle	15,192	19,037	25,127	30,497	36,511	140.3
Meat Sheep	102,667	111,595	131,249	165,373	194,261	89.2
Meat Production						
Total (1000 t)	34,307	38,426	44,993	52,601	59,151	72.4
Pork	26,353	28,543	32,048	36,484	40,377	53.2
Beef	1,803	2,341	3,270	4,154	4,949	174.5
Mutton & Lamb	1,250	1,377	1,609	2,015	2,400	92.0
Poultry	4,542	5,736	7,552	9,347	10,746	136.6
Rabbit	185	204	229	268	306	65.4
Dairy (1000 t)	5,639	5,625	6,089	6,728	7,358	30.5
Milk	5,031	4,978	5,288	5,764	6,294	25.1
Poultry Eggs (1000 t)	10,199	11,796	14,790	16,767	19,540	91.6
.Wool (1000 t)	256	259	279	307	333	30.0
Year -End Inventory						
Large Animals (1000 hd)	134,653	139,649	149,184	158,614	166,495	23.6
Cattle	107,642	112,918	123,318	132,058	139,813	29.9
Horse	10,017	9,967	10,037	10,072	10,193	1.8
Sheep (1000 hd)	109,719	111,679	117,445	127,263	132,690	20.9
Goats (1000 hd)	97,610	105,694	123,084	149,594	170,684	74.9

Source: Chinese Agricultural Statistics Information, 1992, 1993 and 1994. Abstracts of Nationwide Agricultural Statistics,1995 and1996. *Growth rate between 1992 and 1996 was calculated by the author.

(3). Productivity in animal husbandry was enhanced greatly by technological progress. Taking swine as an example, slaughter hogs sharing total year-end inventory rose from 95.1 percent to 119.2 percent or increased 24 percent in four years. It resulted from application of advanced technology.

(4). Per capita meat and egg consumption increased greatly. The "Vegetable Basket Program" aimed at increasing vegetable and other fresh products and "Straw Feed Cattle Program" were implemented, as a result, per capita meat production increased from 29.8 kg in 1992 to 49.5 kg in 1996, and so did egg production per capita from 8.9 kg to 16.3 kg. The shortage of meat and egg supply was reversed.

Productivity of fishery enhanced and output of aquatic products grew greatly.

During the recent four years, fishery production continued to grow at a high rate. Aquatic production reached 28.13 million tons (Table 2.2.6), which increased 12.55 million tons or 80.6 percent as comparing with 1992, an increment of 3.14 million tons annually. Nowadays, aquatic production in China accounted for 30 percent of total world production. It was the first time that fishery production reached such a scale and realized such rates of growth in China. In 1996, aquatic products derived from ocean and fresh water sources were 15.60 million tons and 12.53 million tons (55.46 percent and 44.54 percent) respectively. Aquatic products from capturing and farming were 12.818 million tons and 15.315 million tons (45.56 percent and 54.44 percent) respectively. Compared with 1992 level, salt water products declined 4.5 percent whereas fresh water products rose. Meanwhile, varieties of fish, shrimp, crab, shellfish and algae increased.

In recent years, China has made great progress in fish farming and in the management of aquatic reso
urces. The Ministry of Agriculture began to stop fishing in Donghai (Eastern China Sea) and north part of Huanghai (Yellow Sea) in hot seasons of July and August in 1995, which got favorable comments in the world.

Deep-sea fish harvesting developed rapidly in production scale and working regions since 1985. There were 1131 ocean - going fishing vessels and 850 thousands fish products by 1995. Per capita aquatic products reached 23.53 kg in 1996, which increased 10 kg over that in 1992 because of fishery development and implement of the

Table-2.2.6 Aquatic Products (1000 tons)

Item	1992	1993	1994	1995	1996	1996 Over 1992 %
Total	15,576	18,262	21,464	25,172	28,133	80.6
From Salt Water	9,337	10,760	12,415	14,391	15,599	67.1
(1)Sources						
Capturing	6,912	7,673	8,959	10,268	11,222	62.4
Farming	2,425	3,087	3,456	4,123	4,377	80.5
(2) Products						
Fish	5,176	5,460	6,474	7,581	8,235	59.1
Shrimp &crab	1,274	1,391	1,709	1,848	2,047	60.7
Shellfish	2,044	2,919	3,236	3,927	3,997	95.5
Algae	568	694	745	749	929	63.6
Others	275	290	250	286	391	42.2
From Fresh Water	6,239	7,501	9,049	10,781	12,534	100.9
(1) Type						
Capturing	901	1,019	1,153	1,373	1,596	77.1
Farming	5,338	6,482	7,896	9,408	10,938	104.9
(2) Products						
Fish	6,988	7,152	8,626	10,209	11,883	70.0
Shrimp &Crab	124	155	203	271	364	193.5
Shellfish	105	136	155	208	185	76.2
Others	21	58	64	930	101	381.0

Source: Chinese Agricultural Statistics Information, 1992, 1993 and 1994. Abstracts of Nationwide Agricultural Statistics,1995 and1996. *Growth rate between 1992 and 1996 was calculated by the author.

" Vegetable Basket Project". Since the capacity of aquatic product supply was enhanced and prices were stable , there were few difficulties with residents especially in medium and large cities to buying fish products.

Production conditions improved and modernization promoted in agriculture.

Agricultural input increased and production conditions were improved with the help of development of input industry. Fertilizer application was 38.29 million tons in 1996, increasing by 30.7 percent over that in 1992, or 401kg per hectares application (Table 2.2.7).

Plastic film uses were 1.01 million tons, increasing by 29.4 percent as comparing with 1992, areas covered with plastic film rose to 7.45 hectare. China became a country that used plastic film most widespread.Irrigation developed, effective irrigated area was 50.38 million hectares, where the areas not influenced by flood and drought accounted for 73.8 percent. All of above had taken an important role in sustainable development in agriculture.

Agricultural mechanization developed due to structural adjustments and migration in rural area. Farm machine power was 388.1 million

Table-2.2.7 Fertilizer and Irrigation

Item	1992	1993	1994	1995	1996	1996 Over 1992 %
Fertilizer Applied						
Total (Effective content, mil t)	29.30	31.52	33.18	35.92	38.29	30.7
Average per ha.(kg)	307	331	350	378	401	30.6
Irrigation						
Effective area(mil ha.)	48.50	48.79	48.79	49.37	50.38	3.9
Area not affected by flood and drought(mil ha.)	35.37	34.78	35.27	36.11	37.19	5.1
% of total(%)	72.9	71.3	72.3	73.1	73.8	1.2
by machine(mil ha.)	31.89	31.64	31.53	32.21	32.89	3.1
% of total(%)	65.8	64.8	64.6	65.2	65.3	-0.08
Plastic Film Applied						
Total (1000t)	781	707	887	915	1,011	29.4
Land usage(1000t)	380	375	426	470	562	47.9
Cover area(1000ha.)	4,386*	5,933	6,777	6,493	7,449	69.8

Source: Chinese Agricultural Statistics Information, 1992, 1993 and 1994. Abstracts of Nationwide Agricultural Statistics,1995 and1996.
** Fertilizer applied per ha.;area not affected by flood and drought; proportion of irrigated area by machine and its growth rate are all calculated by the author. * represents 1991's figure.

kw in 1996, increasing 28 percent as comparing with 1992. Machine cultivated area accounted for 57.8 percent, machine sowed area accounted for 57.8 percent and machine harvesting accounted for 21.9 percent (Table 2.2.8). During the same time, rural electricity usage was 1,834.3 billion kwh, an increase of 65.7 percent over 1992 level, and the usage per hectare reached 1921 thousand kwh. It suggested that agricultural modernization became an important agenda again in the process of agricultural development.

Farmers' living standards continued to improve and heads toward the "well-off state" after surpassing subsistence level.

With development in agriculture and rural economy, farmers' income, consumption and life quality enhanced. Per capita net income in rural area increased from 709 Yuan to 1926 Yuan during 1991 - 1996, or a real growth rate of 5.7 percent annually per ha.; rural electricity usage and electricity usage per household are calculated by the author

At the same period, expenditure of rural household rose from 620 Yuan to 1572 Yuan, or annual growth rate of 6.4 percent. According to the comprehensive index of better- off in rural area, the index of farmers' life rose from 58.5 points in 1991 to 77.7 points in 1996, it suggested that average farmers' living standards were surpassed subsistence level and began to step up to better-off. There were 34 counties of being better - off in Jiangsu Province in 1996, accounting for 52.2 percent of all the counties in the Province; and 19 counties near the better-off standards, sharing 29.2 percent of total. It meant that the process of being better-off was speeding up in the developed regions in China.

In 1996, food expenditure accounted for 56.3 percent of total consumption, housing expenditure for 13.9 percent, 8.4 percent for culture, entertainment and services, 7.3 percent for clothing, 4.5 percent for applications and other equipment, 3.7 percent for medical treatments and health, 3.0 percent for traffic and communication and 2.0 percent for others.

Commodity expenditure rose to 65.3 percent, increasing 3.2 percent as comparing with 1991. It implied that consumption structure of rural households was optimizing and life quality was being improved.

Table-2.2.8 Agricultural Machinery and Rural Electricity Usage

Item	1992	1993	1994	1995	1996	1996 Over 1992 %
Agricultural Machine Power						
Total (million kw)	303.08	318.17	338.03	360.60	388.10	28.1
Per Ha. Cultivated Land (1000kw)	3.18	3.35	3.56	3.80	4.07	28.0
Tractors						
Large & Medium (number)	758,904	721,216	693,154	670,308	670,848	-11.6
Power(million kw)	2,630	2,533	2,464	2,401	2,415	-8.2
Small(million number)	7.51	7.88	8.24	8.63	9.19	22.4
Power(million kw)	67.20	70.43	74.00	78.35	83.85	24.8
Level of Agricultural Mechanization						
Cultivated Area By Machine (%)	53.8	55.4	55.2	56.3	57.8	7.4
Sown Area By Machine(%)	17.7	18.3	19.0	20.0	21.4	20.9
Harvested Area By Machine(%)	9.1	10.9	10.5	11.2	12.0	31.9
Rural Electricity						
Total Usage(billion kwh)	110.7	121.9	147.4	165.6	183.4	65.7
Per Household Usage (1000 kwh)	484	530	636	712		
Per Ha. Usage(1000 kwh)	1,160	1,282	1,553	1,744	1,921	65.6

Source: Chinese Statistics Information, 1992, 1993, and 1994; Abstracts of Nationwide Agricultural Statistics,1996. Chinese Statistics Yearbook, 1995; Abstract of National Statistics, 1996. Agricultural machine power per ha; rural electricity usage and electricity usage per household are calculated by the author.

With income increment, farmers' housing was improved and per capita living space was 21.7 square meters in 1996, which increased 3.2 square meters as comparing with 1991. At the same time, household saving in rural area was 767.1 billion Yuan, increasing 2.3 times or 535.2 billion Yuan.

Main Problems and Measure.

Recently, agriculture in China has made great progress, however, there are many problems yet. The problems and measures are as follows:

The foundation of agriculture is not solid and the external environment for agriculture needs improving.

For a long time, Chinese Government has maintained strategy that agriculture is the foundation of the national economy. However, agriculture is still unstable and fluctuated. There were two periods when agricultural production and specially grain production stagnated in the late 1980's and early 1990's. In some developed regions, neglecting agriculture and grain production also appeared while rural industry grew. Consequently, agriculture shrank, inflation was aggravated, and central and local government had to resume addressing agriculture and grain production. Although the growth rate between industry and agriculture was adjusted in recent two years, unbalanced development still exist. During 1991 - 1994, the ratio of agricultural growth rate to industry's was 1: 4.7-5.6, while the most suitable ratio should be 1:2.7-3.0 like in 1995-1996. It implied that the foundation role of agriculture needs further strengthening and never neglecting.

Agricultural and rural structure requiring adjusting and the internal structure needs optimizing.

In the period of 1993-1996, agricultural proportion declined 5.2 percent, while proportion of animal husbandry and fishery increased 6.4 and 2.8 percent respectively and forestry proportion kept the same, but only 4.9 percent. This suggested that the adjustment direction was correct and further adjustment was necessary in agriculture, animal husbandry and fishery sector and also in sub-sectors. Taking agriculture as example, its structure should meet the demand of household consumption and development of animal husbandry, the quality needs improving with quantity increases, and especially rice and wheat with high quality should increase. At the same time, food used and feed used grain should develop distinctively, in other words, feed production directs towards specialization, industrialization and a large scale, particularly, corn with high protein, green corn and forage grass used as feed should be addressed.

It is essential that TVE structure in the rural economic structure. Be readjusted. The industry value is 70 percent share of gross TVE value, while the value of agricultural enterprises and services only accounted for a little part. TVE development should focus on agricultural processing and food industry, so can absorb agricultural comparative advantages to help its survive successfully and develop vigorously. Meanwhile, TVE with inferior goods, heavy pollution and low efficiency should close, merge or shift.

The wide gap of household living between urban and rural and among regions is large and the huge amount of surplus labor waiting for employment in rural area or migration .

Although household living is improved significantly in urban and rural area in recent four years, the differential between them are large and even enlarged in some periods. For instance, the income gap between urban and rural was 1:2.53, 1:2.60, and1:2.47 in 1993, 1994 and 1995 respectively but only 1:2.37 in 1978, and it only declined to 1: 2.27 in 1996. Living standards among eastern, central and western parts of China kept large differences, taking per capita living expenditure as an example, in 1996 it was 1974 Yuan in east, 1451 Yuan in central and 1181 Yuan in west, or 1:0.74:0.60. All of above reflected the rural economic differences in regions. It will take a long time to eliminate or even to diminish the gap, therefore, what should be done at present stage is to prosper rural economy, to make prices of industrial and agricultural products balance, to cut down farmers' economic burden for their incentive to farm, to maintain farmers' rights of self-management, and to prevent harming farmers.

Aggravated contradictions among increased demand for agricultural products, increased population and decreased resources

There were 1223.89 million population in China in 1996, increasing 80.56 million as comparing with 1990, or an increase of 13.427 million annually. During the same periods, cultivated land per capita reduced from 0.0837 hectares to 0.0774 hectares: only one third of the world average. Per capita forests and forage grass area was only one seventh of the world average. Water resources were shortage: 2000 cubic meters per capita, and lack of water was very severe in northern China. There were 450 million labors in rural area, it was large human resources in one hand; and big problem of employment on the other hand. By 1996, although 130 million labors had worked in TVEs, about 70 percent of labor were still engaged in agriculture. Under such conditions, every agricultural labor produced only 1400 kg grain, 160 kg meat and 70 kg aquatic products. There exist large differences in productivity between China and the developed countries, but the production cost is continuing to go up in China, as a result, China is in the unfavorite conditions in international competition.

In a long run, a basic approach to solve above problems is to carry out strategy of flourishing agriculture by education and technology, which will transform traditional agriculture into that applied modern science and technology and will raise farmers' educational level, and will solve agricultural problems finally with high labor and land productivity and high output ratio of resources. At the same time, the strategy of sustainable development should implement to form better resources and favorite environments for generations' survival and development .

Conclusions

Since reform and opening to the outside world, China has made great strides in economic and agricultural development. Recently the national economy is growing rapidly, agricultural productivity is continuing to improve, and transformation from a central planned into a socialist market economy is accelerating, it shows the prosperous future of agricultural development in China.

In reviewing of agricultural development during 1993-1994, I believe following conclusion:

(1). During the four years, although some new problems and contradictions emerged in China's agriculture, over all, the achievements were outstanding. Supply of agricultural products was much better than that before, and farmers living standards improved significantly. Given the direction and speed of development, by the year 2000, China should be well positioned to feed itself in the 21th century.

(2). It needs to develop new institutions and increase vigor to establish market systems in rural areas. To improve the household responsibility system and performance of farmers' cooperatives, the following issues should be addressed.

    1.   Farmers' rights and benefit, and improving their living standards should be the starting points of government policy.

    2.   Solve the surplus labor problem in rural area should receive considerable attentions. The significance and difficulties involved in solving this problem is as the same as that of unemployment in the state enterprises in cities.

    3.   Actual increases in agricultural investment in capital, physical and technology should be attained in order to extend new agricultural technology, to accelerate the process of modernization, to solve low productivity problems in agriculture thoroughly, and to raise agricultural comparative advantages and market competition greatly.

References:

1.   PCR State Statistic Bureau1996 and 1997. and Chinese Statistic Yearbook 1995, and 1996, China Statistics Publication Company,

2.   PRC Ministry of Agriculture, Abstract of Nationwide Agricultural Statistics 1992, and 1993, Agricultural Publication Company, 1996 and 1997.

3.   PRC Ministry of Agriculture, Annual Report of Agricultural Development in China 1995, and 1996, Agricultural Publication Company.

4.   Jiang Jinaping and Cheng Xihuang, Agriculture, Rural Area and Farmers and Modernization Construction, Agricultural Publication Company, 1996.

5.   Jiang Zheming, Strengthen the Agricultural Foundation Role, Further Reform in Rural Area and Promote Complete Development in Rural Economic and Social Development, RENMINRIBAO, July 15, 1996.

6.   Central Committee of the Communist Party of China Held Rural Working Meeting, RENMINRIBAO, October 19, 1993.

7.   Central Committee of the Communist Party of China Held Rural Working Meeting to Arrange Rural and Agricultural Work, RENMINRIBAO, March 24, 1994.

8.   Li Peng, The Ninth Five-Year Plan of the National Economy and Social Development Report on Prospect Targets by 2010, People's Publication, 1996.

9.   PCR State Statistics Bureau, Household Consumption in Rural Area Stepped into New Stage, China Information News, September 3, 1997.

CHAPTER 2.3

Expectations and Strategies for Food Security in China by the Early 21st Century

Mei Fangquan

CAAS. Infomation Institute, Ministry of Agriculture, China

China's food and agricultural development is entering into a historically important stage from now to the 2030s when population will peak while arable land will decrease to its lower level; During this period, exploitation of presently non-arable land resources will require a large investment and the Chinese people will in general have a much better living standard. In this period, the development of China's food and agriculture will be an important asict that attracts a worldwide attention.

Changing trends in China's basic conditlons

It is estimated that by the year 2030, China's population will increase to its peak at about 1.6 billion, per capita arable land will decrease to 0.078 ha., per capita national income 10800 y,uan RMB, and an expenditure on food 1915 y,uan RMB.

Continuous increase of population with its peak in 2030s. During the past15 years. China has her population increased from 987 million in 1980 to1211 million in 1995 with an average of 15 million (1.28%) annually. According to prediction by multiple ways, China will have her population increasing to 1300, 1400-1430, 1470-1540 and1570-1630 million, respectively, by the year 2000, 2010, 2020 and 2030, that will be a period with a biggest population. Thus, it is essential for China to have a strict control on her population increase. The following analyses on food development is based on the above predicted population.

Continuous decrease of arable land to 110 million ha. by the year 2030. In the past, China had a traditional acustomed acreage of arable land decreased from 99.3 million ha.in1980 to 95 million ha. in 1995. However, recent findings from satellite, remote sensing, aerial photography and land survey, indicate this number was incorrect and the actual land is closer to132 million ha.(an increase of 39 % over earlier estimates). It is estimated that this arable land will decrease to 131 million ha. and 125 million ha.. by the year 2000 and 2030; and the per capita arable land will decrease to 0.1 ha. and 0.077 ha., respectively.

Continuous growth of national economy and people's purchasing power double. The nationwide residential consumption increased by 2.63 times, i.e. from 630 Yuan in 1978 to 1659 Yuan rmb in 1995, of which expenditure for food increased from 422 Yuan to 896 Yuan (The Engel's index decreased from 67% to 54%). It indicates a significant improvement in people's living standards. By the year 2000, per capita national income will be 2500 Yuan rmb (calculated with a comparable price in 1990, the same in the follows) and the per capita residential consumption will be 1280 Yuan; and they will be 10800 and 5040 Yuan respectively, by the year 2030 if the per capita national income increases by 5% annually. The expenditure for food (the Engel's index will decrease from 48% in 2000 to 38% in 2030) will be 1915 Yuan. There will be a big increase in the demand for food and the people's living standard will upgrade from better off to wealthy (According to the FAO classification standards, the better off level has the Engel's index 40-49% and the wealthy, 30-39%.).

Growth of Demand for Main Food

By the years 2000, 2010, 2020 and 2030, China will have a total grain demand of 500, 580, 645 and 720 million t, with 30, 38, 43 and 50% of them used as animal feed, respectively. By the year 2020, per capita food consumption will be similar to that in Japan and the main food nutrition at a wealthy level.

Grain demand should be brought into line with total food requirements, even with the consumer goods as a whole for a systematic analysis.

Per capita consumption of various kinds of food increases most rapidly in the period when people's lives become better off after their basic food and clothing problems have been settled. The direct consumption of food grain per capita decreased from 253 kg in 1985 to 234 kg in 1995 and will decrease further to 213 kg by 2000. The proportion of food grains for direct consumption will decrease from 61% in 1993 to 55% in 2000. Meanwhile, there will be a rapid increase in the demand of food from animals, fruits, edible oil and sugar. That shows a significant improvement of quality and nutrition in the diets of the Chinese people, however, food grains is still a basic substance for such an improvement.

Great changes have taken place in quantity and structure of food grain and food consumption during the past 17 years. China had a total grain consumption increased from 300 million tons in 1978 to 465 million t in 1995; per capita food grain consumption, from 300 kg to 385 kg; and the proportion used as animal feed from 12% to 27%. At the same time, there was a trend of rapid increase in consumption of animal food, with per capita consumption of meat, egg and aquatic products increased from 8.86 kg, 1.97 kg and 3.5 kg in 1978 to 28.91 kg, 11.75 kg and 12.09 kg in 1995, respectively; and edible oil, sugar and fruits, from 1.6 kg, 3.42 kg and 6.6 kg to 6.29, 5.42 kg and 30.94 kg. The long-term significant increase in consumption of animal foods and fruits has greatly improved nutritional conditions of the Chinese people.

Major food consumption in 1995:

(Based on the data calculated from China Statistical Yearbook" and "Introduction of National Agricultural Statistics")

	Total amount of consumption (million t)	Consumption/per capita (kg)
Grain	460	235
Meat	50	28.19
Eggs	15	11.75
Milk	0.54	4.05
Aquatic products	22	12.09
Fruits	41	30.94

STRATEGIES FOR FOOD SECURITY IN CHINA

By the year 2000, China's food consumption composition will exceed the comparatively well-off level. Especially the consumption level of animal edible products as compared with prediction data calculated by scientists

Amount of major food demand by the year 2000 (average)

	Total demand (million t)	Demand per capita (kg)	Consumption per capita (kg)
Grain	520	400	213
Meat	60	46	32
Eggs	20	16	14
Aquatic products	30	23	15
Milk	10	8	6
Fruits	46	36	32

in the past . The average energy per capita per day will be 2660 kcal, and protein 72g (high grade protein exceeds thirty percent).By the year 2010, the food structure and level of consumption of residents both in cities and towns in China will be in the mid-stage of comparatively well-off. Food quality would be improved obviously.

Amount of major food demand by the year 2010 (average)

	Total demand (million t)	Demand per capita (kg)	Consumption per capita(kg)
Grain	560-590	400-420	193
Meat	67	48	34
Eggs	26	19	16
Aquatic products	38	24	18
Milk	22	16	14
Fruits	61	44	40

By the year 2020, the Chinese people will have a new developmental stage in nutritional composition of food that would be close to developed countries and regions: The average energy intake per capita/day will be 2600 kcal, protein will be 80g per capita/day (high grade protein by 50% over), fat will be 78g per capita/day (among it, animal fat by 35%).

Amount of major food demand by the year 2020 (average)

	Total demand (million t)	Demand per capita (kg)	Consumption per capita(kg)
Grain	600-645	410-430	173
Meat	75	50	35
Eggs	33	21	18
Aquatic products	48	32	21
Milk	36	24	20
Fruits	75	50	45

By the year 2030, the estimated amount of grain consumption per capita will be 420-450 kg. The direct consumption of grain will be continuously decrease to a value approximate to that of Japan (125 kg per capita). The amount of meat consumption per capita would be steady. Poultry, beef, mutton, aquatic-products and milk will be continuously increased. The feed conversion rate of these animal products is much higher than pork. While by spreading and improving the management of advanced technology, the marketing rate of fattened and cutability of animals and poultry will be increased. Grain consumption per capita will be approximate to that in Japan and Taiwan region.

Amount of major food demand by the year 2030 (average)

	Total demand (million t)	Demand per capita (kg)	Consumption per capita (kg)
Grain	645-720	420-450	140
meat	80	50	35
Eggs	37	24	20
Aquatic products	58	36	24
Milk	56	35	30
Fruits	88	55	50

By the year 2030, major food consumption per capita will be close to that of present day Japan, data is given below.

Comparison of of food consumption between Japan and China

		Grain	Meat	Eggs	Aqnatic products	Milk	Fruits	Vegetables
1990	Japan	125	28	15	40	63	44	114
2020	China	173	35	18	21	20	45	120
2030	China	140	35	20	24	30	50	120

Grain consumption was 440450 kg per capita, based on calculations of equivalent food consumption in Japan in 1990. Since 85% aquatic products in Japan came from marine fishing that did not consume the fish feeds, the actual grain consumption is significantly less than the amount mentioned above. Statistical data analysis of the highest income families of the cities residents in China in the recent 10 years shows that the main grain consumption per capita will approach to the estimated amount per capita in 2020 in the whole country, similar to the consumption per capita and the changing trends of the structural evolution in Japan.

By the years 2000, 2010, 2020 and 2030, 33%, 38%, 43% and 50% of the total grain demands will be used as animal feeds. If we adjust the planting structure, carry out the triple-structure engineering, Speed up development of the integrated production and manage systems of grain crops4eed crops economic crops, planting-breeding-process, production-process-marketing, benefit of high quality feed crops will be 50% higher than that of grain varieties eaten directly by human which are used as feed Up to the first 30 years of the 21 century, increased grain consumption will be basically used as feed. The changing trends of grain consumption demands in the future would be better analyzed from the viewpoint of the benefit of the structure changes.

By the year 2030, the total grain demands will probably be about 645-720 million t with the grain ration 214-224 million t, feed grain 323-360 million t and others (grain for industries and commerce, seeds, and reserve increments) 108-136 million t.

Growth of Production and Supply of Major Food

Grain is important. Major foods in China include grains and animal meet converted from grains. Thus , the grain situation dictates the status of

China's foods. By the years 2010, 2020 and 2030, the grain sowing area can still be 107 million ha, the total grain production could be about 540-560, 570-620, 620-690 million t respectively, just by implementing advanced technology presently available . The analysis of the arable land bearing capacity indicates that China can produce 830 million t of grain annually.

During the past 10 years, the area planted to grain remained nearly 110 million ha. This area should continuously remain above 110 million ha in the long term . By the year 2030, the arable land (statistical area) will be reduced from 130 million ha in 1993 to 123 million ha., but the multiple crop index would be increased from the present 156% to above 160%, the total crop area will approximately be 145 million ha with 74% for planting grain, area sown to grain will be about 107 million ha. (The ratio of the grain planted area: the total crop area equals 74% in 1993).

The potential for increasing multiple crop index in feed crops could be more than 160%. The average multiple crop index all over the country at the present is about 156%. The winter fallow land in the south is 6.7 million ha.. The land for green manure crops can be changed for feed crops. The multiple cropping area can be increased by about 3.33 million ha., giving the total cropping land 153 million ha. alone, significantly more than that in 1993. If the ratio of the grain cropping area remains stable, that would be no less than 110 million ha.

By the year 2010, the total grain production will probably be near 560 million t.
The fertilizer utilization will be increased from 35.95 million t in 1995 to 43 million t (standard fertilizer). The irrigated land will be 57 million ha, with 62% of arable land. Giving the corresponding input and economic management policies, the grain yield will probably increase from 4245kg./ha. in 1995 to 5100-5250 kg/ha., similar to that in early 1990s in Jilin, Liaoning, Beijing, Zhejiang, Jiangsu, Shanghai, Guangdong and Hunan. It may exceed that level if breakthroughs are made in grain production technology and comprehensive management.

The total food grain output may be 570-620 million t by the year 2020 when suitable advanced techniques will be practiced; fertilizer application will be 47 million t; irrigated farmland will be 60 million ha. occupying 67% of total farmland, it is feasible to produce a total grain output of 570-620 million t with a yield of 5400-5625 kg/ha. in all grain producing land, that is that same level in recent 3 years in Beijing, Shanghai, Jilin, Liaoning, Jiangsu and Zhejiang, if measures are taken in combination with policies, input, techniques and management. Both the yield and total output may significantly exceed that level if new breakthrough are made in development of grain crop varieties and various techniques.

China's arable land has a maximum grain producing capability of about 830 million t. Three independent studies on strategies for intermediate-long-term development of food production in China shows that 840 million tons of grains can be produced by China's arable land. The maximum is 830 and 820 million t respectively in the analysis by the Department of Regionalization, Ministry of Agriculture, and by the National Land Management Bureau, by taking into consideration of China's climate, land resources, advanced techniques and materials.

It is possible to achieve total grain production of 645-690 million t and per ha. yield of 6000-6300 kg by the year 2030. In 1995, the true yield was only 3105 kg per ha. (actual survey area). The estimated planting area for developed regions, such as Beijing, Shanghai, Liaoning, Zhejiang, etc. is only 15-20% higher than the actual survey area, and the yield per unit area is closer to the true yield as compared to national average. If 50% of feed grain crops is replaced by high-yielding and high-quality feed crops the per ha. yield of 6000-6300 kg is achievable even by using existing conventional practical techniques alone, by the year 2030.

Major animal food products will continue to grow into the 2030s. The continued growth of grain with basic support and advancements of science and technology will overflow into animal food production and continue to improve the utilization efficiency of food resources. The per capita production of meat, egg, milk and aquatic products will reach, respectively, 44.4, 15.2, 7.1 and 24.2kg in the year 2000, and 47.2, 17.4, 16.7 and 29.0 kg by the year 2010, indicating that the demand for improving diet will be basically satisfied.

During the first 30 years of the 21st century basic demand-supply balance for food grain and major foodstuff can be realized. In order to realize this goal, the proper guarantee of material, technology, capital and effective economic regulatory policy must be attained. Food grain imports can be restricted to 20-30 million tonnes. The shortage of grain and foodstuff lies mainly in feed grain. In view of the comparative advantage of grain in the international markets and the adjustments of grain demand-supply structures, wheat is chosen as the major grain to be imported and the increased domestic grain supply is used mainly as animal feed to increase animal food products and adjust dietary composition.

Strategic Selections for Food Development

In order to attain the basic balance between demand and supply of staple food and to satisfy the ever-increasing demand for higher living standards in the early years of the 21st century, economic system reform and the 2 fundamental changes in economic growth mode must be carried out and at the same time the two strategies, sustainable development and rejuvenating the nation through science and education, must be implemented.

Strengthen Agricultural Infrastructures and Increase Agricultural Inputs. The acreage under effective irrigation should be increased as soon as possible, from 49.37 million ha in 1995 to 53.33 million ha in 2000, 56.7 million ha in 2010 and 73.0 million ha in 2030, bringing 80% of farmland under irrigation. Fertilizer input should be increased from 35.95 million nutrient tonnes in 1995 to 37.50 million tonnes in 2000 and 43 million tonnes in 2010. Meanwhile soil improvements should be accelerated for middle and low productivity farmlands, the goal is to improve 3.33 million ha each year for the next 15 years, so as to secure a stable productivity for the farm lands.

Establish a Macro Management System Combining Production, Supply and Marketing. In order to enforce effective macro regulation, it is urgently needed to set up a perfect and integrated system which takes into account of the national grain and foodstuff as well as an agricultural reserve system, risk foundation system and a price protection system, and to incorporate multi-sector management mechanisms in agriculture, commerce, water conservancy, chemical industry, foreign trade, price, planning etc. into a unified and harmonious system.

Establish a National Food Security and Early Warning System. A perfect and sensitive food security and warning system needs to be developed, which is an essential and powerful tool for macro regulation and plays an important role in the timely prevention, elimination and alleviation of the influences resulting from the total amount, structures and regional imbalance in food grain and staple foodstuffs.

Highlight the Regulation of Price Structure to Promote Stable Growth of Food Production. As China's grain and staple foodstuff prices have reached the level prevalent in international markets, keeping the same price level is required by the long-term strategies for the whole national economy and international trade. Measures must be taken to guarantee farmers' income by regulating the price parities between industrial and agricultural products, especially the price of the means of agro-production. If control over the price of agricultural inputs is difficult to realize, the direct in-kind exchange of such means of production as fertilizers, energy etc. for grains can be used to maintain stable grain prices and agriculture's comparative benefit.

Accelerate implementation of "tri-structure Enginnering" to Increase Overall Benefits of Food Resources. Changing the farming sector from its traditional dual structure of "food crops-cash crops" to a tri-structure of "food crops-feed crops-cash crops". this is characterized by coordinated crop cultivation and animal husbandry development and integration of production, supply and marketing while promoting the quick formation of new and highly profitable industrial agricultural systems. In order to push on the project, comprehensive measures and effective policies are urgently needed and the feed industry should be treated as a relatively independent sector.

Implement Sustainable Food and Agricultural Technology Policy. As most new varieties of food grain and foodstuffs and new technologies for their production create mainly social benefits, effective policies need to be made to give priority support to food and agricultural technologies. At the same time priority support should also be given to those advanced practical technologies which make rational use of agricultural resources, protect the environment and favor high-yield, superior-quality and high-efficiency agriculture combined with sustainable development. Focus should be on strengthening rational location of township and village enterprises and preventing pollution by industries and large farming operation.

Implement Strategy of Sustainable Regional Development. In the eastern district controlling the over and unbalanced use of agricultural chemical, controlling the over utilization of crop land and water area resources are important .

In the central areas with agriculture as the main industry; protection of the rural ecological environment should be strengthened; improvement in soil fertility of medium and low-yielding croplands and increasing the productivity of agricultural resources are important strategies. In the under developed western area, much attention should be paid to water and soil erosion caused by the over utilization of resources, irrational development, over reclamation and over grazing of grassland. This will protect the environment.

From the above analysis it can be concluded that the time from the present to the 2030s is a critical period for China's food and agricultural development. More than 1 billion Chinese people not only can feed themselves but also be able to step forward from the comparatively well-off level into the well-off food consumption period with more balanced diet. With this, Chinese people can make a greater contribution to the world.

Chapter 2.4

Multilateral Institutions in China's Agricultural Production and Trade[1]

Uma Lele,
Albert Nyberg
and Joseph Goldberg[2]

The World Bank,
1818 H Street, NW, Washington, DC, 20433.

China-World Bank Partnership: A Success Story

World Bank-China partnership has grown in strength and scope since China became member of the World Bank Group in 1981. It is now a partnership of global significance. China is the World Bank's largest and one of the best borrowers. With nearly $28 billion committed to China through 184 loans and credits over the 1981-1997 period, the World Bank has become China's largest multilateral lender, although the World Bank's 1996 commitments to China as a share of Chinese external capital inflows that year were only 8 percent of the total. The agricultural sector has been the largest recipient of those commitments (slightly over 25 percent of the total), substantially higher than the sector's share in the overall World Bank lending, which now stands about 15 percent. The importance of agricultural lending to China is explained by two factors. With one fifth of the world's population and one of the fastest growing economies in the world, China has the

---

[1] A paper prepared for a meeting on Agricultural and Sustainable Development: China and Its Trading Partners Symposium at The Texas A & M University System, January 12-15, 1998.

[2] Advisor, Agricultural Research Group, Environmentally and Socially Sustainable Development Vice Presidency; Senior Agricultural Economist, China and Mongolia Department, East Asia and Pacific Region; and Sector Leader, Rural Development and Environment Unit, Europe and Central Asia Region, respectively; The views expressed in this paper are those of the authors and do not necessarily represent the views of the World Bank. We are grateful to Jason Yauney for excellent research assistance.

potential to significantly affect the global food and fiber situation. That will depend on China's future demand growth, the capacity to meet that demand domestically and the extent to which China will enter global markets. Another factor is China's project performance. According to the World Bank's Operations Evaluation Department which routinely audits all World Bank financed projects, the Chinese projects show a the highest success rate regardless of how effectiveness of the World Bank lending is assessed, i.e., the extent to which loan conditions are met, the speed of project implementation, project impact as reflected in return on investments effect on poverty reduction or prospects for sustainability. Agricultural sectors typically faired less well than some other sectors in the World Bank's overall lending, but once again China's success stands out. (World Bank, 1997b, 1995, 1994 OED Reports)

What explains this outstanding performance, what role has the World Bank played in Chinese agricultural development and trade, how can it be expected to contribute in the future, and what lessons does the World Bank-China partnership offer to other World Bank member countries?

This paper explores these issues. It first outlines agriculture's role in China's overall economic development since 1978 when economic reforms began, followed by the challenges China faces in the next 20 years. This section draws on the recent 7 volume World Bank economic report on China. Evolution of the World Bank's lending is considered next, followed by discussion of the issues posed in future Bank lending to agriculture, as China graduates from a low income country, recipient of concessional IDA credits to a full-fledged middle income borrower of World Bank loans. The paper ends with some general lessons from the World Bank-China partnership.

China's Economic Performance since Reforms in 1978

China's performance must be considered no less than an economic miracle. Agriculture has played a central role in that miracle. Starting from being one of the poorest countries in the world, with 60 percent of the 1 billion population living below poverty and earning less than $1 a day, China has experienced one of the fastest rates of agricultural and overall economic growth. Its agricultural growth of 6 percent and industrial growth of over 8 percent per capita, for two full decades (from 1978 to 1997), has also been remarkable for its speed and duration. Its transition to a market economy has been unique for its combination of experimentation and incremental reforms leading to rapid progress in several areas, although agriculture, which was a clear leader in reforms now lags behind other sectors.

Since 1978 China has lifted over 200 million people out of poverty, an unprecedented decline. Again agriculture has played an important role in poverty reduction. By international standards China's social indicators as reflected in close to universal access to primary education, low infant mortality and high life expectancy have been outliers, in view of China's low initial per capita income. (See Table I) China's integration with the world economy has advanced rapidly, leading to a strong external position including rapid export growth and reserves estimated to be well over $100 billion in 19%. China has essentially privatized farming, liberalized markets for many goods and services and intensified competition in industry while introducing modern macro economic management. Both the transitions to urbanization and liberalized economies have taken long time in most industrialized and currently

Table 1 Under-S Mortality Rate in China and other Asian Economies (deaths per 1,000 live births)

Year	China	India	Indonesia	Sri Lanka
1960	173	235	214	140
1965	144	n.r.	n.r.	n.r.
1970	115	n.r.	n.r.	n.r.
1975	85	195	151	69
1980	60	n.r.	n.r.	n.r.
1985	44	n.r.	n.r.	n.r.
1990	44.5	127	111	22

n.r. Notreported

developing countries. It took the *UK* 58 years (from 1780 to 1838) to double per capita incomes, the US 47 years (from 1839 to 1886) Japan 34 years (from 1885 to 1919) and Korea 11 years (1966 to 1977). China has doubled its income twice in periods of 10 years each (1978 to 1996). Whereas liberalization of the economy has also been fraught with many risks and set backs in Eastern block countries, China has telescoped both these in a relatively short period and done so successfully.

Key Features of China's Econoinic Miracle

Three features of China's growth have been significant. First is its regional dimension. if China's 30 provinces would be counted as individual economies, the twenty fastest growing economies in the world would have been in China including many of the coastal provices Although growth has by no means been confined to these, coastal regions, the loss of highly fertile land in them to industrialization is one of the costs to food self sufficiency as discussed later. Second, there have been large cycles in the rates of growth in the provinces with fluctuations in the rates of inflation, although China has not experienced the runaway inflation of Latin American countries or the FSU (the Former Soviet Union). The third is its impressive productivity growth. China's growth has been less dependent on the conventional inputs of capital and labor. Only 37 percent of the growth is accounted for by the growth in capital and another 17 percent by increases in the quantity and quality of labor. Nearly half of China's GDP growth 4.3 percent per year is explained by other factors. Agriculture's performance was even more impressive. The annual rate of agricultural growth averaged 7.4 percent and total factor productivity growth was 6.6 percent during 1978-85 period, i.e., during the period of the household responsibility system. In the 1985-95 period the annual growth rate declined to 5.8 percent, but most of the growth came from additional inputs and investments with total factor productivity growth declining to 1.1 percent (between 1985 and 89). Cereal production hardly increased, and yet nonstaple foods, (livestock, fisheries, vegetables and melons) grew at 9 percent or more alter 1985. These slow downs in total factor productivity and the varying rates of agricultural growth raise important issues about future sources of growth of domestic production and trade discussed below.

Four factors have driven China's growth: high savings rates averaging 37 percent of GDP between 1978 and 1996, one of the highest savings rates in the world, structural changes, reforms and the special conditions in 1978. As other papers in this volume indicate, agriculture has been at the forefront of many of the reforms which have led to the structural changes, e.g., the decline in the agriculture's share in the workforce from 71 percent to about 50 percent in twenty years. Once again it took the U.S. 50 years and Japan 60 years to achieve the same rate of decline. Agricultural reforms consisted of increasing procurement prices for grains, allowing farmers to sell above quota production at market prices, lowering grain quotas, increasing grain imports, and expanding private interprovincial trade. But certainly the most important feature of the reforms has been the household responsibility system, in which collective land was assigned to households for up to fifteen years and local governments took the strong initiative of transferring production decisions and profits from communes to households. Agricultural reforms became the cornerstone of the reforms in the entire economy providing the basis for reform of the industrial structure. Increased rural household incomes increased demand and markets in rural areas, while increased rural savings and investments took advantage of thegreater profitability of rural industry. These changes were supported by changes in political philosophy and attitudes leading to reduced government controls, increased contracting of rural enterprises by the urban industry, reduced production restrictions and lower taxes. The combination of reforms resulted in collective enterprises accounting for 36 percent of the industrial output by 1988 which is where it remains today.

Other factors which explain China's extraordinary performance are the reform of the state enterprises, the substantial growth of the village and township enterprises and more recently privately and individually owned enterprises. The decentrailization, pragmatism and incrementalism. A favored which must also be given credit is the use of "models", local authorities experimenting with policies in specific provinces, prefectures, countries and even firms. If the experiment works, then it becomes a model for others to emulate and is replicated. If the reforms fail, then the cost of failure tends to be contained and limited. Reforms have occasionally been reversed if the government believed that growth was not being served or the stability was jeopardized. This highly decentralized approach to reforms in recent years is in sharp contrast to that followed during the periods of the Great Leap Forward and the Cultural Revolution and seems almost to have been a reaction to the previous experience.

Reform of the Trade System

Although agricultural trade reforms have been the last to be introduced, overall trade reforms set the stage for agricultural trade liberalization. Thousands of corporations now trade internationally, and consistent with a decentralized strategy many are sanctioned by local rather than the central government. Tariffs on imports have been reduced and procurement targets for export products, which applied to over 3,000 in 1979, are zero today. Exporters retain a portion of the foreign exchange receipts, with more liberal laws in retention of foreign exchange by individual traders. Realistic exchange rate also has done a great deal to stimulate exports together with liberalization of restrictions on direct foreign investment creating special economic zones. By 1993, more than 9,000 zones largely in the coastal areas led to substantial growth of foreign investment by foreign entrepreneurs bringing in new technology, services and increasing competitiveness.

China's well-to-do diaspora spread throughout the world has also been an important source of investments and skills which the government has been pragmatic enough to encourage. Nearly 70 percent of the $37 billion invested in China in 1996 through foreign direct investment (FDI) came from the Chinese diaspora. Much of this investment has gone into relatively small scale labor intensive manufacturing, also contributing significantly to employment and income generation and a wider base of business management skills

Challenges for the Future

As would be expected rapid changes have given rise to a new set of challenges: periods of macro economic instability stemming from partially complete reforms; increased insecurity in employment and incomes; mounting environmental pressures associated with population pressure, agricultural intensification and urbanization; rising cost of food self-sufficiency; growing inequality; pockets of stubborn poverty and an occasionally hostile world trading environment

China has demonstrated remarkable capacity to meet these challenges. Based on China's strengths the World Bank has projected annual per capita income growth in China up to year 2020 between 4 to 8 percent, with a strong likelihood of 7 percent growth annually. To realize such high economic growth rates, government policies and the pace of reforms will be critical in several areas: skilled economic management including particularly the management of the government budget and public expenditures, reform of the public enterprises and the banking system, deepening of the capital and other commodity markets. in addition, rapid economic growth in China could be hamstrung by environmental degradation, food insecurity, income and employment insecurity, domestic social instability and a possibly more hostile world economic environment.

Food Security

A combination of rapid economic and income growth, population growth and urbanization is likely to result in major increases in food consumption expenditures, in part reflecting a shift in the consumption from cereals, root crops and tubers to meats, poultry, chicken, marine products, fruits, vegetables, dairy products and oilseeds.

China's ability to feed itself has been a matter of intense debate in recent years. Following China's substantial imports in 1995 Lester Brown in his book "Who will feed China?" predicted alarmingly high rate of growth in food imports of up to 200 million to 370 million tons by 2030 associated with higher rates of Chinese per capita consumption, with adverse effects on the low income food import-dependent countries, particularly in Africa (Lester Brown). Brown's predictions have been refuted by virtually every analyst of the global food situation (Alexandratos 1996 and 1997; Paarlberg 1997; Rozelle and Rosegrant 1997; Pinstrup-Andersen, Pandya-Lorch, and Rosegrant, 1997; World Bank 1997.) and the Government of China (Information Office of the State Council of the People's Republic of China 1996). However, the Brown projections did help focus the attention of the Government of China, the World Bank and others to more critically examine the food security prospects for China. Some analysts have forecasted that China will become a major food exporter (Chen and Buckwell 1991, Mei, 1995). A few others have predicted that China will become a major importer (Garnaut and Ma 1992, Carter and Funing 1991, Brown 1995). The World Bank has projected demand for food grains to rise to almost 697 million tons of trade grain, 608 million tons of milled grain equivalent including 206 million tons of feed grain in 2020, a significant portion of it due to a rapid rate of urbanization and change in the demand patterns related to it, including the shift from cereals to meat, poultry and fish. The World Bank argues that, *provided China invests sufficiently in agriculture*, it will be able to supply most of its domestic food needs with likely increases in food imports to the tune of between 30 million tons to 90 million tons by 2020.

Alternative Quantification of Chinese Production and Trade Scenarios

Rozelle and Rosegrants' alternative simulations also suggest that *only with extraordinary income growth, severe resource degradation and failure to investment in agriculture could China's net cereal imports increase by 85 million tons in 2020 leading to cereal price increases of 10 percent relative to the base scenario.* However, they project increases in government investment of 5 percent annually could make China a net exporter of 31 million tons in 2020, causing prices to decline by 11 percent.

Expanded livestock production could also put pressure on cereal imports in their simulations causing world market prices to increase by 10 percent depending on the technologies China uses in livestock intensification. *Furthermore without further investment in agriculture,* China's meat imports are also estimated to increase to 24 million tons from the current low level of 0.4 million tons. However, with investments and associated improved efficiency in the used animal feed and technologies, net meat imports could be half (0.2 million tons) of those currently.

Importance of Increased Investments and Policy Reforms in Agriculture

It is clear that the *precise levels of Chinese food trade will depend on investments in agricultural research, fertilizer production and import policies, investment in irrigation and pricing and trade policies, but they are unlikely to contribute more than a marginal increment to global cereal trade.*

Agricultural Land

China's agricultural land is exceptionally small relative to its population. Official sources list cultivated and sown areas to be 95 million and 145 million hectares, respectively. However, satellite imagery indicates that the true cultivated area is about 132 million hectares. The 40 percent additional cultivated area implies that fertilizer application rates and yields are 40 percent less than indicated in official statistics and offer considerable potential to increase yields with important implications for food security. Under this scenario, the wheat and corn yields are similar to those in other Asian countries and well below the yields in major developed countries. According to some estimates since 1988 capital construction has removed 190,000 hectares of land from cultivation annually. Other estimates show greater loss of land. This has been compensated by land reclamation of 245,000 hectares annually. However, the land lost in the Shandong and Southeastern coastal provinces is high quality land where multiple cropping is practiced compounding the loss. Reclaimed land tends to be of poorer quality. Besides, the cost of land reclamation has varied between $2,000 to $20,000 per hectare depending on the terrain.

Agricultural Investments

Whereas China increased agricultural investments considerably in the 1960s and 70s, growth in investment has slowed in the 1980s and 90s. In the government's 9th plan strong emphasis has been placed on agriculture in general and grain production in particular. Agriculture still contains a large share of the population, is an important source of rooting out poverty and keeping urban food prices low, an essential

61

ingredient of urban political stability. But there are two quite different routes China could take. One is to continue the current emphasis on increased grain production and self sufficiency through a policy of controlled grain and fertilizer markets and subsidies as China has pursued in the past. The other is to increase production of relatively high value and labor intensive crops such as fruits, vegetables, poultry, piggaries and marine products while relying on the international market for the supply of grains.

Policy Reforms

Although the government has freed many markets, it still controls food grain markets, holds one of the largest stocks of grain equivalent, estimated to be between 260 to 400 million tons, if stocks at the level of prefectures, counties, provinces and the national level are considered. This level of stocks allows it to maintain 95 percent self sufficiency. National and provincial level stocks of 40 million and 90 million are considered necessary. They are acquired by purchasing three quarters of all marketed grain at well below market prices, a level considered to be important to maintain low urban prices for consumers. Government pays provinces through the budget and the banking system for maintaining stocks. Taken together these policies cost China Y 85 billion annually, more than 5 times the resources the government spends on anti-poverty programs. The expenditures constitute three percent income tax on every man, woman and child. As agricultural supplies have far exceeded population and income growth, consumer reliance on government distributed grain has been reduced to a minimum in the last several years. This suggests that the government could gradually liberalize the domestic grain market reducing for instance, government purchases from 75 percent to 25 percent of the marketed grain output by 2020, while also liberalizing international trade in grain, in much the same way that it has liberalized industrial production and trade both domestically and internationally.

Fertilizer Consumption

Fertilizer consumption in China has quadrupled since 1978. But China still uses poor quality fertilizers and there is much scope for improving the efficiency of fertilizer applications. While the rates of fertilizer application are still lower than in Japan or Korea, the marginal return to fertilizer use has declined due to the imbalance in the use of fertilizers with far too much emphasis on nitrogenous fertilizers and far too little on potash. Control of the fertilizer imports and emphasis on domestic self sufficiency may explain the imbalanced use of fertilizers. Policy reforms in the fertilizer sector involve liberalized trade in fertilizers, particularly in the types of fertilizers in which China is not a producer, rationalization of the domestic fertilizer industry to increase efficiency in production, and a more balanced application of fertilizers.

Agricultural Research

Agricultural Research has done much to increase agricultural production. Whereas government expenditures on research increased considerably in the 1970s, research spending as a share of agricultural output has decreased steadily over the last fifteen years and currently stands at between 0.4 to 0.5 percent of agricultural GDP (excluding investments by provinces), while the pressure on the research system to earn resources by commercialization of research results has increased. As a share of agricultural production, China's central government investment levels are a quarter of those in industrial countries if only public sector research is considered and perhaps one-tenth of those if investments in public and private sector research are compared, and perhaps half of those in Brazil. Yet only a small share of the income earned from commercialization of research is allocated to research (Pray 1997). The research by Rozelle illustrates that the number of new rice cultivars developed, released and commercially used in the first half of the 1990s was similar to that of the first half of the 1980s, but 25 percent below the output in the second half of the 1980s (Rozelle, 1996). But it is important to note that yields continued to increase despite the reduction in new cultivar releases. *There is insufficient evidence to conclude that research productivity and output is declining, but it does suggest that greater attention must be paid to this issue because it could reflect a weakening of the agricultural research system. But it could also signal the inadequacy of technology transfer (extension) as research staff state that only 30 percent of the technology developed is regularly applied in farmers' fields and the balance remain "on the shelf," although new variety adoption rates are about 80 percent. However, it could reflect high risk or financially unattractive technology* (World Bank, 1997c). China will need to increase its research expenditures considerably if the challenge of increased food self reliance is to be realized.

A China-CGIAR meeting in November 1997 indicated that Chinese policymakers are aware of the importance of investment in agricultural research, and the need for stronger linkages between the Chinese and the international agricultural research system. Nevertheless, returns to research are long term and difficult to capture and there are many competing demands on public resources. An important stimulus to investment in agricultural research will be the government's continued recognition of the consequences of not investing, namely, far greater reliance on international grain markets at reasonable prices for a large country such as China.

Water Management

Water is one of the biggest constraints to increasing agricultural production particularly given the uneven distribution of water between the north which is drought prone and the south which is under constant threat of flood. Irrigation systems are essential for multiple cropping but are badly in need of repairs and maintenance, and there is much wastage of water due to poor water pricing policies. Substantial increase in investments in water are needed to ensure flood control and improved water management, rehabilitation of dams and containment of dikes together with improved policies to conserve the use of water. Besides industrial uses of water and the depletion in the quality of water due to industrial pollution are some of the most important sources of constraints to water use in agriculture (see below the discussion on environment).

Investment in Port and Grain Handling Facilities

As China liberalizes grain trade leading to average projected grain imports of about 60 million tons, it would need to expand the grain handling capacity of its ports. Only three ports have the capacity to handle bulk grain and delays in offloading entail considerable costs. At today's prices off loading of grain would entail costs of $900 million requiring new deepwater births, high volume rail corridors and bulk loading and unloading facilities.

Environmental Challenges

There are few quantitative estimates of projected environmental degradation associated with intensive agricultural production. Yet environmental cleanup will pose a competing challenge for investment in agriculture. Abundant use of coal to meet the burgeoning demands for energy and rapid growth of cities with associated growth of automobiles and municipal waste have been large sources of air pollution with levels being high in China even by the standards of developing countries. Mortality rates of chronic pulmonary disease, a leading cause of death in China is five times as high as in New York, not only leading to considerable loss of life, but high cost of hospitalization and lost work days estimated to be 7.4 million person years a year.

Acid rain and water use and pollution associated with industrialization and urbanization are other major sources of water constraint. Disposal of industrial waste and organic fertilizer runoffs have been the major source of water pollution. Yet the health impact has been generally contained due to the widespread availability of safe drinking water. Still, water pollution is increasing water shortages and increasing the cost of provision of drinking water by having to move to safer areas.

China's per capita energy demand and increased use of automobiles associated with urbanization will critically shape the environmental prospects in China. Reduction in environmental pollution will call for replacement of outdated and environmentally polluting technologies, increasing energy efficiency and emission standards to that of Europe and the U.S., and pricing polluting industrial inputs to reflect the cost of pollution cleanup. As administered prices give way to market prices, the government would need to consider imposing taxes on such items as coal, based on sulfur and ashe content, investing in alternative technologies in public transport and domestic energy use such as the greater use of gas for cooking and planning and regulating industrial, public and household uses more effectively.

Integrating with the World Market

China's integration with the rest of the world has been remarkably swift, and its trade patterns have been highly diversified, a phenomenon facilitated by import of goods for further processing and the window of Hong Kong. Between 1978 and 1995 the value of exports and imports as a share of GDP tripled, with China becoming the second largest recipient of foreign direct investment after the U.S. Link between trade, FDI and China's high savings rates have been key to China's rapid economic growth. China now accounts for 40 percent of the FDI to developing countries and is the largest FDI recipient among developing countries. In 1995 the $38 billion inflows accounted for 13 percent of industrial output, 12 percent of tax revenues and 16 million jobs mostly to the coastal cities in manufacturing. China wants to shift the investments more to the interior and towards infrastructure projects. Foreign investors are increasingly involved in power generation, railways, ports and highways with local governments selling shares to foreign investors. Agroindustries will benefit from these investments in the interior.

Further reforms will have to entail reduction of import quotas, and tariff and non tariff barriers. China's accession to the World Trade Organization and the associated reduction in trade barriers will also benefit many of China's trading partners, particularly if it includes agricultural trade reforms.

As in other countries, trade reforms involve substantial adjustment costs including reduced protection of state enterprises, a more flexible labor market, less employment and income security leading to concerns about domestic sociopolitical stability. Liberalization of cereal and fertilizer imports also cause concern about access to world markets and likely increases in world prices and domestic food price instability due to China being a large importer. Yet the World Bank's analysis shows that the current administered government price and stocking policies may be more destabilizing to domestic supplies and prices than would be a liberal trade regime at home and abroad. Furthermore, China can enjoy substantial efficiency gains from playing up to its comparative advantage, in the case of agriculture, between the production of grain vs. high value crops, and benefit more from the import of technology from industrial countries. Virtually all the research and technology development is done in industrial countries and increasingly much of it, including in agriculture through research on genetically engineered products, is in the private sector. China can benefit from increased use of imported technologies which have tended to spread rapidly and there is growing interest in the major agricultural exporters to China and multinational corporations in the expanding Chinese market for seeds, fertilizers, cereals, edible oils and other products.

China's external debt position gives it the flexibility to liberalize agricultural policies incrementally in much the same way that it has managed reforms so far. By the end of 1996 China's external debt of $130 billion with debt to export ratio of 85 percent and the debt to GNP ratio of 20 percent was less than half the developing country average and among the lowest in the developing world. Yet in view of the recent developments in East Asia, China needs to show prudence in foreign borrowings, monitor its short and long term debt more effectively and develop a clear framework for external borrowing including developing selected performance guarantees for its foreign borrowers and integrate its domestic public finance needs with the strategy for external borrowing.

China and its Trading Partners

China's growth and industrialization has had three effects on industrial economies. First, they experience a more rapid demand for capital goods and knowledge intensive industries as well as for primary including particularly agricultural products. The World Bank projects an 8 percent annual growth in exports to China with increase in their overall export growth by 0.5 percent, from 2 percent to 2.5 percent due to the Chinese growth. Also relative prices of capital and knowledge intensive industries will increase vis a vis those of labor intensive goods leading perhaps to structural adjustment in industrial countries towards relatively greater production of capital and knowledge good products and services as well as to primary products in which land extensiveness permits some industrial and advanced developing countries comparative advantage, for example in the production of such items as fertilizers, seeds, wheat, corn, soybeans and edible oils and certain types of fruit, vegetables and meats.

World Bank Lending to China

The World Bank's lending strategy has been to assist China on several fronts discussed above through three interrelated instruments: (1) Economic and sector analysis; (2) investment lending; and (3) technical assistance, training and institutional development.

China's large size and global importance, its well demonstrated and sustained political commitment to reforms and a proven track record of economic and social performance has resulted in the China country department being able to attract some of the best talent in the World Bank.

Economic analysis has involved bringing to bear in-house World Bank expertise combined with expertise on China in universities in industrial countries in close collaboration with Chinese government officials and research and educational institutions. The purpose of economic and sector work has been to provide objective analysis of the state of the Chinese economy and its major sectors, the impact of its policies, institutions and investments and areas needing further reforms. It would be fair to say that the World Bank's professional knowledge, access to the highest levels of the government and objective analysis has helped to articulate pros and cons of important macroeconomic and sectoral policy issues, provided quantitative, intellectual support to the reformers in China in a transparent manner and perhaps helped speed up the reforms and their implementation where there already was considerable internal segment of opinion in favor of such reforms. Examples include fiscal and tariff reforms, grain price increases, liberalization of commodity markets and reform of the state enterprises.

Investment lending provides financial and technical assistance to major sectors of the Chinese economy. Although China has become the World Bank's biggest borrower, the Bank's financial commitments dwarf in relation to both the level of domestic savings and external private capital flows reported earlier. That is part of the reason why the World Bank has had a strong and healthy partnership with China with its lending playing an important catalytic role, in several cases developing important "models" which could be replicated elsewhere, once they were demonstrated to be successful on a small scale.

As of 1997 the World Bank (International Bank for Reconstruction and Development-IBRD which lends at near commercial interest rates and International Development Association-IDA, the World Bank's concessional window of lending) had committed $28 billion to China. (See Figures 4.4.1-5). of the total *$18.75* billion were on IBRD terms and $9.2 billion on concessional IDA terms. Forty three percent of the IBRD and IDA lending went to energy and transportation sectors particularly in

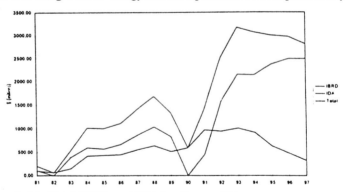

Figure 2.4.1. Total IBRD and IDA lending to China.

the early years of collaboration when private investment was less active, and another 37 percent to agricultural and industrial sectors (See Figure 2.4.2). With $7.7 billion of Bank and IDA lending agriculture received 27 percent of the lending and has been the largest sector for World Bank loans and credits. As Figure 2.4.5 shows agricultural projects have ranged from agricultural research, seed production, rural credit, irrigation, control, soil conservation, afforestation, area development, rural resettlements, fisheries and poverty reduction. Fifty four percent of the concessional $9.2 billion IDA lending went to agriculture followed by education, health, transportation, water supply, urban development, etc. (See Figure 2.4.4). They are the so-called softer sectors, where public sector involvement is crucial to alleviate poverty and create social capital and where returns are long term and benefits, while substantial, are difficult to capture for private investors.

Environment has been a beneficiary beginning the decade of the 90s with nearly $1 billion of World Bank commitments.

The Bank has also provided technical assistance and training through self standing technical assistance projects for developing project planning and implementing capacity. (See Figure 2.4.2)

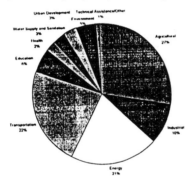

Figure 2.4.2 IBRD and IDA lending by sectors.

Again agriculture dominated in technical assistance projects. But their emphasis has shifted to policy and institutional reforms in the 1990s to such areas as financial, legal and fiscal reforms. Technical assistance credits have amounted to only a quarter of a billion dollars but perhaps have had a very high rate of return. In addition, informally through its staff involved in economic and sector analysis and lending, the World Bank offers considerable assistance for human and institutional development in such areas as project preparation and implementation, economic, social and environmental assessment, procurement and disbursements.

An important part of the World Bank assistance has been in bringing a range of new technologies and ensuring their rapid dissemination. Examples include cotton varieties from the U.S. and citrus from Spain, tubewell irrigation in the semi arid North China Plain which was one of the most successful World Bank projects as well as methods of watershed management, flood control and soil conservation, river basin management, afforestation and reform of the state enterprises.

Another important part of the World Bank's role has been in the areas of formulating strategies, planning and implementation, e.g., in the health sector, improving the location and financing of health sector interventions with greater focus on investments outside the main urban centers, planning and implementation of rural resettlements, and poverty reduction through developing grassroots approaches. Rural resettlement projects have either been components of other projects or self standing projects, some involving resettlement of close to 200,000 households. Compared to other developing countries, China has a good record in resettling populations and creating viable livelihoods.

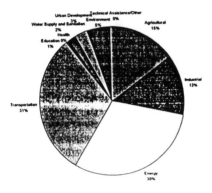

Figure 2.4.3. IBRD lending by sectors.

The World Bank funded poverty projects have not only targeted the remote resource poor regions of China, bypassed by the economic reforms, but demonstrated how, through well conceived projects using existing local institutions, targeting the poorest households within these poor regions can enable reaching households effectively through such interventions as primary education, health, assistance with labor migration, small scale industrialization, rural water supply and agricultural innovations. Prior to designing these projects, the World Bank was able to work with the concerned Chinese authorities in developing a national strategy for poverty alleviation and in bringing it to the attention of the highest policymakers for their endorsement. Bank financed projects have been viewed by the Chinese government as providing important models on decentralized planning and implementation which are being replicated elsewhere in China. The lessons learnt offer considerable potential to increase the efficiency of the nearly $2 billion spent on poverty reduction by the Chinese authorities. In a country where the use of successful models works well, the World Bank has often played an important catalytic role in many sectors, for possible replication elsewhere by the Chinese. In that sense the World Bank's substantive contribution has been larger than suggested by the financial transfers. A recent internal Quality Assessment Review of these poverty projects gave them high marks for effectiveness both due to the innovative work done by the World Bank's task managers and the strong commitment to poverty reduction shown by the Chinese officials at the national, provincial, county and village level leading to highly effective planning and implementation.

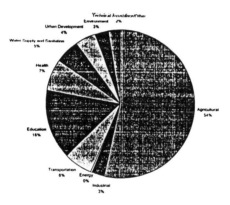

Figure 2.4.4. IDA credits by sectors.

Challenges of Shifting From IDA credits to Bank Lending

Unlike in other developing countries the Chinese Ministry of Finance expects the entities borrowing credits and loans from the World Bank to be directly responsible for paying back the debt, including bearing foreign exchange risks. Such decentralized responsibility and accountability and emphasis on early cashflow explains why World Bank funded Chinese projects perform well and generate high rates of return. Even at the lowest level in China there is considerable appreciation of the importance of making good use of World Bank resources and to ensure quick and high payoff. However, as China graduates from being an IDA country, the borrowing entities and the finance ministry are less inclined to borrow from the World Bank at commercial rates to assist the "softer" sectors, including agricultural research, poverty alleviation, health, education and rural water supply, the returns for which are long term and difficult to capture in immediate revenues. The government will on the other hand seems likely to be willing to borrow for investment in water management, irrigation, production of high value activities such as meat, poultry, fisheries, fruits and vegetables as well as the major grains. Yet it is clear that the opportunity to wrestle with difficult issues with long term payoffs through the development of models which IDA lending provided will be lost with the loss of IDA to China.

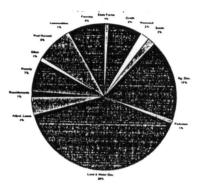

Figure 2.4.5. Distribution of World Bank/IDA lending by types of agricultural projects.

Lessons from the World Bank-China Partnership

A less well recognized role (even within the World Bank) is in learning lessons and broadly disseminating the successful experience of Chinese economic development and China-World Bank partnership to the rest of the developing world, in much the same way that the Chinese replicate successful models of development initiated at the local level in other parts of China. What value can the World Bank add to the economic performance of a country such as China? China has a strong commitment to grow and to eradicate poverty. It has one of the highest savings rates and the highest inflows of international capital, mostly from the Chinese diaspora. It has liberalized its economy pragmatically and consistently. It has a strong domestic economic management capacity and a large base of trained human capital. Most importantly, China has shown a strong capacity to learn from its own mistakes and the mistakes of others, and to adapt the knowledge and experience gained skillfully to its own circumstances. China's internal stability and continuity in the government including low turnover of government officials as well as a significant share of idealism and nationalism has created a context for development. An important question is how replicable are these in other parts of the developing world.

The $28 billion of World Bank loans and credits, while small in overall terms, have not been insignificant in a country whose financial capital needs are considerable. But it is the nonfinancial services the World Bank has provided including its economic and sector analysis, the innovativeness of its projects, contribution to the internal dialogue on policy and institutional reforms, and the bringing in of new technology and know-how on large and myriad small issues in which the World Bank's contribution has been significant. Involvement in project lending allows the Bank a rich hands-on experience, and China's commitment to development enables it to attract the best talent in the world to help address China's challenging problems. The World Bank not only expects this successful partnership to continue, but to learn lessons from it for other parts of the world.

References

Alexandratos, N. 1996. China's Projected Cereals Deficits in a World Context. *Agricultural Economics* 15, p. 1-16.

Alexandratos, N. 1997. china's Consumption of Cereals and the Capacity of the rest of the World to Increase Exports. *Food Policy* 22 (3), p. 253-267.

Brown, Lester R. 1995. *Who will feed China?* New York: W.W. Norton and Company.

Carter, C., and Z. Funing. 1991. China's Past and Future Role in the Grain Trade. *Economic Development and Cultural Change* 39, p. 791-814.

Chen, L.Y., and A. Buckwell. 1991. *Chinese Grain Economy and Policy.* Wallingford, U.K.: C.A.B. International.

Fan, S., "Data Survey and Preliminary Assessment of Agricultural Investment in China." A Report to Food and Agriculture Organization, January 30, 1995.

Garnaut, R., and G. Ma. 1992. *Grain in China: A Report.* East Asian Analytical Unit. Canberra, Australia: Department of Foreign Affairs and Trade.

Information Office of the State Council of the People's Republic of China. 1996. *The Grain Issue in China.*

Lele, Uma, Terry Sicular, and Saeed Rana, Alleviating Rural Poverty: Learning from China, Forthcoming.

Mei, F. 1995. China Can Feed its Population. *China Daily* 29 (April).

Nyberg, Albert, Food Security in China, A paper presented at a Symposium, "Food Production and Environment in China: Challenges for the 21st Century", at the 89th Annual Meeting of the American Society of Agronomy, Crop Science Society of America and Soil Science Society of America, Anaheim, CA, October 28, 1997.

Paarlberg, R.L. 1997. Feeding China: A Confident View. *Food Policy* 22 (3), p. 269-279.

Piazza, Alan and Echo H. Liang, The State of Poverty in China: Its Causes and Remedies.

Pinstrup-Andersen P., Rajul Pandya-Lorch and Mark Rosegrant, The World Food Situation, Recent Developments, Emerging Issues and Long Term Prospects, a paper presented to the International Centers Week,. The Consultative Group on International Agricultural Research, Washington, DC, October 27, 1997.

Pray, Carl, Final Report of the China Agricultural Research Funding Project, September 22, 1997, Unpublished.

Rozelle, S., C. Pray, and J. Huang. "Agricultural Research Policy in China: Testing the Limits." Post-Conference Workshop on Agricultural Productivity and Research and Development Policy in China, August 29, 1996, Melbourne, Australia.

Rozelle, S., and M.W. Rosegrant. 1997. China's Past, Present, and Future Food Economy: Can China Continue to Meet the Challenges? *Food Policy* 22 (3), p. 191-200.

World Bank. 1997. China 2020 Series (7 volumes), Development Challenges in the New Century. Washington, DC.
(2) Clear Water, Blue Skies: China's Environment in the New Century
(3) At China's Table: Food Security Options.
(4) Financing Health Care: Issues and Options for China
(5) Sharing Rising Incomes: Disparities in China.
(6) Old Age Security: Pension Reform in China
(7) China Engaged: Integration with the Global Economy.

World Bank. 1997b. The Annual Review of Development Effectiveness, Operations Evaluation Department. Washington, DC.

World Bank. 1997c. China: Long-term Food Security, Rural and Social Development Operations Division, China and Mongolia Department. Washington, DC.

World Bank. 1995. Annual Review of Evaluation Results 1994, Operations Evaluation Department. Washington, DC.

World Bank. 1994. Annual Review of Evaluation Results 1993, Operations Evaluation Department. Washington, DC.

ANNEX

World Bank Supported Projects in China, by Sector

FY	Project Name	Bank	IDA	Total
		(U.S. $ millions)		
	Agriculture Sector			
82	North China Plain Agriculture		60.00	60.00
83	State Farms I (Heilongjiang)	25.12	44.99	70.11
84	Rubber Development		40.00	40.00
84	Special Fund		60.00	60.00
84	Rural Credit I		50.00	50.00
85	Agricultural Research II		25.00	25.00
85	Seeds		40.00	40.00
85	Forestry Development		42.26	42.26
85	Pishihang-Chaohu Area Development	17.00	75.00	92.00
86	Rural Credit II		90.00	90.00
86	Freshwater Fisheries		60.00	60.00
87	Red Soils		39.95	39.95
87	Xinjiang Agricultural Development		69.63	69.63
87	Gansu Provincial Development	20.00	150.50	170.50
88	Rural Credit III		170.00	170.00
88	Northern Irrigation		103.00	103.00
88	Coastal Lands Development	32.61	60.00	92.61
88	Daxinganling Forestry		56.21	56.21
88	Rural Sector Adjustment Loan	200.00	100.00	300.00
89	Shaanxi Agricultural Development		106.00	106.00
89	Shandong Agricultural Development		109.00	109.00
90	Jiangxi Agricultural Development		60.00	60.00
90	National Afforestation		300.00	300.00
90	Hebei Agricultural Development		150.00	150.00
91	Mid-Yangtze Agricultural Development		64.00	64.00
91	Rural Credit IV	75.00	200.00	275.00
91	Henan Agricultural Development		110.00	110.00
91	Irrig. Agricultural Intensification	147.10	187.90	335.00
92	Tarim Basin		125.00	125.00
92	Guangdong Agricultural Development		162.00	162.00
93	Sichuan Agricultural Development		147.00	147.00
93	Agricultural Support Services		115.00	115.00
93	Taihu Basin Flood Control	100.00	100.00	200.00
93	Grain Distribution	325.00	165.00	490.00
94	Second Red Soils Area Development		150.00	150.00
94	Songliao Plain Agriculture Development		205.00	205.00
94	Xiaolangdi Multipurpose	460.00		460.00
94	Xiaolangdi Resettlement		110.00	110.00
94	Loess Plateau Watershed Rehabilitation		150.00	150.00
94	Forest Resource Development and Protection		200.00	200.00
95	Yangtze Basin Water Resource Development	100.00	110.00	210.00
95	Southwest Poverty Reduction	47.50	200.00	247.50
96	Shanxi Poverty Alleviation Project		100.00	100.00
96	Seed Sector Commercialization	80.00	20.00	100.00
96	Gansu Hexi Corridor	60.00	90.00	150.00
96	Animal Feed	150.00		150.00
97	Heilongjiang Agricultural Development	120.00		120.00
97	Wanjiazhai Water Transfer	400.00		400.00
97	Zinba Mountains Poverty Reduction	30.00	150.00	180.00
97	Xiaolangdi Multipurpose	430.00		430.00
	Subtotal	2819.33	4922.44	7741.77

Part III
MAJOR ISSUES AND CONCERNS

Chapter 3.1

Concerns for the Future of Chinese agriculture

Francis Tuan[1]
Economic Research Services (ERS),
U.S. Department of Agriculture.
Washington D.C.

Since rural reforms were implemented in 1979, agricultural production in China in value terms, measured by the gross value of agricultural output, increased 182 percent, or an annual growth rate of 3.6 percent. Food production also rose significantly. Over the past 17 years, grain output grew 2 per cent per year, oilseeds 8 percent, meat 10 percent, and aquatic products 12 percent.

As China's average annual growth rate of food output exceeded the average annual growth of population, per capita consumption of most agricultural commodities grew considerably for the same period. In some years, imports of grain, oilseeds, vegetable oils, and sugar partially contributed to the growth of consumption, because domestic production of these commodities failed to meet the faster growth of consumption. Nevertheless, the ratio of annual grain imports as a share of China's total grain output, for example, were generally less than 5 percent. During the marketing years of 1986/87 and 1994/95, China was even a net grain exporter (Tuan, 1997). As stated above, China's rural reforms indeed facilitated impressive growth of farm output over the last one and half decades. However, the significant increase was accompanied by rapid growth in per capita consumption of food products induced by significant increases in income and population growth during the same period. Probably all researchers related to agriculture would agree that China, the world's largest developing

---

1 Views expressed in this paper are those of the author's and not necessary those of the U. S. Department of Agriculture.

country (22 percent of the world's population and possessing only 7 percent of world's total cultivated land, according to official statistics), will likely face greater challenges in meeting the continued growth of demand for food as we move into the 21st century. These challenges include continuing to produce more agricultural products, particularly of food output, to further liberalize domestic agricultural commodity markets and distribution and foreign trade, and to formulate and implement agricultural policies that will support and facilitate the development of China's market oriented economy. However, there are many concerns and issues frequently are raised by the Chinese and foreign researchers and by the Chinese government officials as well as those of other countries.

The following sections are devoted to describing and discussing those concerns and issues. I trust that China's Government will pay close attention to these issues if the country plans to achieve the sustained overall development as experienced in the past 17 years and to meet the Chinese people's greater need for food products in coming decades.

Ecological concerns

As we all know, levels of crop production are determined by two factors, planted (or sown) area and yield. Thus, to achieve higher output, we will need either larger area or higher yield, or both. To increase planted area, a country will need to expand cultivated land. To raise yield, farmers will need to either apply more and/or better quality input, or to use inputs more efficiently.

Preservation of cultivated land area

China claims its cultivated land area is only 7 percent of the world's total and is shrinking continuously, particularly over the last 10 or 15 years. In 1996, China officially reported a total cultivated area of 95.0 million hectares, significantly reduced from 99.4 million in 1978. This is equivalent to an average loss of about 250,000 hectares per year over the past 17 years. A recent report issued by China's National Land Administration revealed that cultivated land area in China could, in fact, be as much as 40 percent more than the under-stated official land area. Nevertheless, losing farm land is still a critical issue because the cultivated land areas lost during the last decade has been largely the most productive land.

China's cultivated land area had again been surveyed during the first agricultural census, which was held in January, 1997. China's State Statistical Bureau (SSB) is responsible for conducting the census and is confident that the country's actual cultivated land area will be accurate when the census results are tabulated and published later this year. The SSB does not believe that the new, higher total cultivated area announced by the National Land Administration was a valid estimate because the cultivated area statistics collected by the agency using detailed regional and soil surveys were from different years and not from the same date or a set time frame. Most important, the announcement of China's actual cultivated area will affect the forecasts of most researchers studying the potential of the country's crop production. In general, the larger the cultivated areas China has, the bigger potential the total production of farm output will be.

Potential growth of crop yields

Under-stating cultivated area in China's agriculture related publications has had a long history. Although sampling surveys employed by the SSB in recent years provided good estimates of crop yields, actual crop yields were then adjusted and hence over-stated in order to accurately reflect crop output levels. The implication of such a problem is that China's crop yields have long been inflated. In the past decade, experts specializing in China's agriculture and economists or econometricians trying to model and forecast China's food production argued that China's crop yields, particularly for wheat and rice, already approach world class levels. Further increases in crop yields would be costly and difficult. However, given that China has in fact been under-stating cultivated land by 30 to 40 percent, China's actual crop yields instead leave much room for future growth. This potential growth appears to be more consistent with most Chinese predictions or forecasts that China is capable of advances in crop yields in the future.

Higher crop yields in China can be achieved by increasing use of chemicals, or more effective combinations (particularly potassium and phosphates) and timely application of fertilizers. Timely and appropriate application of pesticides and herbicides or a better pesticide management program is not only critical to yield growth but also beneficial to the rural environment or ecology. In addition, plowing back of crop residues after harvest instead of using large portion of crop stalks as an energy source (particularly in northeast China) should be emphasized and practiced. This practice is vital to maintaining good soil structure for sustainable agricultural development.

CONCERNS FOR THE FUTURE

Better management of water resources

Water availability will be critical for continued yield increase in agricultural production, particularly of rice and wheat. Wheat output kept growing over the past decades in a large part because of the expanded irrigation, pumping ground water in the North China Plains. This continued underground water pumping has reportedly caused underground water table in north China depleting at an unacceptable rate. Some areas in north China already are facing serious subsidence problems. The lack of management for more efficient use of water along the Yellow River also caused river bed at the lower reaches to dry up in summer during the last several years. This also further pressed farmers to pump up more underground water in the North China Plains.

In general, our field visits to northern China's crop fields in irrigated areas indicated that actual water applications to crops were well in excess of the amount needed to ensure plant growth. Since irrigation dominates water use in rural areas, improved efficiency in the delivery and application of irrigation water not only create a new source for other users, but also ensure minimal effect on crop output. Another way to create water availability is to introduce new sources of water in the north. For example, to divert water from the south to the north has been a project that China's Government has been planning for many years. On top of the Gezhou Dam Project, China needs to evaluate carefully where and how to divert the water from the Yangtze River to China's northwest region and/or the North China Plains. These plans certainly will be vital to the future development of China's agriculture. In addition to the serious shortage of irrigated water in the north, water resource in the south has been heavily polluted in the last decade because of rapid growth of rural industries and the water quality now becomes a issue. Therefore, both water quantity and quality have become serious concerns in China's long-term agricultural development.

Pollution caused by rural industrial enterprises

Although management of water resources will be particularly critical in the long-run in maintaining and increasing China's crop yields, especially for winter wheat in the North China Plains, mismanagement of water use caused by the rural industrial enterprises over the past 10 to 15 years and increasing chemical applications could further deteriorate China's rural environment. According to a recent and relatively large scale of survey on China's industrial pollution, which was conducted from January of 1996 through December of 1997 by China's National Environment Protection Bureau, the

Ministry of Agriculture, the Ministry of Finance, and the State Statistical Bureau, over 50 percent of the total organic waste, smog and dust, and solid pollutants released by China's industry came from rural industrial enterprises (World Journal, 1997). There are 1.2 million rural industrial enterprises which are considered major sources of pollutants released from manufacturers which produce non-metal and mining products, textile product, food processing output, metal products, chemical products, and mechanical products. Rural industrial enterprises are currently concentrated in Zhejiang, Jiangsu, Shandong, and Guangdong provinces. The pollutants released by those industries mainly affect the environment and people's health. This situation will have to be effectively corrected and controlled by the Government. The cost associated with the correction of the environmental damage should be shared by those industries, not the society or the Government.

Economic concerns

One of the most critical issues regarding China's agriculture is to understand the changes in food consumption pattern. It is important especially for the government economic planners and international commodity analysts as they try to forecast China's future food needs and their implications for world agricultural trade.

Uncertainty of future demand for food

This is an issue because statistics or information published prior to the institutional reforms (such as abolishing of communes and implementing rural household production responsibility systems), or even before marketing and pricing reforms (such as the contract procurement system and elimination of the urban food rationing system), are generally not appropriate or useful in projecting or forecasting future food consumption in China. In short, there is a lack of useful data base or statistical series that can be based on to make dependable economic projections or forecasts. China's SSB and other government agencies often conduct income and expenditure surveys on urban and rural residents on a regular basis, but surveys are not using random sampling methods. In other words, surveys are not scientifically designed. It implies that survey results have been statistically biased. The biased statistics or parameters used in the models for forecasting future demand may cause huge differences with the end results, particularly when multiplying by China's huge population base. For instance, if annual per capita consumption of rice is estimated at 125 kilograms from a survey rather than the actual 121 kilograms per person. The difference is only 4 kilograms. This difference, however, is multiplied by China's population of 1.2 billion, resulting in an overestimate of almost 5 million tons of rice demand every year for the country as a whole. This small error on the estimate of total rice consumption may be considered by China's economic planners or analysts as trivial because it accounts for only about 4 percent of China's total annual rice production. However, 5 million tons of rice consists of about one-third of the world's total rice trade. Assuming China decides to import the majority of the 5 million tons of rice from the international market, there is no doubt that world market rice prices would skyrocket over night.

Growing demand for feed grain

In general, estimating China's future demand for feed grain is much more crucial than estimating the demand for food grain. This is a focal issue since Lester Brown published his book *Who Will Feed China*, although the issue has been discussed by many other economists for several years. Due to sharp increases in consumption of meat and other livestock products observed in recent years, the derived demand for feed grain is expected to rise continuously during the coming decades, mainly due to rapid economic growth and population increase. Chinese government officials reported in the last year or two that meat output, therefore availability for consumption, is over-stated because of double counting and/or government interference in the process of compilation of meat statistics from local reports. This suggests that the lower per capita consumption of meat reported by the urban and rural household expenditures surveys may be more reliable, despite urban household expenditures surveys' lacking of restaurant consumption of meat and other food items. This implies that the lower meat consumption in China has more room to grow compared with those of neighboring countries or economies in Asia, particularly Taiwan and Hong Kong.

China's livestock industry or sector, particularly hog and poultry production, has a complex structure that mingles back yard and specialized household feeding with modern commercial and confinement types of feeding. How much feed is used for livestock feeding is still and will continue to be a mystery until some comprehensive studies are carried out to carefully estimate China's actual feed or grain-meat conversion ratios. My personal experience with China's livestock sector suggests that China has a low grain-meat conversion ratio in the rural back yard feeding, largely because of substantial non-grain supplements fed to the animals. How long this low grain-meat conversion process can be sustained and what are the actual conversion ratios are critical determinants to the issue of how much total feed grain will actually be needed for the next 10, 20, or 30 years.

In short, China's total demand for grain would then be clearer if we could manage to understand the above two issues regarding food demand or consumption. In general, I do not agree that pre-assumed fixed levels of per capita consumption of grain (about 400 kilograms), as shown in most studies published in China, is an acceptable way of saying that China will not need to import large quantities of grain in the future if China could achieve a total grain production calculated based on that level. The continued structure changes in livestock production in rural areas in the future implies that tremendous amount of feed grain will be required to meet the development of animal feeding for the country.

The issue of surplus rural labor force

Another economic issue related to China's rural economy is the surplus of labor. China's SSB statistics show that there are about 450 million individuals in the labor force in the rural areas of the country. Over the last decade, more than 120 million rural workers or members of farm households had transferred to village and township enterprises. Nonetheless, there are still 330 millon farm laborers remaining in rural areas, with a significant portion of the labor force being considered as under-employed. This, together with the long existing household registration system which strictly prohibits rural workers from permanently moving into urban areas, per household farm land currently averages less than 0.5 hectare. This small size of average per household farm land suggests that China needs to further develop more village and township enterprises, particularly around medium and small cities and townships in the western region, in order to absorb the big pool of excess farm labor and to balance regional development between the coastal and inland areas in China.

Shortage of agricultural investment

While rural industrial enterprise development needs government to increase its investment in the inland regions, government investment, mainly in the form of fixed capital assets and research and development (R&D), is also important to China's agricultural development in the future. The Chinese Government played an important role in agricultural output growth in the past. According to recent studies (Fan, 1996), government investments in agriculture declined sharply after rural reforms were implemented in 1979. The Chinese government believed that savings which resulted from the rapidly rising agricultural incomes of the farm households would provide the main stream of agricultural investment, therefore offsetting the reduction in central government investment. However, over the last 16 or 17 years, farmers have mainly invested in building new houses and invested in rural industrial enterprises rather than farm production or farmland.

Although the Chinese Government resumed higher investments at the end of the 1980's and during the initial years of the 1990's to respond to the slow down of agricultural production, central government investment in 1994 was still 34 percent below that of 1978 when the level of government investments reached 38 billion yuan. Undoubtedly, the downward trend in government investment in agriculture will make it difficult for the sector to achieve a sustainable pattern of growth in the long-run. The most striking phenomenon of the recent trend was the decline in government funding for agricultural R&D, the resulting deterioration of the national agricultural research system, and the slow down in the rate of release of new crop varieties and agricultural technology. There is no doubt that China's Government urgently needs to stimulate and encourage more investment, including foreign capital, in the agriculture sector. More importantly, the experience of other developing countries has also indicated that the financing, organization, and management of investment are critical in promoting future production and growth in agriculture.

Political concerns

To discuss politically related issues or concerns had been very sensitive in the past and is still quite sensitive this days. However, they need to be discussed with sincere and with constructive sense in mind. Two issues, land tenure system and market development, are raised to be discussed in this section.

Land tenure security up for debate

As discussed above, to encourage farmers to invest in agricultural production as well as in farmland improvement will be a critical issue for China's agricultural development in the future. Recent surveys conducted in China indicate farmers are not particularly interested in investing in farmland improvement unless they know they own or they are much better legally protected in using their land for a long period of time. It is obvious that farmers would not be interested in investing in any production if no secured returns are expected. Farmers often complained during the surveys that their collectively allocated land could be asked to return any time if the collectives are reallocating their cultivated land or reassigning the land for other uses. Since 1993, the Chinese Government has implemented a new land policy that aims to prolong the land contract duration in hope that farmers would take longer term action towards land use, especially the investment in agricultural land. Surveys conducted in 1997 in villages showed that farmers had showed no signs in adjusting either their investment behavior or production decision. I believe that a stable but more importantly **legalized** long-term land tenure program that is guaranteed by the Government or a privatized land ownership system would be able be more effectively induce farmers to invest in productive activities as well as in land improvement.Studies on land tenure issues have been carried out both in and out of China. It has consistently been cited that China has a poor land tenure and is one of the areas that has been mentioned as a major reason for slow growth, however has been largely ignored by reformers (Zhang, 1996). Privatization of land obviously goes against the Chinese Government's well. Without state government's legalized guaranteed program, it is clear that farmers will not feel secured in investing in land improvement. Therefore, under what forms of land tenure system, given the collective ownership of land, to ensure the best use of land resource in the long-run will definitely need further research and debates, so that a better and effective land tenure system can be formulated and presented to farmers.

Establishing a unified national market system and improving its linkage with the international market

Since China implemented rural reforms in 1979, agricultural commodity markets, except for grain, cotton, silk cocoons, and tobacco, have generally been liberalized. In an effort to gradually liberalize the grain market, the country's four decade old urban grain rationing system was eliminated in 1993 (although it was partially reinstated in 1994 because of skyrocketing grain prices). However, the fixed price procurement of grain has yet to be completely abolished by the government.

Three years ago, to ensure more production of grain, China's Central Government implemented the "grain-bag policy" (or Governor's grain responsibility policy) in order to force farmers to increase sown area for grain. In other words, all provincial governors are responsible for their province's own production and distribution of grain (and cotton). This policy, unfortunately, is a step backward in attempting to liberalize the grain market because it overlooks the principle of comparative advantage, leading to a misallocation of resources (Crook, 1996). Therefore, China's Government appears either not quite ready to liberalize the grain market or misunderstands the ultimate goal of market liberalization. Consequently, the Government still holds tightly to the mentality of using administrative fiat as the more effective measure in achieving the goal of increased grain production. More importantly, China needs a unified national market for not only vegetables and fruits, but also grain and other agricultural commodities. During the establishment process of the unified national market, using grain as an example, grain consumers in major grain producing provinces or even in the entire country may encounter a relatively significant upward price adjustment because of a sudden reduction of grain output in areas with less comparative advantage in grain production. From both the experiences in other countries and the theoretical point of view, fluctuating levels of commodity prices in a big national market are always lower than those of many small markets divided up from the big national market. In addition, because of the development of unified national market, China's basic infrastructure such as transportation and storage will also be gradually strengthened.

Based on the same reasoning, the rapid economic growth along the coastal areas will increase its demand for output and intermediate input or resources produced in inland provinces. Farm households in inland areas may become richer because the economic development triggered by the growing demand from the coastal provinces will help develop inland economic development which in turn increases the demand for local agricultural commodities and even rural labor force. This chain effect may actually reduce the income differentials between the two regions and eventually eliminate the imbalanced development in China (Lin, 1995). To link China's domestic agricultural market with the international market will also be important and inevitable, particularly if China joins the World Trade Organization (WTO). The linkage should enable China to better use its resource and become more competitive in agricultural commodity trade in the world market and therefore benefiting it's overall economy. If China joins the WTO, China's agricultural production and demand for imported food will be closely related to the changes in world demand and supply of agricultural products. However, China will also face challenges with the fierce competition in international agricultural trade. With surplus labor and relatively abundant natural resources, China needs to take advantage of these endowments and adopt modern technology related to food processing and storage. This will facilitate China to produce more quality agricultural products and compete in the world markets.

Social concerns

After the dismantling of the commune and the implementation of more rural reforms, there are a couple of concerns often raised during the rural surveys and studies. These are the issue of deteriorating rural education system and the worsening income inequality issues.

Strengthening rural education system

A related issue to the rural surplus labor is the deteriorating rural education system, currently lacking of financial support and quality teachers. Strengthening the system, which is basically run by the rural collectives, will be vital for the continuation of China's agricultural modernization. Without good quality or better educated farmers, agricultural production which relies on adoption of new technology will not make continued and expected progress as China's government leaders or planners would like to see. On the other hand, the surplus rural labor force will not be readily available for the transfer from farming to working in the village and township enterprises. To help improve the rural education system China's Central Government and/or Provincial Governments need to step up their roles. In developed countries and many newly industrialized countries or economies, elementary or secondary education systems are enforced and strengthened by the central and provincial level governments. China needs to seriously consider how to provide better education to the rural children, particularly in the areas such as stable financial supports and training of school teachers. In fact, building rural human capital is such an important issue which will not only benefit China's rural development, but also contribute to the country's overall economic growth in the future.

Income differentials between coastal and inland areas

In fact, rural family income differentials between coastal and inland provinces narrowed after rural reform began in 1979. However, improvement extended only into the mid-1980's. Since then, the narrowing of the gap of rural household income between the two areas has reversed and widened. The phenomenon was aggravated in the 1990's, largely because of the rapid growth and concentration of rural township and village enterprises in the coastal provinces. Coastal provinces are also conveniently located to allow trade with neighboring provinces as well as foreign countries. More advanced infrastructure such as transportation, storage, and communication in the coastal areas also contributed to the rapid economic development and created millions of job opportunities for rural labor to shift to work outside of farming. In contrast, even government investment pouring into the inland provinces has not stimulated economic growth because of the lack of infrastructure in the inland area. One result of this imbalanced development is the so called "blind migration" of farmers moving from inland provinces to eastern coastal areas in order to seeking higher incomes. The movement of millions of farmers from the west to the east is creating social and service problems, particularly in big cities and municipalities.

To alleviate these problems, China's Government needs to better plan its investment in the inland region or provinces, not only to balance the overall development of the entire economy and reduce poverty areas, but also to mitigate the growing pressure from "blind migration."

Statistical system in China

As China's rural areas developed and grew rapidly over the past 17 years, statistical agencies at the central and provincial levels have been busy collecting and compiling information and data. The statistics collected and processed have been largely for the use of the Government's planning purposes and for informing how successful the country has developed its economy. In a more market-oriented economy, information and data collection should serve a much broader needs from different parts of the economy. Timely collection, processing and dissemination of data and analysis are critical for the government agencies, banking system, merchants, and farmers in making their decisions. Regarding China's agricultural information and statistical system, I have the following concerns:

The need of a comprehensive information system and market analysis work

While reforms implemented since 1979 lead its rural economy much closer to a market economy, China, the world's largest producer and consumer of many agricultural commodities, urgently needs to set up a comprehensive agricultural information system and a timely, reliable, and objective agricultural situation and outlook analysis program for its markets to operate efficiently.

A comprehensive agricultural information system is important for China because China's agricultural related functions are scattered among different Ministries. Currently, information collected by one Ministry is seldom shared with other government agencies. Exchange of information among government agencies seems to be gaining support by some government program analysts, researchers, and officials. A unified or comprehensive agricultural information system, in which situation and outlook analysis and policy analysis can be performed and based on, is vital to providing the appropriate statistics and information to ensure timely and accurate policy decisions.

An agricultural situation and outlook program is set up to analyze and explain what is happening now (situation analysis) and forecast what will take place in the future (outlook analysis). A situation and outlook program is a basic investment by the government. The program, providing data about production, consumption, stock levels, and trade of all major commodities, benefits the government and private businesses in making decisions. For the government, timely knowledge and understanding of market development and prospects is critical for government planners and for those who implement policy. The same is also true for farmers and agribusinesses who are deciding on investments, on what to plant, on how much input to apply, or when to buy and sell their commodities. Many other businesses, for example banks, also need timely information to make financial decisions. In general, the more information one can obtain and the more up-to-date and reliable it is, the lower the risks involved and the costs of doing business. An effective information system means that markets work better, costs and risks are lower, and output and income are higher (Tuan, 1995).

I see that China is also in an urgent need to build a timely, reliable, and objective situation and outlook program and the program should be based on the continuation of improvements in collecting, processing, and compiling of agricultural and economic information. The United State Department of Agriculture is working together with several Chinese government agencies to set up such a program. The project is expected to last two years and I believe it will benefit not only the Chinese Government but also the United States and world agricultural trade.

Replacing the traditional comprehensive reporting system with random sampling survey methods

China's more market-oriented economy requires more modern and scientific ways in order to collect accurate, timely, and useful information and statistics for the country's planning purposes and for researchers, private businessmen, and various industries. The design of the current statistical reporting, inherited from the Soviet system, was for the use of the traditional planned economy. Not only many new statistical indicators are not in place for the future planning purposes, China is also facing the issue of inaccuracy of data collected through the current reporting system. Of particularly noticeable statistical discrepancies in the last decade were the under-reported cultivated areas and over-estimated livestock/meat production statistics as discussed previously. China conducted the first ever agricultural census in January, 1997. Those two statistical indicators gathered from the census should provide a good check regarding the discrepancies as mentioned above. False reporting of data and statistics have long been a issue in rural areas, particularly after the reporting of statistics is related to the annual evaluation of the cadres' yearend accomplishments or performance (Zhong, 1997). My personal experience from the visits to rural villages and townships in China also indicates the truth and the extend of the false reporting of agricultural production information, including grains.

China's Government also need to provide broader spectrum of statistics for researcher, students, and foreign experts to do analytical reports. Good analytical reports will provide central and local governments useful and valuable suggestions for making future plans. Users of information and data are those who could suggest what other or new data or information are needed to better provide and meet the needs of government, private enterprises, bankers, merchants, and farmers in making critical decisions.

Summary and Conclusions

China's rural reforms implemented since 1979 have stimulated agricultural production, particularly food output. However, income-driven increases in food demand were also rising rapidly. Because of reductions in cultivated land area, deteriorated natural resources, a large population, and other issues as discussed above, together with expected rapid economic growth, China now faces even greater challenges in producing more agricultural products. China's government must seriously investigate and urgently resolve many of the above issues critical to achieving sustained development of its agricultural sector. Understanding and resolving these issues will help not only the country in making timely and correct decisions in the future, but also the world stability.

References

Agricultural Research Institute, Chinese Academy of Agricultural Sciences, "A Macro-Analysis on our Country's Foodgrain Issues," Agricultural Economic Problems, No. 2, 1995, Beijing, China. (Chinese edition)

China State Statistical Bureau, China Statistical Yearbooks, various issues, Beijing, China.

Crook, Frederick W., "How will China's Rice Bag Policy Affect Grain Production, Distribution, Prices, and International Trade?"

Paper presented in the Workshop on Agricultural Policies in China at OECD, Dec 12-13, 1996, Paris.

Fan, Shenggen, "Public Investment in Rural China: Historical Trends and Policy Issues," Paper presented in the Workshop on Agricultural Policies in China at OECD, Dec 12-13, 1996, Paris.

Guo, Shutian, "Revisit Current Chinese Farmers' Problems,"Agricultural Economic Problems No. 10, 1995, Beijing, China. (Chinese edition)

Lin, Justin Y. and Li Zhou, "Our Country's Agricultural Crisis and their Basic Solution," Economic Research No. 6, 1995, Beijing China. (Chinese edition)

Tuan, Francis C., "Food and Agriculture Issues in China" Paper presented at the International Symposium on Stable Supply of Food and Sustainable Agricultural Development in Asian Region, Feb 13-14, 1997, Tokyo.

Tuan, Francis C., "Importance of Agricultural Situation and Outlook Analysis in Developing China's Rural Markets," Paper presented in the Hainan International Symposium on China's Agricultural Reforms: Experiences and Prospects, April, 1995Hainan Province, China.

World Journal, "Rural Industries: Becoming the Main Source of pollution," December 24, 1997, page D6, East America Edition, New York.

Zhang Linxiu, Huang Jikun, and Rozelle, Scott, "Land Policy and Land Use in China," Paper presented in the Workshop on Agricultural Policies in China at OECD, DEC 12-13, 1996, Paris.

Zhong, Funing, "An Analysis on the Inflated Statistics About Meat Output and Its Reasons,"China Rural EconomyChinese Academy of Social Sciences, pp. 63-66, October1997, Beijing China.

CHAPTER 3.2

Land Tenure Security and Farm Investment In China[1]

Anning Wei
World Bank, 1818 H Street, NW,
Washington D.C.20433, U.S.A.

Min Zhu
Bank of China, Beijing, P.R. China

Shouying Liu
Development Research Center (DRC) of State Council of China.
Beijing, China

Reforming property rights in land and tenancy is one of the core problems in developing economies (Bell, 1990). In Africa, the primary concern is to convert indigenous customary land ownership to individual land ownership (Basset and Crummey, 1993; Bruce and Migot-Adholla, 1994). In Latin America, the challenge is to redistribute land and transfer low-efficient large farms into small farms, which is expected to release the social tension caused by high land concentration (Scofield, 1990; Grindle, 1990; Hall, 1990). In some East Asian countries, like Philippines (Bell, 1990) and Thailand (Feder, 1993), the land reform was carried out, or is expected to be carried out, by legalizing the tenure system to increase the tenure security and protect tenants' rights.

In Southern Asian countries like India, Pakistan, and Bangladesh, the land reform has focused on transferring the ownership from landlords to actual

---

[1] The opinions expressed in this paper are entirely those of the authors and should not be attributed to the organizations the authors are associated with, The comments and supports from many colleauges and friends are deeply appreciated, particularly: J. Hao (WCCS), R. Burcroff (the World Bank), F. Tuan, and F. Crook (United States Department of Agriculture), X, Chen, and M.Lu (DRC).

cultivators, which would alleviate rural poverty and promote the social status of the rural subordinate class (Herring, 1990; Jannuzi and Hanstad, 1990; Singh, 1988). In Eastern and Central Europe and former Soviet countries, large scale privatization and restructuring movements are transferring the former state/collective farms to private-ownership-based family and cooperative farms (Euroconsult and Center for World Food Studies, 1994; Brooks and Lerman, 1994; Lerman, Brooks, and Csaki, 1994). All these international experiences in land reform target two basic goals like China's land policy: increase agricultural productivity and social stability.

The international experiences that are the most relevant to China are those of reforming communal land ownership and managing system. In the nature, the current land system under the Household Responsibility System (HRS) in China is very much like the communal land systems prevailed in the indigenous communities of America, and the tribal communities in Asia and Africa.

In the communal land system, land is owned by communities, farmed by family units through the customary tenure arrangement. Land sales and rental usually occur within community. Sales or lease to outsiders are either forbidden or subject to approval by the whole community. Land is not allowed to be collateral in formal credit markets (Binswanger et al, 1993). Communal land systems are deeply embedded in cultural and political systems where private property rights are not well established and protected. As the supply of land is reduced by population pressure, agricultural terms of trade are appreciated by commercialization, and as technology advances, the commune land system can not provide the secured tenure and enforce effectively the newly-emerged land. Therefore, land can not be allocated efficiently, the land market is fragmented, farmers have little incentives to invest in land, and farmers have no access to the credit markets. To overcome these impediments, the commune land system has to be moved in the direction of individual land rights (Feder and Feeny, 1993).

As being well documented, the current land system under the HRS in China has the following shortcomings inherited from the commune land system: The insecurity of tenure, small farm size and fragmented land allocation, the confused ownership structure, and high tenure enforcement costs. In practice to cope with this insecure land tenure to protect themselves[2]. At the same time, in 1993 the government extended farmers' land contract for 30 years and requested local community authorities to secure farmers' current land holdings. It appears straightforward that secured land tenure will motivate farmers to invest in land as well as other farm assets given attractive input and output markets. However, Feder et al. (1992) found the linkage between land tenure security and farm investment was very weak in China. Based on 400 farm household survey, Kuang (1995) argued farmers did not feel current tenure particularly insecure though contracted land had been redistributed periodically due to demographic changes. Actually this is not rare in the international experiences. Bruce and Migot-Adholla (1994) found that in Africa the positive relationship between land tenure security and investment depended on the commercialization of agricultural production. The positive relationship existed more often in the region where the production was highly commercialized. Feder (1993) observed in Thailand that only when land was allowed to use as a collateral to obtain credits could the increased tenure security promote land investment.

This lead to a debate about how to pursue land reform in rural China. Kuang reported that most surveyed farmer opposed the government's 1993 decree that extended farmers' current land contracts for 30 years and froze their current land holdings. Kuang suggested, rather than land security, the current reform should emphasize relieving farmers from the burden of government procurement, intervention on their production and marketing decisions, and chaotic implicit and explicit taxation. At the same time, many authors support the government's 1993 decree and considered further individualization of property rights would sustain healthy agricultural development.

In rural China, land is a production resource as well as subsistence insurance for many farmers, and land tenure arrangement concerns not only individual villager's but also community's welfare, and the state's

---

[2] Currently, four major types of new land tenure are very outstanding across China. They are the dual field system, new cooperative system, large commercial family farms, auctioned "primary" mountain land

food supply system bases on the land tenure. As Liu et al. (1996) argued that current land system in China is the balance of the interests of the state, collectives, and individuals. Further reforms have to reach a new equilibrium of the interests of three groups. Land reform is a complicated process and has multiple dimensions. This study can not conclude the debate about the direction of further land reform, but will establish empirical evidence on the relationship between land tenure security and farm investment based on the unique rural household survey data to contribute this debate.

Farm investment is a very important dimension for land reform to concern in China. China finished its rural economic reform featuring by the Household Responsibility System (HRS) in 1984, and during the period of 1978-84, agricultural output grew at a remarkable average annual rate of 7.7% (in 1980 real price), and in which average annual growth rate of grain production reached 4.9%. However, when the greater agriculture prosperity was expected, the average annual growth rate of grain outputs dropped to 1.3% during the period of 1984-93, and in 1994, the total grain output decreased 11.9 million tons comparing with 1993 (China Statistics Yearbook, 1994; China daily, 3/12/1995)[3]. This brings pressure to the urban retail grain price which soared 70-100% across the country's major cities in 1994. The grain import shot up 34% in 1994 too (Wall Street Journal, 3/10/95). But, public investment on agricultural fixed assets slipped from 0.45% of GDP in 1978-84 to 0.35% of GDP in 1985-93, and the agriculture-related household fixed asset stock dropped from 8.7% of GDP in 1985 to 5.2% in 1993 (China Statistics Yearbook, 1994; China Agriculture Statistics Yearbook, 1994). While farm households has became the main source for agricultural investment since 1984, farmers' investment grew very slowly since the early 1980s, and was stagnant or even declined in many years. The major portion of farmer households' investment went to tractors, draft animals, etc., land related investment is especially low (Rural Statistic Yearbook of China, 1993; West and Pan, 1993; Feder et al. 1992). Land abuse has been observed widely in rural China (Huang, 1993). If considering rapid rises in agricultural input prices (there were no reliable indicators available though they were deeply felt by farmers), the situation of agricultural investment becomes more worrying.

---

[3] This implies an almost zero growth of per capita grain output which is still significantly lower than international level.

"Who will feed China?", Lester R~ Brown's (1994) question woke up the concern of international communities about the possible threat that China's- demand may cause to world supply. While many scholars argued, based on the pace of the progress of agricultural technologies and using more precise consumption data, that Mr. Browns' pessimistic assessment was without a solid foundation, the prospect of agricultural investment remains unclear. Improving farm households' investment through land tenure reform will 6bviously increase agricultural investment, and therefore contribute to China's food security and sustain the long-term agricultural development.

The present study is organized as follows. The next section describes the household survey data used for the study. The third and fourth sections discuss the situation of land tenure security and farm investment for the period of the study. The fifth section specifies econometric models and reports the estimation results. The last section concludes the study.

Data

A survey was conducted in April 1994 in rural China by the Rural Development Department of Development Research Center and the State Statistical Bureau. They both are the central government agencies of China. The survey was to investigate the evolution of land system in rural China during the period of 1988-93. A similar survey had been conducted in 1988 for the period of 1978-88~. The new 1994 survey covered 80 villages located in four provinces, i.e., Zhejiang (30 villages), Henan (10 villages), Jilin (25 villages), and Jiangxi (11 villages). The survey issued two types of questionnaires, the questionnaires for village leaders to collect information at the village level, and the questionnaires for farm households to collect information about family farms. In total, 80 village and 800 household questionnaires were collected. Zejiang province is located in southeastern China with very developed rural industries and a very productive fanning sector. Rice is the main grain product and two rice crops are produced each year. Another rice province under the survey is Southern province Jiangxi. Like zhejing, Jiangxi is

---

[4] The proceedings of International Conference of Land System in Rural China (1992. Beijing), entitled as "The Transition of Land System in Rural China" by Task Force of Land System in Rural China, summarize the findings of this survey.

rice production has two crops each year. Located in central China, Henan province has the climate suitable for one wheat and one corn production per year. All these three provinces have small farms. The fourth province, Jilin, is in Northeast China with relatively large family farms and corn as the main agricultural crop. Among all four provinces non-farming activities are developed only in Zhejiang. Table 3.2.1, most of it was produced by Liu. et al. (1996, p.25), summarizes the basic characteristics of the four provinces based on the survey data.

Table 3.2.1 The Characteristics of the Sampled Provinces, 1993

	Zhejiang	Henan	Jilin	Jiangxi
Number of the Sampled Villages	30	10	25	12
Number of the Sampled Households[1]	300	100	200	200
Arable Land per Household	2.2	6.3	17	9.6
Grain Yield (kg/mu)	748	573	455	475
Grain Quota (kg/mu)	98	83	164	167
Per Capita Income (yuan)	1711	945	861	901
The Share of Non-Farming Income in Total Income (%)	67.0	2.4	13.3	13.2

[1] This row is produced by the present author.

Basically, the sample coverage represents different development levels of rural China and the main categories of climate and grain crops. The research results from this survey may have national implications.

Three studies have been done on this survey data. Liu et al. (1996), based on the questionnaires of village leaders, demonstrated that the diverse land tenure systems emerging during the process of rural reform are the equilibrium of the balance of the interests of three groups: the state, local collectives, and individual. The study warns that the reform of current land tenure system is much more difficult than that of the HRS, since the current property rights dilemma is intrinsically more complex due to the fact that the potential conflicts of interest among those three groups are greater. Similarly using the village level data, Carter et al. (1996) examined the determination of land tenure system in rural China. They found the state grain procurement quota and nonagricultural income are two important factors shaping the use rights, tenure security, and transfer rights. Yao and Carter (1996) used a part of household data (Zhejiang and Jiangxi two provinces) to analyze the productivity implications of various land tenure components. These three studies paved the way for the present research to pursue. Some of their results, especially those from village data, will be used in some analyses followed

Patterns of Land tenure Security

Before the economic reform initiated in 1978, land in China was legally owned by the village collectively. In practice, however, land is owned and operated by the state, through a three-tier system, commune-production-brigade-production team. A production team is typically a village. The HRS reorganized this system by distributing village land equally among the residents (either by head or by labor) and replacing 100% government procurement by a fixed quota of grain submission to government (an explicit land tax) plus a contracted grain procurement (an implicit land tax).

The fundamentals of land system under the HRS remain confused. In terms of landlords, the former production team, the former brigade, the former commune, and finally the state all claim explicitly and implicitly that they have the ownership over land and have the rights in receiving land tax and rent. This also creates situations of redistributing land across various governmental bodies (Huang, 1993; He, 1993).

The HRS retains the collective nature of land ownership from the old system, each villager is guaranteed the right to continuously use and collect income from equally share of village land. The land a villager leases is the realization of his/her use rights over one share of the whole village's land, but not the use rights over a specific piece of land. Further, the amount of land a villager leases in general is independent of his/her human and physical capital stock. Land is a production input as well as a social security belt for farmers.

Naturally, when demographic changes (marriage, death, birth, and leaving) happen, it will be demanded to redivide land among villagers. The redistribution pressure is especially intensive in the villages where the ratio of land to population is low and the non-farming employment opportunities are scarce. Tenure is not secure for farmers who lease the land from villages.

1988 and 1995 surveys measured the tenure security in the above sense at the village level for the period of 1978-93. 1988 survey covers 1978-88, and 1994 survey covers 1988-93. Table 3.2.2 reports the frequency of land redistribution at the village level.

For the both time periods, about 70 per cent villages had redrawn the boundaries of the land-holdings of their farmers at least once. Comparing columns 3 with 4 of Table 3.2.2., the structure of redistribution frequencies for the two periods (1978-88 and 1988-93) are surprisingly similar: about 30-35 per cent villages never redistributed their land among villagers, about 35-45 per cent villages did once, about 15-20 per cent twice, about 8-10 per cent three times or more. This means the insecurity of land tenure did not change very much for 14 years.

Table 3.2.2 Frequency of Land Redistribution at the Village Level From 1978-1993 (Numbers are percentages) [1]

Frequency	1978-1988 [2]	1988-1993 (1994 Survey, 80 villages) [3]				
		Weighted Average	Zhejiang	Henan	Jilin	Jiangxi
Never	34.8	30	46	0	30	18.2
Once	37.1	42.1	30	30	69.2	36.4
Twice	19.8	16.3	20	70	0	0
Three Times	8.3	4.1	3.3	0	0	18.2
More than Three times	0	4	0	0	0	27.2

[1]Because of dismal errors, the summation may not be exactly 100.
[2](1988 survey 253 villages) Extracted from He (1993, p.39)
[3]Extracted from Carter et al. (1996, p. 27)

For the results of 1994 survey, as noted by Carter et al. (1996), the land redistribution did not occurred often in Zhejiang province mostly because that non-farming opportunities are relatively abundant in rural area, farmers do not value land being a income source and subsistence insurance as much as the farmers in other provinces. In Jilin province where land is relatively not scarce, the frequency of the redistribution is also low. Henan and Jiangxi are short of non-agricultural employment opportunities in rural regions, and population pressure on land is high. There, land has been redistributed very often.

The household questionnaires of 1994 survey also investigated tenure security at the household level by asking the surveyed farmers how frequent land-holdings were changed for the period of 1988-93. The survey covered two types of changes in farmers' land-holdings: (1)

Farmers voluntarily increased and decreased their land inventories, and(2) land-holdings had to be adjusted because of demographic changes. Only the second type of land adjustment is directly related to tenure security. Fortunately, demographic changes are responsible for about 80% of tenure changes in rural China (He 1993, p. 39), the discussion based on the second type of land adjustment covers the base of land security problem of rural China.

Out of 800 households, 345 households reported their 370 adjustment incidents due to demographic reasons. These adjustments were done through three ways. First, original land arrangements are thoroughly reshuffled within the village and the village land is redistributed among the residents. Second, only those families that had demographic changes will go through the adjustment, extra land will be taken away, and the shortage will be compensated. Third type is that landholdings remain the same, but the land related fees and implicit/explicit taxes will be adjusted to reflect demographic changes. Table 3.2.3 describes the frequency distribution of these three methods at the household level.

Table. 3.2.3 The Distribution of Land Adjustment according to the Methods

	Frequency
redistribution	213
adjustment according to demographic changes	148
adjust burden only	1
other	8
	370

It is clear that adjusting land-related duties without changing landholdings is not a common .practice. Thorough-redistribution is the most important method. But, the marginal adjustment according to demographic changes is also a significant way. These last two methods have different implications for tenure security. Redistribution changes the position, the number of parcels, and even the quality of a household's landholding, the marginal adjustment may retain most of them.

Patterns of Farm Investment

Insecurity of land tenure may thwart farmer's investment tendency. At the same time, the próblems with other components of land ownership may also hurt farmer's investment incentives. Liu et al. (19%) decomposed the property rights of farm land into four basic parts: residual income rights, unencumbered use rights, rights to secure procession, and transfer rights. The HRS reforms individualized residual income rights. But for the other three rights the HRS still exposed farmer to the intervention of local collectives and the state. Farmers do not have fill use and transfer rights, which may discourage their investment activities, especially those land-embedded non-transferable investment.

This situation become more worrying if considering the fact that, as documented by Feder et al. (1992), started from 1983, household farms have replaced the state and collectives to become the main source of agricultural investment. However, farms' investment was only 3-5 per cent of their expenditure. 1994 household survey investigate household farms investment activities for the period of 1988-93. Two types of farm investment are of particular concerns, one is land-related investment, the other is investment on fixed capital.

Land-related investment include construction of irrigation well, irrigation ditch, pumps, and drainage facilities, activities of flattening land, soil improvement, green manure and manure. Farmer's investment in this category was recorded by two item, i.e. capital expenditure and labor input. For some activities, labor work is enough, some need only capital, others may need the both. Land-related investment is embedded in the particular parcel of land, and is not physically transferable. Without a well-functioned land market to compensate previous tenant's land-related investment, which is the case of contemporary China, tenure insecurity has immediate threat on farmers' land-related investment. Investment on fixed capital include green house, barns, animal houses, and various machinery. Comparing with land-related investment, this type of assets is more independent of tenure security and other land property rights.

For both categories of farm investment, farmers' input is rather modest Table 3.2.4 characterizes the structure of farm investment. For land related investment. It shows green manure and manure are the two type of investment occurred the most often. Other type of investment which have no immediate impact on yield or take more capital happened rarely.

Table 3.2.4 Activities of Farm Investment (1988-93)

Land-Related Investment			Investment of Fixed Capital		
Investment activities	Frequency of House-hold	Percen-tage	Investment Items	Frequency of House-hold	Percen-tage
irrigation well	20	2%			
irrigation ditch	2	0%	green house	6	0.6%
green manure	300	30%	barn	133	12.4%
flatten land	14	1%	animal house	214	20.0%
manure	514	51%	tractor	66	6.2%
drainage	130	13%	planter	24	2.2%
soil improve-ment	11	1%	harvester	5	0.5%
pond	8	1%	pump	68	6.4%
			thresher	298	27.9%
			rice planter	1	0.1%
			dryer	0	0.0%
			plant protection	250	23.4%
			other	4	0.4%
total	999	100%	Total	1069	100.0%

The contents of investment on fixed capital were more diversified than those of land related investment. Table 3.2.4 shows that threshers, plant protection machines, and animal house are three main investment items. Investment on operational machinery is minimal, which reflect the resource endowment in rural China, land and capital are scarce and labor is abundant.

The above 999 items of land related investment were reported by 602 households, the 1069 investment items of fix capital were reported by 515 households. Table 3.2.5 below shows that for 6 years, most households did only one type of land-related investment and most households did only one or two types of fixed capital investment. Table 3.2.6 reports per household farm investment for the period of 1988-93. For 6 years, in average, each individual family farm spent only 302 Yuan and 42 working days on land related investment, and only 1933.2 Yuan on fixed capital investment.

Table.3.2.5 Intensity of Farmer's Investment Activities

How Many kind of Investment A Household Conducted for the Period	Frequency of Households	
	Land-Related Investment	Investment of Fixed Capital
1	342	161
2	156	215
3	78	93
4	22	40
5	0	4
6	4	0
7	0	1
8	0	1
Total Households	602	515

Converted working days into the cash value by multiplying it with daily agricultural wage rate, then all farm investment items can be summed up to give rise a total farm investment per household, which is 2618 yuan, dividing it by 6 leads to annual farm investment per household, 436.3 yuan. This is only 6% of total income of a average farm household in 1994.

For the four sampled provinces, Zhejiang has the highest annual household income and grain yield as indicated by Table 3.2.1. It is expected that Zhejiang will have highest farm investment. But Table 3.2.6 shows it is not true. For the per household land -related investment Zhejing has the second lowest capital expenditure, and the lowest labor input. For the investment of fixed capital, Zhejiang is also the second lowest. In all those investment items, Zhejiang is far behind Jilin which is number one in every category.

Two important factors distinguish Zhejing from other sampled provinces. Zhejing's non-farming income hold about 70% of total farm income, that is five times more than that of Jilin and Jiangxi and thirteen times more than that of Henan. Off-farm employment opportunities are easily available. Also, farm size in Zhejiang is extremely small, 2.2 mu per household, that is about one third of that of Henan, one fifth of that of Jiangxi, and one twelfth of Jilin. These suggest that nonfarm employment opportunities may have negative and farm size may have positive impacts on farm investment.

Table.3.2.6 Farm Investment in Four Sampled Provinces (1988-93)

	Land-related Investment (602 households)				Fixed Capital Investment (515 Households)	
	Number of House-hold	capital (Yuan)	labor (working day)	capital per labor	Number of House-hold	capital (Yuan)
Zhejiang	227	351.0	42.8	8.2	269	1697.2
Henan	99	416.1	45.7	9.1	71	2526.6
Jilin	123	750.8	148.5	5.1	80	3377.7
Jiangxi	153	23.1	71.6	0.3	95	941.4
average		360.1	72.1	5.0		1933.2

Jilin has the largest farm size, which also lead the investment in every category. Jiangxis land-related investment in terms of capital and fixed capital investment are far below the second lowest province Henan. But Jiangxi's farm size and the share of nonfarming income are not the lowest. But, land tenure in Jiangxi is extremely insecure. Table 3.2.2 tells for the period of 1988-93 for all sampled provinces, in average, 30% of villages experienced none land redistribution and 8% of villages experienced three times or more. However, in Jiangxi only 18.2% experienced none land redistribution, but 35% of villages experienced three times or more. This indicates that land tenure security may have strong positive impacts on farm investment.

Table 3.2.6 clearly proposes following hypothesis for the followed econometric test:
• Farm size has positive impacts on farm investment;
• Availability of nonfarming employment opportunities has negative impacts on farm investment;
• Land tenure security has positive impacts on farm investment.

Farm investment determination model and estimation results

Feder et al. (1992) conducted a study to search for the determinants of farm investment and residential construction in post-reform China. By solving a farm's profit maximization problem, the following econometric model was proposed to identify the determinants of farm investment:

$$I = a + b_1F + b_2C + b_3T + b_4S + b_5H + e \qquad (1)$$

Where, I: investment, F: farm Size, C: credit, T: tenure security perception, S: initial capital stock, H: household characteristics. a and b are coefficients, and e is error term.

The similar specification was also used to explain farmer's housing construction behavior. In their study, the econometric estimation did not show direct links from credit and tenure security to investment, only farm size had significant impacts on investment.

While the above Feder et al.'s work advanced the research in the field significantly, there are four limitations: (1) land tenure security was measured by people's perception toward the possibility of contract disruption and the security of using the same land after the contract expires. This measurement missed the aspect that how often actually the tenured land has been disrupted, which may be more important for telling tenure security . (2) The farm investment did not include land-related expenditures, such as green manure and manure, irrigation, and so on. (3) The impacts of other property right components, such as use rights, was not included. (4) The impact of nonfarming employment opportunities was not considered.

Using the household data of Zhejiang and Jiangxi, Yao and Carter (1996) estimated green manure, one of land related investment as shown by Table 3.2.4, determination equation. Their model specification is similar to Feder et al.'s (1992) and found tenure security, transfer rights, and use rights have significant impacts on the green manure.

The present study will continue Yao and Carter's work by expanding the scope of sample into all four provinces and the definition of land-related investment into all 8 kinds of investment expenditure. Also, Feder et al.'s model specification was adopted with certain modifications. The following model will be estimated:

$$I = a + b_1F + b_2N + b_3T + b_4A + b_5L + b_6G + b_7D1 + b_8D2 + b_9D3 + e \quad I=1,2, ...N. \quad (2)$$

The definitions of I, F, and T, as well as a, b and e, are the same as in equation (1). For i-th household, N is the non-farming job opportunities,

A is agricultural income earned by a household, G is the state grainprocurement quota, L is the number of labor in the family, D1 thorough D3 are dummy variables to distinguish Zhejiang, Henan, and Jilin from Jiangxi, respectively..

Differing from Feder et al's (1992), and Yao and Carter's (1996) work, the specification of (2) omits the components of age and education of household heads, because the implications of age and education is not clear. Aged people with little education may not necessarily invest less than younger and better-educated people, since longer farming experience may suggest a more urgent need for invest more. This may be the reason that these two variables are rarely significant in Feder et al's (1992), and Yao and Carter's (1996) work.

Grain procurement quota (G) is introduced into the model since it is the key concern of the government's intervention on farmer's land using rights. Farmers' freedom in making production and marketing decisions is limited by the government's grain procurement quota and related intervention about grain planting acreage and marketing channels. For i-th farm household, G is the procurement quota per mu land, measuring farmers' freedom in using land. Agricultural income (A) is included into the model. Farmers' investment activities are related to expected return. Suppose input prices are constant, higher agricultural income may induce higher investment in the near future.

Equation (2) did not include the measurement of land transfer rights. On the one hand, the household questionnaire did not provide reasonable measurement of the transfer rights perceived by an individual, on the other hand, without a functional land pricing mechanism, the freedom of transferring land may not have clear impacts on farm investment.

Non-farming job opportunity (N) will be represented by the ratio of nonfarming income to total income for i-th household. High values are usually related to better non-farming working opportunities.

Measuring land tenure security at farm level is a challenge. Naturally the frequency of land adjustment is a good approximate. An individual farm that experiences twice land adjustments has more insure land tenure than the other one that experience only once has. However, for the data available for this study, this type of measurement suffers two shortcomings. On the one hand, it does not have enough variation since

330 out of 440 household had only one time land adjustment; on the other hand, it ignores the differences in adjustment methods. For tenure security, redistribution has much profound implications than the marginal adjustment due to demographic changes. These concerns lead to the second measurement which may be more precise and practical. We choose those 330 households that had only one land adjustment, and divided them into two groups: one group had redistribution, the other had only demography-related partial adjustment. Therefore, N will be a be 0 or 1 dummy variable to distinguish redistribution from partial adjustment. When N is 0, i-th farm had one time redistribution; when N is 1, the farm had a partial land adjustment.

The richness of household questionnaire about farm investment provide an unique opportunities to define farm investment in different ways to assess the impacts of tenure security on a particular type of investment, land related or fixed capital. In the case of land related investment, a farm has a cash expenditure and a labor input which
can be estimated separately. In total, five estimation will be conducted, (1) land-related capital investment, (2) land-related labor investment, (3) land-related investment, (4) fixed capital investment, and (5) farm investment. (3) is the summation of (1) and (2), and (5) is the summation of (3) and (4). Table 3.2.7 below reports OLS estimation results.

The number of observations for each function is different. Though totally there are 800 household observations, for each variable there are always some households that have no data. To make a full data set for estimating a particular function, only those household observations that has all required variables are assembled. Therefore, there are 281 observations for land-related investment function, 188 for fixed capital investment function, and 170 for farm investment function. Appendix provides a table showing the structure of these three data sets, i.e. each province contribute how many, and for each province how many observations had only partial land adjustment, and how many had thorough land redistribution.

The residual diagnose for all five estimation did not find the heteroscadesticity problem which is a usual concern of cross section data. The diagnostics have been done on the residuals to check the heteroskedasticity problem since the cross section data is being used. The calculated Lagrange multiplier values for all five estimation in Table 3.2.7 are greater than the critical values, therefore, the null hypothesis of heterosckedasticity has been rejected in all cases.

Table.3.2.7 Estimation of Farm Investment Determination Model

	land related invest ment	land related capital investme nt	land related labor investme nt	fixed capital Investme nt	farm investme nt
Ag. Income	0.03 (3.31)* **	-0.00 (-0.61)	-0.00 (-0.45)	0.06 (1.81)*	0.31 (4.56) ***
Non-ag. Employ- ment	-903.59 (-2.01) **	-85.18 (-0.83)	-60.66 (-2.10)**	-321.85 (-0.44)	-622.67 (-0.71)
No. of labor	-181.31 (-2.90) ***	-16.01 (-1.13)	-0.09 (-0.02)	128.59 (1.29)	-202.89 (1.65)*
Farm size	10.67 (0.75)	7.14 (2.11)**	0.01 (0.01)	53.22 (1.83)*	83.91 (2.14)**
Tenure Security	346.96 (1.88)*	100.42 (2.41) ***	54.03 (4.58) ***	-871.13 (-3.37) ***	-347.23 (-1.07)
Grain Procuremen t Quota	-1.36 (-1.72) *	0.23 (1.30)	-0.10 (-1.92)**	0.04 (0.03)	-0.82 (-0.48)
D1	271.94 (0.73)	374.44 (4.47)** *	3.51 (0.14)	1414.2 (2.30)**	2556.4 (3.45) ***
D2	62.83 (0.27)	441.65 (8.60)** *	-26.09 (-1.78)*	2535 (6.47)** *	3094.5 (6.49) ***
D3	1991.5 (6.55) ***	607.51 (8.84)** *	40.04 (2.05)**	3377 (6.57)** *	5522 (7.97) ***
Constant	1330 (3.89) ***	-65.81 (-0.85)	84.30 (3.85)** *	-671 (-1.03)	-897.08 (-1.16)
Observation	281	281	281	188	170
Adjusted R^2	0.40	0.46	0.25	0.39	0.57

1: t-ratios are in brackets. *, **, ***: Indicate statistical significance

But the diagnoses of residuals showed that there was a multicorrelation problem. Very likely farm size and grain procurement quota are positively related. Since OLS estimator in the presence of multicollinearity remains unbiased and efficient, and major undesirable consequences is that variance of coefficient is exaggerated, Kennnedy (1993, p.181) suggested "don't worry about multicollinearity if the R^2 from the regression exceeds the R^2 of any independent variable regressed on the other independent variables". All five estimation fit in this case and the "do nothing" strategy was adopted.

The values of adjusted R^2 range from 0.25 to 0.57 for the five estimation, and four of them have the values 0.4 and above. For the cross section data, these are good fitness, which means the model can explain farmer's investment behavior reasonably well.

For the function of land-related investment, six out of ten coefficients are statistically significant. Dummy variable of tenure security has positive coefficient and statistically significant. This means that land tenure security does have positive impacts on land related investment. The farmers who had only marginal land adjustment are more willing to invest in land than those who had to go through land redistribution. Land tenure security matters here.

The coefficients of both agricultural income and nonfarming job opportunity are statistically-significant, but that of agricultural income is positive and that of non-farming job opportunity is negative. As the theory predicts that high farming income leads to more investment and better nonfarming job opportunities lead to less investment. The number of labor in the family is negatively related to land investment, as suggested by negative statistically significant coefficient of the variable. Since some investment could be substituted by labor work.

Land use rights, measured by grain procurement quota per mu land, also has a negative statistically-significant coefficient . A high grain procurement quota is typically related to more serious restrictions on farmer's land use rights, which reduces farmer's investment incentives, therefore leads to less land-related investment.

The coefficient of farm size is not statistically significant, it seems farm size does not have direct impact on farmer's decision on land-related investment in general.

The above relationships between land-related investment and various determinants were explained by other two models in more details, i.e. the functions of capita' and labor expenditure on land-related investment.

Land tenure security has more obvious impact on capital expenditure and labor expenditure for the land-related investment, as indicated by more Farm size is important for cash investment on land, but not for labor investment on land. More labor in a family will lead to more labor investment on land, but not necessarily more cash investment. Grain procurement quota is only statistically significant for the labor land investment function.

In all three land-related investment functions, land tenure security demonstrates significant positive impacts on land-relate investment. However, the estimation of fixed capital investment function is just opposite. The coefficient of land tenure security is statistically significant, but is negative, which means that the more secure is land tenure, the less is fixed capital investment. The farmers whose land had been only partially adjusted tend to invest less in land than the farmers whose land had been thoroughly redistributed. The land tenure security has opposite impacts on the land related investment and fixed capital investment. A possible explanation is that most farm investment requires capital which is very scarce in rural China. When land tenure is being secured, farmers are willing to reduce their investment on fixed capital and shift more resources on land-related investment. This explanation need to be investigated further.

Agricultural income and farm size are the other two variables that are important for fixed capital investment. More agricultural income and larger farm size lead to more fixed capital investment. Farm investment function summarizes the farmers' behavior of land-related as well as fixed capital investment. Since tenure security has opposite implications for the two types of investment, not surprisingly land security is not statistically significant in this function. When farm investment was considered in general, quantitatively it is hard to. find direct link between tenure security and farm investment. Aggregation wipes out the clear impacts of land tenure security on particular investment.

To farm investment, agricultural income and farm size are positively related, and the number of labor is negatively related.

Dummy variables, D1, D2, and D3, are statistically significant in most of five functions, and have positive sign. D1 represents Zhejiang, D2 Hena. and D3 Jilin. Farms in these three provinces invest more than those in Jiangxi in cash land-related investment and fixed capital investment, the positive sign is consistent with this.

Summary and Conclusions

Since 1983, household farms have become the main agricultural investment force in rural China. China's food security and sustainable agricultural development very much depend on farms' investment. The household responsibility system (HRS) reform in early 1980s has individualized farm's residual income rights, and introduced an effective incentive mechanism into the farming system. At the same time, the HRS retains the commune nature of the land property rights, land tenure is not secure for most farms, and farmers' land use and transfer rights are not fully guaranteed. Some researchers and policy makers argued that tenure security is the primary obstacle that hinders farmer's investment, but some other researchers suggested that nonfarming job opportunity and government intervention on farming operation are more influential on farmer's investment decision. This leads to a debate about how to reform China' current land system. This study, based on a survey conducted on 800 rural households in 1994, is to establish empirical evidence on the linkage between land tenure security and farm investment.

A survey conducted in 1988 found that for the period of 1978-88, about 70% of landholdings had been redistributed at least once, a 1994 survey had the same finding for the period of 1988-93. This means tenure security did not improve for all 16 years in China.

Farmers' investment activities were very modest for the period of 1988-93. Most farms engaged in only one or two types of land-related investment, one or two types of fixed capital investment. The most common land-related investment are manure and green manure, irrigation items were minimal. Threshers, animal house, and plant protectors are the three major fixed capital expenditures. In terms of total farm investment, the amounts were extremely small. An average farm household spent only 6% of its total income on the investment each year.

For the four sampled provinces, Zhejiang, Henan, Jilin, and Jiangxi, Zhejiang has the highest household income and land productivity, but has the second lowest farm investment. The abundant nonagricultural employment opportunities and extremely small farm size in Zhejiang may limited farmer's investment incentives. At the same time, Jiangxi that experienced the most frequent land redistribution records the lowest farm investment. These observations encourage more rigorous econometric analyses.

Based on empirical work done by Feder et al. (1992) and Yao and Carter (1996), an econometric model was constructed to examine the linkage between farm investment and land tenure security, as well as other possible determinants.

The estimation shows that tenure security appears to have no explicit impacts on farm investment, which is the same as the findings of Feder et al. (1992). But, this is only an illusion of a quantitative analysis based on aggregated data. When disaggregated farm investment into two categories, land-related investment and fixed capital investment, land tenure security demonstrates consistently significant positive influence on farmer's land-related investment activities. The farmers who have more secured land tenure tend to invest more on land than those who have less secured tenure.

Interestingly, based on disaggregated data, the econometric work showed that land tenure security has significant negative impacts on farmers' fix capital investment, The farmers who have more secured land tenure tend to invest less on fixed capital than those who have less secured tenure. Possible explanation is that when land tenure is secured, resource-scarce farm households may shift investment from fixed capital investment to land-related investment.

The opposite implications of land tenure on land related investment and fixed capital investment may explain why the direct linkage between land tenure and farm investment as a whole could not be confirmed empirically though in the theory and daily observation such a linkage has often been suggested.

Since 1993, the terms of farm households' land contracts have been extended by 30 years. While the empirical research based on the data before 1993 will be still useful for justify or unjustfy this government policy, a new survey to investigate farmers' investment behavior under new contract will be very helpful to understand the importance of tenure security

References

Basset, Thoms J. And Donald E. Crummey, edited, 1993. Land in African Agrarian Systems. The University of Wisconsin Press. Wisconsin, USA.

Bell, Clive. 1990. Reforming Property Rights in Land and Tenancy. The World Bank Research Observer. Vol. 15:143:166.

Binswanger, Hans P., Klause Deininger, and Gershon Feder. 1993. Power, Distortions, Revolt, and Reform in Agricultural Land Relations. The World Bank policy research working paper. WPS 1164.

Brooks, K. and Z. Lerman, 1994. Land Reform and Farm Restructuring in Russia. World Bank Discussion Paper 233, The World Bank, Washington, D.C..

Brown, Lester R. 1994. Who Will Feed China? . WorldWatch, Vol. 7: 5, pp. 10-19.

Bruce, John W. and Shem E. Migot-Adholla, edited, 1994. Searching for Land Tenure Security in Africa. Kendall/Hunt Publishing Company, Iowa, USA.

Carter, M. R. Liu, S., M. Roth, and Y. Yao. 1996. "An Induced Institutional Innovation Perspective on the Evolution of Property Rights in Post-Reform Rural China.". Department of Agricultural and Applied Economics, University of Wisconsin. USA.

Chen, Junshen. "On the rural labor surplus and related policies" Economic Daily 1/28/95

Euroconsult and Center for World Food Studies 1994, Farm restructuring and land tenure in reforming socialist economies. The World Bank Discussion Paper 268. The World Bank, Washington, D.C...

Feder, Gershon. 1993. The Economics of Land and Titling in Thailand. In The Economics of Rural Organization; Theory, Practice, and Policy edited by K. Hoff, A. Braverman, and J.F. Stiglitz. Oxford University Press.

Feder, Gershon and D. Feeny. 1993. The Theory of Land Tenure and Property Rights. In The Economics of Rural Organization; Theory, Practice, and Policy edited by K. Hoff, A. Braverman, and J.F. Stiglitz.

Oxford University Press.

Feder, Gershon, L.J. Lau, J. Y. Lin and X, Luo. 1992. The Determination of Farm Investment and Residential Construction in Post-Reform China. Economic Development and Cultural Change. xx:1-:26.

Grindle, Merilee S. 1990. Agrarian Reform in Mexico: A Cautionary Tale. In Agrarian Reform and Grassroots Development: Ten Case Studies edited by Rory L Prosterman, Mary N. Temple, and Timothy M. Hanstad. Lynne Rienner Publishers. USA and UK.

Hall, Anthony L. Land Tenure and Land Reform in Brazil In Agrarian Reform and Grassroots Development: Ten Case Studies edited by Rory L Prosterman, Mary N.Temple, and Timothy M. Hanstad. Lynne Rienner Publishers. USA and UK.

He, Daofen. 1993. Transition in Agricultural Land System at Village Level. In The Transition of Land System in Rural China edited by . Task Force of Land System in China. Beijing University Press. (In Chinese)

Herring, Ronald J. 1990. Explaining Anomalies in Agrarian Reforms: Lessons from South India In Agrarian Reform and Grassroots Development: Ten Case Studies edited by Rory L Prosterman, Mary N. Temple, and Timothy M. Hanstad. Lynne Rienner Publishers. USA and UK.

Hoff, K. 1993. Designing Land Policy: Overview. In The Economics of Rural Organization; Theory, Practice, and Policy edited by K. Hoff, A. Braverman, and J.F. Stiglitz. Oxford University Press.

Huang, Qinhe. 1993. Review on China's Agricultural Land Policy and Current Situation. In The Transition of Land System in Rural China edited by . Task Force of Land System in China. Beijing University Press. (In Chinese)

Jannuzi, F. Tomasson and James T. Peach, 1990. Bangladesh: Strategy for Agarian Reform. In Agrarian Reform and Grassroots Development: Ten Case Studies edited by Rory L Prosterman, Mary N. Temple, and Timothy M. Hanstad. Lynne Rienner Publishers. USA and UK.

Johnson,D.G.,1988, "China's Economic Reform" Economic Development and Cultural Change 36

Kennedy, P. 1993. A Guide to Econometircs (Third Edition). MIT Press. Cambridge.

Kung, J. K. 1995. "Equal Entitlement versus Tenure Security under a Regime of Collective Property Rights: Peasant's Preference of Institution in Post-reform Chinese Agriculture". Journal of Comparative Economics. 21:82-111.

Lerman, Z., K. Brooks., and C. Csaki. 1994. Land Reform and Farm Restructuring in Ukraine. World Bank Discussion Paper 270. the World Bank, Washington, D.C..

Liu Shouying, 1994,"The Structure and Changes of China's Land Tenure System," The Report on China's Agrarian Institutions, No.2, China's Land Tenure Research Group, Development Research Centre,the State Council, China.

Liu, S., M. R. Carter, Y. Yao. 1996. "Dimension and Diversity of Property Rights in Rural China: Dilemmas on the Road to Further Reform". Department of Agricultural and Applied Economics, University of Wisconsin. USA.

Luo, Yushen and Hongyu Zhang. 1995. The Institutional Innovations of Agricultural Land System Post Household Responsibility System. Economic Research, 1995 issue 1. Beijing, China. In Chinese.

McMillan,John, John Whalley, and Lijing Zhu,1989,"The impact of China's economic reforms on agricultural productivity growth," Journal of Political Economy, 97

Migot-Adholla, S., P. B. Hazzel, B. Blarel, and F. Place. 1993. Indigenous Land Rights Systems in Sub-Saharan Africa: A Constraint on Productivity?. In The Economics of Rural Organization; Theory, Practice, and Policy edited by K. Hoff, A. Braverman, and J.F. Stiglitz. Oxford University Press.

Scolfield, Rupert W. 1990. Land Reform in Central America In Agrarian Reform and Grassroots Development: Ten Case Studies edited by Rory L Prosterman, Mary N. Temple, and Timothy M. Hanstad. Lynne Rienner Publishers. USA and UK.

Task Force of Land System in China, 1993. The Transition of Land System in Rural China . Beijing University Press. (In Chinese).

West, Loraine and Minlin Pan, 1993. "The Role and Function of Households' Investment in Agricultural Development in China". Symposium of Agricultural Development in China in 1990's. Edited by Chinese Agricultural Economics Association. China's People University Press, 1993. Beijing, China.

Yao, Y. and M. R. Carter. 1996. "Land Tenure, Factor Proportions, and Land Productivity: Theory and Evidence from China." Department of Agricultural and Applied Economics, University of Wisconsin. USA.

Appendix: The Structure of the Data for Three Farm Investment Functions

total farm investment function					
	Zhejiang	Henan	Jilin	Jiangxi	Total
No. of observations	54	53	17	46	170
in which					
Household who had redistribution	36	42	0	46	124
Household who had adjustment	18	11	17	0	46
land-related investment function					
	Zhejiang	Henan	Jilin	Jiangxi	Total
No. of observations	58	80	54	89	281
in which					
Household who had redistribution	39	69	0	89	197
Household who had adjustment	19	11	54	0	84
fixed capital investment model					
	Zhejiang	Henan	Jilin	Jiangxi	Total
No. of observations	59	53	28	48	188
in which					
Household who had redistribution	36	42	0	48	126
Household who had adjustment	23	11	28	0	62

CHAPTER 3.3

No Impasse for China's Development[1]

Jian Song
State Councilor, and Chairman
State Science and Technology Commission
China

The central topic is "Food and Water - A question of Survival", a subject of world' s major concern, which is intensified greatly at and after UN World Food Summit held in Rome in November 1996. One hundred and eighty six national delegations, many led by Heads of State or Government, attended the Summit which concluded with a Declaration on World Food Security and Plan of Action that pledge to halve the number of hungry people from the current 850 million by the year 2015. [1]

All politicians as well as scientists are still struggling to better understand this crucial situation of the world and its consequences. Eight hundred and fifty million people are facing food scarcity today, while the world' s farmers are to feed 96 million more people added to the planet each year.

The per person world grain harvest has decreased from 415 kg in 1985 to 360 kg in 1996. The farmers can expect little help from fishermen, the seafood catch is not growing since 1990.

The world has fallen into a bitter controversy on a fundamental, moral and ethical principle. While almost unanimously recognized that access to adequate food is an unquestionably universal human right, the leaders of numerous countries and religious communities speak against population growth control, even though the ever increasing demand for food and water is beginning to press against some limits of the planet's carrying capacity.

But I would not like to and can't be involved into this world-wide controversy, due to lack of deep study on the latter and the sense of etiquette either. Thus I prefer to accept the verbiage, as many politicians often define it, population, food and water, environment together

---

[1] Speach delivered at "Forum Engelberg - 8th conference" 1821 March 1997 in Engelberg, Switzerland

formulate a big problem of a complex mega-system.China's recent rapid development has caught world-wide attention. But Mr. Lester Brown, the Director of the US based World Watch Institute, made a new warning to China: "Who will feed China" in the next century [2]. I paraphrase it, if the Chinese would not be able to feed themselves, then they would starve the world. This was not the first warning, but a new one that has drawn a great attention of the Chinese science community and leadership. In fact, it has been a central problem for the country and investigated by the intellectuals of several generations since the founding of the People' s Republic in 1949, when the population size was only less than 600 million.

In the background of dramatic growth of the world population after the Second World War, China's population was doubled between the 1950s and the 1990s. The situation has become a cutting edge of concern for scientists and politicians of this country (Figure 3.3.1). It took decades, however, for China to reach a consensus that the nation' s future depend upon whether we are able to control population growth within some limit. To solve this complex problem mingled-together with resources and environment, is really a terrifying challenge to China for the next century. It is a consensus for the Chinese science community that the curb on population growth holds the key to this multi-faceted issue. Today, China feeds 1.2 billion population, 22 percent of the world' s total, with the produce from only 7 percent of the world' s arable land. None has the right to complain of too many people in China, this is the legacy of history. No one is supernumerate, all men are created equal, as our faith tells us.

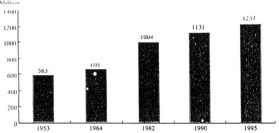

Figure 3.3.1 China's population 1950s - 1990s.

But in the next century if China' s population doubles again in a matter of thirty years, that would surely lead to something terrible! It would be a population explosion in its true sense. Fortunately, a number of Chinese scientists started research on this vital issue as early as the 1950s.

AGRICULTURE IN CHINA 1949-2030

Based on the previous study on the theory of demography, starting from Thomas Malthus' "First Essay on Population" [3], many Chinese scientists have long been engaged in studying the theory of population growth dynamics in general and particularly for China' s case, with the help of concepts and methodology of system analysis and control theory. We have made the following statements which now are agreed among demographers world-wide [4].

- The population growth rate and size of a country are wellcontrollable parameters of the system.
- For each nation or region, there exists a critical value of Total Fertility Rate (TFR, average child bearings per female), denoted by β_{cr} ,which defines the stability of population dynamics. This critical value can be calculated from the standard date of census.
- If a country would like to stop population growth, TFR should be controlled below its critical value $\beta_{cr.}$ In the terminology of mathematics, the first eigen-value of the equations of population dynamics then becomes negative. On the contrary, if TFR is kept greater than its critical value the eigen-value becomes positive, then the population size will be increasing intrinsically.
- The critical value of TFR for China was estimated as 2.10 for the late 1980s. Due to the long "time constant", the inertia of population system, commensurate to the life expectancy of the population, if China keeps for long its TFR at 2.0, near the "replacement level", it would take 70 years from now on to stabilize the population at 1. 7 billion. For example, if China's TFR is well below 1.5, i. e. in average a couple have only 1. 5 children, then the population size would be stabilized within 30 years at a much lower level.

In the early 1980s, based on an in-depth investigation, we published a projection on China's population growth for 100 years ahead, and proposed an action program to reduce growth rate. It immediately caught the eye of the top government leadership.

We are gratified that a proper family planning policy was formulated in China in the early 1980s. It is encouraged to practice deferred marriage and child bearing. Each urban family is strongly suggested to have one child, and every rural family at most two, while no restrictions for minority nationalities at present, whose population stands at 8 percent of the national total. This policy would lead to a TFR 1.6 in the long-run.

As a result of 15 years' efforts from 1980 to 1995, China's crude birth rate dropped. From 2.24 % to 1.72 %, and natural increase rate, from 1.56 % down to 1.05 % , with TFR reduced from 2.20 to 1.9, substantially lower than the critical value 2.1 (Figure 3.3.2). The government of China is determined to follow this as a long-term policy, a policy now understood and well accepted by most of the people

In December 1996, China' s State Statistical Bureau published a projection on population growth for the period 2000 - 2050, based on the data of 1995 sampling survey. If the TFR is kept from now on at the present level of 1.9, the total population would be 1.27 billion in 2000, peak at 1. 48 billion in 2030, and decline to 1.42 billion in 2050. However, if the TFR is dropped to 1.6 in the first decade of the 21st century, the population size would be stabilized at only 1. 377 billion in 2020, dropping to I - 234 billion in 2050. (Figure 3.3.3)

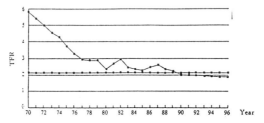

Figure 3,3,2TFR of China 1970-1996

This situation makes both scientists and politicians of the country out of anxiety with a sigh of relief. It is safe to conclude that China's population will peak within a half of century, at a level very likely around 1.6 billion and begin to decline afterward. To be on a safer side, let us use the 1.6 billion as the base for the following discussion on China's food security. According to FAO, food security should have three specific aims: ensuring production of adequate food supplies, maximizing stability in the flow of supplies, and securing access to available supplies on the part of those who need them. Undoubtedly, to produce enough grain and achieve "self-sufficiency" will be the decisive factor for China's food security in the future

The term "self-sufficiency" doesn' t mean to produce everything needed. But for China it really means to produce enough grain to secure food supply near the "well-to-do" level for the whole population. In other words, for a population of 1. 6 billion in the mid-21st century, it needs 640 million tons of annual grain harvest to guarantee an average 400 kg per citizen.

AGRICULTURE IN CHINA 1949-2030

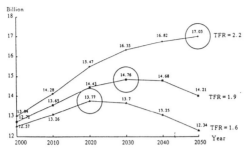

Figure 3.3.3. TFR vs. Population in different periods to 2050.

According to the study and analysis made by Chinese and international agro-scientists, this is a must and reachable and good grounded target. [5] The harvest of 1996 hit an all-time high of 480 million tons, more than 400 kg per capita. China should increase harvest by 160 million tons to reach 640 million, 30 percent more than the present yield, within 30 to 40 years to come.

Such demand requires an average annual growth of one percent in food supply. We see this as a target surely attainable from either economic or technical point of view, based on the past performances of the country' s agriculture and the potential of resources.

• An average annual growth rate of 3.1 percent of crop yield registered in the past 46 years. (Figure 3.3.4)
• The still big potential to increase per hectare yield. There are 90 million hectares, two thirds of the existing cultivated land, are low- and medium-yielding. The improvement of technologies may increase harvest by more than 30 percent.
• Popularization of the available generic technologies can make substantial contribution to the national grain yield. For example, rice yield can definitely surpass the present over 5 tons per hectare to reach 6 tons (US), 7 tons (Korea), or 8 tons (Australia). That is an increase of more than 30 percent from present level.
• Biotechnology science, new hybrid and trans-genetic traits and varieties are the key to the future for production increase, as indicated by many scientists. - 20 million hectares of wasteland are cultivable and awaiting reclamation -for agricultural purposes.
• The 250 million hectares of grassland still remain largely unattended.
In spite of the waning of ocean fish-catching as Dr. Brown elaborated in Rome recently, two million hectares of in-land waters and three million square kilometers of shallow sea in China can provide enormous water

surface for development of aqua-culture industry. The aqua-culture yield
has seen an annual growth of 20 percent in this decade ensuring 20 kg of
fish food per capita. It is expected to have quadrupled aqua-culture yield
within next two decades

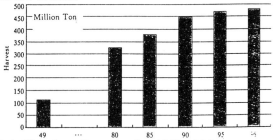

Figure 3.3.4. China's grain production 1949-1996

Quite recently, scientists from the Chinese Academy of Sciences
estimated theoretically possible maximum grain yield for China based on
analysis of regional ecological conditions. It could reach 1,026 million
tons a year. Even under extremely adverse weather conditions, the yield
can reach 534 million tons, almost 20 percent higher than today. If
irrigation improved, it could easily reach 818 million tons, 510 kg per
capita for a 1.6 billion population. [8]Having examined food supply
perspectives of China, now I come to the water problem, the most critical
factor for China's future agriculture.

In the recently published "Comprehensive Assessment of the Freshwater
Resources of the World", the- United Nations reported that up to 1.2
billion people, or one fifth of the world's population, face medium-high to
high water stress. The report warned that by 2025 when the world's
population exceeds 8.3 billion, two thirds of them would be affected by
moderate-to-severe water shortage, unless supplies are used more
efficiently and more resources are developed. A FAO report identifies
China as one of the 26 most water-deficient countries of the world, and
300 out of the 600 cities are suffering from water shortage. Indeed, the
annual runoff water is estimated at about 2.8 trillion cubic meters, 2,250
cubic meter per capita, only one forth of the world's average.

Uneven distribution of water resources is another feature for the country.
(Fig-5) The north part of the country with 64 percent of cultivated land
are afforded only 17 percent of water. 250 million hectares of grassland
in the north and northwest China can be converted to good farmland if
irrigated. Most geologists and meteorologists explained that the draught

of the north and northwest China is caused by the Himalayan cycle of mountain building started in the Cenozoic Era some 60 million years ago, which blocks the monsoon winds from the Indian Ocean.

In discussing solutions, some scientists, especially theoretic physicists seriously suggested to level off or split up, by atomic explosions, part of the Himalayas to let in the humid winds from the Indian Ocean. This may be just a courageous imagination. Though human activities may lead to global warming, it is beyond human ability now to change the climate of Central Asia and North China.

So we have to look for other ways around. Hydrologists are studying the possibility and making proposals to divert water from the abundant South to the thirsty North. Every year, up to one trillion cubic meters of water flow down the south provinces along the Yangtze river into the Pacific Ocean, without being well utilized. One proposal is to channel water from the Yangtze to the North.

Three diversion routes have been extensively studied. The Western Route is to chop up a 75-km tunnel in the Qinghai-Tibet Plateau and transfer 20 billion cubic meters of water from the upper reaches of the Yangtze to the Yellow River flowing down across north provinces. This will be a huge project, but is quite possible to be launched next century. [9]

For the Eastern Route, it is to pump water from the Yangtze at its lower reaches, and transfer it northward through about 1,000 km along an ancient artificial waterway, the Grand Canal, developed about 1,500 years ago. The amount of water to be diverted is estimated at 14 billion cubic meters a year in average.

However, a high-level investigation committee favors the construction of the Middle Route first, which may supply the North with 20 billion cubic meters of water annually at affordable operational cost. The water is taken from middle reaches of the Yangtze, and may flow to Beijing by natural gravitation without pumping.

Nevertheless, the most important solution to the problem of water shortage, at present and for many years to come, is yet to popularize water-efficient technologies in agriculture, such as the shift to dripp irrigation. According to projections by the China Agenda 21 and experts at the Water Resources Ministry, in 2030 a total of 160 - 220 billion cubic meters of new water resource should be found to satisfy the country's demand, mainly for agricultural and industrial growth. Now, China's total capacity of water supply from 86,000 dams, feeder channels, surfacewater and groundwater extraction projects is around 600 billion cubic meters. The actual water utilization is 525 billion cubic meters, of which 80 percent going to irrigation. China' s current agricultural water use-efficiency is around 10 percent. If the efficiency reaches the level of Japan (30 percent), the loss of 84 billion cubic meters can be avoided. If it reaches half of that of Israel (80 percent), 126 billion cubic meters could be saved. So a large-scale application of water-efficient irrigation technologies would save 100 billion cubic meters annually. On the other hand, many new projects of reservoirs building, river diversion and groundwater exploitation, in addition to the water transfer plans from the Yangtze to the North discussed above, could increase the capacity of water supply by 100 billion cubic meters per year.

According to the analysis of scientists and agronomists, the above listed newly developed and saved water resource of 200 - 226 billion cubic meters a year would be available by 2030. This will be able to sustain the agriculture and well-to-do living conditions for the 1.6 billion population of China.

Let me summarize and respond to the warning raised by many friends world-wide. Chinese science communities and the government have reached the consensus that China can feed itself, and will be able to get to the level of middle developed countries in the first half of the next century. However, everyone admits that this is a tremendous challenge, and an uphill battle for the whole nation. We must do our utmost to control population, persisting on the right family-planning policy, to raise the agriculture level, and to develop new water resources. All will depend heavily on the application and advancement of science and technology. With the successive efforts of contemporary and the following generations, China will have a bright future. No impasse ought to happen. No dooms day would come.

References
1 Rome Declaration of World Food Security, World Food Summit, 13 - 17 November, 1996, Rome, Italy.
2 Lester Brown, *Who Will Feed China, World Watch*, September/ October, 1994, pp. 10 - 19.
3 Thomas R. Malthus, *First Essay on Population*, London, 1798.
4 J. Song and J. Y. Yu, *Population System Control*, Springer-Verlag, Berlin, 1988.
5 T. C. Tso, *Population & Food in China*, A Lecture at Maryland State University, March 7, 1996.

6 D. Gale Johnson, *China's Future Food Supply: Will China Starve the World?* 1995, Background Paper provided by Dr T. C. Tso at the International Seminar on Agricultural Development of China, 7 - 11 June, 1995, Beijing, China.

7 Robert L. Paarlberg, *Feeding China: A Confident* View, 1995, Remarks prepared for meeting of New York Council on Foreign Relations research group on China and the Environment, Washington, D.C., April 24, 1995.

8 Shi Yuan - Chun, *Prospective Analysis of Potentials for the Grain Production in China, Science and Technology Daily,* 21 October, 1996.

9 The Changjiang (Yangtze River) Water Resource Commission, *The Western Route Pr 'ect of Water Transfer from South to North,* 1995.

Part IV

TOWARD 2030

Chapter 4.0

China's Grain Economy: Past Achievements and Future Prospect*

Justin Yifu Lin

Peking University and
Hong Kong University of Science and Technology

China is an extremely land-scarce economy on per capita basis. In the 1990s, cultivated land per capita is only about 0.1 hectare, merely 40 percent of the world average. [1] Moreover, China is one of the driest countries in the world. Water runoff is below the world average, only about one-third is exploitable, and annual precipitation has a very uneven distribution. The water shortage in the North and in the Northeast is very acute, while flood often occurs in the South. Due to the poor natural endowments and climate conditions, agricultural production failures often frequented China before the last several decades of 20th century. Observers in the West once referred to China as the "land of famine" (Mallory 1926). China's success to feed itself in the recent decades is a celebrated achievement in both Chinese and world's history.

---

[1] Unless otherwise indicated, the statistical figures in the paper are taken either from various issues of *China Statistical Yearbook* or *A Statistical Survey of China*, published by the State Statistical Bureau. This chapter draws heavily on Justin Yifu Lin "How Did China Feed Itself in the Past? How Could China Feed Itself in the Future" 1998 Distinguished Economist Lecture, CIMMYT, 1998.

Despite its past success, recently China's ability to feed itself in the future becomes a great international concern. Chinese economy is expected to have a dynamic growth in the coming decades (Lin et al. 1996). On the one hand, rapid economic growth is expected to drastically shift China's comparative advantages away from agriculture (Anderson 1990). On the one other hand, urbanization, marketization, and dietary diversification will significantly increase China's food demand. Moreover, Chinese population may increase another 40 percent to reach 1.6 billion in 2030. Because of the sheer size of China's economy, small changes to the gap between China's food demand and domestic supply will have large effects on world agricultural trade. Whether China's food production can keep up its increasing need in the coming decade is an issue of both national and international significance.In the chapter, I will first discuss how China succeeded in feeding its large population in the past. I will then discuss whether China has the potential to meet its future food demand and what measures are required if China attempts to continue feeding itself in the future.

Lessons from Past Achievements

China has been the most populous nation in the world for millennia. Despite the government's efforts to curb birthrate, China, like many other developing countries in the world, experienced a rapid population growth in the modern times. In 1952-96, Chinese population doubled from 575 million to 1.2 billion. Because virgin land in China has virtually depleted after thousand years of continuous population expansion and cultivation, every increase in population was unavoidably accompanied by correspondent reduction in per capita cultivated land. In the 1950s, the per capita cultivated land was already less than a half acre (0.45 acre or 0.18 hectare in 1952). The per capita cultivated land dropped to merely a quarter acre in the 1980s (0.1 hectare).[2] However, in spite of the reduction in

---

[2]For a long time, China's official figure for cultivated land was 96 million hectares. It is widely believed that the official figure underestimates the actual land area by as much as 30 percent. A recent agricultural survey confirmed the suspicion. The Chinese government may revise the figure to 120 million hectares in the coming release of the survey findings.

per capita available agricultural resources, the Chinese people's nutrition improves substantially in the past decades, as Figure 4.0.1 show. By the 1990s, the per capita daily intakes of calorie, protein, and fat in China all surpassed the averages. The Chinese diet is largely dependent on food grains. According average intakes of developing countries and reached or exceeded the world's to an estimate, about 90 percent of the caloric content and 80 percent of the protein intake in the Chinese diet come from grains (Smil 1981). Therefore, grain production holds the key to China's food situation.

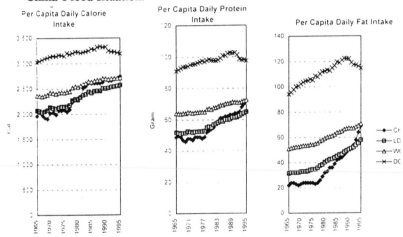

Figure 4.0.1 Nutrition of people. Source: FAO, FOOSTAT (1997)

China changed from a net grain exporter in the 1950s to a net grain importer for most years after the 1960. The net export or net import, nevertheless, never exceeded five percent of domestic grain production (Figure 4.0.2). China has been virtually a grain self-sufficient economy. The improvement in Chinese population's nutrition intakes in the past decades was a result of increase in domestic food production. In effect, China's grain output tripled from 164 million tons to 490 million tons from 1952 to 1996. As a result, per capita grain availability increased from 285 kg to 400 kg in the same period (Figure 4.0.3).

Despite the effort to claim new land, total cultivated land in China has been declining since the 1950s due to the competing usage for industrial development, residential construction, and so on. The increase of cropping intensity index from 130 in the 1952 to 158 in the 1996 offset part of the decline in cultivated land. However, the grain sown acreage after reached the peak of 136 million hectares in 1956 declined continuously to 110 million hectares in the 1990s. The increase in grain output comes totally as a result of increase in yield. As shown in Figure 4.0.4, the grain yield climbed three steps in the past four decades, from less than 2 tons/hectare in the 1950s and 60s to 2 tons/hectare in the 1970s, then to 3 tons/hectare in the 1980s, and finally to 4 tons/hectare in the 1990s. The average annual growth rate was 2.75 percent for the 44 years in 1952-96. This is one of the most celebrated records in the world agricultural history.

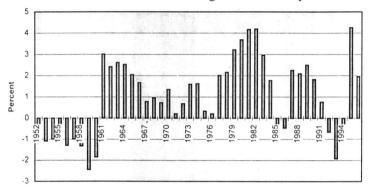

Figure 4.0.2 Net grain import as percentage of domestic output. Source: Ministry of Agriculture, Planning Bureau (1989)

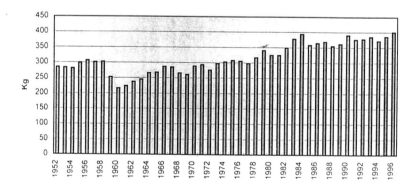

Figure 4.0.3 Per capita grain output. Source: Ministry of Agriculture, Planning Bureau (1989)

In addition to the remarkable performance in agriculture, Chinese economy has also experienced a dramatic transformation in the same period. Between 1952 and 1996, the GDP in China increased 25 times (Figure 4.0.5). Especially, after the late 1970s, Chinese economy has become the fastest growing economy in the world, with an average annual GDP growth rate of 9.9 percent in the period of 1978-96. In the 1950s, China was predominantly an agrarian economy. The share of agriculture in the national economy has dropped from 58 percent in 1952 to less than 25 percent in the 1990s

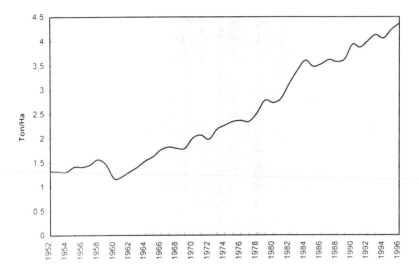

Figure 4.0.4 Grain Yield. Source: Ministry of Agriculture, Planning Bureau (1989)

An important issue that confronts most developing countries is how to develop agriculture rapidly in order to meet the increased food demand brought on by explosive population growth and to support urban industrialization. From the above prospective, China's achievements have been very remarkable. However, in the process of reaching the above achievements, China also has made many mistakes, and paid a high price for those mistakes. China's experiences provided many useful lessons for other developing countries.

At the founding of the People's Republic of China in 1949, China had a predominantly agrarian economy, with 89 percent of the population residing in rural. Heavy industry was a major characteristic of a developed country's economic structure at that time. In order to strengthen national power, the Chinese government adopted a Stalin-type heavy-industry-oriented development strategy in 1952. The goal was to build as rapidly as possible the country's capacity to produce capital goods and military materials. Agriculture in effect was treated as a supporting sector.

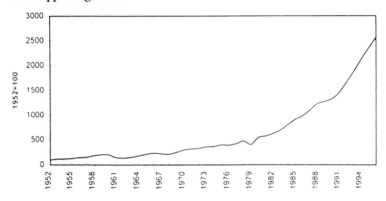

Figure 4.0.5 GDP Growth in China. Source: State Statistical Bureau, various issues.

Capital was extremely scarce at that time and the voluntary saving rate was far too low to finance a high rate of investment in heavy industry sought by the development strategy. To facilitate rapid capital expansion, a policy of low wages for industrial workers evolved alongside the heavy-industry-oriented development strategy. The assumption was that through low wages, the state-owned enterprises would be able to create large profits and to reinvest the profits for infrastructure and capital construction. The practice of establishing low prices for energy, transportation, and other raw materials, such as cotton, was instituted for the same reason.

To implement the low-wage policy, the government needed to provide urban population with inexpensive food and other necessities, including housing, medical care, and clothing. The government instituted a restrictive food rationing system in 1953, which was kept in practice until the early 1990s.[3] Meanwhile, in order to secure the

---

[3]In addition to grain, edible oils, pork, and sugar are included under rationing.

low- price food supply for urban rationing, a low-price compulsory grain procurement policy was imposed in rural areas in 1953. The domestic grain trade in China was virtually monopolized by the state. The industrial development strategy also resulted in a great increase in demand for agricultural products because the increase of urban workers, the need to expand agricultural export to earn foreign exchanges for importing equipment for industrial projects, and the increase industrial demand for raw material. Under this conditions, agricultural stagnation and poor harvests would not only affect food supply but also have an almost immediate and direct adverse impact on industrial expansion.

As the government was reluctant to divert resources from industry to agriculture, the government pursued a new agricultural development strategy, which relied on mass mobilization of rural labor to work on labor-intensive investment projects, such as irrigation, flood control, and land reclamation, and to raise unit yields in agriculture through traditional methods, such as closer planting, more careful weeding, and the use of more organic fertilizer. Collectivization of agriculture was the institution that the government believed would perform these functions

The independent family farm was the traditional farming institution in rural China. The typical farm not only was small, but also fragmented. At the beginning, the government promoted a "mutual aid team" system, in which 4 or 5 neighboring households pooled together their farm tools and draft animals, and exchanged their labor on a temporary or permanent base. The movement was surprisingly successful. It encountered no resistance from the farmers and was carried out relatively smoothly. The main rationale of the collective movement was to pursue the economies of scale and to mobilize rural labor for construction of agricultural infrastructures. This initial success greatly encouraged the government leadership and led the government to take a bolder approach. The size of farm was enlarged rapidly through alterations of farming institution in a short sequence from the mutual-aid teams, to primary cooperatives, to advanced cooperatives, and finally to communes in 1958, which had an average size of 5000 farm households and 10,000 workers.

Billions of man-days were mobilized as expected. The communal movement, however, ended up with a profound agricultural crisis between 1959 and 1961. The gross value of agriculture measured at the constant prices of 1952 dropped 14 percent in 1959, 12 percent in 1960, and another 2.5 percent in 1961. Most importantly, grain output was reduced 15 percent in 1959, another 16 percent in 1960, remained at the same low level for another year, and did not recover to the level of 1952 until 1962. A careful study of recent released demographic data leads to the conclusion that the crisis resulted in about 30 million excess deaths and about 33 million lost or postponed births in 1959-61(Aston et al. 1984).[4]

The government did not abolish communes after the crisis but delegated agricultural operation and management to a much smaller unit, the "production team," of 20-30 neighboring households. This system remained the basic farming institution from 1962 until 1979 when the rural reform replaced the production team system with a household-based farming system, called the household responsibility system reform.

The government, nevertheless, became more realistic and, for a number of years immediately after the crisis, gave the priority of development to agriculture. The government's policy started to emphasize modern inputs. China's irrigated acreage increased gradually from 30.55 million hectares (29.7 percent of total cultivated area) in 1962 to 44.97 million hectares (45.2 percent of total

---

[4] There has been many attempts to explain this crisis. The conventional hypotheses for the sudden reduction in grain output were three successive years of bad weather, bad policies and bad management in the communal movement at that time, and incentive problems due to the unwieldy, large size of the communes. Lin (1990) found that empirical data does not support the conventional hypotheses and proposed that the main cause of the agricultural collapse was the deprivation of the peasants' right to withdraw from the collectives. This switch in the form of organization changed the incentive structure for the peasants and consequently undermined agricultural productivity. See further discussion in the latter of this section. Lin and Yang (1997) also found that urban-bias in the grain distribution during the famine year had a much larger power in explaining the cross-section differences in death rate during the famine period than the food availability.

cultivated area) in 1978; but, as Table 4.0.1 shows, most of this increase came from the spread of powered irrigation rather than the construction of labor-intensive canals and dams. The utilization of chemical fertilizer was also accelerated, from a very modest 4.6 kilograms per hectare in 1962 to 58.9 kilograms in 1978. Equally impressive was the expansion in the utilization of electricity, a 17.5 folds increase between 1962 and 1978.

Table 4.0.1: Use of Modern Inputs

	Irrigation			Tractor-Plowed Area	
Year	Total Irrigated Area (M. ha)	Irrigated Area in Total Cultivated Area (%)	Powered Irrigation in Irrigated Area (%)	Total Area (M. ha)	Share in Sown Area %
1952	19.96	18.5	1.6	0.14	0.1
1957	27.34	24.4	4.4	2.64	2.4
1962	30.55	29.7	19.9	8.28	8.1
1965	33.06	31.5	24.5	15.58	15.0
1978	44.97	45.2	55.4	40.67	40.9
1984	44.64	46.1	56.4	34.91	30.9
1995	49.28	51.9	65.6	n.a.	n.a.

Year	Chemical fertilizer		Electricity M.KWH
	Total Amount M t.	Per hectare kg/ha	
1952	0.08	0.5	50
1957	0.37	2.3	140
1962	0.63	4.6	1,610
1965	1.94	12.4	3,710
1978	8.84	58.9	25,310
1984	17.40	120.6	46,400
1995	35.94	239.7	71,200

Source: State Statistical Bureau, *A Statistical Survey of China, 1996.* Beijing: China Statistics Press, 1996; and Ministry of Agricultural Planning Bureau, *Zhongguo noncun jingji tongji ziliao daquan, 1949-86* (A comprehensive book of China rural economic statistics, 1949-86), Beijing: Agricultural Press, 1989.

However, the most celebrated change was the establishment of an agricultural research and promotion system for modern varieties. As a matter of fact, the agricultural research is an area that the Chinese government can be proud of. The Chinese Academy of Agricultural Sciences was founded in Beijing in 1957, and, concurrently, each of the 29 provinces in the mainland also established its own academy of agricultural sciences. Each national and provincial academy consists of several independent research institutes. Most prefectures also founded prefecture research institutes. In addition, agricultural research was conducted in a few research institutes of the Chinese Academy of Sciences and in some universities.

This system of agricultural research was disrupted, from 1966-1976, during the Cultural Revolution. The Chinese Academy of Agricultural Sciences, as well as many provincial and prefecture academies, were reorganized, and many research scientists were sent in small groups to work on farms during this period. The agricultural research system was restored after the end of Cultural Revolution in 1976 and many counties also established their own agricultural research institutes at that time. The agricultural research institutes were funded by governmental budgets at their corresponding levels. The Ministry of Agriculture and the State Science and Technology Commission, however, also provided grants to research projects at institutes in lower levels.

The division of labor among the nation-, province-, prefecture-, and county- levels of research institutes was rather broad, and there were considerable overlaps. The research institutes in the Chinese Academy of Agricultural Sciences emphasized basic and applied research with national significance. They were also responsible for technical supervision and coordination of provincial programs. The institutes in provincial academies stressed applied research in accordance with the ecological conditions of its province. Prefecture institutes mainly engaged in selection and adaptive research. County research institutes were primarily responsible for extension work. Each research institute sets its own research agenda under the supervision of its corresponding level of government and institutes at higher levels. In addition, the Ministry of Agriculture and the State Science and Technology Commission also initiated some centrally-orchestrated nationwide research programs on specific crops and problems.

Variety improvement has been the core of China's agricultural research program from the very beginning. In the early 1950's, emphasis was given to the selection and promotion of the best local varieties. Meanwhile, new varieties of rice, wheat, cotton, corn and other crops were also imported from abroad. A major breakthrough in rice-breeding occurred in 1964. In that year, China began full-scale distribution of fertilizer-responsive, lodging-resistant dwarf rice varieties with high-yield potential, two years earlier than IRRI's release of IR-8, a variety which launched the Green Revolution in other parts of Asia. At about the same time, hybrid corn and sorghum, improved cotton varieties, and new varieties of other crops were also released and promoted. The high-yielding varieties were promoted and adopted in production rapidly. A second major breakthrough in rice-breeding occurred in 1976. In that year, China became the first country to commercialize the production of hybrid rice. Some researchers heralded the innovation and commercial development of hybrid rice as the most important achievement in rice breeding in the 1970's (Barker and Herdt 1985, P. 61). By 1979, the percentage figures for area sown with high-yielding varieties were 80 percent for rice, 85 percent for wheat, 60 percent for soybeans, 75 percent for cotton, 70 percent for peanuts, and 45 percent for rape (Ministry of Agriculture, Planning Bureau 1989, pp. 248-9).

The planned economic system in a socialist economy is known for its ability of mobilizing resources for specific projects. For that, the national concerted effort for the innovation and commercialization of hybrid rice is a good example.[5] However, the above-described

---

5. Rice is a self-pollinating plant. For commercial hybrid rice production, it is necessary to have a genotype that is male sterile (a rare genetic characteristic), cross it with "maintainer lines" to produce off-spring with male sterility but other desirable characteristics, and then cross again with "restorer lines" to produce seeds with normal self-fertilizing powers. The first male-sterile variety was discovered in 1970 by Yuan Longping, a devoted breeder at a high school, and his assistant. In 1971, the search for "maintainer lines" and "restorer lines" became a concerted nationwide program involving more than twenty research institutes in several provinces. The first maintainer variety was discovered by Yuan and another researcher in Jiangxi Province in 1972, and the first restorer variety was discovered by a breeder in Guangxi Province in 1973. A hybrid combination with marked heterosis was bred in 1974. Regional production tests were conducted simultaneously in hundreds of counties in 1975. In 1976, hybrid rice began to be commercially released to farmers.

decentralized research system also followed very remarkably underlying economic principles. According to the theories of technological innovation in a market economy, the allocation of resources should reflect the market size of the technology (Griliches 1957, Schmookler 1966) and the scarcity of factor that the technology is replacing (Hayami and Ruttan 1970). Lin (1991a) argued that theoretically the principles for effective technological innovation in an economy without markets is the same as that in a market economy. Lin (1991b, 1992a) also found empirically that the pattern of research resource allocation in a research institute and the farmers' adoption of new technology in a locality could be explained very well by the market size and factor scarcity in that locality.

The efforts in research and the application of new technologies are expected to contribute significantly to agricultural output growth. The yield profile in Figure 4.0.4, which measures the changes in land productivity, provides an indirect evidence of research' and technologies' contribution to grain production. An empirical study, that directly measures the contribution of research to production, find 20 percent of the agricultural output growth in the period of 1965-93 is attributable to research-induced technological change (Fan and Pardey 1997). Another empirical research focusing on China's rice production in the period of 1970-90 also confirms the primacy of technological change in explaining the yield improvement (Huang and Rozelle 1996). However, when we look at the total factor productivity (TFP) instead of the land productivity and look the productivity profile for the whole period from 1952 to 1996 instead of just a sub-period, the picture is quite different. The studies by Fan (1997) and Wen (1993) show that the TFP throughout the 1960s and 1970s was lower than that in the 1950s and did not rise above the 1952 level until the agricultural reform started in 1979. For reference purpose, Fan's estimate of TFP in 1952-95 is replicated in Figure 4.0.6.

Two puzzles exist in Figure 4.0.6: (1) Modern technologies, such as high-yielding varieties, chemical fertilizers, and the more reliable powered-irrigation system, are expected to improve agricultural productivity. Then why, in spite of the widespread application of modern varieties and intensive use of modern inputs in the1960s and

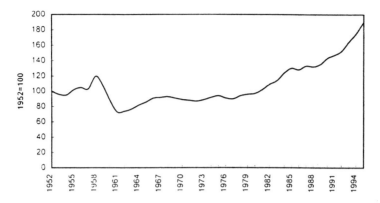

Figure 4.0.6 Total factor productivity in Agriculture. Source Fan (1997)

1970s, the TFP was even lower than the level reached in the 1952, at that time the modern technologies were still not available or rarely used? (2) Why the TFP increased in the period from 1952 to 1958, when the farming institution was changed from the household system to the collective farming and the TFP increased again in the period after 1979 when the farming institution was changed from the collective system to the household system? As discussed, when China adopted the heavy-industry-oriented development strategy in the early 1950s, the government simultaneously promoted collectivization as a strategy for the development of agriculture. Due to the time and spatial dimensions and biological nature of agricultural production, effective supervision of labor efforts in an agricultural collective is very costly (Lin 1988) The success of an agricultural collective depends inescapably on a tacit promise of self-discipline established by the collective members. However, a self-enforcing agreement can be sustained only if the members of the collective have the right to quit the collective when the other members do not honor their promises (Telser 1980). At the beginning of the collective movement, the government actively encouraged farmers to join the collectives but also followed the principle of voluntarism, allowing individual farmers to quit a collective if the collective failed their expectations. Consequently, the self-enforcing agreements in most collectives could be sustained and the overall agricultural performance was improved due to the economies of scale in the collectives. However, there was a built-in danger in the initial success of a collectivization movement. Due to the differences in their time preferences, abilities, and other endowments, some members of a collective might take advantage of the low supervision in the collective and attempted to evade the responsibilities stipulated in their self-enforcing agreement. Consequently, the disintegration of some collectives was inevitable, even though the overall performance of the movement was successful. The collapse of some collectives was an effective discipline mechanism itself (Holmstrom 1982). It made a potential violator of the self-enforcing agreement realize that honoring the agreement was to his advantage. Encouraged by the initial success, however, zealous political leaders of a collectivization movement might interpret differently some individual members' exit from the collectives. These individuals were viewed as the enemies of the movement. To prevent the further collapse of other collectives, compulsory measures were taken. The collectivization was thus changed from a voluntary to a compulsory movement, and the safety-valve was removed. In a compulsory collective, the incentives to work depends on the effectiveness of supervision (Lin 1988). Because the supervision of production in agricultural collective is too costly to be effective, farmers' incentive to work is low and agricultural productivity is depressed to a level lower than in the individual household farming.

The change from a voluntary movement to a compulsory movement in China occurred in 1958 when the government pushed to replace the advanced cooperatives with the gigantic communes. Lin (1992b) estimated that the productivity in a compulsory agricultural collective was about 20 percent lower than that in a household farm. The studies by McMillan et al (1989), Fan (1991), and Huang and Rozelle (1996) confirm Lin's finding. Therefore, in spite of the favorable impact of modern technologies, the TFP in the 1960s and 1970s was lower than that in the 1950s. And the TFP increased in the period of 1952-58 because the exit right was respected during that period and the TFP also increased after 1979 because the farming institution changed from a compulsory collective system to the individual household system.

Prospect and Policy Options for the Future

While China's ability to feed itself in the last four and a half decades was highly claimed, really remarkable achievements in Chinese agriculture did not occur until the beginning of reform in 1979, as the figures in table 4.0.2 show. Major elements of the reform included the replacement of collective team system with the household responsibility system, the expansion of rural product and factor markets, the liberalization of agricultural prices, except for grain and cotton (Lin 1997a). Among these reforms, the change to the household farming system had the largest productivity impact (Lin 1992). Between 1979, the beginning of farming institutional change, and 1984, the completion of this change, we see the largest annual growth rate in agriculture's TFP as well as in total grain output and per capita grain output (See Table 4.0.2). However, the farming institutional change's effect on agricultural production was once and for all. Its effect would have been depleted by 1984. After 1984, while the TFP growth rate is still substantially higher than the pre-reform period, the annual growth of grain output declined substantially with an average annual growth rate of 1.55 percent in 1984-96, which was even lower than the average annual growth rate of 2.41 percent in the pre-reform period of 1952-78. As a result, in 1984-96, the annual growth rate of grain output per capita was 0.14 percent,the lowest since 1952. The poor performance in grain production was mainly due to the government's continuous intervention in the grain production and marketing. As the government has liberalized the prices and marketing of most other agricultural products, the production of grain was not as profitable as other products and farmers do not have enough incentives to increase grain output (Lin and Li 1995).

Table 4.0.2: Average Annual Growth Rate of TFP, Grain, GDP and Consumption

Period	Agricul tural TFP*	Grain Total	Per Capita	GDP	Consumption Level Index Nation	Urban/ Rural	
1952-96	1.51	2.52	0.77	7.7	4.5	4.7	4.0
1952-78	-0.25	2.41	0.40	6.1	2.2	2.9	1.8
1978-84	5.10	4.95	3.70	9.3	7.7	4.5	9.0
1984-96	3.91	1.55	0.14	10.2	8.1	8.5	6.4

TFP figures are for the period of 1952-95.
Source: TFP figures are taken from Fan (1997), other figures are taken from various issues of *China Statistical Yearbook*.

The rural reform is a part of China's overall reform. The success of the household responsibility system in 1978-84 encouraged the Chinese government to take bolder measures in reforming the overall economy. As a result, the annual GDP growth rate increased from 9.3 percent in the period of 1978-84 to 10.2 percent in 1984-96. The consumption level of the nation as a whole and especially the consumption level of the urban population also have a corresponding acceleration (see Table 4.0.2).

When China started the reform in 1978, the Chinese government adopted a gradual, incremental approach. The experience in China shows that this approach is very effective in maintain social stability and stimulating economic growth (Lin et al, 1996a). Because of the inherited nature of this approach, there are many remaining issues in the economic system (Lin 1997b). However, Chinese economy is expected to maintain the existing dynamism for several other decades. This is because the driving forces for economic growth in an economy are capital accumulation, improvements of resource allocation, and technological changes. China has one of the highest saving rate in the world, reaching 35 percent of GDP per year. In the transition from a planned economy to a market economy, the rooms for improving resource allocation are very large. More importantly, the current technological level in China is very low. Because of the advantage of backwardness, the room for low-cost, technological catch up is very large. China's GDP is likely to maintain close to 10 percent annual growth rate foranother two or three decades (Lin et al. 1996a).

Currently China's per capita nutrition availability is above the world's average but is still some distance away from the developed countries' standard, especially the intakes from animal products are substantially lower (see Table 4.0.3). Certainly, how much animal products will be on a person's daily diet is determined by many economic and non-economic factors other than income level alone. However, judging from the experiences in Hong Kong, Korea, and Japan, when China's income increases, the demand for animal products will have a proportionally large increase. This implies that, although per capita direct consumption of grain may decline as income rises, the increase in indirect demand through the consumption of animal products will outweigh the decline in direct consumption demand.

In addition, the demand for grain will also increase due to population

Table 3: Average Per Capita Daily Nutrient Availability and Sources, 1995

Item	China		Developed countries		Hong Kong	
	Total	From animal products	Total	From animal products	Total	From animal products
Energy (K. Cal	2741	506	3191	861	3285	1048
Protein (Grams	72	24	98	55	109	78
Fat (Grams)	69	44	114	63	137	72

Item	Korea		Japan	
	Total	From Animal Products	Total	From Animal Products
Energy K.Cal	3268	511	2887	596
Protein (Grams)	85	35	96	53
Fat (Grams)	82	38	80	36

Source: FAO, FAOSTAT.

growth. Currently China has 1.22 billion population. China may add another 30 to 40 percent by 2030. Even maintaining at the current consumption level, China's grain demand will rise substantially.

However, as income and population grow, the cultivated land will decline due to construction of houses, roads, factories, and facilities. Since 1988, 190,000 hectares of farmland have been taken over each year for those purposes. Environmental degradation may further reduce the amount of land available for cultivation. New land will be claimed for farming, but it is of poor quality and is less productive (World Bank 1997, p. 66).

Income growth will also increase the demand for vegetables and fruits, which will compete with grain for limited cultivated land. Figure 4.0.6 shows, the areas that were sown to grain decline from 80.3 percent of the total sown acreage in 1978 to 73.8 percent in 1996 while the areas sown to vegetables and fruits increased from 2.5 percent to 7.7 percent in the same period. Similarly, the demand for fresh water fish will also reduce the land available for grain and other crops.

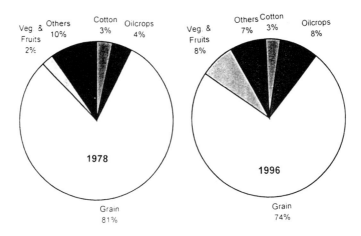

Figure 4.0.7. Structure of sown acreage. (left 1978; right 1996)
Source : Ministry of Agriculture, Planning bureau (1989) and State Statistical Bureau (1997)

Moreover, as income grows, China's comparative advantages will change. Not only land but also labor, water, and other resources will be shifted away from the production of grain to other more profitable agricultural and non-agricultural products. Recently there were many studies about China's future grain demand and supply (Brown 1995, Garnaut and Ma 1992, Huang et al. 1997, and World Bank 1997).[6] Most studies differ little with respect to their projections of China's grain demand because their similar assumptions about population and income growth. However, the projections on grain

---

[6] For a review of these studies, see Fan and Agcaoili-Sombilla (1997).

144

supply differ greatly, leading to confusing estimations about the magnitudes of China's future grain import demand.[7]

Leaving the issue of demand aside, the magnitude of China's future grain imports depends on how much China can produce and will produce domestically, which in turn depends on: (1) the technological potential for further yield-improvement, (2) the government's investments in research and extension to bring out the technologies that can tap the yield potential, (3) the infrastructure investments to maintain soil fertility and improve irrigation system, and (4) the farmers' incentives to adopt new technologies and to maintain soil fertility.

1. Technological Potential for Yield Improvement

Figure 4.0.8 shows the 1995-97 average yields of rice, wheat, and maize of China, the world average, the country with the highest yield in the world, and the highest experimental yield on a area larger than 0.1 hectare that has ever obtained in China.[8]

Among China's three main grain crops, rice has the highest yield. Rice is also China's only grain crop of which yield is among the world's top. In 1995-97, China's average rice yield was 6.2 tons/ha, which was 63 percent higher than the world's average of 3.8 tons/ha and was only 25 percent lower than the world's highest yield of 8.3 ton/ha, achieved in Puerto Rico. However, China's actual rice yield was only about 35 percent of the highest yield that has ever obtained on China's experimental field. Like the case in the world, among the three main grain crops in China, wheat has the lowest yield. The yield in 1995-97 was only 3.7 tons/ha, which was 48 percent higher than the world's average of 2.5 tons/ha and 57 percent lower than Netherlands's 8.7 tons/ha.

---

[7]Brown estimates that China will require to import from 207 million tons to 369 million tons of grain in 2030, which he thinks will drain world supplies, force grain prices up, and deny other low-income countries access to world grain market. However, his prediction is based on misinterpretation of data and erroneous assumptions. Brown's projection aroused many public concerns and awareness of the issues both in China and in the world. His study, nevertheless, cannot sustain academic examinations and can be discarded (Alexandratos 1996).

[8] The figures for the average yields of 1995-7 are obtained from FAO's FAOSTAT. The highest experimental yields are taken from Lin et al 1996 b.

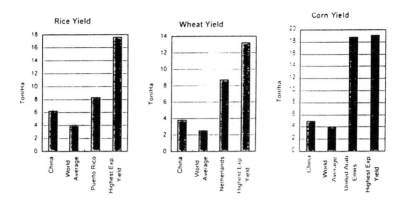

Figure 4.0.8. Yield variations for rice, wheat and corn. Source: FAO, Faostat (1997)

The highest ever recorded experimental yield in China was 13.2 tons/ha. The gap between the actual yield and the highest experimental yield is 9.5 tons/ha, or 257 percent of the current yield level.

China's maize's yield in 1995-7 was 4.9 tons/ha. It was only 22.5 percent higher than the world's average of 4.0 tons/ha and just 26 percent of the world's highest yield of 18.8 tons/ha achieved in the United Arab Emirs. The highest ever-observed experimental yield in China was 19.1 tons, which was only slightly higher than the actual yield in the United Arab Emirates. The gaps between China's actual yield and the world's highest yield and between China's actual yield and China's highest experimental yield are two different measures of the potentials for yield improvements in China. From these measures, we can conclude that there are still large potentials for yield improvements in China.

2. The government's investments for agricultural research.

A large portion of the observed yield gaps between the actual yields on the farm and the highest yields on the experimental plots or in the world are attributable to some technical constraints.[9] Table 4 reports

---

[9] Not all the yield differences are due to technical factors. Socio-economic factors, for example, the purity of seeds and the timing of applying fertilizers and water, can also contribute to the yield differences between actual yield

the results of an expert survey of over 2000 agronomists and plant breeders in China in 1992 for their judgments about the technical constraints that contributed to the observed yield gaps between China's actual yields and the highest potential yields for the three major grain crops.

Table 4.0.4 : Technical Constraints

	Rice	Wheat	Maize
Yield Gap due to Technical Constraints (tons/ha)	8	9	11.5
Contributing Factors (%)	100.0	100.0	100.0
1. Crop Characteristics	52.0	52.5	49.0
2. Environmental Conditions	32.1	32.3	35.2
3. Soil Conditions	7.4	6.2	6.9
4. Pests, Diseases, Weeds	2.2	2.4	2.7
5. Weather Conditions	5.3	6.6	5.2

Source: Lin, Justin Yifu et al., 1996 b.

According to the survey, technical constraints resulted in a total of eight tons in the difference between the actual yield and highest experimental yield for rice, 9 tons for wheat, 11.5 ton for maize. The differences in crop characteristics, such as plant architecture, photosynthetic ability, the duration of maturity and so on, had the large impacts, contributing to about half of the yield losses due to the technical factors for all three crops. The second most important technical factors are the differences of environmental conditions--for example, duration of sun light, the accumulated temperature, and the humidity--between experimental fields and the average fields. It contributed to about one-third of the yield loss arising from technical factors. Soil conditions, weather conditions, and pests, diseases, weeds contributed to the rest of yield losses (Lin, et al. 1996b). Potentially, technical constraints can be overcome by breeding seeds with the desirable traits and by other new technologies. They are the directions for yield-improvement research.

In the past, China's agricultural research system was remarkable. After the agricultural crisis in the 1959-61, the Chinese government has relied on modern technologies for increasing grain yield and output. Since 1961, the number of research personnel increased 7.3

percent and annual expenditure increased 5.7 percent per year. China now has the largest team and experimental yield of agricultural research personnel.[10] China's agricultural research intensity (ARI) (measuring agricultural research investment relative to agricultural GDP) in the early 1960s was 0.41, much higher than the developing countries' average of 0.24 in the same period of time. Throughout the 1960s to the 1980s, ARI was maintained at the same level (Fan and Pardey 1997). With these efforts, China was the world leaders in many areas of grain research in the 1960s and 1970s. However, looking ahead there are several alarming signs in China's agricultural research system.

As part of the overall market-oriented reform, there is a reform in agricultural research funding policy in the 1990s. The government reduced fiscal appropriation for agricultural research, shifted funding from institutional supports to competitive grants, and encouraged research institutes to commercialize their technologies, using part of the proceeds to subsidize their research. The research institutes' real income from technology commercialization increased substantially. However, the proportion that was used to subsidize research was far from enough to compensate for the reduction in fiscal appropriation (Scott, et al. 1997). As a result the ARI declined to 0.34 in the 1990s (Huang 1997).[11] New agricultural technologies, such as genetic engineering, requires heavy research investments. The reduction in research funding will hurt China's agricultural research capacity in the long run.

China's agricultural research system has another serious problem, that is, the lack of appropriate compensation for scientists. A highly qualified plant breeder earns approximately the same salary as an unskilled manual worker. This compensation system deters promising young students to pursue advanced studies in agricultural sciences and discourages many talented agricultural scientists trained abroad to return and work in China. As Chinese economy moves towards a market system, such a deterrent effect will become more

---

[10] It should be admitted that only 5-6 percent of the researchers in China hold a postgraduate degree while in the national systems in other less developed countries the proportion is 60-70 percent (Fan and Pardey 1997).

[11] Actually, China's ARI in the 1980s was already a quarter lower than the average of developing countries due to the substantial increases in other developing countries, especially those in Asia (Fan and Pardey 1997).

agricultural scientists trained abroad to return and work in China. As Chinese economy moves towards a market system, such a deterrent effect will become more and more serious. In effect, the agricultural research system has started to lose research staff since the mid 1980s. In 1986, the total number of agricultural scientists in the national system, excluding those in the universities, was 232,683. It dropped 15 percent to 196,708 in 1996.[12]

3. The investments in agricultural infrastructure.

With the increase of population, land and water available for agriculture will become increasingly scarce by definition, if the comparison is made in per capita term. Population pressure may also lead to growing environmental degradation, including erosion and salinization, and reduce land suitable for cultivation. The breeding of higher-yielding and environmental stress-resistance varieties may compensate the decline in cultivated land. However, it is also important to protect the resource base through erosion and water control and soil fertility enhancing measures. In the pre-reform period, Chinese government was known for its ability to mobilize agricultural labor in slack seasons for environmental-improving projects, such as upgrading irrigation system, salinization control, reforestation, and terracing. After shifting to the individual household farming system, mobilization of rural labor for such projects become harder. The government's investment in agricultural infrastructure projects for upgrading or maintaining agricultural resource base becomes increasingly important.

When the reform started at the end of 1978, the government committed to increase fiscal expenditure on agriculture from 13 percent of budget at that time to 18 percent in the pursuing years. However, as figure 4.0.9 shows, the share of expenditure on agriculture did increase for one year but declined sharply after 1980 due to the success of rural reforms in bringing out remarkable agricultural growth. The share of budgetary expenditure for agricultural infrastructure has a similar pattern of changes. After 1985, due to the stagnation in grain production, the share of government's fiscal expenditure on agriculture recovered somewhat but declined again in the 1990s when the growth in grain output resumed. The right-hand panel of figure 4.0.9 shows the expenditures of government's investment in agricultural infrastructure projects, measured in the 1978 constant prices, reduced from

---

[12] The number of research scientists are provided by Jikun Huang.

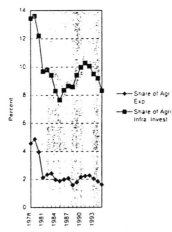

Figure 4.0.9A. Agriculture in government budget

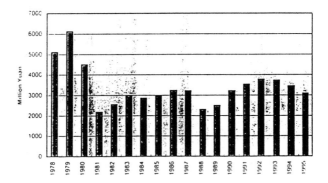

Figure 4.0.9B Real Agr. infrastructural investment
Source: Statistical Bureau, various issues.

the height in 1979 to about half of that level afterwards. Probably because of
the increase in environmental stress arising from the population pressure and economic activity and because of the reduction in infrastructure investments, Chinese agriculture's ability to resist flood, draught, and other unfavorable weather impacts has declined. Figure 4.0.10 shows the percentage of sown area that was hit by natural disasters and had a 30 percent or more output reduction than the yield in a normal year. It is clear that there is an increasing trend since the 1970s. Without government's substantial investments in agricultural infrastructure projects to offset environmental stresses arising from population pressure and intensification of economic activities, Chinese agriculture system's ability to resist unfavorable weather impacts will decline. The hope to have a stable and sustained yield improvement and output growth will be jeopardized.

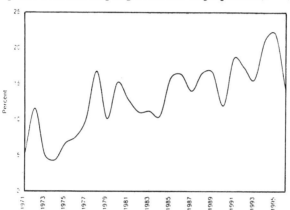

Figure 4.0.10. Percentage of sown area affected by disasters
Source: Ministry of Agriculture, Planning Bureau (1989) and State Statistical Bureau, various issues.

4. The farmer's incentives to adopt new technologies and to maintain soil fertility.New technologies provide the hope to increase grain output through raising yields instead of expanding sown areas. However, new technologies will work only if farmers adopt them in the production. The adoption of new technologies is an economic decision. In a market economy, whether farmers will adopt a new technology depends on the costs and benefits of the adoption. The costs of adoption are predominately determined by the availability of extension services and seed production and distribution system. The benefits of adopting new technologies depends on the magnitude of yield increase or cost reduction and the level of output price. After

shifting to the household farming system, Chinese farmers have become more sensitive to the costs and benefits of technology adoption (Lin 1991c).

Before the reform, the administrative network was quite effective in extending new technologies and making them available to farmers. There are evidences that the original extension network is disrupted after the shift to the household farming system, which will have an adverse effect on the costs of adopting new technologies (Lin 1991c). Moreover, although the government has liberalized most markets, it still controls domestic grain trade. Currently the government purchased about three quarters of all marketed grain (World Bank 1997). The government's procurement prices will have a large effects on the profitability of using new technologies. The government's predominate concern is the welfare of urban consumers. It often suppressed the procurement prices at well below the market prices and raised the prices only when domestic grain supply threatened the food security. To increase farmers' incentives for adopting new technologies, the government needs to strengthen the extension network on the one hand, and liberalize grain prices, making production of grain as profitable as other crops, on the other hand.

Most farm land in China has been cultivated continuously for thousands years. It is remarkable that the soil fertility did not decline. The wide application of organic fertilizers and use of green manure have contributed to the maintenance. However, measures for maintaining soil fertility are a kind of long-term investments. Farmers will have incentives to adopt those measures when their long-term returns are secure. When the household responsibility system was first adopted in the late 1970s and early 1980s, the collective land was leased to individual household for only one to three year. That contract duration was too short to recoup soil fertility maintenance and other long-term investments. In response to the need, the government later allowed the contract to be extended to 15 years and now to 30 years. However, frequent redistribution and relocation of land among households in a village are often found in many areas. To increase the incentives for maintaining soil fertility and investing in soil fertility enhancing activities, the government needs to adopt effective policy measures to assuring farmers' tenure security of land contracts.

An issue similar to the land is the maintenance of water resources and the rights to use of water. Water is even more scarce than land for agricultural production in China. However, water usage is very wasteful. Most irrigation and drainage systems were old and poorly maintained because of fragmented responsibilities among levels of government and inadequate budgets. The water charges are too low. On the one hand, the low charges reduce farmers' incentives to conserve water in production. On the other hand, the charges are not sufficient to cover maintenance costs, reducing the government's incentive to invest in and maintain the irrigation system. For improving the incentives of farmers' water conservation and government's public investments, it is necessary to reform the existing low-charge policy, raising the prices to a level that can give investments in irrigation projects appropriate returns and is sufficient to maintain the working condition of the systems.

Conclusion

In the past, the keys for China's success to feed itself were agricultural research, modern technologies and individual-household based farming system.

Chinese government still commits to the policy of producing enough grain domestically to meet its food need in the future (Huang 1997b). The potential to achieve so is there. Some policy reforms are required to tap the potential. As stated by Johnson (1994), if China has a grain problem in the future, it must be caused by a series of policy failures. However, we do have reasons to be optimistic. The past experiences show that, whenever grain production became a threat to China's food security, the government would adjust its policies and give agricultural sector appropriate supports.

If in the future Chinese government gives up the policy of grain self sufficiency, it does not mean that China will lose the ability to feed itself. Just the opposite! With the trade liberalization, China is likely to export more of labor-intensive, high value-added food stuffs, in addition to labor-intensive manufactured products. In value terms, China may be a net exporter of food (Lu 1996). Moreover, the improvement in resource allocation will increase Chinese people's income and their household budgets for diets. Chinese people will feed themselves even better.

References:

Alexandratos, Nikos. "China's Projected Cereals Deficits in a World Context." *Agricultural Economics*, Vol. 15, No. 1, September 1996, pp. 1-16.

Anderson, Kym. *Changing Comparative Advantages in China: Effects on Food, Feed, and Fibre Markets*. Paris: Development Centre Studies of OECD, 1990.

Ashton, Basil; Hill, Kenneth; Piazza, Alan; and Zeitx, Robin. "Famine in China 1958-61." Population and Development Review. Vol. 10, December 1984, pp. 613-45.

Barker, Randolph and Herdt, Robert W. *The Rice Economy of Asia*, Washington, D.C.: Resource for the Future, 1985.

Brown, Lester. *Who Will Feed China? Wake up Call for a Small Planet*. New York: W.W. Norton and Co. 1995.

Fan, Shenggen. "Production and Productivity Growth in Chinese Agriculture: New Measurement and Evidence." *Food Policy*, Vol. 22, No. 3, 1997, pp. 213-28.

Fan, Shenggen. "Effects of Technological Change and Institutional Reform on Production Growth in Chinese Agriculture." *American Journal of Agricultural Economics*, Vol. 73, N. 2, 1991, pp. 266-75.

Fan, Shenggen and Agcaoili-Sombilla, Mecedita. "Why Do Projections on China's Future Food Supply and Demand Differ?" EPTD Discussion Paper No. 22, Washington, D.C.: IFPRI, 1997.

Fan, Shenggen and Pardey, Philip G. "Research, Productivity, and Output Growth in Chinese Agriculture." *Journal of Development Economics*. Vol. 53, 1997, pp. 115-37.

Garnaut, Ross. and Ma., Guonan. *Grain in China*, East Asia Analytical Unit, Department of Foreign Affairs and Trade, Australia, 1992.

Griliches, Zvi. "Hybrid Corn: An Exploration in the Economics of Technical Change." *Econometrica*, Vol. 25, 1957,pp. 501-22.

Hayami, Yujiro and Ruttan, Vernon. "Factor Prices and Technical Change in Agricultural Development: The United States and Japan, 1880-1960." *Journal of Political Economy*, Vol. 79, September-October 1970, pp. 115-41.

Huang, Jikun. "*Qiantian woguo nongye keyan touru zhengce*" (on China's agricultural research investment policy) *Nongye Jishu Jingji*, 1997 a, No. 2, pp. 10-3.

Huang, Jikun. "Agricultural Policy, Development, and Food Security in China: A Report Submitted to FAO," Center for Chinese Agricultural Policy, Beijing, 1997b.

Huang, Jikun and Rozelle, Scott. "Technological Change: Rediscovering the Engine of Productivity Growth in China's Rural Economy." *Journal of Development Economics*, Vol. 49, 1996, pp. 337-69.

Huang, Jikun; Rozelle, Scott, and Rosegrant, Mark. W. "China's Food Economy to the twenty-first Century: Supply, Demand, and Trade" 2020 Vision Discussion Paper No. 19, Washington, D.C.: IFPRI, 1997.

Holmstorm, Bengt. "Moral Hazard in Teams." *Bell Journal of Economics*, Vol. 13, No. 2, Autumn 1982, pp. 324-40.

Johnson, D. Gale. "Does China Have A Grain Problem?" *China Economic Review*, Vol. 4, No. 1, pp. 1-4.

Lin, Justin Yifu. "Institutional Reforms and Dynamics of Agricultural Growth in China." *Food Policy*. Vol. 22, No. 3, 1997, pp. 201-12. (a)

Lin, Justin Yifu. "Issues in China's Economic Reform: Roots and Options" Shui On Centre for China's Business and Management and School of Business and Management, Research Reports on Doing Business in China: Current Issues, Hong Kong University of Science and Technology, 1997. (b)

Lin, Justin Yifu. "Hybrid Rice Innovation: A Study of Market Demand Induced Technological Innovation in a Centrally Planned Economy." *Review of Economics and Statistics*, Vol. 74, No. 1, February 1992, pp. 14-20.

Lin, Justin Yifu. "Prohibition of Factor Market Exchanges and Technological Choice in Chinese Agriculture". *Journal of Development Studies*, Vol. 27, No. 4, July 1991, pp. 1-15. (a)

Lin, Justin Yifu. "Public Research Resource Allocation in Chinese Agriculture: A Test of Induced Technological Innovation Hypotheses." *Economic Development and Cultural Change*, vol. 40. No. 1, October 1991, pp. 55-73. (b)

Lin, Justin Yifu. "The Household Responsibility System Reform and the Adoption of Hybrid Rice in China." *Journal of Development Economics*, Vol. 36, 1991, pp. 353-72 (c).

Lin, Justin Yifu. "Collectivization and China's Agricultural Crisis in 1959-61." *Journal of Political Economy*, Vol. 98, No. 6, December 1990, pp. 1228-52.

Lin, Justin Yifu. "The Household Responsibility System in China's Agricultural Reform: A Theoretical and Empirical Study." *Economic Development and Cultural Change*, Vol. 36, no. 4 (Supplement), April 1988, pp. S199-S224.

Lin, Justin Yifu and Li, Zhou. "Current Issues in China's Rural Areas." *Oxford Review of Economic Policy*. Vol. 11, No. 4, 1995, pp. 85-96.

Lin, Justin Yifu and Yang, Dennis Tao. "Food Availability, Entitlements and the Chinese Famine of 1959-61." Peking University, China Center for Economic Research Working Paper, No. E1997011, 1997.

Lin, Justin Yifu and Wen, Guanzhong James. "China's Regional Grain self-sufficiency Policy and Its Effect on Land Productivity." *Journal of Comparative Economics*. Vol. 21, 1995, pp. 187-206.

Lin, Justin Yifu; Cai, Fang; and Li, Zhou. The China Miracle: Development Strategy and Economic Reform, Shanghai People's

Publishing House and Shanghai Sanlian Sudian, 1994 (for Mainland China); The Chinese University of Hong Kong Press, 1995 (for overseas), 1996 (English edition), Tokyo: Nihon Hyo Ron Sha, 1996 (Japanese edition); and Seoul: Baeksan Press, 1996 (Korean edition); Ho Chi Minh City: Saigon Times, 1997 (Vietnamese edition); Paris: Economica, 1997 (French edition). (1996 a).

Lin, Justin Yifu; Shen, Minggao; and Zhou, Hao. *Agricultural Research Priority in China: A Demand and Supply Analysis of Seed Improvement Research for Major Grain Crops in China.* Beijing: China Agriculture Press, 1996.

Lu, Feng. "Grain versus Food: A Hidden Issue in China's Food Policy Debate." Working Paper No. No. E1996003, China Center for Economic Research, Peking University, 1996.

Mallory, Water. *China: Land of Famine.* New York: American Geographical Society, 1926.

McMillan, John, Whalley, John, and Zhu, Lijing. "The Impact of China's Economic Reforms on Agricultural Productivity Growth." *Journal of Political Economy.* Vol. 97, August 1989, pp. 781-807.

Ministry of Agriculture, Planning Bureau. *Zhongguo nongcun jingji tongji daquan, 1949-1986.* (A comprehensive book of China's rural statistics), Beijing: Agriculture Press, 1989.

Rozelle, Scott; Pray, Carl; and Huang, Jikun. "Agricultural Research Policy in China: Testing the Limits of Commercialization -led Reform." Department of Economics, Stanford University, 1997 (Mimeo).

Schmookler, Jacob. *Invention and Economic Growth.* Cambridge, MA: Harvard University Press, 1966.

Smil, Vaclav. "China's Food: Availability, Requirements, Composition, and Prospects." *Food Policy,* Vol. 6, No. 2, May 1981, pp. 67-77.

State Statistical Bureau. *China Statistical Yearbook,* various issues. Beijing: China Statistical Press.

State Statistical Bureau. *A Statistical Survey of China*, various issues. Beijing: China Statistical Press.

Telser, Lester G. "A Theory of Self-enforcing Agreements." *Journal of Business*, Vol. 53, No. 1, January 1980, pp. 27-44.

Wen, Guanzhong James. "Total Factor Productivity Change in China's Farming Sector: 1952-1989." *Economic Development and Cultural Change*, Vol. 42, No. 1, October 1993, pp. 1-41.

杵　砧

CHAPTER 4.1

Food Security Strategy for the Early
Twenty-first Century[1]

Albert J. Nyberg
Sr. Agricultural Economist
World Bank

Historically, China has been famine prone and as recently as the late-1950s and early-1960s, famine claimed millions of lives. For this reason Government officials remain concerned about grain balances and why they have pursued grain self-sufficiency policies and a program of large domestic grain storage. Such a strategy is costly and inefficient, but ensures China's autonomy in domestic grain availability. Officials question the reliability of the international market to meet their needs if China experienced large grain shortages. But domestic markets also are distrusted and government intervention pervades grain marketing and trade. Recent policy reform has brought domestic input and output prices to near international price levels and multiple farmgate prices are being consolidated. But marketing reforms incorporating greater private participation have been short-lived, with frequent retrenchment China's economic growth over the past two decades has been remarkable. But it's agricultural growth has lagged behind, and within agriculture, output of high value commodities such as livestock and aquatic products, fruits and vegetables grew at annual rates of 9.0 percent or more, while cereal output grew by only 3.2 percent.

Rapid income growth has altered food consumption patterns. Per capita (direct) consumption of grains has declined, and consumption of higher quality fruits, vegetables, and livestock products has increased, particularly among urban consumers. Overall grain utilization has increased primarily through indirect consumption (the conversion of grain to livestock products) magnifying the importance of feed grains. As

---

[1] This is a summary of a Wold Bank study entitled "China Long -Term Food Security", undertaken by the author and several associates in the late 1996- early 1997.

incomes grow, consumption will continue to shift toward higher value food items. But, future consumption growth will likely be less than current income elasticities suggest because consumption of livestock products is already high relative to income levels. China is the world's largest cereal consumer, ranking fifth in per capita cereal consumption, and cereals will continue to supply the bulk of calories, as they do in most other countries. Expected changes in consumption and production structures will require China to rely more on the international market to balance cereal supply and demand. But, as a large and agroclimatically diverse country, China will be substantially self-sufficient in food. To be otherwise would incur prohibitively expensive transport costs. Estimates of annual import requirements have fluctuated between 20 and 50 million tons[2] for the early decades of the 2000s.

The following discussion focuses on the issue of National *chronic* food insecurity; although it also addresses *transitory* food insecurity that may occur through drought or other seasonal calamity. The importance of household food security is appreciated but this and other poverty issues are too complex incorporate in this work.

Consumption and Demand

CONSUMPTION

The two principal data sources for estimating food consumption and demand are the State Statistical Bureau's (SSB) annual household consumption and expenditure surveys and FAOs food balance sheets. SSB urban survey results clearly show that per capita consumption of grain has declined over the past decade—when income has been rapidly increasing, implying that grains may have become an "inferior" food.

These data also indicate that per capita meat consumption increased only marginally over the same period. Rural household consumption *trends* are similar but at different *levels*. The FAO methodology derives consumption as a residual from production and net trade, adjusted for nonfood uses and losses. The two consumption estimates are compared

---

[2] See Fan. Shenggen, and Mercedita Agcaoili-Sombilla, "Why Projections on China's Future Food Supply and Demand Differ?" paper presented at a conference on *Food and Agriculture in China; Perspectives and Policies*, Beijing China, October 7-9, 1996.

in Table 4.1.1. The major differences in the two estimates are in meat products and vegetables; other differences could be attributed to measurement error.

Table 4.1.1: China Per Capita Food Consumption

	SSB Household Survey (1992)	FAO Balance Sheet (1992-94 avg.)
	------------------------(kg/capita)------------------------	
Grain (cereals)	235.91 [a]	224.2
Pork/Beef/Mutton	20.27	27.7
Poultry	2.35	5.0
Aquatic Products	9.20	13.5
of which Fish	2.00	6.0
Vegetable Oil	6.12	6.4
Vegetables	116.85	86.6
Fruit	38.80 [b]	28.5

(a) Deriving overall grain consumption by combining rural and urban consumption of individual grains provides a national per capita estimate of 216kg. (b) Urban consumers only. *Sources*: State Statistical Bureau, China Statistical Yearbook, 1994 (Table 8.6) and FAO, Food Balance Sheets, 1992-94 Average, FAO Statistical Series No. 131, Rome 1996.

The SSB data for meat and poultry consumption are inconsistent with production, implying losses that are not credible. Alternatively, the FAO estimates appear to overstate meat production—hence, consumption is also overstated. Scholars have argued that both estimates overstate cereal consumption by as much as 15 percent.[3]

---

[3] Garnaut, Ross and Guonan Ma; *Grain in China,* East Asia Analytical Unit, Dept. of Foreign Affairs and Trade, Canberra, Australia, 1992.

The *balance sheet* approach indicates the average Chinese diet is very high in cereal and meat consumption and is more consistent with residents of higher income countries. Daily caloric consumption at about 2,730 is 95 percent of the level of Malaysians whose per capita GNPlevels are 550 percent greater.[4] Meat consumption, at 32.7/kg per capita, is essentially the same as in the Republic of Korea where per capita GNP is fifteen-fold that of China. Rice and wheat comprise 85 percent of direct cereal consumption, but coarse grains are important in some rural locations, particularly areas unsuited for rice or wheat production. Otherwise, coarse grains are consumed indirectly in the form of meat and livestock products and used in industrial and food manufacturing. Currently, grain-to-meat conversion rates are very high (low feed/meat ratio), because most livestock subsist on grazing, green fodders, and household kitchen wastes—except for commercial poultry raised in cages. Grass and forage fed livestock production can be expanded, but most future incremental meat production, will require more grain-intensive production modes.

CONSUMPTION PROJECTIONS

Long-term food consumption projections are difficult not only because of uncertain base consumption levels but also because of unstable income elasticity coefficients. As a result, projections often imply consumption levels and caloric intake of unrealistic proportions. Projections for indirectly consumed items, such as feed grain, are even more complicated because feed conversion ratios have not been empirically derived and are based on approximations and estimates. Alternative approaches—such as time-lag and income-lag comparisons with other nearby Asian countries—also are inappropriate given China's extraordinarily high per capita consumption levels and relatively lower income levels.

The following consumption estimates were based on SSB household survey data with future consumption changes derived from the Global Trade Analysis Project (GTAP) model[5] using the income elasticities

---

[4] Based on Atlas method, and 230 percent greater if based on Purchasing Power Parity estimates.

[5] The GTAP model is a global general equilibrium model maintained at Purdue University by a consortium of research and policy institutions including the World Bank, World Trade Organization, European Commission, OECD and numerous bylateral agenciesThe model and its underlying database are

listed below. An average annual GDP (income) growth rate of 7.0 percent between 1995 and 2020 was assumed. Utilization coefficients, developed from China's Input-Output Tables,[6] allocated grain production to direct consumption, manufactured/processed products, and alternative uses, such as livestock feed. A large proportion of cereal consumption is in manufactured /processed products. Estimated per capita and total consumption for 2020 are given in Table 4.1.2

TABLE 4.1.2: ESTIMATED *DIRECT* FOOD CONSUMPTION, 2020

	Income Elasticity	Per Capita Estimate (kg)	Total (million tons)
Rice	-0.05	66.5	94.6
Wheat	0.07	51.5	73.9
Coarse grains	-0.18	23.1	31.9
Meat	0.62	51.3	73.6

With very large increases in meat consumption, China's feed grain requirements will expand rapidly. Pork is currently the major meat component, 70 percent of which is produced under "backyard" conditions using very little grain or manufactured feeds. However, much of the incremental production will necessarily come from more "commercial" operations; otherwise, the animal numbers would be unsustainable high. The level of model aggregation hides the details of meat consumption (pork, poultry, etc.) and future consumption growth will likely be greater in more feed efficient poultry.

The model estimated feed needs using a single feed conversion ratio. However, to more accurately determine feed grain requirements, they

documented in Hertel (1997) and is widely used for analysis of agricultural and general trade policy worldwide.

[6]   SSB, Department of National Economy Accounts, "Input-Output Table of China," China Statistical Publishing House, 1987 and 1992.

were estimated separately for poultry, pork, and "other red" meat under the assumption that technology would improve feed conversion rates by 2020. The total estimated cereal requirements were 608 million tons of trade grain. Details are contained in Table 4.1.3. Grain requirements for food and manufacturing will be about 368 million tons, for feed grains 206 million tons, and 34 million tons for other uses.

TABLE 4.1.3: ESTIMATED CEREAL REQUIREMENTS FOR 2020

	Rice [a]	Wheat	Coarse Grain	Total
		(million tons)-		
Cereals for food	94.6	73.9	31.9	200.3
Feed Grains	46.3	21.6	138.2	206.1
Manufactur. products	56.3	68.6	43.1	168.0
Post-harvest loss	10.4	7.2	9.7	27.3
Seed requirement	1.1	3.5	1.6	6.2
Total Trade	208.6	174.8	224.5	607.9

/a   Total paddy equivalent is 297 million tons.
/b   Post-harvest losses assumed to be 5% of production.

Some coarse grains will continue to be used as a food grain and some fine grain will be used as feed. Wheat bran, as a flour milling by-product will continue to go into feed. About 12 percent of rice production is now used to feed livestock. This is due to the relatively low price of hybrid rice, as well as transport constraints to moving coarse grains into the rice areas. The model suggests that rice used in livestock feed will increase to 22 percent of production in 2020, but better market integration would improve food and feed grain allocations. These estimates of China's future grain requirements are greater than other recent studies have estimated. This derives primarily from the large estimate of grain used in manufacturing.

Total cereal requirement estimates are sensitive to several factors. China's large population means that small changes in per capita consumption of food grains and meat result in large changes in total estimates. Small changes in income growth, income elasticities, or price elasticities also will result in significant changes in total cereal requirements. Also, small changes in feed conversion ratios will produce large changes in total estimates. *Thus, any estimate should be carefully examined to ensure the assumptions are understood and acceptable*

Production Constraints and Solutions

Agricultural growth since 1978 can be divided into two periods: during Household Responsibility System (HRS) implementation (1978-85), and the post-implementation period (1985-95). During the initial period agricultural growth averaged 7.4 percent and total factor productivity (TFP)[7] growth was 6.6 percent.[8] Subsequently, growth declined to 5.8 percent, most of which was created by additional inputs and investments; the TFP growth rate declined to 1.1 percent, raising concern about future growth in food production.

The various components of agriculture and food production grew at very dissimilar rates, particularly since 1984/85. Cereal production increased the least, reflecting its relatively low profitability. Nonstaple foods, livestock, fisheries and vegetables /melons grew at rates of 9.0 percent or greater after 1985 (Figure 4.1.1).

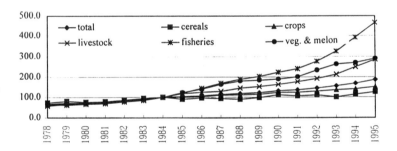

FIGURE 4.1.1: INDEX OF GROSS OUTPUT OF AGRICULTURAL COMPONENTS

---

[7] Total factor productivity is defined as a ratio of output to the weighted sum of inputs. It is typically expressed as a ratio of index numbers measuring the change in output relative to the change in total inputs.

[8] Wen, G.J., "Total Factor Productivity Change in China's Farming Sector: 1952-1989", Economic Development and Cultural Change 42(1), October 1993, pp. 1-41.

FOOD SECURITY STRATEGY

ACCOUNTING FOR PAST GROWTH

Various scholars have identified investments and institutional innovations as factors contributing to rapid growth. To assist in identifying additional constraints and solutions to agricultural growth and food security, factors contributing to growth during the reform period were reviewed and evaluated in a dynamic multisector output response model. The results are summarized in Table 4.1.4

Table 4.1.4. Contribution to Crop Production Growth

Factor	Southern China 1978-84			1984-95	Northern China 1978-84		1984-95	
	Rice	Other Grain	Cash Crops	Rice	Wheat	Corn	Wheat	Corn
	percent							
Research Stock	1.61	5.03	8.51	1.65	3.30	5.84	3.43	6.07
Irrigation Stock	0.13	0.22	0.00	0.39	0.43	0.46	0.47	0.50
Institutional-	2.10	2.71	2.23	0.00	3.86	0.00	0.00	0.00
Input/Output Price	0.44	1.18	0.99	-0.11	0.86	1.94	-0.75	-1.14
Land Prices	-0.03	0.00	-9.14	-0.01	-0.04	1.83	-0.01	0.40
Labor Prices	-0.96	-2.95	1.29	-0.21	-1.30	-5.90	-0.08	-0.36
Environmental Factors								
Disaster	0.06	0.20	0.02	-0.07	0.30	0.36	-0.02	-0.03
Erosion/Salinize	0.00	-0.37	-0.86	0.00	0.01	0.13	-0.02	-0.35
Residual	1.20	1.35	11.55	-1.28	0.21	-1.17	-0.09	-0.19
Growth Rate	4.54	7.37	14.59	0.36	7.63	3.48	2.12	4.89

Growth was separately analyzed for southern and northern China using different crops to account for the different cropping patterns and agroclimatic environment. Factors included in the analysis were the HRS institutional reform, investments in research and water control infrastructure, input-output price ratios, land and labor prices, and environmental factors. Land use/land markets and agricultural extension were excluded from the analysis because of the lack of change in the former and inability to quantify changes in the latter. Thus, their contribution to growth is undetermined and by default is in the residual.

These analyses confirmed that the HRS was important. However, this was a one-time event and will not contribute to future growth. Other important contributors to growth were investments in agricultural research, which was the engine of agricultural growth during the reform period. Also, irrigation infrastructure investment was an important contributor to agricultural growth, and may have contributed more than the coefficients indicate as many of the investments were for rehabilitation and compensation infrastructure (to compensate for irrigated land lost to urban and industrial growth) rather than *capacity expansion*

Agricultural Research Investment

During the post-revolution period China's research system created new technologies, developed new crop varieties, and improved agronomic practices which improved the total factor productivity in agriculture and expanded crop production in the face of declining output/input price ratios. The growth accounting analysis is consistent with conclusions of other scholars[9] who found that growth in Chinese agricultural productivity was overwhelmingly attributable to research investments. Research in any given year builds on past research and adds to the foundation for future research. Research investments accumulate and last for several years before fully depreciating. Thus, research *stock* is more important than investments in any particular year (except for continuity of research projects).

Although China's agriculture has been well served by research, three issues create concern for the future. National agricultural research budgets have quadrupled, in real terms, over the past three decades and the number of research scientists employed has increased sixfold,[10] but these numbers are deceptive because: (a) funding per scientist has declined, and funds per scientist influences output (new varieties) more than total budget; (b) research costs have generally increased more rapidly than inflation and genetic research is turning to more costly

---

[9] Huang, Jikun, Mark W. Rosegrant, And Scott Rozelle, "Public Investment, Technological Change and Reform: A Comprehensive Accounting of Chinese Agricultural Growth." International Food Policy Research Institute, Washington, D.C., July 1995.

[10] Fan, S., "Data Survey and Preliminary Assessment of Agricultural Investment in China." A Report to Food and Agriculture Organization, January 30, 1995.

biotechnology; and (c) research topics have expanded to incorporate pest resistance and drought tolerance along with the former higher yield and shorter season foci.

TABLE 4.1.5 CENTRAL GOVERNMENT INVESTMENTS IN AGRICULTURAL RESEARCH

Period	Research Scientists	Agricultural Research Intensity /a	Annual Average Expenditures constant 1990 prices	
			Total Y million	per Scientist Yuan
1965-69	10,166	0.33	464	46,001
1970-74	10,618	0.41	720	68,518
1975-79	19,319	0.49	1,022	52,729
1980-84	33,111	0.44	1,404	42,482
1985-89	50,330	0.40	1,763	35,336
1990-94	61,835	0.39	2,063	33,276

/a   See footnote 11.

Equally important is the decline in agricultural research intensity[11] (ARI) that has occurred over the past 15 years. Between 1975-79 and 1990-94 the ARI declined from 0.5 percent to 0.4 percent (the coefficient excludes research funded from Provincial budgets). This ratio is slightly below the average of other developing countries but is only one-fourth of the investment ratio of developed economies. The defacto decline in the research budget is exacerbated by research *commercialization* of the past decade. Research institutes have been encouraged to commercialize their activities and earn a portion of their income through technology sales. Ideally, the institutes would license the technology to manufacturers and commercial entities, but weak protection for intellectual property rights made this less attractive than the

---

[11] Agricultural research intensity is the term applied to the ratio of agricultural research expenditures over agricultural gross domestic product.

commercialization option. However, commercialization evolved into unrelated activities with higher capturable returns; including manufacturing, restaurants and hotels, and commercial trade. Diverting researchers from their primary task seriously detracts from the institutes' research objectives especially since only about 15 percent of the commercial revenues are allocated to the institutes' research budgets.[12] *Agricultural research is a public good which should be fully funded from the public treasury.*

In 1982, the Chinese Academy of Science initiated a *competitive grants* program which allocates funds directly to research scientists. Conceptually, competitive grants are an efficient method of allocating research funds, but may limit funding to established scientists and institutes and make it difficult for newcomers, including better trained younger scientists, to participate. Competitive public research funding has shifted resources to national level institutes at the expense of prefectural institutes. Whether this is efficient or not, yet remains to be determined—prefectural institutes introduced over 50 percent of the nations new rice cultivars between 1990 and 1994.

Chinese counterpart agencies have access to technology generated by the Consultative Group on International Agricultural Research (CGIAR) institutions[13] but China benefits only marginally from agricultural technology developed by private international corporations such as transnational seed companies. Greater internationalization of research and technology through increased cooperation with international public and private institutions would be mutually beneficial. *Given the overwhelming contribution of research to agricultural growth, real investments in research must be better managed to ensure that efficiency and long-term food security objectives are met. Research must be sufficiently funded to maintain total factor productivity growth.*

---

[12] Rozelle, S., C. Pray, and J. Huang; "Agricultural Research Policy in China: Testing the Limits," Post-Conference Workshop on Agricultural Productivity and Research and Development Policy in China, 29 August 1996, Melbourne Australia.

[13] CGIAR is a loose-knit group of donors that support 18 institutes engaged in international agricultural research.

Otherwise, declining production growth will necessitate increased imports

Fertilizer and Food Security

The rapid increase and widespread use of chemical fertilizers has been a major factor in the remarkable increase in grain and food production. Domestic production provides 70-75 percent of the nutrient supply using both ingenious local technologies and modern international technologies. Fertilizer use has quadrupled since 1978, spurred on by the availability of fertilizer responsive crop varieties. Fertilizer used in 1995 was valued at Y 125 billion and represents the major cash input in crop production. The apparent declining effectiveness of incremental fertilizer application has led to concern that the potential for further yield increases, using existing cultivars, is limited. The Chinese Academy of Agricultural Sciences (CAAS) has identified the following causes of declining fertilizer effectiveness: (a) unbalanced and underapplication of nutrients, especially underuse of potash (K_2O); (b) poor quality of fertilizers; (c) poor application methods; and (d) poor distribution of fertilizers.

Unbalanced Supply and Use of Nutrients

The importance of proper nutrient balance is well known and generalized nitrogen:phosphate:potash ratios of 100:50:25 have been recommended for several years, but 1995 application ratios were 100:47:16. (Korea applies N, P, and K in a 100:120:80 ratio and Japan applies 100:40:40.) An analysis of the 1995 macro nutrient balances, based on input and uptake for the 17 most important crop and livestock products is summarized in the table below.

TABLE 4.1.6. NATIONAL NUTRIENT BALANCES (million tons)

	Nutrient Uptake (crop & animal)	Assessed Losses	Nutrient Input /a	Balance
Nitrogen	19.52	10.14	32.52	2.86
Phosphate	6.83	3.49	12.51	2.18
Potash	17.50	1.85	12.62	-6.73

/a Includes chemical fertilizers, crop residues, animal feeds, organic wastes, atmospheric deposition, and nitrogen fixation.

The results indicate that about 3 million tons of nitrogen and 2 million tons of phosphates applied were not used by crops and livestock. The excess nitrogen and phosphate, valued at Y 18 billion, was wasted and possibly contributed to environmental pollution. Conversely, potash removal was 6.7 million tons greater than that applied. The under-application of potash diminished the efficiency of nitrogen and phosphate uptake and crop production was lower than if balanced fertilizers had been applied. Although some nutrient uptake is provided by soil nutrient reserves, long-run sustainability requires crop and animal nutrient uptake be replaced by either natural processes (atmospheric deposition, biological nitrogen fixation, etc.) or by fertilizer application.

Research projects and field trials undertaken by Chinese agricultural research entities—often in collaboration with international institutes— indicate how yields respond to balanced fertilizer application (Table 4.1.7).

TABLE 4.1.7. YIELD RESPONSE TO POTASH IN BALANCED FERTILIZER TRAILS

Crop	Potash Applied (kg/ha)	Yield without K	with K	Increment	Average Yield Response (kg output/ kgK$_2$O)
		----------------------(kg/ha)--------------			
Rice	98	6,038	7,020	982	10
Wheat	98	2,790	3,900	1,110	11
Corn	113	5,048	6,570	1,522	13
Tomato	165	23,318	30,773	7,455	45
Br. Bean	120	2,145	3,233	1,088	9
Water melon	150	31,290	38,430	7,140	47
Citrus	n.a.	40,148	53,783	13,635	
Pineapple	375	22,530	28,590	6,060	16

Source Stauffer, Mark D. And James D. Denton, "Importance of Plant Nutrients in Increasing Agricultural Productivity–Chinese Experience," presented at the Fertilizer Association of India's Annual General Meeting, December 7-9, 1995, New Delhi.

The potential increases in crop productivity from balanced application of fertilizers range from 16 to 50 percent. The under-application of potash, despite clear evidence of its benefits, may stem from the limited amount of potash available from local sources—and it is costly to produce.

Fertilizers quality

About 50 percent of China's supply of nitrogen and phosphate consists of low grade ammonium bicarbonate (ABC), single superphosphate (SSP), and fused magnesium calcium phosphate which is produced in small scale factories. Ammonium bicarbonate as a nitrogen nutrient source is unique to China. Its manufacture involves a relatively simple technology using widely distributed anthracite coal deposits as feedstock and permitted China to rapidly expand production capacity and nitrogen fertilizer application. But, its manufacture is energy inefficient (mitigated by an abundance of anthracite deposits), highly pollution prone, and the product decomposes in storage. Similarly, phosphate fertilizer production remains primarily low analysis SSP which can be manufactured from widely distributed low grade rock using simple and low-cost technology. But not all of the contained phosphate is water soluble and therefore is not immediately available for plant uptake. However, SSP provides traces of the minor element sulfur, in which Chinese soils are deficient. The poor quality of the low grade materials is not only in their low nutrient content (less than 20 percent) which makes them costly to transport, but also in their handling and agronomic efficacy. As application volumes increase, it becomes more important to improve production and transport efficiency by shifting to higher analysis fertilizers.

Fertilizer Supply and Underapplication

The average application rate of plant nutrients is officially reported as 240 kg/ha. This appears to overstate the actual application as recent satellite imagery indicates that the cultivated land base is about 40 percent greater than the land statistics indicate and there is "double counting" of about 10 percent of the single-nutrient fertilizers that are blended into compound fertilizers. Correcting for these statistical anomalies reduces the application rate to a more modest figure of 155 kg/ha. China's average fertilizer application rate is therefore below the East Asian developing country average and far below Japan and Korea where yields are much higher. This low application rate may result from the limited supply availability. New factories using modern technology and producing high analysis, high quality products are under construction, and others are planned, which will meet incremental needs but will not immediately replace the low grade fertilizers. New modern factories require large investments and long gestation periods for construction, which may require a decade or more. Planning guidelines call for nitrogen self-sufficiency and 75 percent self-sufficiency in phosphates by 2000. The increased production is to come from large scale plants that would cost in excess of $5.0 billion. But delayed construction and increasing demand will prevent achievement of these objectives.

Fertilizer imports have averaged 8.6 million tons (nutrient content) annually over the past decade. Urea has been the major import, followed by potash and compounds. SINOCHEM has the exclusive right to import fertilizer; as such it leverages its position to obtain very favorable prices. However, it is inflexible and has not responded to demand changes. China must be prepared to import large quantities of potash as local supplies can meet only a small fraction of requirements.

It is estimated that 50 percent of the nitrogen applied to irrigated fields is lost by evaporation, because fertilizer is broadcast. Applying the fertilizer directly to the root zone by using fertilizer drills or slow release fertilizer tablets would reduce this problem. The nitrogen in ABC is particularly volatile.

Land

Cultivated land per capita is low, but land *per se* is not a limiting food security constraint given the potential; for increased yields under balanced fertilizer application, for expanding agricultural land through land reclamation, and for long-term investments in land if farmer incentives are improved.Official sources list China's cultivated and sown areas as 95 and 145 million hectares, respectively. However, satellite imagery indicates the true cultivated area is approximately 132 million hectares. If crops are distributed proportionally on the more than 37 million hectares of cultivated land not in the official estimates, individual crop yields are 40 percent lower than officially reported.

The land base is steadily eroded by residential and industrial encroachment and infrastructure construction. Since 1988, capital construction has removed 190,000 hectares of agricultural land from cultivation annually. (Some recent studies suggest the actual area removed may be greater.) This is about one-third of the annual decrease in cultivated area and is the only land that is *permanently* alienated from agriculture. These losses have been concentrated in Shandong and the southeastern coastal provinces where high multiple cropping compounds the land loss. Land reclamation, averaging 245,000 hectares annually, has partially mitigated these cultivated land losses (though newly reclaimed land is less productive than that lost). Some cultivated land is lost by transferring it to alternative agricultural uses such as pastures, forests and fish ponds. Other fluctuations in cultivated land represents marginal land which is cultivated or lies fallow depending upon prices and profitability.

An estimated 13.5, 2.4, and 2.0 million hectares of barren, tidal, and wastelands,[14] respectively, are *potentially* reclaimable although infrastructure (irrigation works, roads, etc.) and farm buildings would reduce the cultivable area by 25 to 30 percent. Almost 50 % of these areas are located in Xinjiang and Heilongjiang where water shortages would constrain reclamation and some are located in remote areas or are low in natural fertility—but could still be productive pasture or forest. About 25 percent of the land lies within, or south of, the Yangtze River basin and would be more amenable to irrigation. One-third of the potentially reclaimable tidal lands are in the coastal provinces abutting the Bohai Gulf (Liaoning, Tianjin, Hebei, and Shandong). Food production on reclaimed tidal lands would be concentrated in aquaculture and rice. The cost of reclaiming these lands varies from a low of Y 15,000/ha for barren land to Y 150,000/ha of tidal land.

The cropping index ranges from less than 100 in some northeastern and northwestern provinces to over 250 in Hubei. Short frost-free periods limit the potential for increasing the index in the northern-most and high altitude provinces, except for green house production of vegetables, and

---

[14] Barren land may be either virgin land or abandoned cultivated land. Tidal lands are those which are inundated during high tide and reappear during low tide, and wastelands are primarily those areas where rocks and other waste materials from mining operations have been discarded.

water shortages limit multiple cropping in other northern locales. However, improved water control, shorter season varieties, and appropriate production incentives will stimulate additional multiplecropping in the southern and middle latitude provinces.

Water Resources

Water scarcity is China's most limiting agricultural production resource, particularly in the northern corn and wheat regions. Consequently, efficient use of existing supplies is particularly important. Investments in improved water control and delivery systems would conserve water and modestly improve yields.

Only about 30 percent of the water diverted into irrigation canals is actually delivered to crop root zones. The losses are due to delivery inefficiencies, which are the responsibility of public entities, and to inefficient on-farm water use. *Delivery improvement will require lining the canals and other investments but on-farm efficiency could be improved with appropriate water pricing.* Efficiency gains in delivery and application must be pursued, but achievable gains are limited. Even with improved efficiency, productivity growth will require incremental supplies.

Incremental Supplies

Annual average water runoff in China is considerably below the world average and only about one-third of the runoff is exploitable. Water exploitation was only 60 percent of the potentially exploitable supply in 1993. Capturing more will require large investments in storage, diversion works, recycling, pumping, and conveyance systems.

Major increases in the supply of exploitable water to northern China must await the completion of one or more of the three components of the South-North Transfer scheme from the Yangtze River. Prefeasibility studies on the eastern and middle routes conclude that 20 and 25 billion m^3 per year could be transferred to the more arid north with respective investment costs of roughly $10 and $15 billion. Also, available irrigation supplies could be expanded up to 25 percent by reusing drainage water mixed with fresh water.

FOOD SECURITY STRATEGY

Water Consumption

Industries, municipalities, rural residents, and irrigated agriculture are the major users of captured water. Irrigation water for agriculture is aresidual, determined by alternative uses that are expected to grow rapidly during the next two decades, especially municipal and industrial uses. Increases in *industrial* water requirements are closely linked to industrial growth which is projected to be sustained at an average of 6 percent through 2020. *Municipal* water consumption is primarily determined by urban population (although it is rationed in some northern Chinese cities), which is projected to grow by 4.0 and 3.0 percent, respectively, during the 1990s and 2000s. Water demand by *rural households* is projected to stabilize as small per capita increases will be balanced by a declining rural population.

Irrigation water requirements are a function of the effectively irrigated area, efficiency of the delivery systems, cropping practices. However, as a residual user, crop requirements may not be met. It is more efficient, in terms of output, to fully irrigate all the land possible with the water available and idle (or cultivate as dryland) the balance. However, this is unacceptable on equity grounds and, in reality, water deficits would be shared with consequent production reductions.

Water Conservation Investments

Government investments in water conservation fluctuated over the past four decades but consistently claimed about 70 percent of agricultural sector investments. Real investments were relatively low during the 1980s, reflected in the marginal increase in irrigated area during the decade. Both investments and irrigated area increased sharply in the 1990s; between 1989 and 1995 irrigation investments trebled and irrigated area expanded by 10 percent. The marginal cost, in constant 1990 yuan, of this expansion was 10,000 Yuan per ha, representing very efficient investments.

Infrastructure

Only one-third of cereal production is commercially marketed outside the village of production. But with increasing urbanization and increasing quantities of cereals fed to livestock the marketed surplus will rapidly increase. However, lack of transportation and supporting infrastructure are serious constraints to efficient grain marketing and distribution. This is particularly true for bulk grain transportation, intermodal transfers, handling, and transit storage. The inability to rapidly move large quantities of grain limits the development of rational grain markets and creates market fragmentation.

Bulk Grain Logistical Systems

To market grain more efficiently China needs more dedicated bulk handling systems. Bulk logistical systems create synergism and benefits that would be unavailable if introduced as individual elements. But such systems are costly and justifiable only on high volume corridors. Insufficient deep-water ports, dedicated grain berths, and port handling facilities make external grain flows unnecessarily costly and lack of domestic bulk transport facilities makes it difficult to transfer grain between the hinterland and the coast. Existing bulk services are limited to a shuttle system which transfers imported wheat from Tianjin to Beijing and low volume transfers from Dalian to inland northeastern cities.[15]

Ports

Fourteen major ports handle 98 percent of China's internationally traded grain. But only six ports have *specialized grain handling berths*; although additional specialized berths are under construction at several others. High-capacity bulk unloading equipment is available at *some* major import terminals, but for most ports bulk unloading consists of low-capacity grab cranes, conveyor hoppers, and often improvised portable conveying equipment that cannot be operated in inclement weather. *The lack of dedicated all-weather high-capacity off-loading equipment increases port transfer costs as discharge and demurrage costs range from $4,000 to $12,000 per day per vessel (depending upon size and age of the vessel).* Daily loading charges are generally double the discharge cost. Grain vessels servicing China's grain imports are primarily of 45,000 to 50,000dead weight tons (dwt), but few berths can accommodate vessels of this size and partial off-loading is often required before berthing China has no grain berths capable of handling 80,000 dwt

---

[15] Additional systems are under construction in the northeast, Yangtze River, and southwest areas under a Bank Group-financed project.

("Panamex") vessels. Transocean shipment on smaller vessels is more costly than on larger vessels, therefore, ocean freight costs on China destined grain are higher than for countries receiving the larger vessels.[16]

Importing grain on 45,000 - 50,000 dwt vessels and slow off-loading at poorly equipped ports cost an estimated $10.00 *extra* per ton. If grain imports treble in the next two decades, this could cost $400 million *extra* per year. However, this figure represents *potential savings* if adequate investments are made in deep-water berths and port handling equipment.

To efficiently accommodate future grain flows, China must add a new bulk deep-water grain berth of 4.0 to 5.0 million tons capacity biennially over the next two decades. Berth and bulk handling investment costs would be roughly $60 to $85 million per berth, depending upon existing ancillary facilities.

Grain flows through *inland ports* averaged 6.5 million tons per year during the 1990s. All these flows are bagged grain using slow and costly intermodal transfers. Inland waterways are intensively used in other parts of the world as a low-cost transport mode for bulk grain but have been neglected in China. This neglect has stemmed, in part, from the administratively determined rail tariffs which, contrary to other countries, are below barge tariff rates. *Market-determined tariff rates for all transport modes would increase the demand for bulk barge transport along the major inland waterways.*

Rail Transport

Despite large investments and improved efficiency in the rail system, limited capacity still constrains feed grain shipments from Northeast China to southern destinations. Grain specific investments are difficult to justify because grain comprises only 4 percent of rail freight.

---

[16] *Source:* Lloyd's Maritime Information Services, Inc. Also, during a recent 90 day period the average *cost per ton* for various size bulk grain shipments between the US Gulf Coast and China were: 30,000 dwt—$31.00; 55,000 dwt—$25.00; and 80,000 dwt—$16.00. Fertilizer shipments are frequently in even smaller ships (15,000—25,000 dwt) with per ton rates in excess of $40.00.

Nevertheless, future grain requirements of bulk users (food millers and manufacturers, feed millers, and industrial manufacturers), will increase.

Special purpose bulk wagons would increase efficiency through rapid loading/unloading and increased turnover per wagon. Bulk rail wagons are efficient only on high volume corridors, preferably with backhaul opportunities, with bulk loading/unloading facilities at railheads. Individual bulk hopper wagons with a turnover capability of about 4 million ton/km annually (based on 1,400 km average distance of grain shipments) cost about $50,000 each. *A $625 million investment in 12,500 bulk wagons would handle 50 percent of the grain traffic,* but ancillary loading and unloading facilities would require further investments to reap the full benefits of bulk transport.

Storage

Grain storage requirements increase as grain production and domestic trade expand, but much of the storage facilities are constructed on-farm. The State adds about 1.0 million m² or 1.7 million tons of storage space annually, mainly long-term storage for reserves. Virtually all of these additions are flat warehouse storage which does not lend itself to rapid turnover. China could reduce government storage by relying on nonstate enterprises to market and distribute grain if more flexibility were permitted for spatial and temporal price variation.
High throughput transit storage is crucial to efficient intermodal interfacing in a bulk grain logistical system. Transit storage requirements at intermodal transfer points are highly site specific and must be carefully calculated to avoid dead storage.

Policies and Institutions

Most grain policies are control oriented, urban biased, focused on the short term and impede long-term food security. Managing and supporting the grain component of the food sector is complex because it directly involves multiple ministries and substantial market intervention. During the late 1980s and early 1990s, the government gradually liberalized grain policy and reformed grain institutions, but rapid price increases in late 1993 led to retrenchment in 1994. Subsequent reform has been modest and focused on decentralization. The major objectives (b) increase control of production, pricing and marketing—government quota procurement of 50 million tons plus 40 million tons of additional

of grain policy are to: (a) maintain 95 percent grain self-sufficiency[17]; \\
of grain policy are to: (a) maintain 95 percent grain self-sufficiency[18];
procurement to control 70-80 percent of the marketed grain; and
(c) continue controlling stocks and international trade.

Governor's Responsibility System (GRS)

In 1995, the Central Government delegated responsibility for balancing
local supply and demand, portions of pricing and marketing control and a
portion of grain responsibility (including financial responsibility) to the
provinces. Market transactions between Government Grain Enterprises
(GGE) in surplus and deficit provinces replaced centrally "planned"
interprovincial grain transfers. GGEs from deficit areas are prohibited
from procuring grain in the countryside of grain surplus areas, but must
buy from wholesale markets at county level or above. Surplus producing
provinces must maintain three months supply of grain stocks and deficit
provinces—six months supply. Provinces are free to subsidize farmers'
inputs and consumers' purchases as their revenues allow. Private traders
cannot procure grain in the countryside until State grain quotas have been
filled.

The State retained the grain quota (below market price grain which must
be delivered to Grain Bureaus), national stock responsibility, and the
international grain trade monopoly. The net effect of the GRS policy was
to decentralize responsibilities, replacing the nationwide policy with
provincial policies, but the policy content changed only marginally.

Quota Procurement and Farm Income

In 1995, 142 million tons of grain was marketed outside the village of
production—representing 30 percent of production. Government grain
enterprises procured 65 percent of this quantity—purchasing 46.2 million

---

[17] Information Office of the State Council, "The Grain Issue in China,"
October 1996.

[18] Information Office of the State Council, "The Grain Issue in China,"
October 1996.

tons of quota grain and an additional 46.3 million tons at negotiated prices. Quota grain prices averaged 60 percent of the free market price, and negotiated prices averaged 90 percent. These below-market prices imposed a total implicit tax of Y 40.7 billion on grain producers. *Recent grain price and production increases have shown that farmers respond to price. Thus, artificially low average prices depress production and marketed surpluses and hamper food security.* As incomes of grain farmers are typically below that of aquatic, livestock, fruit, and vegetable producers, the tax is discriminatory as well.

TABLE 4.1.8. GOVERNMENT GRAIN PROCUREMENT AND VALUE, 1995

State Procurement			Price			
Quota	Negotiated	Production	Quota	Negotiated	Market	Implicit Tax
---------------- (million tons) ----------------			------------------ (Y/ton) -----------------			
Rice [/a]	18.4	10.3	185.2	1,107.5	1,729.0	1,897.5	16.3
Wheat	17.1	14.1	102.2	1,080.0	1,528.0	1,688.0	12.7
Corn	9.3	15.0	112.0	855.0	1,385.0	1,580.0	9.7
Soy beans	1.0	4.2	13.5	1,814.0	2,422.0	2,711.0	2.1
						Total	40.7

/a   average of indica and japonica (unmilled).

The Government proposes to retain procurement quotas to control grain stocks but 1997 quotas will be procured at market-determined prices and consumption subsidies will be discontinued. Unifying prices at the "market" level will improve farmer incomes but the purchase price will *remain state-determined*, albeit by competition between various GGEs. Nongovernment grain enterprises participate in the market only after the State satisfies its procurement needs.

State Grain Enterprises

The Grain Bureaus (GB) have two conflicting roles: to perform regulatory and policy functions and to operate commercially profitable enterprises. This "two-track" system allows the enterprises to mix staff, activities, grain stocks and financial accounts. Quota grain is sold as higher price negotiated grain, reserve grain is used as commercial stocks, and consolidated accounts facilitates reallocating commercial losses to policy losses and claiming government reimbursement.

The GB is not a pure monopsony, but it procures grain at below-market prices before competitive procurement is permitted. They have little incentive to be efficient because their losses are reimbursed; they have minimal competition for raw materials and only modest competition in marketing (GBs procure about 70 percent of marketed surplus). Most consumers now purchase the major share of their grain needs from the private sector, but private retailers purchase most of their supplies from GB enterprises.

Pricing policy, combined with enterprise inefficiency, is costly. IOU procurement is prohibited, but there are reports of delayed or closed procurement due to lack of funds or storage space.[19] GB net losses continue to accumulate, rising to Y 19.7 billion in 1996. Government consumption subsidies for "grain and edible oil" averaged Y 20 billion over the past five years[20].

Grain policy and enterprise inefficiency costs Y 80 billion annually (implicit farmer tax plus losses and consumption subsidies) making reform imperative. Full market competition must replace the monopoly/monopsony the GB now enjoys. To achieve and maintain competitive marketing efficiencies government enterprises must be subject to the same constraints, incentives, and standards as private traders, collective farms, and grain processing companies. Competition could result in *some* GGEs declaring bankruptcy and exiting the industry leaving some of the 4.1 million GB workers unemployed; but many of the GBs would remain in business—albeit with fewer workers. The cost of social assistance to all 4.1 million employees would be far less than the current inefficiency costs of Y 80 billion.

---

[19] Xinhua news agency reports in various newspapers.

[20] An estimated annual subsidy of Y 3 billion for cotton was subtracted from the price subsidy for grain, edible oil and cotton listed in *China Statistical Yearbook,* 1996.

Grain Reserves

Governments in the major grain exporting countries have divested their grain storage activities because the cost of holding reserve stocks exceeds the perceived benefits of grain price stability. Reserves are maintained as commercial stocks by more efficient and competitive private enterprises. Food grains in these countries comprise a small portion of agricultural incomes and consumer expenditures, and price instability is *relatively* unimportant. *Nevertheless, there are economic arguments for price stabilization in developing countries where food grains are a major cost component of consumer expenditures.*[21]

The State Administration for Grain Reserve (SAGR) manages reserve stocks on a noncommercial basis, but pays Provincial Grain Bureaus to handle the grain on its behalf. *This separation of the management and operational functions has hindered the Government's ability to effectively stabilize prices.* Price stabilization releases can be triggered by a 20 percent price increase, "within a short period of time," with State Council approval. But activating releases have been difficult because the reserve grain is physically combined with commercial stocks and is sometimes sold and thus unavailable when needed. Also, the GBs are reluctant to release stocks because storage fees are an important income source.

The State manages an extraordinary quantity of grain to maintain price stability a reserve target of 40 million tons and 90 million tons of commercial and semi-commercial grain. Additional reserves are maintained by provincial authorities and other jurisdictions, and farmers are known to retain large stocks. Information on reserve stocks and stock releases are unavailable; thus, reserve costs and efficiency cannot be evaluated. *But price stability could surely be effected with smaller reserves*[22].

---

[21] Rice price stabilization by Indonesia's Bulog (Badan Urusan Logistik) is estimated to have contributed almost one percentage point to GNP growth in its early years, 1969-74 (Timmer 1996).

[22] Indonesia's Bulog stabilized domestic rice prices during the past three decades by relying heavily on private-sector traders, performing only a marginal market role. Bulog's *maximum* annual rice purchase was 25 percent of production with carryover stocks are 1.0 million tons.

Large reserves successfully protect against *possible* transitory insecurity, but less costly alternatives should be sought. Futures hedging is an efficient management tool when annual imports are *required* to meet consumption requirements but inappropriate as insurance for *possible* imports. Futures options are relatively inexpensive *theoretical* alternatives, but there are too few market transactions in these instruments to be a viable option for China. However, *greater reliance on the international market would be more efficient than maintaining stock for the most severe contingency, although improved port and handling capacity would be necessary.*

TABLE 4.1.9. COST OF MAINTAINING WHEAT RESERVES VS. IMPORTING

Year	Purchase and Carryover to Following Year /a ($/ton)	Unit Value of Imports ($/ton)
1990		172
1991	220	118
1992	157	142
1993	185	129
1994	170	131
1995	172	174
1996	223	229

/a   Assumptions: Interest–10%; Storage losses—5%; $18/ton/year storage cost.

From 1990 to 1996 the average cost of internationally procured grain would have been $35/ton less than procuring (at an international price equivalent) and storing domestic grain for subsequent use. *Proposed policies* call for: (a) independently maintained reserve storage with separate staff and accounts, (b) new SAGR warehouse construction in readily accessible grain deficit locations, (c) "first in, first out" inventory management, and (d) adding future reserves grains are available, not procured according to *Plan. Implementation of these proposals should permit SAGR to more effectively implement policy decisions and perform its stabilization role.*

Input Pricing and Marketing

Fertilizer marketing remains a planned component of the input supply sector. The fertilizer supply market is basically a single channel, monopoly structure operated by the China National Agricultural Means of Production Corporation (CNAMPC). CNAMPC takes domestically manufactured fertilizer from the manufacturers and receives imported fertilizer at the ports for onward distribution. Only 10 percent (estimated) of the domestically manufactured fertilizer is traded outside this channel. Prices and marketing margins received by CNAMPC are government controlled; imported fertilizer, which typically has higher nutrient content, is higher priced.

The existing marketing structure has provided farmers with nitrogen and phosphate fertilizers. But it has failed to deliver the types and combinations of fertilizer nutrients recommended and needed. Also, it is insufficiently flexible to respond to the large and rapid changes in fertilizer quantities and composition required in the next century. Independent dealers, whose incomes depend upon helping farmers optimize their fertilizer input, effectively perform this service in many countries of the world—and they should be an improvement for China.

International Trade Policy and Institutions

China is the world's largest fertilizer importer and in some years, the world's largest grain importer. Imports in both commodities are "planned" up to six months before the beginning of the year and implemented by monopoly trading companies. Given China's importance in the international market and monopoly status of these companies, they are able to impose considerable leverage in pricing and delivery terms, typically importing fertilizer and grain at c.i.f. prices very near international prices levels.

The monopoly trading companies and "planned" import-export systems are effective, but operate with uncertain efficiency. *Planned grain trading has exacerbated fluctuations in rice, wheat and corn supplies over the past decade (measured by either production or GB procurement—plus net imports).* More competitive and open trade regimes might result in higher farmgate corn prices, lower cost of wheat flour to consumers, and more stabilizing responses to import requirements.

Grain Price Trends

After a long period of stable subsidized cereal prices Government decided, in 1993, to reduce its fiscal costs by increasing administratively set prices and moving toward market determined prices. Domestic prices began to trend upward in 1993 and increased sharply in 1994 before leveling off in 1995 and declining in 1996. By the end of 1996 wholesale cereal prices had returned to the levels of early 1994. Throughout the period price movements were relatively more smooth and without the monthly volatility experienced in the international market - particularly rice. But in 1994, domestic (wholesale) prices of corn increased 65 percent, rice 75 percent, and wheat 60 percent—while international prices of corn and rice were declining and wheat prices were stable. Both domestic and international price peaked in early 1996 with domestic prices above international prices.

The combination of administratively determined procurement prices and planned external trade has hindered internationalization of the domestic market. As domestic prices are now higher than international prices, domestic market liberalization and international integration are difficult. Reducing quota and negotiated prices would hurt rural incomes and maintaining domestic prices above international levels would require import protection with either quotas or tariffs. Current price differentials implicitly subsidizes producers and, unless passed on to consumers, will require large government subsidies. Moreover, higher domestic grain prices provide incentives for GGEs to import rather than procure domestically. Thus, China faces the prospect of unsanctioned cross border trade—particularly with adjacent rice exporting countries.

Water Pricing Policy

Water is the most limiting resource in food production. But current low water prices encourage overuse and contribute to low water use efficiency and subsequent scarcity. The Water Financial Directive of 1988 called for marginal cost pricing of water by 1997. But water prices charged to most farmers remain below supply costs, ranging from zero to Y $0.40/m^3$ in the Yellow River Basin. The marginal value of water in agriculture ranges from Y 1/m³ for grains and Y 4/m³ for vegetables.[23] In the few locations where rational irrigation water pricing has been implemented, water demand has declined without affecting yield levels. *Water charges should be immediately increased to cover supply costs and a timetable established for achieving marginal cost pricing. However, increasing prices may be difficult to implement without ensuring irrigation services and infrastructure are adequately maintained and system performance improved.*

Land Use Policy

The family farm management structure (household responsibility system) reintroduced in the late 1970s and early 1980s radically altered incentives and rapidly increased labor and land productivity. Cultivation contracts for farmland were recently increased from 15 to 30 years. But incentives remain inadequate to stimulate investments needed to improve land productivity. Because land might be redistributed before contract expiration, due to rural population and household growth, tenure remains insecure.

Many farmers, particularly in the southeast coastal provinces, now engage in part-time agriculture earning a major portion of their income from non-farming activities. They have minimal incentive for productivity enhancing investments or practices (terracing, soil amelioration, and degradation-prevention measures) and are less interested in agricultural productivity than in retaining their land use right as a social safety net.

Other farmers have begun to produce more profitable commodities such as fruits, vegetables, livestock and aquatic products. That leaves farmers who produce grain, oilseeds, cotton, etc., at a serious income disadvantage—not because of scale diseconomies but because of an inappropriate land and labor resource combination. Rural incomes will diverge even further unless grain producers can gain access to additional

---

[23] World Bank. "China–Yellow River Basin Investment Planning Study" (Two Volumes). Report No. 11146-CHA. Agriculture Operations Division, China and Mongolia Dept., EAP Region, June 30, 1993.

land. If these farmers had access to the land of the part-time farmers mentioned above, for which financial compensation could be made, national productivity would improve. *The various land use experiments underway, including shareholding and leasing of village contracted land, need to be expanded and institutionalized to facilitate the rapid development of a land use market.*

Food Supplies

Domestic Grain Projections

Water shortages will increasingly limit agricultural growth—particularly in the northern corn and wheat belts. A physical constraints model was developed to evaluate the impact of various water options on agricultural output in the nine water regions.[24] The model was estimated using three different yield growth rates[25] (as a proxy for investments in research) under alternative irrigated area and water availability scenarios.Estimated base case grain output for 2020 was 636 million tons (Table 4.1.10). Based on rainfall and water runoff statistics, the model confirmed that irrigation water will be limited in the Northern corn and wheat producing regions but will remain sufficient in the Southern rice producing regions. *The implication is that rice may become an important feed grain as well as a food grain.* In many locations in southern China, low quality hybrid rice is cheaper than coarse grains and is already used as a feed grain. As farmers gain experience in corn cultivation and new varieties adapted to southern climates are developed, corn production will expand in the south

---

[24] The nine water regions, I - IX, are sequentially, Northeast, Haihe, Huai/Shandong, Yellow, Yangtze, South, Southeast, Southwest, and Northwest. A description of· the regions and their component provinces is contained in Annex 5 of the study entitled "China's Long Term Food Security".

[25] A continuously increasing yield growth rate of 1.0 percent annually was the high rate; a yield growth rate of 1.0 percent until 2010 with subsequent growth rates of 0.5 percent was the base rate; and the low rate was a yield growth rate of 1.0 percent up to 2000 with subsequent growth rates of 0.5 percent. The assumptions underlying the model and alternative scenarios evaluated are described in Annex 5 of the study entitled "China's Long-Term Food Supply".

But producing rice as feed grain will likely continue—despite its lower feeding value—as hybrid rice yields are double the corn yields in this region.

TABLE 4.1.10. GRAIN PRODUCTION ESTIMATES FOR 2020

	High	Base	Low
		(million t)	
Rice	313	298	283
Wheat	151	144	137
Coarse grains	203	194	186
Total	667	636	606

*Rice is paddy equivalent..

The rate of industrial growth is the most influential factor in determining future irrigation water availability—and food production. The faster industry grows, the less water remains for agricultural use. The location of industrial growth affects agriculture differentially. Rapid industrialization in water regions I and II would exacerbate irrigation water shortages; alternatively, industrialization in the water surplus areas of southern and southeastern China would not deprive agriculture of water, but the impact of transferring cultivated land to industrial use would be greater as the multiple cropping index is higher in the south.

Results from the model clearly indicate the need for investments in water development, saving, and recycling to meet municipal and industrial requirements and efficiently provide agriculture with incremental water supplies. In 2020, the marginal agricultural impact of the proposed South-North water transfer schemes (Eastern and Middle routes) would be about 15 million tons of grain, 10 million tons of vegetables, and various tonnage of other crops with a 1995 value of Y 57 billion ($6.7 billion), yielding a crude IRR of 20 percent for agriculture.

Besides greater investment in water resources, the model also indicates the need to invest in: (a) land reclamation sufficient to maintain the cultivated and irrigated land base; (b) agricultural research to ensure crop yields continue to grow and multiple cropping expands; and (c) agricultural extension to transfer research results into farmer actions.

The extension issue is particularly important to improve fertilizer nutrient balance. Unless fertilizer nutrient balance is improved, grain supply deficits will likely double.

In the base case scenario of investments and water availability, [26] cereal supply is estimated to be 636 million tons of rough grain in 2020 assuming the cultivated land base remains constant. In addition, 486 million tons of fruits and vegetables and 50 million tons of oilseeds would be produced. However, this will fall short of the anticipated needs of 695 million tons of grain. About 60 million tons of wheat and feed grains will need to be obtained from the international market. About one-half of the imports will be feed grains. If investments are inadequate to maintain the resource base and increase productivity, reduced production will require even greater imports.

International Grain Supplies

During the 1990s, world grain production has averaged over 1.7 billion tons. Coarse grains comprised the largest share, followed by wheat and rice. About 11 percent of production enters international trade. Wheat is the most widely traded grain, followed by coarse grains and rice. China accounts for about 20 percent of world production (Table 4.1.9). China's participation in the world market has been modest as both a net importer and a net exporter over the past 15 years, with neither accounting for more than 5 percent of international trade. However, the world market has the potential to supply China with substantially larger quantities of grain in the future

Potential Grain Export Supplies

The potential supply of grain available from the world market is substantially larger than the current world trade of 200 million tons.

---

[26] Eleven of the variables in the model (total water availability, industrial demand for water, return flows of municipal and industrial water, usable water returns, drainage water reuse, water use efficiency, effective irrigated area, rainfed multiple cropping index, cultivated area, rainfed and irrigated yield growth) were included as low, medium, high options. Three other variables (two South-North water transfers routes, and balanced fertilizer application) were included as yes-no options.

Table 4.1.11 World Grain Production and Trade
(Annual Average 1992/93-1996/97)

Commodity	World Production (million tons)	China Production (million tons)	World Trade (million tons)	China's Share of World Production (percent)
Coarse Grain	839.7	100.6	90.3	12.2
Wheat	551.1	115.0	98.7	20.8
Rice (milled)	367.0	128.7	12.1	35.8
Total	1,759.6	344.3	201.1	19.8

Source: United States Department of Agriculture, Foreign Agriculture Service, "Grain: World Markets and Trade," FG 10-96, October 1996.

However, stagnant world import demand since the late 1970s has postponed development of this potential. The traditional grain exporters accounted for 90 percent of the exports in the first half of the 1990s.[27] Domestic per capita consumption is already high in these countries, and except for Thailand, population growth is low; therefore, production growth is primarily a function of their ability to export. In the face of reduced export demand they reduced grain crop areas by 34.5 million hectares from the highs of the early 1980s to keep from accumulating stocks. The production potential of this land was roughly 115 million tons. *These lands could be returned to production almost immediately if there were incentives to do so, as indicated by increases in 1996 cereal plantings and harvest.*

Other potential sources of export supplies include the former Soviet Union which reduced grain crop areas by 27 million hectares over the same period due to reduced demand, and to lower input availability. Total production declined by 86 million tons between 1990 and 1995. *With*

---

[27] The traditional grain exporters and their share of total world exporters were: the United States (42%), the European Union (22%), Canada (11%), Australia (7%), Argentina (6%), and Thailand (2%). However, the United States overwhelmingly dominates the course grain market.

proper incentives, some of this production could return. Argentina also has the potential to expand grain production and export. Argentina is sparsely populated with vast areas of grassland that could be converted to grain production if export markets and prices provided adequate incentives. The pasture is flat delta land which is primarily used for cattle grazing. Only a small portion of the land receives fertilizers, pesticides or herbicides. Current grain yields in Argentina reflect the low input use and averaged 2.76 tons/ha from 1990 to 1995 compared to 4.84 tons/ha in the United States. Even at current low yields, the production potential of the pasture land in Argentina is large.

Grain stocks are rebuilding but remain low—limiting immediate expansion of exports. Current export grain stocks in the major exporting countries are about 100 million tons, some 40 million tons above the previous low stock levels of 60 million tons in 1995/96 (USDA 1997). Under these stock conditions world trade could immediately expand by roughly 40 million tons, but over the longer term the potential for increased grain supplies is large. *For example, world grain production increased by 7.5 percent in 1996/97 in response to higher prices, but production in the five largest exporters increased by 20 percent and by 40 percent in Argentina.*

World Grain Prices and Price Volatility

World grain prices have historically declined over long periods relative to overall consumer prices, however, they have also increased sharply over short periods of two to three years. Real average annual prices for wheat, maize and rice prices declined by 50 to 60 percent between the 1970s to the 1990s (World Bank, 1997). But, prices increased sharply between 1971 and 1974, when real prices of all major grains increased by 50 to 100 percent, created in large part by a rapid rise in energy and fertilizer prices. Between 1993 and early 1996 real prices of wheat and corn increased by 32.8 percent and 45.7 percent (Table 4.1.13), due to unusually low world stock levels combined with a drought in the southwest wheat producing regions of the United States. The price increases were greater for importing countries than the numbers suggest because the United States and the European Union had been subsidizing grain export prices to increase export market share. When prices began to rise and export supplies dwindled, the subsidies were terminated.

TABLE 4.1.12. WORLD GRAIN PRICES, SELECTED PERIODS 1991-96
($ nominal)

Period	Wheat	Rice	Maize
1991 (annual avg.)	128.66	314.40	107.40
1993 (annual avg.)	140.24	270.00	102.10
1996 (Jan-Jun avg.)	231.40	349.60	183.00
1996 (Dec avg.)	175.70	319.20	117.70

Source: World Bank, IEC.

Real grain price declines are forecast over the next 10 to 15 years according to recent studies by the World Bank, FAO and IFPRI.[28] These forecasts assume that historical yield increases will continue and that population growth slows as forecast by the United Nations. However, the forecasts do account for increased demand due to rising incomes.

Future grain price volatility is likely to increase from the levels of recent decades because of policy changes in the major exporting countries. These changes include both those made due to the Uruguay Round GATT Agreement on Agriculture and those made unilaterally in the United States and the European Union for budgetary reasons. Policy changes in the major grain exporting countries were largely responsible for the sharp declines in world grain stocks from 465 million tons in 1986 to 245 million tons in 1995. These changes likely will keep world grain stocks low in the future and lower stocks could lead to greater price volatility because the buffer against a poor harvest is less. The level of grain stocks are important because they reflect the ease with which grain exports can be expanded. During the mid-1980s, the major grain exporting countries had carryover grain stocks of 260 million tons while world grain trade was less than 200 million tons. This large reserve of grain could be used to expand exports without significant price increases. In contrast, current world grain stocks are about 100 million tons and could not meet a significant increase in exports. *If China were to increase grain imports*

---

[28] Islam, Nurul, ed. Population and Food in the Early Twenty-First Century: Meeting Future Food Demand of an Increasing Population." International Food Policy Research Institute, Washington, D.C., 1995

to expand exports without significant price increases. In contrast, current world grain stocks are about 100 million tons and could not meet a significant increase in exports. *If China were to increase grain imports significantly and rapidly, its actions would increase world grain market prices. But if imports were gradually increased over a number of years, the price impact would be marginal.*

Market Access

Concern about market access has often led to policies aimed at a high degree of self-sufficiency in traditional food-importing countries, stimulated by the poor policy record of the major grain exporters. In 1996 the European Union imposed a wheat export tax in an attempt to reduce exports and prevent domestic prices from rising. During the 1970s the United States embargoed either soybeans or grain in three successive years to prevent higher consumer prices. Targeted export embargoes have also been used for political reasons, as in 1980 when the United States embargoed grains, other foods, and agrochemical exports to the former Soviet Union in protest of the invasion of Afghanistan. But this 16-month embargo was largely ineffective because alternative supplies were available.

Exporting countries often limit food exports if prices rise sharply, so unless an importing country has a long-term agreement, supplies from individual countries are not assured. Further, the supplies are most likely to be unavailable when prices rise sharply—precisely when market access is most important. The best defense against such action by importing countries is to diversify imports among competing suppliers and to enter into long-term agreements that specify market access conditions.

Scenarios

With appropriate incentives grain output in the exporting countries (plus the former Soviet Union—FSU) could increase by 200 million tons and would bring them to their previous maximum output levels. However, most regions of the world are net importers and will require increased imports over the coming decades. Thus, the real issue is whether the traditional exporters (plus FSU) could meet all the incremental import requirements without large price increases.

In the early 21st Century most of the incremental worldwide demand will be met from domestic production—Japan, the Republic of Korea, and

Taiwan (China)[29] excepted. Nevertheless worldwide import requirements will triple between 1990-94 and 2030, requiring production in the exporting countries to grow by 1.1 to 1.4 percent per year—readily attainable if investments in research are maintained.

Food Balance Options and Conclusions

At the national level China has the potential to remain food secure over the next 2 to 3 decades if various reforms are implemented and a number of investments are made in agriculture and infrastructure. It is assumed that China will rely more on market forces to signal investment, production and consumption decisions and once it joins WTO, will integrate more fully into the international marketplace. Within this framework domestic grain production will increase, but irrigation water shortages will limit expansion requiring increasing reliance on the international market for incremental grain needs.

With a more open trade regime and competitive market, both China and China's trading partners will benefit through exploiting their comparative advantage. Increased incomes and urbanization will change diet composition; direct grain consumption (per capita) will decline and consumption of livestock products (indirect grain consumption), and manufactured cereal products will increase.

Options

The best estimate of China's 2020 cereal demand is about 608 million tons (697 million tons of unmilled trade grain), but *actual* requirements depend upon factors such as income and population growth rates, and changing food preferences. A *major portion of the incremental needs can be met from domestic resources if investments are made to* :
- Improve agricultural research and extension to maintain total factor productivity growth. It is critical to rapidly achieve balanced fertilizer applications.
-

---

[29] Alexandratos, Nikos; "China's Consumption of Cereals and Capacity of the Rest of the World to Increase Exports," *Food Policy* (forthcoming)

- Develop water resources to enable aggregate irrigation water availability (including improved efficiency, recycling and reuse) to increase by an average of 0.5 percent annually between 1995 and 2020. This will doubtlessly require water transfers from the Yangtze to the Yellow river basin.
- Reclaim and develop land to maintain the current stocks of arable and irrigated land.
- Develop dedicated-integrated bulk handling port and rail facilities to transport larger quantities of domestic and imported grains.

A series of policy reforms centered on less government intervention in the cereal sector, market determined prices, and open competitive marketing and trading regimes for inputs and cereals must accompany these investments. With these investments and reforms, China will likely produce 90 percent of grain requirements, and rely on international suppliers for the balance—about 60 million tons of grain by 2020—but if investments fall short, more imports will be required. The major grain exporting countries can readily supply this amount, but unless China invests heavily in port facilities and bulk logistical systems, imported grain will be extremely costly because even greater investments would be needed to improve the current outmoded handling systems—covered rail wagons, bag handling equipment, and flat storage.

Domestic Grain Supplies

Government would like to maintain 95 percent grain self-sufficiency, but to limit future imports to 5 percent of consumption requirements would be inefficient and costly. To achieve that objective, import tariff protection of about 25 percent would be required and domestic wholesale prices would be about 50 percent greater than international prices.

Constraints and Options

Pricing and Marketing. *To improve food security, grain production, and marketing efficiency, China must decrease government intervention and increase reliance on market forces to determine prices. Market competition for both inputs and outputs is essential to ensure efficient farmgate prices and marketing margins.* To achieve and maintain marketing efficiency both GGE and nongovernment marketing agencies must operate under the same constraints, efficiency incentives, and commercial standards. To ensure markets are integrated and farmers receive proper price signals, the Government must discontinue the spatial and temporal monopoly/monopsony privileges of the GGE and eliminate below market price State procurement. It must also relax countryside procurement constraints placed on GGEs from deficit areas and on private traders. Instead, the Government would procure grain for poverty groups, military forces, civil servants, strategic reserves, etc., at competitive "market" prices and transfer their policy functions to other noncommercial agencies. Similarly, CNAMPC would lose its monopoly fertilizer marketing rights and compete with other marketing agencies.

To improve water use efficiency, including irrigation, the Government must increase water prices to cover supply costs and establish a timetable to increase prices to long-run marginal cost levels. Also, market-determined freight tariff rates would encourage more efficient transport modes to develop, such as bulk barge carriers.

Price Stabilization and Strategic Reserve Programs are Costly. *Price stabilization is possible with considerably less market participation than the 70-80 percent of marketed surpluses the GGEs currently procure.* If the State wishes to stabilize prices and maintain strategic grain reserves these activities should be managed and operated by a government agency which is completely separate from commercial GGE operations. Price stabilization programs operate with price bands which generally follow the long-term international price trends; otherwise, costs will become exorbitant and unsustainable. But during periods of volatile prices, large short-term costs may be incurred even though long term trends are followed. The price stabilization programs in India and Indonesia rely on the private sector to carry out most marketing operations with government procuring only a small share of marketed surpluses.

Changes in Land Policies could Boost Food Production. Changes in land policies could provide incentives to increase land productivity and food security and to improve rural income distribution by: (a) allowing rural residents to either, lease their land cultivation rights *to* others (when off-farm activities provide most of their income) or lease cultivation rights *from* others when underemployed on their own farms (which also would improve rural income distribution); and (b) encouraging the private sector to invest in land reclamation. The rural land experiments underway should be evaluated and consolidated, legalized and publicized to enable farmers and land developers to understand the alternatives and opportunities.

International Trade Policies are too Rigid. The monopoly trading corporations procure and import fertilizer and cereal grains at c.i.f. prices very close to international prices. But farmers have *not* been provided with the appropriate balance of fertilizers and, because supply-demand conditions change between planning and execution, planned international grain transactions has exacerbated domestic supply volatility. Planned trade and trade monopolies should be discontinued so that open trading could support rapid responses (e.g., facilitate appropriate imports and permit timely exports to prevent unnecessarily large corn stocks from accumulating in the northeast provinces) and ensure that imported commodities are procured and handled efficiently.

Research and Resource Investments. Agricultural research is an excellent investment, but research must be prioritized and research budgets and agricultural research intensity should be increased. China has had a dynamic agricultural research system, but over the past 15 years agricultural research intensity has declined raising doubts about its ability to sustain long-term growth in total factor productivity.

Research indicates that *balanced fertilizer* application would increase land productivity by 12 to 15 percent on average—producing an additional 50 to 60 million tons of cereals. Whether farmers' constraints to applying balanced fertilizers lie within the agricultural extension system or in the fertilizer market structure, they must be identified and resolved.

China's fertilizer industry will require large investments and long lead times to develop. China has sufficient feedstock material to develop large scale ammonia-urea manufacturing plants to meet incremental nitrogen requirements and replace a portion of the low analysis ABC plants. Also, the richer phosphate deposits are being exploited but earlier planned investments in accompanying TSP manufacturing plants are lacking.
Water is the most limiting food production resource—therefore is an investment priority. Increasingly, water will be diverted from agricultural to urban and industrial uses and large investments will be required to maintain and increase *irrigation water* supplies. A first step is to improve *efficiency* by treating and recycling municipal waste-water, lining canals, and improving on-farm application. Municipalities and industries must invest in wastewater treatment to permit environmentally safe recycling.
Water resources development typically is very costly with high social

returns but low financial returns. Except for small pond/reservoir based irrigation systems, only the government has the capacity to undertake *new* investments. This includes water transfers from the Yangtze to the Yellow River Basin which will require long planning and implementation horizons and huge investments.

Satellite imagery indicates that China's cultivated land base is about 40 percent larger than official statistics indicate. This means that fertilizer application rates and crop yields are 40 percent below reported levels, which represents tremendous potential for increasing productivity and production. Newly reclaimed land mitigates the land lost to urban encroachment and infrastructure construction, but it is less productive than the prime agricultural land it replaces. *Continued investments in land reclamation are required to maintain the land base.*

Increasing urbanization and increasing cereal-based livestock production will increase the quantity of grain marketed, handled and transported. Investment in improved handling infrastructure is essential to efficiently move these increased quantities of grain. This includes dedicated bulk grain handling logistical systems involving inland and coastal waterway, rail, transit storage and intermodal interfacing to rapidly and efficiently move the larger volumes.

Investment Summary

To efficiently produce 636 million tons of cereals and import another 60 million tons, large investments will be required in addition to normal maintenance and capital replacement. Water resource development involving South-North transfers, improved irrigation canals and distribution systems, and municipal-industrial wastewater treatment will require the largest investment. Other large investments will be needed in transportation-handling infrastructure, including ports, bulk rail and waterway transport and ancillary bulk handling equipment and in fertilizer manufacturing. Investment and recurrent budgets for agricultural research and extension also need to be increased. Rough estimates of *incremental* investment requirements are indicated in Table 4.1.13.

Nonstate Investments Incentives

Given the large variety and size of investments required to provide future food security, incentives for nongovernment investments are needed.

- International *seed* companies might invest in research in China if permitted and if intellectual property rights were strengthened, but most agricultural research will need to be funded from domestic public resources.
- International *fertilizer* companies might invest in China's domestic industry. This would partially resolve the fertilizer investment constraint and ensure the latest mining and manufacturing technology was employed.

TABLE 4.1.13. ESTIMATED INCREMENTAL INVESTMENT REQUIREMENTS, 1995-2020($ MILLION, CONSTANT 1995 PRICES)

Water Resources:		64,000
South-North Water Transfer	25,000	
Irrigation Distribution Improvement [a]	39,000	
Municipal-Industrial Wastewater Treatment	n.a.	
Infrastructure:		1,385
Ports	600	
Bulk Rail Wagons	625	
Bulk Barges	90	
Inland Terminals, with bulk handling equipment	70	
Fertilizer Manufacturing Plants	5,000	5,000
Agricultural Research [b]	25,660	25,660
Agricultural Extension	n.a.	
Total		96,045

[a]   To maintain the trend of the past decade investments in Water Conservancy should increase by about $120 million annually; that is, incremental 2020 expenditures should be $3.0 billion more (in 1995 terms) than 1995 expenditures.

[b]   To maintain the trend of the Reform period, agricultural research investments should increase by five percent annually; that is, 2020 investment expenditures should increase by $1.75 billion (in 1995 terms). This may or may not be sufficient to maintain TFP growth.

- *Land* reclamation and long-term *productivity enhancing* measures require incentives for nonstate sectors to invest. These might include extended cultivation rights or the privilege of leasing out cultivation rights and compensation for land reclamation companies through land sales or leases— perhaps for nonagricultural use. Longer term farmer contracts may induce them to terrace or undertake other land improvement activities.

- *Domestic marketing and transportation infrastructure* incentives for nonstate investors are difficult to design, but large domestic and international corporations may be prepared to invest in bulk transit storage and bulk barge or rail wagons if dedicated to their operations. With a more open grain market, international grain corporations might invest in ports and bulk handling equipment. Without major private sector grain market participation, transport and logistical investments in high volume grain corridors would likely depend on the public treasury.

The International Market

International grain stocks have been rebuilt to about 100 million tons from the low levels of 1995/96. This provides a small margin for increased trade. But, the major grain exporting countries could rapidly produce and export much larger quantities of grain if they expected that growth in export demand would be maintained.

International grain prices are expected to continue their long-term decline. But short-term volatility is likely to increase because the exporting countries have discontinued or reduced government storage; thus, there is a smaller buffer to mitigate drought or calamity-induced shortages. Therefore, it would be to China's advantage to enter into long-term grain contracts with grain traders from exporting countries.

China's 14 major grain ports have a theoretical import capacity of 23 million tons. While these and other ports could handle additional quantities of grain it would be very costly to import beyond design capacity as marginal costs increase rapidly beyond that point. Thus, it is crucial to construct deep-water ports capable of handling 80,000 dwt vessels and to widely link intermodal bulk grain logistical systems with port facilities.

REFERENCES

Alexandratos, N. 1997. China's Consumption of Cereals and Capacity of the Rest of the World to Increase Exports. Food Policy, FAO, Rome. Forethcomng 1997.

Anderson, K., B. Dimaranan, T. Hertel and W. Martin. 1996. Asia-Pacific Food Markets and Trade in 2005: A global, economy-wide perspective. Discussion paper No. 1474, Center for Economic Policy Research, London.

Aubert, C. 1996. China's Grain Distribution and Price Policies. Workshop on Agricultural Policies in China, OECD Headquarters, December 12-13.

Barichello, Richard (University of British Columbia, Vancouver, Canada). 1996. The Nature of State Trading in Indonesia: The case of BULOG. International Agricultural Trade Research Consortium, Annual Meeting, Washington, DC, December 15-17.

Brown, L. R. 1995. Who Will Feed China? Wake-up Call for a Small Planet. W. W. Norton & Company, New York.

Cao, Mingkui, Shijin Ma & Chunru Han, 1994. Potential Productivity and Human Carrying Capacity of an Agro-Ecosystem: An Analysis of Food Production Potential of China. Agricultural Systems Vol. 47: 387-414.

Chern, W. 1997. Survey of Food Demand Elasticities in China?

China, State Statistical Bureau, China Statistical Yearbook, various issues.

Chu, A.C.P. 1996. Sustainable Agriculture: Integrating pastoral and arable land in Southern China—a model to increase arable land—what can the New Zealand experience contribute? Workshop on Agricultural Policies in China, OECD Headquarters, December 12-13.

Crook, F. W. 1996. How will China's 'Rice Bag Policy' affect grain production, distribution, prices and international trade? Workshop on Agricultural Policies in China, OECD Headquarters, December 12-13.

Du Ying (Research Center for Rural Economy, Ministry of Agriculture). 1996. Perfection and Reforms of Chinese Food Production and Marketing Policies. This paper was presented at the International Symposium on Food and Agriculture in China: Perspectives and Policies, October 7-9, Beijing, China.

Environment and Production Technology Division, International Food

Policy Research Institute. 1995. Data Survey and Preliminary Assessment of Agricultural Investment in China—A Report to Food and Agriculture Organization, United Nations (January 30,).

ERS/USDA. 1996. Long Term Projections for International Agriculture to 2005. ERS staff paper no. 9612. Economic Research Service, United States Department of Agriculture, Washington, D.C.

Evenson, R.E., and D. Jha. 1973. The Contribution of Agricultural Research Systems to Agricultural Production in India. Indian J. Agr. Economics 28:212-230.

Fahlbeck, E., and Z. Huang. 1996. The Property Right Structure of Farm Land and Family Farming in China—An Analysis of Developing Options. Workshop on Agricultural Policies in China, OECD Headquarters, December 12-13.

Fan, S. 1991. Effects of Technological Change and Institutional Reform on Production Growth in Chinese Agriculture. Amer. J. Agr. Economics, 73: 266-75.

Fan, Shenggen. 1996. Research Investment, Input Quality, and the Economic Returns to Chinese Agricultural Research. China Workshop - Global Agricultural Science Policy for the 21st Century.

Fan, S., and Philip Pardey. 1996. Research, Productivity and Output Growth in Chinese Agriculture. J. Development Economics (in press).

FAO. 1995. The State of Food and Agriculture. Rome.

Feng Lu. 1996. Feed Demand and its Impact on Grain Economy in China. Workshop Agricultural Policies in China, OECD Headquarters, Paris, December 12-13.

Food Studies Group. "Global Cereal Markets and Food Security." International Development Center, Queen Elizabeth House, April 1996.

Glenshaw, Peter A. (World Bank, Washington, DC, USA) 1996. The Role of Fertilizers in Food Security—The Case of China. Presented by Sulfur Institute at Fifth Biennial Symposium: "Sulfur Markets - Today and Tomorrow."

Griliches, Z. 1958. Research Costs and Social Returns: Hybrid corn and related activities. J. Political Economics 66:419-31.

GTAP, GTAP Database version 3.3. 1996. Global Trade Analysis Project. Purdue University, Indiana.

Guanzhong, James Wen (Baruch College/City University of New York).

1993. Total Factor Productivity Change in China's Farming Sector: 1952-1989. Economic Development and Cultural Change 42:1-41.

Hertel, T. W. 1997. Global Trade Analysis: Modeling and Applications. Cambridge University Press, Cambridge.

Hong, Y. 1996. Sources of Productivity Disparities in Regional Grain Production in China. China Workshop—Global Agricultural Science Policy for the 21st Century, p. 87.

Huang, J. China's Integration into the Global Economy: Longer Term Implications for Agriculture and the Agro-Food Sector. Center for Chinese Agricultural Policy, Chinese Academy of Agricultural Sciences.

Huang, J., S. Rozelle and M. W. Rosengrant. 1995. Supply, Demand and China's Future Grain Deficit. Food, Agriculture and Environment Discussion Paper, International Food Policy Research Institute, Washington,

Huang, J., S. Rozelle and M. W. Rosegrant. 1997. China's Food Economy to the Twenty-First Century: Supply, Demand and Trade. Food, Agriculture and the Environment Discussion Paper 19, IFPRI.

Huang, J. and L. Zhang. 1996. The Macroeconomic and Public Expenditure Implications of Grain Policies in China: Some selected issues for discussion. Workshop on Agricultural Policies in China, OECD Headquarters, December 12-13.

Huang, J., and H. Bouis. 1996. Structural Changes in the Demand for Food in Asia. Food, Agriculture and the Environment Discussion Paper 11. International Food Policy Research Institute.

Huang, J., M. Rosegrant and S. Rozelle. 1996. Public Investment, Technological Change and Reform: A Comprehensive Accounting of Chinese Agricultural Growth. China Workshop - Global Agricultural Science Policy for the 21st Century, p. 57.

Huang Yanxin (Policy and Law Dept., Ministry of Agriculture, PRC). 1995. Current Demand for Grain in China and Forecast Until 2000. Conference on "Grain Market Reform in China and its Implications," East and West Center, Honolulu, September 16-19.

Islam, N., and S. Thomas. 1996. Foodgrain Price Stabilization in Developing Countries: Issues and Experiences in Asia. Food Policy Review 3, International Food Policy Research Institute, Washington, DC.

Kalirajan, K. P. and Yiping Huang 1996. An alternative method of measuring economic efficiency: the case of grain production in China. The University of Adelaide, Chinese Economy Research Unit, February 14, 1996.

Lin Bao, Lin Jixiong and Li Jiakang. 1996. The Significance of Balanced Fertilization, Based on Results from Long-Term Fertilizer Experiments. In: Balanced Fertilizer Situation Report #2, Soil and Fertilizer Institute, Chinese Academy of Agricultural Sciences. Beijing, China.

Lin, J., J. Huang and S. Rozelle. 1996. OECD Forum for the Future—conference on China in the 21st Century. Long-term Global Implications, China's Food Economy: Past Performance and Future Trends. Paris.

Lin, J. Y. 1992. Rural Reforms and Agricultural Growth in China. Amer. Econ. Review 82:34-51.

Lin, Justin Yifu and Minggao Shen (Dept. of Rural Economy, Development Research Center of the State Council, PRC). 1994. Rice Production Constraints in China: Implication for Biotechnology Initiative.

Lin, X., C. Yin, S.D. Xu. 1996. Maximizing Sustainable Rice Yields Through Improved Soil and Environmental Management. Proceedings of the International Symposium November 11-17.

Mei Fangquan. Sustainable Food Production and Food Security in China.

Mei Fangquan. Prospects and Strategic Options of Food and Agricultural Production in China.

Potash & Phosphate Institute and Potash & Phosphate Institute of Canada. 1996 Interpretive Summaries: North American and International Research Projects.

Pray, C. E., and Z. Ahmed. 1990. Research and Agricultural Productivity Growth in Bangladesh and Research, Productivity and Incomes in Asian Agriculture. In: R.E. Evenson and C.E. Pray, (eds.). Cornell University Press, Ithaca.

Rae, Allan N. 1996. China's Development, Food Consumption and Livestock - Feeds Policy Interactions. Workshop on Agricultural Policies in China, OECD, Paris, December 12-13.

Rosegrant, M. W., M, Agcaoili-Sombilla and N. D. Perez. 1995. Global Food Projections to 2020: Implications for Investment. Food, Agric. and the Environ. Discussion Paper No. 5. IFPRI.

Rosegrant, M., M. Agcaoili-Sombila and N. Perez. 1997. Global Food Projections to 2020: Implications for Investment. 2020 Vision Discussion Paper Series No. 5, International Food Policy

Research Institute.

Rozelle, S., C. Pray and J. Huang. 1996. Agricultural Research Policy in China: Testing the Limits. China Workshop—Global Agricultural Science Policy for the 21st Century, p. 3.

Rozelle, Scott, Carl Pray and Jikun Huang. 1996. Agricultural Research Policy in China: Testing the Limits of Commercialization-Led Reform.

Schwartz, L. L., A. Sterns, J. F. Oehmke, and R.D. Freed. 1989. Impact Study of the Bean/Cowpea Collaborative Research Support Program (CRSP) for Senegal. Draft—Dep. Agric. Econ., Michigan State University.

Scobie, M., and I.R. Posada. 1978. The Impact of Technical Change on Income Distribution: The case of rice in Colombia. Amer. J. Agr. Econ. 60:85-92.

Simpson, J.R. 1996. Animal feed requirements and availabilities in China—Projections and policies. Workshop on Agricultural Policies in China, OECD Headquarters, December 12-13.

Song, H. 1996. China's Agricultural Infrastructure Construction and Investment Policy. Workshop on Agricultural Policies in China, OECD Headquarters, December 12-13.

Stauffer, Mark D. and James Beaton (Potash & Phosphate Institute of Canada). Importance of Plant Nutrients in Increasing Agricultural Productivity—Chinese Experience.

Tang Renjian. The Reform on Grain Policy: Difficulties and Objectives—Reflections on China's Grain Problem.

Tang Renjian. 1996. The Reform of the GCS in China: Present Situation, Target and Ways. Beijing, China.

The People's Republic of China. 1996. Food Security in China. Contributed to the World Food Summit.

Thirtle, C., and P. Bottomley. 1988. Is Publicly Funded Agricultural Research Excessive? J. Agr. Econ. 31:99-111.

Timmer, C. Peter. 1990. How does Indonesia Set Its Rice Price? The Role of Markets and Government Policy. International Conference on the Economic Policy Making Process in Indonesia, Bali.

Timmer, C. Peter. 1991. Food Price Stabilization: Rationale, Design and Implementation. In: D.H. Perkins and Michael Roemer, (eds.), Reforming Economic Systems in Developing Countries. Harvard Institute for International Development,

Timmer, C. Peter. 1996. Does BULOG Stabilize Rice Prices in Indonesia? Should It Try? Bulletin of Indonesian Economic

Studies 32: 45-74

Tin Nguyen, Cheng Enjiang and Christopher Findlay (The University of Adelaide, Chinese Economy Research Unit). 1996. Land fragmentation and farm productivity in China in the 1990s.

USDA. 1996. Long Term Projections for International Agriculture to 2005. ERS staff paper No. 9612, Economic Research Service, United States Department of Agriculture, Washington, D.C.

Wang, Zhi and J. Kinsey. 1994. Consumption and Saving Behavior Under Strict and Partial Rationing. China Economic Review 5:83-100.

Wen, G. J. 1993. Total Factor Productivity in China's Farming Sector: 1952-1989. Economic Development and Cultural Change, 42:1-41,

World Bank. *Commodity Markets and the Developing Countries.* International Economics Department, various issues.

World Bank. 1997. China—Forward with One Spirit: A Strategy for the Transport Sector. Report No. 15959-CHA, Infrastructure Operations Division , China and Mongolia Dept., EAP Region.

World Bank. 1993. China—Grain Distribution and Marketing Project Staff Appraisal Report. Report No. 11671-CHA, Agriculture Operations Division, China and Mongolia Dept., EAP Region.

World Bank. 1993. China—Yellow River Basin Investment Planning Study." (Two Volumes) Report No. 11146-CHA, Agriculture Operations Division, China and Mongolia Dept., EAP Region, June 30.

World Bank. 1991. China—Managing an Agricultural Transformation, Grain Sector Review. Report No. 8652-CHA; Agricultural Operations Dept., Asia Regional Office.

Wu, Harry X. and Christopher Findlay. 1996, Grain supply and demand in China and its implications. Prepared for the International Symposium on Food and Agriculture in China: Perspectives and Policies, October 7-9, Beijing, China.

Wu, H. X., and C. Findlay. 1996. China's Grain Demand and Supply: Trade Implications. Workshop on Agricultural Policies in China, OECD Headquarters, December 12-13.

Wu, Y., and H. Yang. 1995. Growth and Productivity in China's Agriculture: A Review. Working Paper Series, 95/2, Chinese Economy Research Unit, The University of Adelaide, Australia.

Zhang, L., J. Huang and S. Rozelle. 1996. Land Policy and Land Use in China. Workshop on Agricultural Policies in China, OECD Headquarters, December 12-13.

Zhang, L., J. Huang and S. Rozelle. 1996. China's Agro-Environmental Policies and Measures for Sustainable Agriculture. Workshop on Agricultural Policies in China, OECD Headquarters, December 12-13.

Zhu, D. Qu and Xing Naiquan. New Development on Agricultural R & D Policy in China.

CHAPTER 4.2

Agricultural Policy, Development and Food Security in China

Jikun Huang
Center for Chinese Agricultural Policy

Food security was perceived as not only supply problem, but also a poverty issue. It was recognized that three conditions need to be met: ensuring adequacy of food supply or availability, ensuring stability of supply and ensuring access to food at the household level, particularly by the poor. China's long pursuit to produce enough food for its growing population has highly recognized by international communities. Having over fifth of the world population with only one-fifteenth of the world's arable land, China has been able to achieve its highly self-sufficiency in food.

China was a net exporter of food and even grain in 1950s. Although China began to import grain more than its exports since 1960s, the shares of grain net imports to total domestic consumption were marginal and reached about three percent by the early reform period (1978-84), declined to about one percent only in 1985-90. Except for a record level of nearly 20 million tons of grain net import in 1995, China in fact was net grain exporter with annual net export of more than 5 million tons in 1992-94. While there might be a slightly growing trend in China's grain imports in the coming decades, China has developed its position as a net exporter of food (food here includes both grain and non-grain foods) in value terms during the reform period as China also exports significant amount of foodstuffs with a high added value (i.e., livestock products and other processed foods). Starting from almost balance trade (not in quantity but in value term) in food sector in 1980, net export in food was 2.3 billion US dollars in 1985 and reached the peak level of 3.8 billion US dollars in 1995 (Lu, 1996).

On the other hand, these is also a growing concern of the future food security in China. Firstly, while there has been a general increase in food production in China over last several decades, the year-to-year fluctuations of food supply and prices have also been significant during the transition period of the economy. Market stabilization and food price inflation have been among the major concerns of government policy since late 1980s. Chinese government considers maintaining a comparatively high level of food self-sufficiency as well as stability of food supply and consumer's prices as a matter concerning national security and stability: "Only when the Chinese people are free from food availability and stability of food supply worries can they concentrate of and support the current reform, thus ensuring a sustained, rapid and healthy development of the economy" (The State Council, 1996). To this end, a series of measures are adopted recently in order to stabilize domestic food supply and market through both administrative and economic interventions on factor market, rural infrastructure development, food distribution and marketing system, national and local food reserve schemes, price regulation, international trade, etc.

Secondly, although the national economy has been keeping a strong growth, it remains uneven across regions. Farmer's income in the central and eastern parts of China continued to experience faster growth, while western and southwest regions followed a slower growth pattern as in the previous years. The income distributions among regions, between the rural and urban areas, and within region have been worsening (MOA, 1996).

Thirdly, and more importantly, availability of supply will be a major component of China's food security in the coming decades. Although the rates of food production growth in worldwide exceeded that of population growth in recent decades has implied increased per capita food availability, and real international food and feed grain prices have declined for several decades further implying that supply increases have exceeded increase in effective demand, the situations differ from country to country.

China is still facing great challenge of feeding its growing population using its declining and limited resources of land, water and other resources for food production. The importance of availability of food supply in China is due not only to its a large proportion of the world's population and consumption, but also to rapid industrialization which led to a competition of resource uses between agriculture and non-agriculture on the supply side, and to strong income growth, rapid urbanization, and increasing population which will lead to significant increase in demand for agricultural products on consumption side.

Because grain (both food and feed grains) is a major component of food economy in China, the focus of the food economy and food security issues has mainly been related to grain economy by both the government and scholars. Grain fundamentalism, the provision of adequate cereal grain supplies at low stable prices for urban residents, historically has been an overriding concern of government and recurring food shortages, particularly the famine of the early 1960s, exacerbated the concern for assured and secure grain supplies. This concern, coupled with rapidly increasing incomes and associated consumption pattern shifts, rapidly industrialization and urbanization, domestic infrastructure and transport constraints, and domestic grain price spikes in 1994/1995 as well as Brown's 1995 prediction of China's massive imports of grain in the coming decades was galvanized and addressed worldwide attention recently.

The sheer size of China's economy, its rapid growth, its gradual progress toward a market-oriented and increasing integration into the global economy, its urbanization, the shifting of comparative advantage in the economy from agricultural to the other sectors, dietary diversification, and diminishing agricultural land, water and labor resource to support its growing population, will make it a crucial player in the future development of world markets for inputs and outputs of food and agricultural products, agribusiness, and industry. Small adjustments in the country's food supply and demand, any changes in demand for agricultural inputs as well as the modes or approaches selected by Chinese government in dealing with its food security (i.e., how will China be willing to reliance on trade to meet domestic demand) will have large implications for world agricultural trade. This makes China's long-term food security an issue of both national and international significant.

The purposes of this paper are to review the food security situations, the performance of food and agricultural sector, and the role of food policies in improving the food situations in China, and to identify key issues related to food security which need further interventions. In the next section, previous achievements and sources of growth in agricultural production and food security in China are reviewed. In the third section, current government's policy and programs and their impacts on agricultural production and food security are analyzed. An outlook of

China's food economy then is made in the fourth section. The major challenges and constraints to agricultural production and food security in China are discussed in the fifth section. In the final section, China's experience and lesson in improving its food security are discussed, the issues and constraints are summarized for further study on China's food policies.

Food Security and Performance of Food and Agricultural Sector

Food Security: An Overview

China, with more than 1.2 billion people in 1995, is the world's most populous country. China is highly acclaimed for its ability to feed its over-fifth of the world population with only about 7 percent of the world's arable land. The Chinese experience demonstrate the importance of technological development, institutional changes, incentive structural improvement, rural development and other conducive policies in improving food security situation in a country with limited natural resources.

Availability of Food

China has experienced substantially increase in per capita food consumption over last several decades. Average per capita food availability increased from less than 1700 k calories in 1960 to over 2700 k. calories per day in 1993 (Table 4.2.1), nutrient availability is based on food balance sheet which may under estimate grain for feed use, therefore the figures in Table 4.2.1 are large than those presented in the following several tables which are based on food nutrition intake survey. This increase in food availability was almost achieved exclusively through increases in domestic production. During the same period, protein intake and fat consumption per capita per day also increased from 42 grams to 70 grams, and 17 grams to 58 grams, respectively. These exceed average nutrient availability in countries with a per capita GNP comparable to China's.

In additional to absolute nutrition levels, nutrient sources are the other important indicator of the economic well-being of the population. The average diet of the people in low-income countries constitutes a high percentage of nutrients from crop sources, while a typical developed country population has a mixed diet with a relatively high percentage of

Table 4.2.1. China's average per capita daily nutrient availability and sources, 1960-93.

	Nutrient Availability			Nutrient Sources					
				Crop Products			Animal Products		
Year	Energy	Protein	Fat						
	(K. Cal)	Grams	Gram	Energy (%)	Protein (%)	Fat (%)	Energy (%)	Protein (%)	Fat (%)
1960	1676	42	17	97	93	76	3	7	24
1970	2087	53	23	96	93	67	4	7	33
1980	2470	64	32	94	90	60	6	10	40
1990	2547	70	54	89	81	53	11	19	47
1993	2757	70	58	86	na	na	14	na	na

Sources: data for 1993 are estimated by the Ministry of Agriculture , the others are from Wang, Jensen, and Johnson, 1993.

nutrients from animal sources. Table 4.2.1 shows that the change in the nutrition composition of diet has been consistent with the increasing income of the Chinese population. Traditional diets have typically been based on cereals, vegetables, and small quantities of meat and fish. With rapid increase in per capita income, urbanization, and market expansion, the consumption of more expensive non-cereal food items, particularly the average consumption of livestock products and fish have increased over time (Huang and Bouis, 1996).

Aggregate Household Food Security

The status of aggregate household food security also improved significantly over time. According to FAO's WFS (1996), the household food security situation in China from the beginning of the 1970s as measured by the aggregate household food security index (AHFSI) and the level of food inadequacy have followed closely those of average availability at the national level. The status of AHFSI increased from a low level of 70 percent in 1969-71 to a relative high level of nearly 80 percent in 1990-92, while the food gap (food inadequacy) declined from being slightly higher than 14 percent to being about 3 percent over the same period.

Household Food Security by income group

Tables 4.2.2 to 4.2.4 show the household nutrient intake and sources by income group based on food and nutrition survey conducted by the Chinese Academy of Preventive Medicine (CAPM) and the State Statistical Bureau (SSB) in 1990, and food consumption and expenditure survey by SSB for the years of 1985-92.

Table 4.2.2 shows that while the average per capita energy intake in all sample provinces reached levels more than 2200 k.calories, there is a wide range of nutrient intake by income group. The 25 percent of the population with annual per capita income less than 500-700 yuan (varying among provinces) their daily energy intakes were below 2200 k.calories. Energy intake by the poorest (bottom 10 percent) was only about 82 percent level of the sample average. A similar pattern is also evidenced for protein and fat intakes. Table 4.2.2 also shows that the nutrient intakes are strong linkage with income. To meet recommended daily energy intake level of 2200 k.cal, the income of the poorest 10 percent should be doubled. Table 4.2.3 gives some indicators which could have impacts on household food security over time for the rural population in general and the poor in particular. The improvement in the household food security is evidenced by two major factors: the strong income growth and reduction of poverty population in rural area. Annual per capita rural real income raised by nearly 4 times during reform period (1978-95). In additional to the growth in agricultural sector, the growth in rural non-agricultural income has been the other major source of rural income growth. Ratio of agricultural (crop, forest, livestock, fishery) to non-agricultural income declined from 5.6 in 1978 to 1.7 only in 1995 (Table 4.2.3).

Poverty is among the important causes for undernutrition and therefore food security, improvement in food security situation is also evidenced from a significant reduction of poverty in China. World Bank early estimates show the number of absolute poor to have declined from roughly 260 million in 1978 to 96 million in 1985, or from about one-third to 12 percent of the total rural population (World Bank, 1992). By 1995, the figure further decreased to 65 million, only about 7.6 percent of total rural population (MOA, 1996).

A Brief Summary of China's Approaches

The China's experience demonstrates the importance of technology development, incentive system, institutional reform, rural economy development and other conducive policies in increasing the availability of food for its increasing demand from the growing population and income. Technology has been the engine of China's agricultural economy growth (Stone, 1988; Huang and Rozelle, 1996). China's technological base grew rapidly during both the pre-reform and reform periods.

For example, development of shorter growth season, photosensitive varieties that allowed for more intensive cultivation of the land dramatically raised multiple cropping index and increased annual grain production at given land in 1960s and 1970s. Hybrid rice, a breakthrough pioneered by Chinese rice scientists in the 1970s, increased yields significantly in many parts of the country, and rapidly spread to nearly one-half of China's rice area by 1990. Other grains enjoyed similar technological transformation (Stone, 1988).

Institutional arrangement and government food policies also played important role in China's food production and availability of food for the whole society. Prior to the economic reform (before 1979), this was achieved by centralized plan economy in the expense of economic efficiency. China had adopted an industrial-biased development strategy since the early 1950s. To implement this development strategy, the first concern was accumulation of capital for its industrialization. This was done by keeping low wage for industrial workers with low food prices. Under the socialist plan economy, low wage and low food price system were obtained by prohibiting free market operation and state compulsory agricultural procurement, and food rationing scheme in the market and distribution side, and planned economy in the production side.

The industrial-biased development strategy and this strategy oriented other macro- and sector-specific policies have several implications to the food security at the national and household levels. Equity on income distribution was achieved at expense of efficient lost within rural and urban areas, but more favor to urban consumers.

Table 4.2.2. Per capita daily nutrient intake by income group in China sample provinces, 1990.

Income group	Sichuan	Ningxia	Hebei	Zhejiang	Guang dong	Beijing
Energy intake (K.cal)						
Average	2335	2402	2227	2460	2425	2309
Poorest 10%	1889	1819	1970	1971	2129	1960
10% - 25%	2068	2142	2093	2217	2174	1855
25% - 50%	2271	2319	2201	2314	2191	2091
50% - 75%	2485	2480	2256	2559	2583	2371
75% - 90%	2606	2642	2284	2711	2532	2605
Richest 10%	2852	3074	2559	2983	2797	2972
Protein intake (g)						
Average	59.6	68.5	69.1	63.9	60.7	69.1
Poorest 10%	48.1	54.9	59.4	51.0	51.9	57.9
10% - 25%	52.5	64.6	64.7	57.0	53.0	55.0
25% - 50%	55.4	65.4	67.9	59.6	58.4	62.0
50% - 75%	61.1	68.7	70.2	65.6	64.5	70.3
75% - 90%	66.6	72.5	71.4	70.7	64.7	79.2
Richest 10%	73.6	84.5	80.8	79.2	71.8	90.5
Fat intake (g)						
Average	36.0	33.7	34.0	33.1	39.1	45.5
Poorest 10%	25.5	21.5	27.4	20.2	28.9	35.5
10% - 25%	27.7	25.6	29.2	26.4	32.7	35.4
25% - 50%	31.3	31.1	31.9	30.3	36.8	41.1
50% - 75%	37.0	35.6	35.0	35.1	42.5	48.5
75% - 90%	43.4	41.0	37.1	39.8	42.0	52.6
Richest 10%	57.0	48.9	46.1	47.6	50.6	63.9
Annual per capita income (yuan)						
Average	558	643	658	991	1027	1270
Poorest 10%	251	230	204	299	433	488
10% - 25%	347	321	349	508	624	705
25% - 50%	443	457	499	739	809	1012
50% - 75%	579	686	692	1058	1075	1346
75% - 90%	756	975	948	1457	1433	1748
Richest 10%	1115	1395	1453	2163	2033	2633

Source: Chen et al, 1994.

Table 4.2.3   Income, equity, food budget share and poverty in rural China, 1978-1995

	Real per capita income index		Agri/non -agri income ratio	Gini coefficient	Frequency of poverty (%)	Food budget share (%)	
	Rural	Poorest 20%				Rural	Poorest 20%
1978	72	na	5.6	0.21	32.6	68	na
1980	100	100	3.4	0.24	27.2	62	66
1985	189	165	3.0	0.26	12.0	58	63
1990	218	177	2.9	0.31	10.1	59	63
1995	272	193	1.7	0.33	7.6	59	na

Sources: Computed based on rural food and consumption survey data, SSB and Statistical Yearbook of China, 1996.

Urban resident's food supplies, and to less extent of rural food deficit region, were guaranteed by the urban food rationing (food stamp) system. This system provided a minimal amount of energy and protein requirements for various urban residents based on their age, sex, occupation and allocation. In rural food surplus area, food consumption was determined by the local production and the state food procurement which were fixed for three to five years based on the previous production levels. Rural household food consumption, therefore, was strong linkage with the local production level.

The low economic growth as a result of economic inefficiency prevented large reduction of rural poverty and farmer's income growth. In 1978, China remained one third of total population under the poverty line.

After 1979, China began to adopt institutional and economic reforms to shift from a socialist to a more open, market-oriented economy. Since then, remarkable progress has been achieved in the performance of agriculture and food production as various reform measures gradually liberalized the institutional and market structure of production and consumption. The growth rate of gross domestic product (GDP) during the reform period is roughly double that of the immediate pre-reform period of 1970 to 1978 (Table 4.2.4). GDP has expanded at an average annual rate of more than 9 percent over the last 15 years. The growth rates of food and agricultural production nearly tripled and per capita consumption was four times higher (Table 4.2.4).

Given the limited natural resources, particularly arable land, China realized from very beginning of the reform that agricultural growth is a necessary but not a sufficient condition for booming rural economy and raising farmer's income. Efficient gain from the institutional reform resulted in increasing surplus of rural labor. Internationally, a number of approaches have been followed to deal with agricultural labor surplus, these include migration from rural to urban areas, employment generation in the rural areas (within both the agricultural sector and in the non-agricultural sector), and a combination of these two approaches. The recent history of many developing countries has shown the importance of expanding industry in rural areas to generate employment for increasing agricultural surplus labor. Among them, China is one of the most successful examples of this kind of development.

Table 4.2.4. The annual growth rates (%) of China's economy , 1970-95

	Pre-reform	Reform period	
	1970-78	1978-84	1984-95
Gross domestic products[a]	4.9	8.5	9.7
Agriculture	2.7	7.1	4.0
Industry	6.8	8.2	12.8
Service	na	11.6	9.7
Foreign trade	20.5	14.3	15.2
Import	21.7	12.7	13.4
Export	19.4	15.9	17.2
Population	1.8	1.4	1.4
GDP per capita	3.1	7.1	8.3
Consumption per capita	1.6	8.0	6.8
Rural	0.6	9.2	4.9
Urban	3.7	5.2	7.6

a: Figure for GDP in 1970-78 is the growth rate of national income in real term. Growth rates are computed using regression method GDP and per capita consumption growth rates refer to the value in real terms. Source: SSB, Statistical Yearbook of China, various issues; A Statistical Survey of China, 1996.

In China, the rural township and village enterprises (TVEs, including all enterprises in rural area own by township, village, and below village levels such as private enterprises and others) expanded at a remarkable rate after rural reform started in 1979 and now play a substantial role in

China's economy and farmer's income. By 1994, rural TVEs accounted for about three-quarters of rural gross output values. TVEs dominate many industrial sectors such as textiles, farm machinery and equipment, other simple machinery, construction materials, food processing, and a variety of consumer goods.

Stability of food supply and access of food to the poor are the other dimensions of national food security. In this regard, the government has developed its own strong disaster relief program and large-scale food-for-work schemes. It was proven that China's own capacity to natural disasters has been quite adequate (WFP, 1994). The major constraints in stabilization of food supply in China are the poor market infrastructure and internal transportation.

Remained Issues

Although nutrition improvement has taken place in the past, malnutrition remains, particularly in the inland and the poor. Children are among the most vulnerable and affected groups. According to a recent survey conducted by Chinese Academy of Prevention Sciences, there has been also significantly improvement in nutrition security. On the average, about 15 percent of all children under five are underweight, about 30 percent stunted and 3 percent wasted, though these levels are considered as relatively low comparing to the average of the developing countries. About 25 percent of the population are considered to be at risk of Vitamin A deficit. Iron deficiency anemia is present especially in rural areas.

The income gaps among regions (inland and coastal regions) and between urban and rural have been continually increasing. Within rural area, the income gap between the poor and the rich has also been expanded since the late 1970s. The real income growth of the poor is general lower than that of the whole population (Table 4.2.3). Gini coefficient rose from 0.21 in 1978 to 0.31 in 1990 and 0.33 in 1995.

China has been successful in developing new technology, changing institutional structures, mobilizing labor and other resources to support agricultural growth. However, new technology is uncertain and subject to investment policy and on going research institutional reform. Current technology extension system is deteriorating, addition efficient gains from further institutional change are limited. Current food domestic prices are close or even higher than world market prices and protection of domestic food production could be very cost and difficult if China want to meet WTO's requirement for entry. Resources are shifting away from agriculture to non-agricultural sectors. Expansion of resource frontiers is problematic due to limited availability and costly development. A more detail of these issues as well as other constraints and challenges will be discussed in the rest part of this paper.

Changing Role of Agriculture in Economy

Agriculture has made important contributions to the development of the national economy in terms of gross value added, employment, capital accumulation, welfare of urban consumers, and foreign exchange earning. Growth in the food and agricultural economy provided the foundation for the successive transformations of China's reform economy. However, the importance of this role in national economy has declined over time. Agriculture contributed more than 30 percent to GDP and half to export earning before 1980, but fell to only about 20 percent to each by early 1990s (Table 4.2.5). It provided 81 percent of the labor employment in 1970 but only 52 percent in 1995. The decline in agriculture's importance is even more marked in international trade. The share of primary (mainly agricultural) products in total exports was 50 percent in 1980 (Table 4.2.5). By 1995, that share was only 14 percent. The share of food export as the percentage of total export fell from 16 percent in 1988 to about 7 percent in 1995, the share of food imports fell from 15 percent to 5 percent in the same period.

The declining importance of agriculture is a historical phenomenon common to all developing economies. China is densely populated. Average farm size was less than one hectares as early as the 1950s. With continued population growth on very limited land resources, the country's comparative advantage will rapidly shift away from land using economic activities such as agriculture to more labor intensive manufacturing industrial activities (Anderson 1990). Within agriculture, cropping is the dominant sub-sector, contributing 80 percent of the gross value of agricultural output in 1978 (SSB). This declined significantly to less than 60 percent in 1995. During the same period the shares of livestock and aquatic output values more than doubled. However, it is interesting to note that the decline in agricultural comparative advantage in the economy has not been fully reflected in self-sufficiency of agricultural products in China.

Table 4.2.5. Changes in structure of China's economy, 1970-95.

	1970	1980	1985	1990	1993	1994	1995
Share in GDP							
Agriculture	40	30	28	27	20	20	20
Industry	46	49	43	42	47	48	49
Services	13	21	29	31	33	32	31
Share in employment							
Agriculture	81	69	62	60	56	54	52
Industry	10	18	21	21	22	23	23
Services	9	13	17	19	21	23	25
Share in export							
Primary products	na	50	51	26	18	16	14
-Foods	na	17	14	11	9	8	7
Share in import							
Primary	na	35	13	19	14	14	18
-Foods	na	15	4	6	2	3	5
Share of rural Population	83	81	76	74	72	71	71

a: GDP data in 1970 is not available, national income was used in 1970.
Sources: SSB, Statistical Yearbook of China , various issues; MOFERT, Almanac of China's Foreign Economic Relations and Trade, various issues .

Rice self-sufficiency levels declined somewhat from 104 percent in the early 1960s to about 100 percent by the early 1990s (Huang, 1997). For other major grain commodities over the last four decades, there is no clear indication of a long term declining trend in the self-sufficiency levels. On the other hand, meat, aquatic products, vegetables and fruits have enjoyed export expansion over the last two decades (Lu, 1996).

Growth of Food and Agricultural Production

The growth of agricultural production in China since the 1950s has been one of the main accomplishments of the country's development and national food security policies. Except during the famine years of the late 1950s and early 1960s, the country has enjoyed rates of production growth that have outpaced the rise in population. Even between 1970 and 1978, when much of the economy was reeling from the effects of the

Cultural Revolution, grain production grew at 2.8 percent per annum (Table 4.2.6). Oil crop grew at 2.1 percent per year and outputs of fruits and meats grew even faster (3-7 percent).

After 1978, decollectivization, price increases, and the relaxation of trade restrictions on most agricultural products accompanied the take off of China's food economy in 1978-84. Grain production increased by 4.7 percent per year; the output of fruits rose by 7.2 percent (Table 4.2.6). The most spectacular growth was enjoyed by oil crops, livestock and aquatic products which expanded in real value terms by 14.9 percent, 9.0 percent and 8.8 percent annually, respectively.

However, as the take-off efficiency gains from the shift to the household responsibility system were essentially reaped by the mid-1980's, the growth rate of food and agriculture decelerated (Table 4.2.6). This declining trend was most pronounced among grains and oil crops, where prices and marketing continued to be highly regulated. In contrast, for other crops, livestock and poultry, and fishery products, where price and market liberalization was more advanced, and demand for these commodities was rising, growth rates generally have exhibited increased or steady rates throughout the reform period (Table 4.2.6). While dropping below the rate of growth generated in both the pre-reform and early reform period, production of rice, other grains, and cash crops has continued to expand after 1985. By 1995, grain production reached a level of 466.6 million metric tons (rice in paddy form, or equivalent to 411 million metric tons, mmt, if rice measured in milled rice form), an increase of 162 mmt or 53 percent more than in 1978. Within cereal grains, rice production increased by 35 percent, and the productions of wheat, maize and soybean all are nearly doubled during the reform period of 1978-95.

The production of oil-bearing crops reached 22.4 mmt in 1995, more than four times in 1978 with annual growth rate of about 15 percent in the early reform period (1978-84) and 4.4 percent in the post reform period (1984-1995). Sugar-bearing crop production reached 79.4 mmt, tripled during 1978-95.Livestock, fishery, and fruits generally have exhibited more significantly and accelerate growth throughout the reform period (Table 6.2.6). Past studies have already demonstrated that there are a number of factors which have simultaneously contributed to agricultural production growth during the reform period. The earliest empirical efforts focused on measuring the contribution of the implementation of the household responsibility system (McMillan et. al. 1989; Fan, 1991; Lin, 1992). These studies concluded that most of the rise in productivity in the early reform years was a result of institutional innovations, particularly rural household responsibility system which restored the primacy of the individual household in place of the collective production team system as the basic unit of production and management in rural China.

More recently, Huang and Rozelle (1996) and Huang et. al. (1996) showed that since the household responsibility system reform was completed in 1984, technological change has been the primary engine of agricultural growth. Improvements in technology have by far contributed the largest share of grain production growth even during the early reform period. The results of these studies show that further reforms outside of de-collectivization also have high potential for affecting agricultural growth. Price policy has been shown to have had a sharp influence on the growth of both grain and cash crops during the post-reform period. Favorable output to input price ratios contributed to the rapid growth in the early 1980s.

However, these new market forces are a two-edged sword. A deteriorating price ratio caused by slowly increasing output prices in the face of sharply rising input prices was an important factor behind the slowdown in agricultural production in late 1980s and early 1990s. Rising wages and the higher opportunity cost of land have also held back the growth of grain output throughout the period, and that of cash crops since 1985.

Trends in environmental degradation, including erosion, salinization, and loss of cultivated land show that there may be considerable stress being put on the agricultural land base. Erosion and salinization have increased since the 1970s. These factors have been shown to affect output of grain, rice, and other agricultural products in a number of recent studies (Huang and Rozelle, 1995; Huang et.al., 1996).

Agricultural Development Strategy, Policies and Food Security

Government Targets for Agriculture to 2000 and 2010

Food self-sufficiency has been and will continue to be the central goal of China's agricultural policy. The Ninth Five-Year Plan for 1996-2000 and National Long Term Economic Plan envisage continued growth in

Table 4.2.6 Growth rate (%) of agricultural economy by sector and selected agricultural commodity 1970-1995

	Pre-reform	Reform period	
	1970-78	1978-84	1984-95
Agricultural output value	2.3	7.5	5.6
Crop	2.0	7.1	3.8
Forestry	6.2	8.8	3.9
Livestock	3.3	9.0	9.1
Fishery	5.0	7.9	13.7
Grain production	2.8	4.7	1.7
Rice	2.5	4.5	0.6
Wheat	7.0	7.9	1.9
Maize	7.0	3.7	4.7
Soybean	-1.9	5.1	2.9
Cash crops			
Oil crops	2.1	14.9	4.4
Cotton	-0.4	7.2	-0.3
Rapeseed	4.3	17.3	5.4
Peanut	-0.2	10.8	5.2
Fruits	6.6	7.2	12.7
Red meats	4.4	9.1	8.8
Pork	4.2	9.2	7.9

Growth rates are computed using regression method . Growth rates of individual and groups of commodities are based on production data; sectoral growth rates refer to value added in real terms . Source: SSB , Statistical Yearbook of China , various issues; A Statistical Survey of China, 1996. MOA, Agricultural Yearbook of China, various issues.

agricultural production and farmer income at four percent annually, nearly maintaining food self-sufficiency levels and eliminating absolute poverty. The plans strive to achieve the following key targets by 2000 and 2010:

• Increase grain production to 490-500 mmt (rice is measured by paddy grain) with an annual increase of 10 mmt by 2000; develop sustainable growth in grain production through increases in public investment in agriculture and science and technology with a target of 560 mmt grain production by the year of 2010. Increase meat production and aquatic products by 10 million tons each by 2000;

• Increase agricultural output value (in 1990 prices) by four percent per year and maintain a similar rate of growth throughout the first decade of 21st century;

• Increase poverty alleviation funds and eliminate poverty among remaining 70 million absolutely poor by the year of 2000.

The strategy for achieving the above targets and goals includes measures to deepen rural economic and institutional reforms; improve incentives for farmers and the local government to invest in agriculture, particularly land; stimulating sustainable agricultural development through focusing on increasing the rate of regeneration of renewable resource and regulating the exploitation of non-renewable resources; provide incentives through input and output prices to increase the multiple cropping index; take new measures to strengthen the application of scientific methods to the sector; undertake structural adjustment in the rural economy; optimize agricultural production linkages; strengthen anti-poverty programs; and further open China's agriculture sector to foreign investment and improve the efficiency of foreign capita use in agriculture.

Public Financial and Investment Policies

To implement the industrial-biased development strategy, the first concern was the accumulation of capital for industrialization. The solution taken by China for the primitive accumulation was through agriculture, which accounted for nearly 60 percent of national income and employed more than 80 percent of the country's labor force in the early 1950s.

On the other hand, China also recognized the importance of agriculture and the rural sector in the development of the whole economy. Table 4.2.7 shows that the state financial expenditure on agriculture in real terms had general increased prior to the rural economic reform initiated in 1979. However, they declined in the early 1980sand did not recover to the levels of the late 1970s until the early 1990s.

Table 4.2.7. Government investment (billion yuan in 1985 price) in agriculture, 1965-95

Year	Financial expenditure			Index of Government expenditure bias	Water control investment	Agricultural research expenditure
	Total	Agriculture	% of agri.			
1965	60.3	7.1	12	26	1.3	357
1970	85.9	.5	8	19	2.3	401
1975	108.3	13.1	12.	32	3.4	539
1978	142.2	19.3	14	41	4.5	700
1980	143.4	17.8	12	34	3.2	791
1985	184.4	15.4	8	23	2.0	1104
1990	212.9	19.0	9	26	3.0	1043
1992	249.5	21.4	9	20	5.5	1323
1994	238.7	25.7	11	42	6.9	1458
1995	244.4	25.8	11	42	6.5	1474

Values are in real 1985 price. Government investment (expenditure) bias is constructed by comparing the relative size of its public expenditure in the economy with its contribution to the national income. Here we estimated the government investment (expenditure) bias using an approximation by roughly dividing the ratio of government expenditure in agriculture to total government expenditure by the ratio of net income of agriculture to total national income. If this index value equals 1, IB = 1 (or 100 percent), the investment policy is neutral, encouraging neither agricultural nor non-agricultural productions. If IB < 1, there is an anti-agriculture bias in the government expenditure (investment) policy. Otherwise, IB > 1 implies a pro-agricultural production bias.
Source: SSB, Statistical Yearbook of China, various issues; and SSTC.

The decline in government expenditure in agriculture in 1980s has addressed great attention on the sustainability of agricultural production growth and domestic food supply in the future. The investment policy was reviewed and investment was increased in the early 1990s (Table 4.2.7). Both the Ninth Five-Year Plan (1996-2000) and China's Long Term Plan to 2010 envisage the government increasing investment in agriculture, including rural infrastructure investment, and loans and credits for agricultural production. Among them, irrigation and water control are listed as the first priority of the government investments in the coming decades.

Research and Technology Development Policies

Developed from almost nothing in 1950s, China's research system grew very rapidly after 1960s. It has been successful at producing a steady flow of new varieties and other technologies since 1950s. Farmers used semi-dwarf varieties developed in China several years before the release of Green Revolution technology elsewhere. China was the first country to develop and extend varieties of hybrid rice. Chinese-bred corn, wheat, and sweet potatoes technologies were comparable to the best varieties in the world in the pre-reform era (Stone, 1988)

Since the 1980s, China has implemented a series of technology policies. Reform has attempted to increase research productivity by shifting funding from institutional support to competitive grants, supporting research focusing on problems that will be useful for economic development, and encouraging applied research institutes to earn their money by selling the technology they produce (Rozelle, et. al., 1996).

While competitive grants programs have probably increased the effectiveness of China's agricultural research system, the reliance on commercialization revenue to subsidize research and make up for falling budgetary commitment has weakened the system. Empirical evidence demonstrates the declining effectiveness of China's agricultural research capabilities (Rozelle, et. al., 1996) in early 1990s.

In realizing the recent weakening of research system and the importance of science and technology in raising agricultural productivity, the Chinese government set up several programs to stimulate agricultural technology development and farmers' adoption of new technologies. Both the Ninth Five-Year Plan and the Long Term Plan for 2010 foresee that China will rely largely on introducing new technologies to raise agricultural production, particularly new varieties of crops and livestock.

AGRICULTURAL POLICY DEVELOPMENT

Land Tenure Policy

China initiated its rural economic reform in 1979. While keeping the ownership of collective land, agricultural land within the production team was equally distributed to the team's household based on family's population or combination of population and labor. The importance of this reform on the growth of agricultural production in the early reform period has been shown in a wide range of publications in the literature. The implications of equity distribution of land to farmers for food security and poverty are also obvious.

The other recent significant change in institutional arrangements in agricultural land is the renewal of the land contract system. The new land contract introduced in 1994/95 extended the term by 30 years (or 50 years) from the expiration of the original contract. While the regime of collective land ownership remains unchanged, land-use rights can be transferred with payment. This change of policy was designed to overcome the problem of farmers being increasingly unwilling to invest in agricultural production, especially grain production, because of unclear land titles and the small scale of their holdings. It also has implications for land consolidation and the commercialization of agricultural production. A recent land tenure survey conducted by author in 8 provinces in China indicates that there would be a potential gain in agricultural production resulted from efficient use of land and other inputs and farmer's willingness to invest in agricultural land if a well defined land use right tenure system were developed. To what extent of this impact will have on agricultural production is still needed further investigation.

Food Price and marketing policies

Price and market reforms have been key components of China's development thrust as it gradually shifts from a socialist to a market-oriented economy. Although the process has been characterized by cycles of deregulation and reinstatement of controls, there has been a decisive trend towards liberalization. Moreover, nominal protection rates, while in general high, have declined rather than increased with rapid economic development, as predicted by the political economy of agricultural protection (Huang and David, 1995). China's case is unique because of its socialist history. The state not only balance the interests of producers

228

and consumers, the state itself often considers its own interest as a direct consumer of agricultural products in its own commercial enterprise, and prior to the reform period, also as a producer of agricultural products. It is ironic that in 1993, a greater degree of price liberalization has occurred on consumer and producers when the consumer is the general public. However, the grain marketing was partially re-centralized (re-imposing the state grain quota procurement) after 1994 as a result of high food price inflation in late 1993 and continued in 1994-95. This partial re-centralization of the grain procurement system reflects the importance of the other two central goals of government food policy: stabilization of food prices and the political economy of the urban consumer. Table 4.2.8 present estimates of overall efficiency of government grain price and marketing policy during 1978-92. As Table 4.2.8 shows, grain as a whole has been heavily taxed in most periods studied. However, it differs largely among commodity and over time (rice farmers have been consistently taxed, wheat and other grain productions in China were heavily subsidized before 1990. (see Huang and David, 1995, for detail). On the other hand, consumers in China have been heavily subsidized. The annual CSE due to the price policy for grain as a whole raised from 1.5 billion Yuan in 1979 to about 40 billion Yuan in the early 1990s.

Table 4.2.8 also shows that the average annual cost (or efficient lost) of government grain marketing was about 10 billion Yuan measured at the 1980 price during 1978-92, a figure almost equivalent to government total expenditure on explicit grain price subsidies in the same period and much more than the government total expenditure in the whole agricultural sector. It is also important to note that the cost (or efficiency) of centralized state grain marketing has declined (or improved) significantly over time, from more than 15 billion Yuan in the late 1970s to about 8 billion Yuan in the early 1990s, indicating a remarkable gain being realized from grain market reform. This indicates that a further liberalization of grain market should be pursued and the recent retrenchment of grain price and marketing policies in China may have raise welfare lose of the country as a whole.

Recently, there are the other two important developments in the agricultural marketing system: "two-line operation system" in grain and the provincial governor's "Rice Bag" responsibility system. The former is designed to separate policy operation of state grain procurement from market operations. The system was tested in 1994 and the government decided in July 1995 to implement the reform over all the country in

1996. This reform is an outcome of the "inability" of the State grain bureaus to stabilize grain markets. Their full commitment to profit maximization while still retaining some monopsony and monopoly power in the grain market led to contradictory incentives from their separate State mandated market trading activities. The " two-line operation system" in grain marketing may increase market efficiency and contribute to the stabilization of food price and market, but the degree of gain will largely depend on the actual implementation of the policy. The provincial governor's "Rice Bag" responsibility system seeks to set up a regional food security and grain balancing mechanism where provincial governors and governments are held responsible for balancing grain supply and demand and stabilizing local food market and prices in their respective regions.

Table 4.2.8. Estimates of the distribution of the annual burden and benefits (-cost) of price and marketing policies on grain, million yuan at constant 1980 price, 1978-92.

| Period | Public burden | | | | Benefits | | or cost (-) |
| | Domestic trade | | International trade | | | | |
	Output	Input	Output	Input	Producer	Consumer	Cost
1978-79	5436	2152	904	374	-8195	1593	-15468
1980-84	13976	1454	2674	1030	-3571	10398	-12308
1985-89	12398	508	1165	943	2048	7984	-4982
1990-92	11450	144	1312	823	-34200	39836	-8093

Figures are annual average for the period indicated.
A: Public Burden, based on estimate of public expenditure for implementing price policy; producer cost based on estimate of the PSE due to price and marketing policy; consumer cost based on estimate of the CSE due to price and marketing policy; cost refers to residual on the cost of bureaucracy.

This policy has contributed to increase in output, higher degree of stabilization in grain production, and significant reduction in agricultural price fluctuations in the short-run. However, such a policy is not without cost. Although its impacts on the integration of national grain market are minimum and not as large as most people have expected (Rozelle, et al,

1996; Yu and Huang, 1997), its impacts on the efficiency of resource allocation, diversification of agricultural production and farmer's income should not be ignored.

Input Price and Marketing Policies

Most of agricultural input prices and market have been liberalized since 1980s. But the process of price and market reform for fertilizer, one of the most important agricultural inputs, has not been completed after one- and -a-half decades. With the persisted increase in agricultural input prices, especially fertilizer price, the Chinese government issued a series of new measures to control input prices in order to balance its supply and demand.

A market retrenchment policy on fertilizer also was attempted in 1994, but it was much weaker than the grain quota procurement policies. The state readjusted the factory price of chemical fertilizer. The new fertilizer factory-gate target band was applied to control price inflation before reaching the wholesale level. Further efforts were also made to control fertilizer marketing by trading more through the state main channel. In late 1994, the State Council called a national work conference on reform of the circulation system for agricultural inputs. The targets of this policy were to strengthen macro control and market management, reduce marketing costs, and stabilize fertilizer prices.

Measures adopted by the government in 1995 and early 1996 included: a) Implementing provincial governor's responsibility system for fertilizer pricing and marketing (the provincial government would be responsible for meeting excess demand for agricultural inputs and maintaining the stability of input prices); b) Efforts to continuously control fertilizer distribution (the maximum amount of fertilizer that the government allowed to be sold directly through the non-government channels by the factories was limited to 10 percent of total production); c) Trade control (Fertilizer imports were continuously licensed and managed by appointed trade agents).

China depends heavily on the international market for fertilizer, importing about one fourth of its requirements. Domestic potassium production is extremely limited and about 90-98 percent of the potassium used are imported. Trade policy has strong impacts on domestic fertilizer price and market.

The estimates of the nominal protection rate (NPR) overtime shows that, prior to 1985, government's policy of promoting domestic production of agricultural inputs subsequently raised the domestic price paid by farmers above their world price, more than 100 percent in the early 1970s (Huang and David, 1995). That rate of penalty declined over time, mainly because the depreciation of the yuan raised the border price in domestic currency as domestic fertilizer prices have raise significantly in the 1980s. It was not until the early 1990's farmers received a significant price subsidy close to 30 percent. But as all price controls at the retail levels were lifted in 1993, the NPR for fertilizer rose again to more than 20 percent after 1993. Such input price policy has not compensated for the low output procurement price policy on many major food commodities. A recent World Bank estimate that average nominal protection for agricultural crops was -40 percent in 1993-94, when the implicit tariffs on prices of agricultural inputs, the effective rates of protection is even lower, -43 percent, suggesting that input price policies generally exacerbate the policy bias against agriculture.

Grain Reserve Policy

In order to ensure food security for China's 1.2 billion people under current economic environment, a new grain reserve system has been established in 1990 as a major government intervention on stabilization of food supply and domestic market prices. Government maintain large volumes of stocks. Precise grain reserve remains confidential but the estimates from various sources indicate that the State might maintain a "special" reserve of about 40 million tons of grain procured by the State Grain Bureau from domestic production and imports (USDA). As a part of disaster preparedness measures, the government encourages the establishment of village-level food "help-each-other" reserves. Villagers contribute a portion of their harvest to these reserves. The government also operates a large-scale food-for-work program as one of the main instruments for helping the poor and, at the same time, for constructing basic infrastructure needed for development. Meantime, a state grain market risk fund system was also established in 1994 by various levels of the governments to stabilize their local grain market.

Foreign Exchange and Agricultural Trade Policy

Significant changes in the nature and extent of government trade interventions occurred during the reform process. China's Open Door policy has contributed greatly to rapid growth and also led to increasing reliance on trade (domestic and international) to meet consumers' need. The historical overvaluation of the domestic currency due to the trade protection system has undervalued agricultural incentives. This distortion in price incentive depress agricultural production, but the degree and extent of distortions have been declining over time as a result of foreign exchange reform.

China's exchange rate policy during the reform period has clearly been successful in affecting substantial depreciation in real exchange rate (Huang and David, 1995). Whereas real exchange rates remained constant, and even appreciated over three decades prior to the reform period, it rapidly depreciated during the reform period except for a couple of years after 1985 when high domestic price inflation occurred. Within 15 years, the real exchange rate depreciated by more than 400 percent. The success of the exchange rate adjustments stemmed mainly from the productivity effects of the economic reforms and technological innovations in agriculture, foreign trade, and industry which contributed to the relatively low inflation. China was second only to Indonesia in pursuing aggressive adjustments in the exchange rate in the region over the past two decades. The favorable trends in the real exchange sharply increased export competitiveness and thus significantly contributed to the phenomenal export growth record (i.e., non-grain food products) and consequently the spectacular performance of the country in the 1980s.

The trade has also been liberalizing. Prior to the reform period, the allocation of imports and exports including foreign exchange were strictly based on administrative planning and undertaken by 12 foreign trade corporations (FTC). The process of trade policy reform has involved the introduction of greater competition in international trading and the gradual development of instruments for indirect controls. In 1984 the foreign trade system was decentralized considerably. Provincial branches of national FTCs were allowed to become independent and each province allowed to create its own FTCs. By 1986 there were more than 1200 FTCs. By the early 1990s the number was more than 3000. Foreign trade in what are considered strategic products such as food grains, textiles and fiber, and chemical fertilizer, however, continue to be restricted to specialized national trading corporations to have stronger controls on the level of exports and imports of these products.

Anti-poverty policy

Central and local governments have both sustained their strong commitment to poverty alleviation efforts. Dozens of line ministries and agencies play a role in poverty alleviation in China, including Ministries of Civil Affairs, Finance, Agriculture, Labor, Domestic Trade, Water Resources, Public Health, and the State Education Commission, the State Science and Technology Commission, the Agricultural Bank of China, and the People's Bank of China. To better organize this increasing complex poverty reduction work and in order to institutionalize the anti-poverty program, the State Council in 1986 authorized establishment of a task force and executive agency structure to be replicated at all administrative levels. At the central level is an inter-ministerial task force called the Leading Group for the Economic Development of Poor Area (LGEDPA). The anti-poverty executive agency, the Poor Area Development Office, reports directly to the State Council via the LGEDPA. This institutional set-up strengthens the leadership of the national anti-poverty program, and facilitates policy implementation.

After the mid-1980s, while continuing pre-existing rural social and relief services, China began to adopt a new poverty alleviation strategy to emphasize on economic development programs in the poor areas. This strategic policy encourages the leaders and farmers in poor areas to effectively develop the local economy with the assistance of the government and in line with market demands. This program has been referred to as moving from "blood transfusion" to "blood making" policy.

Infrastructure improvement has been emphasized through the Food-for-Work program administered by the State Planning Commission. More recently the Government's declared objective is to eliminate absolute rural poverty by 2000, substantially reduce relative poverty, and make available adequate nutritionally balanced food in all parts of the country. To this end, new poverty alleviation initiatives have been introduced. The measures included the extension and strengthening of assistance to the poorest of the poor residing in the worst physical environments; the integration of production, education, health, family planning and transport program into comprehensive local intervention packages; and initiation of a "State 8-7 Plan--- A new strategy or plan for assisting the poor" in 1994. A total of 592 counties will be assisted under the plan.

Rural Enterprise Development Policy

In China, the TVEs expanded at a remarkable rate after rural reform started in 1979 and now play a substantial role in China's rural economy. The gross output value of TVEs in real terms increased at an annual rate of 23.5 percent during 1978-95, increased from 63.2 billion yuan in 1978 to 2479.1 billion yuan in 1995. By 1995, rural TVEs accounted for about three-quarters of rural gross output values and about 40 percent of the country's export earnings (SSB, 1996). As the number and size of TVEs increased in rural areas, their development maintained momentum by absorbing surplus rural labor. The sector employed 128.6 million rural laborers in 1995, 4.5 times more than in 1978, with increase of more than 6 million annually (Huang, 1997). The development of TVEs in rural China not only generates employment for increasing labor surplus in rural areas, but also raise farmer's income, promotes rural urbanization and market development, and stimulates structural changes in the rural economy.

Policy has played an important role in setting up and encouraging the development of TVEs in China. In the early-1950s, rural industry was mainly staffed by farmers holding part-time jobs in commercialized handicraft industries. Many of these workshops were very small in scale, either because their technology or the organization of their work offered no economies of scale. During the commune era (1958-78), these workshops and individual handicraft workers were organized into a large number of commune, brigade and team enterprises. A variety of activities were profitably undertaken in this environment and were mainly arranged through local government, including the production of building materials and some farm inputs, subcontracting of various sorts from urban industry, and some simple processing of agricultural outputs.

Economic reform since 1979 has largely altered the operation of the commune-run enterprise system. The state own enterprises' (SOE) monopoly and monopsony powers were eroded during this period and opportunities in the urban economy were becoming more accessible to TVEs. With the recognition of individual and private enterprises by the central government beginning in 1984, employment in these below-village units expanded by 34.4 million during 1984-90 (Huang, 1997). Having increased at an average annual rate of 25%, below-village TVEs employment growth accounted for 85% of total TVEs employment growth during 1984-90.

TVEs emphasize the development of labor-intensive industry and have significant flexibility to respond to changing market conditions. A main feature of TVEs' development has been their attempt to accommodate the country's increasing demand for a greater variety of consumer goods using increasing surplus labor in rural area. TVEs development is also closely linked with that of urban industry in terms of products, technologies, staffing, and facilities, especially in the early stages of the development: subcontracting production from urban large industry, directly or indirectly hiring urban retired technicians in the TVEs, and purchases retired urban equipment. In 1978-84, transferring urban industry assets and retired equipment to rural area account for 35-45 percent of new assets in the TVE.

Prospects for China's Food Economy

Issues and Debates

One of the most closely watched debates by researchers of China's agricultural economy, particularly the food economy, whether inside or outside the country, address the question: Will China be able to produce most of what it needs to feed itself in the 21st century? The preponderance of evidence produced by those who have seriously addressed this debate favors the viewpoint that China essentially be able to feed itself even though imports of grain most likely will rise over the next several decades. For example, Yang and Tyers (1989) have forecast that China will import around 50 million t of grain per year in the late 1990s. Rozelle et. al., (1996) predict that China will need up to 40 million t annually to meet domestic needs in the first two decades of the next century. While China will need to import substantially more grain than in the past, most international food trade and production specialists believe such a rise in demand can be met by current suppliers without long term price increase and threatening world food security. On the other hand, Garnaut and Ma (1992) forecast that China will face up to a 90 million t shortage in grain in 2000. Brown (1995) argues that China's production will fall between 216 and 378 million t short of meeting demand. According to Brown's analysis, the imports flowing into China to meet this shortfall, financed by the foreign exchange earnings of its booming export sector, will drain world supplies, force prices up, and deny less able nations grain stocks needed to feed their populations.

The range of net import predictions is perplexing. The consequences of China emerging as either a major importer or exporter could be enormous for world grain markets and prices. China is a country experiencing rapid development and transformation. The dynamic nature of the economy and continuing reform requires that the projection should be frequently updated to better reflect changes in policies.

The goal of this section is to report our projection results which incorporate the most recent government policies based on a projection framework developed at Center for Chinese Agricultural Policy (CCAP). In this model, addition to price response of both demand and supply, a series of important structural factors and policy variables are accounted for explicitly, including urbanization and market development on the demand side, and technology, agricultural investment, environmental trends, and competition for labor and land uses on the supply side.

Recent policy trends as discussed in previous sections and other factors that will influence future supply of and demand for food are considered in the projection. These trends provide input for the growth assumptions for key parameters employed in the simulation procedure. But it should be kept in mind that there are also some other important factors which have not been included in the projection, such as availability of water for agricultural use and possible decline in return to agricultural research. A summary of key assumptions used in the projection is provided in Huang et al. (1996).

Food Requirement: Results of Baseline Projection

According to the analysis, per capita food grain consumption (all figures in this section are measured in trade grain, rice in milled form, not unprocessed grain) in China hit its zenith in the late 1980s and early 1990s. Food grain consumption per capita falls over the forecast period (Table 4.2.9). The average rural resident will consume greater amounts through the year 2000, before reducing food grain demand in the first decade of the next century. This decline in the rural areas occurs at a time when income elasticities, although lower than the late 1990s, are still positive. As markets develop, rural consumers have more choice, and will move away from food grains. Urban food grain consumption declines over the entire projections period.

In contrast, per capita demand for red meat is forecast to rise sharply throughout the projection period (Table 4.2.10). China's consumers will more than double their consumption by 2020, from 17 to 43 kilograms per capita. Rural demand will grow more slowly than overall demand, but urbanization trends will shift more people into the higher-consuming urban areas (in 1991 an urban resident consumed about 60 percent more red meat than his/her rural counterpart). While starting from a lower level, per capita demand for poultry and fish rise proportionally more.

The projected rise in meat, poultry, fish, and other animal product demand will put pressure on aggregate feed grain demand (Table 4.2.11).

Table 4.2.9. Projected annual per capita food grain consumption under alternative income growth scenarios in China, 1996-2020

Alternative	Per capita food grain consumption (kg)			
scenario	1996	2000	2010	2020
Base Line				
National	224	223	214	203
- Rural	244	246	243	239
- Urban	178	177		
Low income growth				
National	222	220	211	202
-Rural	242	243	240	237
-Urban	176	175	172	167
High income growth				
National	225	225	215	203
- Rural	246	249	246	240
- Urban	177	177	173	165

Source: Author's estimates.

In the baseline scenario, demand for feed grain will increase to 109 million t by the year 2000. By the year 2020, the projected grain needed for feed will reach 232 million t. At this rate of growth, feed grain as a proportion of total grain utilization will move from 20 percent in the early 1990s to nearly 40 percent in 2020.When considered with the projected population rates, aggregate grain demand in China will reach 450 million t by the year 2000 (Table 4.2.12), an increase of 17 percent over the level of the early 1990s (386 million t). Although per capita

Table 4.2.10 Projected annual per capita consumption of meat and fish under alternative income growth scenarios in China 1991-2020

Alternative	Per capita meat consumption (kg)			
scenario	1991	2000	2010	2020
Baseline				
Red meat	17	23	32	43
--Rural	15	20	26	33
--Urban	24	30	40	52
Poultry	2	3	5	8
--Rural	1	2	3	4
--Urban	4	6	8	12
Fish	6	10	17	28
--Rural	4	6	9	14
--Urban	12	18	28	43
Low income growth				
Red meat	17	22	27	34
--Rural	15	18	22	27
--Urban	24	28	34	42
Poultry	2	3	4	6
--Rural	1	2	2	3
--Urban	4	5	7	9
Fish	6	9	14	20
--Rural	4	6	8	10
--Urban	12	16	22	30
High income growth				
Red meat	17	25	36	53
--Rural	15	21	30	41
--Urban	24	32	46	65
Poultry	2	4	6	10

--Rural	1	2	3	5
--Urban	4	6	10	16
Fish	6	11	21	40
--Rural	4	7	12	19
--Urban	12	20	35	61

Source: Author's estimates

food demand is falling in the later projection period, total grain demand continues to increase through 2020 mainly because of population growth and the increasing importance of meat, poultry, and fish in the average diet. By the end of the forecast period, aggregate grain demand will reach 594 million t (Table 4.2.12).

Availability of Grain Supply: Results of Baseline Projection

Baseline projections of the supply of grain shows that China's producing sector gradually falls behind the increases in demand. Aggregate grain supply is predicted to reach 426 mmt (in trade weight) by the year 2000. This projection implies a rise in grain output of only about 10.6 percent over the early 1990s.

Table 4.2.11. Demand for feed grain under alternative population and income growth scenarios in China, 1996-2020.

Alternate scenarios	Demand for feed grain (millions tons)			
	1996	2000	2010	2010
Baseline	92	109	158	232
Low population growth	92	108	153	218
High population growth	93	110	163	242
Low income groth	90	103	139	189
High income growth	95	116	181	286

Production is expected to rise somewhat faster in the second and third decades of the forecast period. Mostly as a result of the resumption of investment in agricultural research during the forecast period, aggregate grain production is expected to reach 486 million t in 2010, an increase of 14 percent during the preceding 10 years; production will reach 570 by 2020, an even higher percentage increase for the decade (17 percent over the 2010 level).

Under the projected baseline scenario, the gap between the forecast annual growth rate of production and demand implies a rising deficit. Imports surge in the late 1990s to 24 million t (Table 4.2.12). After peaking in 2010 at 27 million t, grain imports remain at 25-26 million t level through 2020.

Alternative Scenarios

To test the sensitivity of the results to changes in the underlying forces behind the supply and demand balances, a number of alternative scenarios are run, altering the baseline growth rates of the key variables, including income, population, and investment in technology. The results, shown in Table 4.2.12, indicate that low population growth rates would reduce grain demand by 6 million t, 17 million t and 32 million t in 2000, 2010 and 2020, respectively, compared to the baseline, with total grain imports falling to zero by 2015. With high population growth, annual imports increase to about 30-50 million t in 2000-2010. Low income growth causes a decline in projected grain demand from 594 million t to 550 millon t, resulting in slight exports of grain after 2015. With rapid income growth, projected imports would raise significantly.

Perhaps the most important result shown in Table 4.2.12 is the very large impact of investment in agricultural research on production and trade balances. This is hardly surprising given the large contribution that agricultural research, and the technology it has produced, has made to agricultural productivity in recent years. Increases in the rate of growth in investment in agricultural research and irrigation from a baseline level of 3.5 percent to an alternative of 4.5 percent per year are projected to shift China from an import to an export position by 2020. If, instead, growth in annual investment in the agricultural research system and irrigation fell only moderately, from 3.5 percent per year (as forecast under the baseline projections) to 2.5 percent, by 2020 total production would only be 517 million t, imports under such a scenario would reach a level of 76 million t.

This level of grain imports could be expected only if there was continued decline in the growth of agricultural investment, and if the government did not respond with countervailing policy measures as import levels rose. However, agricultural research and irrigation investments have already recovered in recent years. As grain prices have risen in response to short term tightening of grain supplies, government policy makers

Table 4.2.12 Projections of grain production, demand, and net imports (million tons) under various scenarios with respect to population, income and technology, 2000-2020.

Alternative scenario	2000			2010			2020		
	Demand	Production	Net import	Demand	Production	Net Import	Demand	Production	Net import
Baseline	450	426	24	513	486	27	594	570	25
Low population growth	444	426	19	496	486	11	562	570	-8
High population growth	445	426	29	528	486	42	622	570	52
Low income growth	440	426	15	490	486	4	550	570	-20
High income growth	460	426	34	537	486	51	648	570	78
Low investment rate	450	418	32	512	462	50	593	517	76
High investment rate	454	429	22	514	507	7	596	606	-10
High income & high investment	460	429	31	538	507	31	650	606	43
Low income and Low investment	440	418	22	489	462	27	548	517	31
High income , high investment & low population growth	454	429	25	520	507	13	613	606	7
Low income, low investment & high popultion growth	445	418	26	503	462	40	573	517	56

Source: Author's estimates

have responded with promises of greater investments in agriculture. While most of the investments have been targeted at irrigation, improvements in the operations of research institutes have also been announced.

In addition to domestic investments, the government could also look to the international area for technological products that would allow China time to redevelop its agricultural research system. China is planning to initiate a technological transfer program to introduce 1000 kinds of advanced agricultural technologies from abroad. Such moves would reduce the expected decline in the growth rates of grain production, and also decrease the expected level of imports even if growth in public investments slowed.

Because high income growth may offer a high investment in agricultural technology and infrastructure, simulation results of this combined scenario indicate that the grain import could be reduced by 35 million mt (from 78 million mt to 43 million mt) by 2020 if high income growth scenario were accompanied by the high investment assumption. The import level would fall further to only 7 million mt by 2020 if the high income and high investment scenario were combined with a likely low population growth assumption (Table 4.2.12).

On the other hand, a scenario with low growths in income and investment and high growth in population could raise China's grain import from a baseline of 25 mmt to 56 mmt by 2020.

Simulations also show that production, demand, and imports are insensitive to small changes in price trends. Output price trends do affect China's grain balances, but the effects are small. For every 0.5 percent increase (decline) in the annual projected grain price trend, imports fall (rise) by 2-3 million t. Similar magnitudes are observed with changes for the price of fertilizer; by increasing (decreasing) the projected growth of fertilizer prices by 1 percent, imports increase (decrease) by 4 million t. Hence, if rising fertilizer prices are higher than grain output prices, the change in China's output to input price ratio means more imports will be required to meet the nation's projected deficit (at least through the medium run when higher imports would force prices up, offsetting part of the deteriorating part of the deteriorating output-to-fertilizer price relationship).

Finally, assuming a constant response of production to erosion and salinity as the level of environmental deterioration increases, slight increases in their trends (e.g., an increase on growth rate of 0.2 percent per year from 0.2 to 0.4) have little impact on output (a decline of only about 7 million t in 2020). Extrapolating from these results, substantial impacts would not be found until the erosion and salinity rates accelerate to growth level 5 times greater (or to 1 percent per year increases in erosion and salinity). Even at this level of environmental stress, projected grain imports in 2020 only rise to 60 million t.

Summary and Implications

In the past food self-sufficiency has been the central goal of China's agricultural policy, and it will continue to be in the future. Incorporating the future's likely agricultural policies and trends of economic development into a projection model, our projections show that under the most plausible expected growth rates in the important factors, China's imports will rise steadily throughout the next decade. By 2000, imports are expected to reach 24 mmt. Increasing imports arise mainly from the accelerating demand for meat and feed grains. After year 2000 grain imports are expected to stabilize. Supply growth is sustained with the on-going recovery of investment in agricultural research and irrigation.

There is considerable range in the projections, however, when baseline assumptions are varied in both the short-and long-run. Different rates of agricultural investment and income growth create some of the largest differences in expected imports, but this is what should be expected. While there are a few scenarios where projected levels of imports are somewhat large, from both the view point of China's own domestic needs, and relative to the size of current world market trade, there are factors which may keep China from becoming too large a player in the world market. First, world grain prices would certainly rise in the face of large Chinese imports, a tendency which would dampen Chinese grain demand and stimulate domestic supply. Second, there may be major foreign exchange constraints to importing such large volumes of grain-either government policy makers will not allocate foreign exchange for additional grain imports, or exchange rate movements will discourage imports. Third, limitations on the ability of China's ports and other parts of the nation's transportation and marketing infrastructure to handle large quantities of grains may constrain import levels. Finally, there are many political economy factors that will make China's leaders react to increasing grain shortages and food security. National defense, pride, and ideology will necessarily put a premium on maintaining a rough balance between domestic demand and supply.

Issues and Challenges

On the basis of the projection results presented in the last section, it appears that China will neither empty the world grain markets, nor become a major grain exporter if the government policies are on the right direction. It does seem likely, however, that China will become a more important player in world grain markets as an importer in the coming decades. Net annual grain import of 25 to 30 mmt is a likely level of China's annual grain imports in the coming decades and is also a figure very likely to be acceptable to Chinese government as this represents about 4-5 percent of domestic grain demand only.

However, there is also widely variation of grain net imports to China under various scenarios. Relaxing any major assumptions on policies as well as changes in other factors which have not been considered in this projection work such as the competition for agricultural water use, trade-off between importing feed grain and meat, protection of domestic grain production as domestic price raised, declining return to investment in agriculture and research, impacts of grain and fertilizer market reform on grain production, and so on, are expected to have significant impacts on China's food supply and demand. In the context of the broad definition of national food security, the availability of food supply is only a part of the story. China could face the formidable challenge of feeding its growing population (an annual increase of nearly 14 million in 1990s) and pursuing its high level of food security in the coming decades as there are growing concerns on efficient uses of the country's limited agricultural resources, difficult in overcoming the technical constraints on food production, and more complicate in formulating conducive food and agricultural policies to increase domestic agricultural production and achieve a higher level of national food security.

Resources Constraints

Declining Trend of Limited Agricultural Land

China is an extremely land scarce country. Total cultivated land was about 95 million ha in the early 1990s. Measured on a per capita basis, it has been less than 0.1 ha since the 1980s (declining from 0.13 in 1970 to 0.08 in 1995). Although these official data could be underestimated by about 30 percent (Crook, 1994), this should not imply that China's grain yields are low and China could easily reduce its future grain deficit or become a grain exporter by "realizing" this underestimation of cultivate land. Potential yield gain is very limited for those "un-reported land". Based on our most recent household survey conducted in Zhejiang, Jiangsu, Fujian and Yunan provinces in the late 1996 and the early 1997, un-reported lands mostly are under the categories of marginal, fragile, "corner", and newly reclaimed land, and mainly belong to "private plot" planted to vegetable and other minor crops for home consumption. This implies that there might be a underestimate of crop food consumption (i.e., vegetable) of rural household if the consumption data is based on national production and food balance sheet. This also implies that under-reported grain area or over-estimate of grain yield might be much lower than the aggregate figure (about 30 percent) if production data are more reliable.

It should be noted that land use efficiency in China has been very high and the potential for land reclamation and expansion of the land base is modest. This situation would be further galvanized by increasing competition for land use from non-agricultural sectors, continuing land lost through environmental degradation, and declining irrigation water supplies which might endanger the continued agricultural use of large land tracts.

Water Shortage for Agricultural Use

China is one of the most water-short countries in the world. The water shortage in China, and particularly in North China Plain and Northwest China, has been becoming more acute and water supply and demand balance has deteriorated. Rising demand for urban and industrial water supplies poses a serious threat to irrigated agriculture as the marginal benefits of water are generally believed to be higher in non-agricultural uses. Such without specific allocation mechanism much of the water would be used in alternative uses.

Attempts to improve agricultural irrigation through technical and engineering projects have been attempted by the government in recent decades, particularly before 1980. However, since 1980, the cost of new irrigation investment has increased rapidly. Existing irrigation systems have performed poorly. Many water projects have degraded through waterlogging, salinization, and mining of groundwater aquifers in China. Growth in agricultural productivity in irrigated areas has also slowed.

Shifting Labor to Non-Agricultural Sector

Increased employment opportunities in the non-cropping and off-farm sectors brought about by greater employment availability in rural non-agricultural sectors such as industry and services have led to large shifts in labor use patterns. After putting ever increasing amounts of labor into grain production in the 1950s through 1970s, labor use on all crops fell substantially from 1975 to 1994. Rice farmers use less than half as much as in the pre-reform era. Labor used to produce wheat and soybeans fell similarly (Huang and Rozelle, 1997). Most labors remained in agriculture are either aged or unskilled farmers.

Technical Constraints

Research and Technology

While research and technology have played critical role in the growth of food production in the past, recently falling budgetary commitment to research has weakened the system. A recent initiated national seed program focuses on maintaining a stable increase in crop yield and total agricultural production. However, there are a number of issues that need to be resolved before the program can be successful. These include barriers to entry, allowing the market to set prices, phasing out of subsidies, and development of intellectual property rights.China's grain yields are among the highest in the world. More research effort in future will be required to achieve a similar productivity gain as in the past.

Fertilizer

Agricultural chemical use in China is among the highest in the developing countries. Marginal gains from additional fertilizer use may be modest. More important issues of fertilizer use in China will be unbalanced use of fertilizer and inefficiency of the state monopoly fertilizer market or distribution system. China depends heavily on the international market for fertilizer, importing about 25 percent of its requirements; domestic production is extremely limited and 95 percent of the potassium used is imported. Further reform in fertilizer pricing, market/distribution, and trade are needed to improve the system.

Infrastructure

China's limited grain handling and internal transport capacity will continuously have its impacts on the stability of supply in the future. Rail transport, which is the primary mode of freight movement, is efficient but has insufficient capacity to meet transport needs. Shipment of grain (i.e., maize) from Northeast China to Southern China is more expensive than the transportation cost of import grain from abroad in South China. The volume of grain stored in deficit areas is greater than would be necessary under improved transport conditions. Any short term fluctuation of food production in a particularly area often causes fluctuation in local prices and difficult for farmer to sell their product (in case of good harvest as in 1996-97) such hurt farmer's production incentives.

Food and Agricultural Policies

China's government recognized that government policies are important in dealing with effective supply of food, stabilization of food market and access of food to poor. However, there are still many constraints and challenges facing Chinese government in dealing with food production and food security in future.

Land Tenure System

China should continue to reform its land contract system. Recent modification of land tenure may not have much impact on agricultural production and land investment in the short run, since a proper institutional and legal framework needs to be established before the market can work efficiently. Currently, legal framework related to land markets as well as other factor and product markets is not well adapted to rapidly expanding , increasingly market-based economy.

Water Management

The historical importance of managing water is facing new challenges. While individual household responsibility system reform initiated in late 1970s had increased efficiency of agricultural production and accelerated production growth, the system also brought about a number of problems in water management. The breakdown of the commune led to a decay of water management system. Current water management is commonly believed to be one of the main reasons for increasing water shortage problem, but which to date largely has been ignored by reformers.

As a result of poor management, unclear water property rights, and inadequate maintenance, irrigation systems in China were showing signs of structural deterioration and declining productivity. Unsuitable operation has resulted in damage to the irrigation system. The conflicts for water use has expanded beyond agriculture, occurring not only among farmers within the same irrigation district, but also between government departments over the right to manage water, in particular, to collect water fees and give out water withdrawal permits (Cao, 1992).

Investment Policies

Reform has accelerated the growth of agriculture since the late 1970s, but the new institutional arrangement has not provided the incentive for both public and private to invest in agricultural infrastructure, research and extension. Investment in agricultural research fell in the late 1980s thereby weakening the basis for future productivity grains, at least over the medium term. During the rural reform, the government faced the major challenge of improving the educational levels and technological capabilities of millions of farmers. Given the importance of technology in agricultural production, strengthening research, education and extension should be assumed high priority. Several programs (i.e., Harvest Plan) have been initiated, but these programs are facing the constraint of increasing shortage of funding.

Food Price and Marketing

After nearly two decades of reform, farmers are still required to deliver specified quantities of grain to the Grain Bureau at the lower prices. In addition to the 50 million tons of quota procurement, State marketing agencies are charged with capturing 80 percent of the marketed grain surpluses to ensure that government can exercise substantial control over the open market as well. Despite procurement of 50 million tons of grain at below market prices, less than one-third of these grains are sold at concessional prices to urban consumers. The balances are sold or stored for later sale at "free market" prices creating windfall gains for the Grain Bureaus which are competing with non-state traders.

Provincial Governor's Grain Responsibility System (Rice-Bag) policy initiative made provincial governors responsible for balancing provincial grains supply and demand-with urban residents as the focus of concern. These responsibilities were transmitted to lower jurisdictions and appear to create inefficiency of resource allocations.

Input Price and Marketing

While food price inflation was controlled and prices were stabilized in 1996, agricultural input prices kept raising at a rate higher than the overall price inflation. Increase in agricultural input prices was mainly driven by chemical fertilizer. This was taken by the reformer as an indicator of failure of the fertilizer liberalization policy, which led to the market retrenchment policy on fertilizer in 1994.

However, there was little evidence to indicate that the recent market retrenchment policy on fertilizer had its desired effect. Prices of fertilizer continued to increase at a remarkable rate. The actual retail price of urea reached as high as 1600 yuan per ton in most of the country in 1994 and more than 2200 yuan per ton in late 1995, compared to the 1400 yuan per ton "ceiling price" set by the government. Several questions are raised: Whether recent fertilizer price inflation is related to private sector's participation in the market or related to the state monopoly market and trade policies? Or is fertilizer price inflation due to manipulation by private traders or supply deficit due to domestic production and trade policies?

Trade

External trade policy has also undergone a cycle of liberalization and retrenchment. In 1992 import/export responsibility was partially delegated to provincial authorities for all international grain transactions. But quantity licensing and control policy was re-emphasized for major food commodities (grain).

Other important issues include trade-off of importing feedgrain and livestock, trade management system reform, and cooperation between production, domestic and international trades. As consumption of livestock products increase, it become more important for Chinese officials to know whether importing livestock products may be more efficient than growing/importing feedstuffs and producing livestock domestically. There has also been often observed conflicts between domestic and international trades, production and marketing arrangements, and input and output policies because of lack of coordination among the Ministry of Agriculture (agricultural production), the Ministry of Commerce (domestic agricultural trade and marketing), the Ministry of Foreign Economic Relation and Trade (agricultural export/import), and the National Production Materials Authority.

Poverty and Equity

Great achievement in the poverty alleviation has been made since the late 1970s. But further reduction of poverty has been proven to be difficult due to wide dispersion of the poor population in remote upland areas and the increasing income gap between the poor and developed coastal areas. The quick reductions of poverty through agricultural growth were largely exhausted by end-1984. Ministry of Agriculture estimated that the population in absolute poverty dropped by less than 30 million after 1985 and remained 65 million in rural area in 1995, accounting for 7 percent of the rural population. Most of these residual poor have remained trapped in more remote upland areas where agricultural productivity gains have proven far more problematic. More attention also needs to be focused on the problems of poor people outside the poor regions. The challenges are not only to eliminate absolute poverty, but also relative poverty.

While economy has been successful in keeping high growth, the growth also persisted its uneven pattern across regions and among income groups. The income gap among regions and between the rural and urban has not been narrowed.

Conclusion Remarks

China is the world's most populous country, and is highly acclaimed for its ability to feed its over-fifth of the world population with only about 7 percent of the world's arable land. Despite this extremely limited natural resource and doubling its population in last four decades, per capita daily availability of food, the status of household food security and nutrition all have been improved significantly. The increases in per capita availability of food were almost exclusively through increases in domestic production. The growth of agricultural production since the 1950s, particularly after 1970s, has been one of the main accomplishments of the country's development and national food security policies. China's experience demonstrate the importance of technology changes, institutional reform, incentive system, rural development, and sector specific policies in ensuring adequacy of food supply through increases in domestic production.

China's research system has been successful at producing a steady flow of new varieties and other technologies since 1950s. China's robust growth in the stock of research capital has in significant part been responsible for these dramatic changes. Improvements in technology have by far contributed the largest share of grain production growth, such are major sources of increases in food availability in China. Institutional reform and changes in incentive structure also brought about significant growth in agricultural production. Investment in irrigation is another important determinant of China's agricultural growth in recent decades.

China's experience show that the strong rural economic growth is a key factor for significant reduction of absolute poverty. Broad participation in rural economic growth brought about a tremendous reduction in absolute poverty during the period 1978-85. The country has also succeeded in creating jobs in both farming and non-farming, particularly rural industry, for rural surplus labor since early 1980s. Diversification of production and development of rural enterprises (TVEs) stimulate the growth of farmer's income and improve farmer's living standard.

Grain reserve system, strong disaster relief program and large scale of food-for-work scheme contributed substantially to the stability of food supply and access of food to the poor. China's experience also demonstrates the important relations between economic growth, political and social stability, and food security in a large country where the economy is undergoing a fundamental adjustment and growing rapidly. Maintaining the relative high level of food self-sufficiency is desirable for ensuring not only domestic but also global food security.

In the past food self-sufficiency has been the central goal of China's agricultural policy, and it will continue to be in the future. The Ninth Five-Year Plan for 1996-2000 and National Long Term Economic Plan envisage continued growth in agricultural production and farmer income, and maintaining a high level of food self-sufficiency and eliminating absolute poverty. The strategy for achieving the above targets includes measures to provide more incentive structure and mechanism for both public and private to increase productivity enhanced investments in agriculture (investment in land, irrigation, research, extension etc.), to increase the terms of trade through price and marketing reforms, to raise farmers' income through promotion of rural industrial development and non-farming employment, to strengthen anti-poverty programs by raising poverty alleviation funds and emphasizing on economic development program as well as infrastructure development, and to enhance the state grain reserve system as a major government intervention on stabilization of food supply and market prices, so on. Projection results presented in this paper which incorporate the most recent government policies and economic trend discussed above show that China will neither empty the world grain markets, nor become a major grain exporter. It does seem likely, however, that China will become a more important player in world grain markets as an importer in the coming decades. Increased grain imports on the part of China will, for the short term, benefit grain-exporting countries, especially those dealing with wheat and maize. Net grain imports of 25 to 30 mmt is a very likely level of China's annual grain imports in the coming decades. The increase in net grain import is due mainly to the accelerating demand for meat and feed grains as well as from the continued slowing of supply as a result of reduced agricultural investment in 1980s. China's grain economy is becoming increasing animal feed-oriented one. After the year 2000, however, imports are expected to stabilize as the domestic supply growth rate is sustained with the recovery of investment in agricultural research and irrigation.

On the other hand, China could face the greater challenge of feeding its growing population and pursuing its high level of food security in the coming decades if future policies would not be in the right direction. In the worst case scenarios China will need 70-80 million tons. Other scenarios are less dramatic. Although it is not the case that China does not have the capacity to feed its people, rather it is a matter of economics and how much the Chinese government is willing to invest in these areas. There is reason for optimism, but there is also reason to worry.

There are growing concerns on efficient uses of the country's limited agricultural resources, difficult in overcoming the technical constraints on food production, and much complicate in formulating more conducive food and agricultural policies to increase domestic agricultural production and levels of national food security in the coming decades.

China is land scarce and water short country. This situation would be further galvanized by increasing competition for land and water use from non-agricultural sectors, continuing lost through environmental degradation, and efficient lost due to poor land and water management. Agricultural labors are shifting to non-agricultural sector with increased employment opportunities in the non-cropping and off-farm sectors. The unbalanced use of fertilizer, inefficiency of the state monopoly fertilizer market and distribution system, and to some extent fertilizer import trade policy are the other key issues which have not been addressed much attention by policy makers.

Recently falling budgetary commitment to research has weakened the system. Development of efficient seed industry is not without problem as appropriate supporting institutions (i.e., eliminating entry barriers, allowing the market to set prices, phasing out of subsidies, and development of intellectual property rights) need time to create. Poor land/water management and unclear land/water property rights were showing signs of structural deterioration and declining productivity. A better incentive system is needed to encourage both public and private investment in agriculture. China's constraint on limited grain handling and internal transport capacity will be one of predominant factors shaping China economy.

Great achievement in the poverty alleviation has been made since the late 1970s. But further reduction of poverty has been proven to be difficult due to wide dispersion of the poor population in remote upland areas. Although nutrition improvement over the last two decades has taken place, malnutrition remains. In sum, China is capable to solving its food problems in general and there are great roles which the China's government could play in improving national and global food security. In this regard, a full knowledge about the present policies and their impacts on food production, stability of supply, trade, and household food security are essential in designing a more conducive policy measures and institutional framework to improve the level of food security. The understanding of China's agricultural policies and their impact on the economy will not only be the interest of China, but also the rest of the world as China became integrated into the world market.

References

Anderson, Kym. 1990. Comparative Advantages in China: Effects on Food, Feed, and Fibber Markets, Paris: Organization for Economic Cooperation and Development.

Brown, Lester. 1995. Who Will feed China? Wake-up Call for a Small
    Planet. The Worldwatch Environmental Alert Series, Worldwatch
    Institute.
Carter, Colin and Zhong Funing. 1991. "China's Past and Future Role in
    the Grain Trade." *Economic Development and Cultural Change*
    39(July 1991):791-814.
Crook, F. 1994. "Could China Starve the World? Comments on Lester
    Brown's Article." *Asia and Pacific Rim Agriculture and Trade
    Notes*. ERS, United States Dep. Agric., September (1994):17-19.
Fan, S. 1991. "Effects of Technological Change and Institutional Reform
    in Production Growth of Chinese Agriculture." *American Journal
    of Agricultural Economy*, May.
FAO [Food and Agriculture Organization of the United Nations], 1996.
    Success Stories in Food Security, WFS 96/TECH/11, A Technical
    Paper Prepared for the World Food Summit, November, 1996,
    Rome.
Garnaut, Ross, and Guonan Ma. 1992. *Grain in China: A Report*.
    Canberra, Australia: East Asian Analytical Unit, Department of
    Foreign Affairs and Trade, 1992.
Huang, Jikun. 1997. "China's Rural Enterprise Development and
    Implications for Developing Countries". A Paper presented at
    International Policy Consultant Workshop, Hanoi, March 17,
    organized by World Bank and FAO.
Huang, Jikun and Christina C. David. 1995."Price Policy and
    Agricultural Protection in China," A report to FAO.
Huang, Jikun, Mark Rosegrant, and Scott Rozelle. 1996. "Public
    Investment, Technological Change and Agricultural Growth in
    China." *IFPRI 2020 Discussion Paper*, 1996, IFPRI,
    Washington D.C..
Huang, Jikun and Scott Rozelle. 1995. "Environmental Stress and Grain
    Yields in China." *American Journal of Agricultural Economics*.
    77(1995):853-864.
_____. 1996. "Technological Change: Rediscovering the Engine of
    Productivity Growth in China's Agricultural Economy." *Journal
    of Development Economics* 49(1996):337-369.
Huang, Jikun, Scott Rozelle and Mark Rosegrant. 1996. "China's Food
    Economy to the 21th Century: Supply, Demand, and Trade",
    IFPRI 2020 Discussion Paper, 1996, International Food Policy
    Research Institute, Washington D.C..
Li Peng, 1996. GUANYU GUOMINJINJI HE SHEHUI FAZHAN
    JIUWU JIHU HE 2010 NIAN YUANJING MUBIAO

GANGYAO DE BAOGAO [A Report of National Economy and Social Development for the Ninth Five-Year Plan and 2010 Long Term Goals], People's Press, Beijing.

Lin, Justin Yifu. 1992. "Rural Reforms and Agricultural Growth in China." *American Economic Review*, 82(1992):34-51.

Lu, Feng. 1996. "China's Grain Trade Policy and Food Trade Pattern" A paper presented in Conference on Food and Agricultural Policy Challenges for the Asia-Pacific and APEC, October 1-3, 1996, Manila, the Philippines.

McMillan, J. J. Walley and L. Zhu. 1989. "The Impact of China's Economic Reforms on Agricultural Productivity Growth." *Journal of Political Economy*. 97(1989):781-807.

MOA. China's Agricultural Development Report, 1995, 1996.

MOFERT [Ministry of Foreign Economic Relations and Trade], Almanac of China's Foreign Economic Relations and Trade, and China Custom Statistics.

Rozelle, Scott, Albert Park and Jikun Huang. 1995. "Dilemmas in Reforming State-Market Relations in Rural China." Working Paper, Food Research Institute, Stanford University, 1995.

Rozelle, Scott, Jikun Huang and Mark Rosegrant. 1996. "Why China Will Not Starve the World" *Choices* First Quarter, 1996:p18-24.

Rozelle, Scott Carl Pray and Jikun Huang. 1995. "Agricultural Research in China: Growth and Reforms." Paper Presented at Global Agricultural Science Policy for the Twenty-first Century, Melbourne, Australia, 26-28 August 1996..

State Council, 1996. The Grain Issues in China, Information Office of the State Council, Beijing, China.

SSB, ZGTJNJ. *Zhongguo Tongji Nianjian [China Statistical Yearbook]*. Beijing, China: China Statistical Press, 1980-95.

SSTC [State Science and Technology Commission]. *Zhongguo Kexue Jishu Ziliao Ku, 1985-90; 93* [China Science and Technology Statistical Yearbook, 1985-90; 93]. Beijing, China: State Science and Technology Commission, 1991; 1995.

Stone, Bruce. 1988. "Developments in Agricultural Technology," *China Quarterly* 116(December 1988).

World Bank. 1992. China: Strategies for Reducing Poverty in the 1990s. A World Bank Country Study. World Bank, Washington D.C..

WFP [World Food Program] 1994. "China's National Experience with Food Policies and Program," WFP Committee on Food Aid Policies and Programs, Rome.

Yang, Yongzheng and Rodney Tyers. 1989. "The Economic Costs of

Food Self-Sufficiency in China." *World Development*
17(1989):237-253.

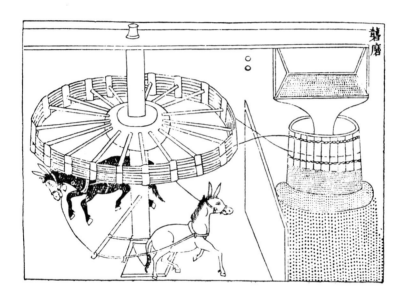

CHAPTER 4.3

Agricultural Investment in China: Measurement and Policy Issues

Shenggen Fan
International Food Policy Research Institute

Agricultural production growth in China has outperformed growth in most of the countries in the world for the last several decades despite several fluctuations. Output in the agricultural sector has grown about 4% a year from 1950 to 1995.[1] Food shortages and massive hunger once predicted by numerous foreign observers were abated and instead replaced by adequate food supply. The health and nutrition condition of a large portion of the population was improved, although sizable pockets of poverty still exist in many parts of the country.

Agricultural investment played an important role in the past rapid growth because of the priority afforded by the Chinese government to increase production. Improved technologies and the provision of rural infrastructure facilities, all of which were made also possible by the availability of investments, not only increased agricultural productivity but reduced poverty in many parts of China. The latter issue has become very important in the more recent years. Poverty alleviation is now drawing equal attention from the Chinese government as that of promoting further productivity

---

[1]There have been some arguments that China's agricultural production growth may be overestimated due to the methodology used in the aggregation of total output by the State Statistical Bureau. Fan (1997) recently used a more appropriate approach (the Tornqvist-Theil index) to measure agricultural production growth in China. The more appropriate approach would lower the estimate of annual production growth from 4.45 to 3.72, or by 0.73 percentage point from 1952 to 1995. Nevertheless, this newly estimated growth rate is still impressive and is one of the highest growth rates among all countries in the world during the same period. To be consistent with other statistical indicators in the paper, we still use the growth rates reported by the State Statistical Bureau.

enhancement in the agriculture sector. As China's economy develops, the agriculture sector faces new problems and challenges. On the one hand, food demand, particularly that for processed foods and high value products such as meat and dairy products, are continuing to grow at high rates. On the other hand, the comparative advantage of Chinese agriculture is rapidly deteriorating as resources like labor, land, and water move out of the agricultural sector. Further increases in agricultural production will be more difficult to achieve in the future. There is little doubt that investment in agriculture will remain the prime mover for China to achieve the twin goals of productivity growth and poverty alleviation. Notwithstanding this, however, trends of growth in agricultural investment has declined in the past one-and-half decades. To make the situation even worse, there is currently no formal mechanism in the government machinery that identifies priorities to allocate investment, which would likely then lead to their inefficient use. Based on this urgent issue, this study is an initial effort to analyze the historical trends and major policy issues related to agricultural investment in China. The paper starts with a review of new challenges facing Chinese agriculture in Section "New Challenges Facing Chinese Agriculture" from which we deduce that increased agricultural investment will still be a critical component in successfully meeting emerging challenges in the sector as other more traditional sources of production growth continue to get depleted. In Section "Historical Trends of Agricultural Investment", an analysis is undertaken on recent trends of agricultural investment using the data that has been recently constructed through the compilation of information from various sources. Historical trends are usually helpful benchmarks in understanding recent developments. An analysis of major policy issues related to agricultural investment is undertaken in Section on "Policy Issues" from which it could be noted that the task does not end in the search for more investment. The financing, organization, and management of agricultural investment is crucial for China's agricultural sector to achieve its national development goals more efficiently especially for a long period of time. Conclusions of the study are presented in the final Section.

New Challenges Facing Chinese Agriculture

The recent and ongoing institutional changes and market reforms within China have had substantial impacts on agricultural production growth. After the first round of institutional reforms (particularly the household production responsibility system) were instigated in 1979, Chinese agricultural output grew by 6.62% per annum from 1979 to 1984. This is a tremendous growth compared with the growth rate of 2.83% per year during the pre-reform period of 1952 to 1978. Aside from the direct effects of the reforms themselves, the increased use and improved quality of agricultural inputs as well as the new technologies and production practices have all contributed to the rapid growth. However, there are many reasons to be concerned about the ability of China to achieve the long-term goal of attaining food security and alleviating poverty since production intensification has had some negative consequences to the country's natural resources.

Over the past 40 years, China's population has grown by 1.8% per annum reaching over 1.2 billion in 1995. This rate of growth has slowed down to 1.1% in the recent years, but further reductions are more difficult. This implies that demand for food will continue to grow from population increases alone. National income grew by 10% per annum in the last 15 years. This rapid growth is expected to continue in the future. As a result, structure of food demand in China will change. The demand for meat, fish, and more convenient, higher quality food products will increase, while food grains and lower quality food products will decline in relative if not absolute terms. Recent evidence suggests that the growth-promoting effects of these early reforms have begun to moderate. In fact, the rate of production growth in agriculture has declined to about 5.0% per year since 1985.

The effects of the reforms have varied significantly across regions within China. The coastal area gained the most, while inland areas where the majority of poor people reside gained very little. More than 300 million people in China, mostly concentrated in the remote rural areas, still receive incomes of less than $1 per day.

The amount of resources available to agricultural production, particularly fertile land and water, has declined due to rapid industrialization and urbanization. Further increases in food production must come from higher yields through more intensive use of fertilizers, pesticides, irrigation, and high-yielding varieties. But the detrimental effect of the excessive use of these inputs to increase production has led to the degradation of natural resources. Moreover, poor farming practices have caused vast environmental problems that have many economic consequences.

Policymakers in China have recognized the important role of agricultural investment in meeting these challenges. But designing policies to mobilize resources to support agricultural investment and, more importantly, using these resources more efficiently are major concerns of policymakers at various levels of the government machinery. Relevant policy issues concerning the efficient allocation of public investment are discussed in more detail in Section on policy.

Historical Trends of Agricultural Investment[2]   The disparate nature of agricultural investment, both regionally and institutionally, poses substantial difficulties in assembling such data for policy analysis.   The institutional volatility that has characterized China over the past several decades presents additional difficulties in developing a historical perspective on these matters.   Nevertheless, data from officially published and unpublished materials were drawn together in an attempt to construct quantitative estimates of agricultural investment in China that are in accord with international standards.   In this paper, agricultural investment is defined in terms of the actual expenses that are used to increase long-term agricultural production capacity.

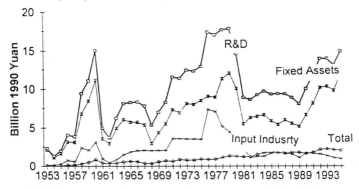

Figure 4.3.1 Government investment in agriculture, 1953-1995

All investment expenditures reported in the text have been deflated to 1990 constant yuan using the consumer price index. We will focus our attention to public investment in agriculture. However, a discussion of investment by collectives and individual farmers is also undertaken. A clear understanding of the historical development of non-government investment in agriculture could provide great assistance to policymakers in their use of public investment as an instrument to attract other sources of investment.   Investments from all sources are then aggregated to analyze the trend of total investment in Chinese agriculture. Finally, a comparison of the trends of investment in the different regions is presented and analyzed.

---

[2]The sources of the data used in this study include various publications from the State Statistical Bureau, Ministry of Agriculture, Chinese Academy of Agricultural Sciences, United States Department of Agriculture, and other published and unpublished sources. For the detailed sources of the data, refer to Fan (1996)

Public Investment in Agriculture

In order to avoid double counting and to be consistent with the internationally recognized definition, the three items are included in the estimate of total public investment in agriculture, that is, public investment in agricultural fixed assets, in agricultural input industry, and in agricultural R&D.[3] These investments are typically used by the government to promote long-run production growth and poverty reduction in rural areas.

Investment in Fixed Assets

Public investment in fixed assets refers to construction and purchase of fixed assets in monetary terms. It includes three components: capital construction investment, investment in technical upgrading and transformation, and other fixed assets investments. Capital construction investment refers to investment in new projects or expansion of existing facilities to increase new production capacity. Technical updating and transformation investment refers to investment in projects to renew, modernize or replace existing assets and related supplementary projects (excluding major repairs and maintenance projects). Other fixed assets investment refers to investment in fixed assets by government-owned units valued over 50,000 yuan (in current price) which are not included in investment of capital construction and technical updating and transformation according to the government regulations.

Capital construction investment is by far the largest among the three components, particularly in agriculture. It accounts for more than 80% of total government fixed assets investment. This investment is mainly used by the government to improve irrigation and drainage systems nationwide, to reclaim land in more remote areas, and to purchase large size machines and tools on state farms.

Public investment in fixed assets grew steadily during the first five-year plan (1953-57), averaging 2.2 billion yuan (Figure 4.3.1). During the Great Leap Forward (1958-60), however, public investment in fixed assets jumped three times, reaching an annual average investment of 8.8 billion yuan. This

---

[3]Researchers in China have different definition about public investment in agriculture, which usually includes the following items: government investment in fixed assets (long term), state financial funds for supporting agricultural production (short term), agricultural credit funds, township enterprise funds for supporting agriculture, agricultural development funds, poverty reduction funds, and foreign capital investment in agriculture. It is obvious that some items are not investment rather than day-to-day operation and management expenses, some items are double counting, and some important items such as state investment in agricultural R&D, and state investment in agricultural input industry are not included.

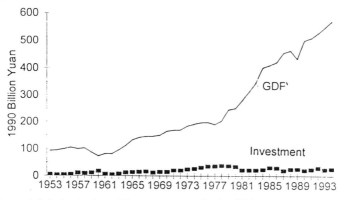

Figure 4.3.2 Agricultural investment and AgGDP, 1953-1994

unsustainable growth in investment was mainly a reflection of the government policy to reach an unrealistic goal of catching up with the growth trends in western countries in a relatively short period of time through large investments in all sectors of the economy. The adjustment made in the following three years reduced the investment to the level prior to the Great Leap Forward. From 1963 to 1979, investment in fixed assets grew smoothly except for a period of two years (1968-69) when the investment was substantially lower because of the effect of the Cultural Revolution. In 1979, investment in fixed assets reached more than 12 billion yuan. In 1981, two years after the introduction of the household production responsibility system, public investment in fixed assets declined sharply by more than 45% in a single year. This slow pace of investment in fixed assets continued until 1989. This is mainly due to the government's belief that savings out of rapidly rising agricultural incomes of the rural households from rural reforms should provide the main source of agricultural investment, thereby, offsetting the reduction in central budgetary allocations. From 1989 to 1992, investment again had a rapid growth, almost doubling in three years. This short period of rapid growth was the government's response to a relative slow down of agricultural production growth after 1985. Since 1992, investment has been sustained at the level of 10 billion yuan per year.

Investment in Agricultural Input Industries
In the Chinese national accounting system, public investment in agricultural input industry is counted as part of industry investment. This particular investment has been primarily channeled to the production of farm machinery and chemicals (fertilizers, pesticides and plastics).

Table 4.3.1 Government investment in agriculture, 1953-1995

Year	Fixed assets	Industry input	R&D	Total
1990 Yuan (in billions)				
1953-57	2.22	.39	0.08	2.69
1958-60	8.80	2.51	0.55	11.85
1961-65	4.66	1.14	0.46	6.26
1966-76	6.56	3.21	0.67	10.45
1977-85	8.16	3.07	1.36	12.59
1986-90	5.80	1.66	1.74	9.21
1991-95	10.14	1.33	2.11	13.58

Source Fan 1996

Until 1976, investment in the agricultural input industry showed an increasing trend and jumped by about five times higher during the Great Leap Forward compared to the previous period. After 1976, this trend was reversed. Since 1977, a dramatic decline occurred in the level and proportion of industrial funds received by these agriculture-support industries, in both absolute and relative terms. This low investment in the agricultural input industry resulted in the stagnation of development of domestic fertilizer and machinery supply capacity. In 1995, China had to import more than 3.7 billion yuan (or 20 million metric tons) of fertilizer from the international market. The foreign exchange used to purchase fertilizer was 30 times the total government investment in agricultural input industry in the same year.

Investment in Agricultural R&D
Public investment in agricultural R&D is accounted for in the total national science and technology budget. The sources of agricultural R&D investment are from different government agencies. Science and technology commissions at different levels of government allocate funds to national, provincial, and prefectural institutes primarily as core support. These funds are primarily used by institutes to cover researchers' salaries, benefits, and administrative expenses. Project funds come mainly from other sources including departments of agriculture, research foundations, and international donors. Recently, revenues generated from commercial activities

---

[4]The data reported here were taken from Fan andPardey (1992) and various publications from Government Science and Technology Commission and Government Statistical Bureau. Research expenditures and personnel numbers include those from research institutions at national, provincial, and prefectural levels, as well as agricultural universities (only research part).

(development income) have become a particularly important source of revenue for the research institutes. The research expenditures reported in this study include only those expenses used to directly support agricultural research.

Overall, government spending on agricultural R&D has increased dramatically during the past four decades, but not without substantial year-to-year swings. Investment in agricultural research was quite modest during the first five-year (1953-57) plan, averaging 80 million yuan per annum. During the Great Leap Forward period (1958-60), expenditures on agricultural research increased dramatically, to 550 million yuan per year. The readjustments in the following three years reduced research expenditures to 460 million yuan per year. During the Cultural Revolution period, research expenditures increased very little. Not until 1977, have expenditures in research grown in a more stable and balanced pattern. However, total agricultural research expenditures have recently stagnated. There has been virtually no increase of research expenditures in real terms since 1992, despite the call from both government and academia to increase government investment in agricultural research.

Agricultural Investment of Rural Collectives and Individuals[5]
Prior to 1979, non-government investment in agriculture was mainly through

Table 4.3.2 Fixed asset investment by rural collectives and farmers

| | Collectives | Collectives | | Farmers invest |
	Farmers	+ Farmers	fixed assets as % of total production expenses
	1990 Yuan (in billions)			
1952-79	n.a.	n.a.	11.55	n.a.
1980-85	8.12	11.05	19.17	n.a.
1986-90	5.03	11.89	16.92	11.08
1991-94	6.98	10.36	17.34	6.07

Source Fan 1996

Aggregate investment in fixed assets by rural collectives and farmers prior to 1980 was reported by the Ministry of Agriculture (1990). But the separate investment farmers and collectives was not reported. Table 4.3.2 presents only investment in fixed assets for agricultural production purpose which is measured in monetary terms for the purpose of increasing long-term production capacity.

Because of the centralized production system organized by collectives, individual farmers' investment in direct agricultural production was minimal. Investment by collectives and farmers rose steadily from 1953 to 1979, reaching an annual average investment of 26 billion ....yuan.

This level, however, dropped sharply after the introduction of the production responsibility system which was introduced in 1979. During this transition period, the production organization was transformed from collectives to individual farmers. Collective investment dropped quickly, but farmers investment did not increase. Between 1980 and 1985, farmers' investment in agricultural fixed capital formation increased from 1.65 to 10.89 billion yuan. But since 1985, the trend has reversed. Farmers' investment has stagnated or even declined, while investment by collectives have shown an increasing trend.

Total Investment in Chinese Agriculture
Total investment in Chinese agriculture was about 2.7 billion yuan in 1953 (Table 4.3.3). From this level, investment increased to the peak of 38 billion yuan in 1978 with substantial fluctuations. After 1978, agricultural investment in China began to decline, to a lowest level of 20 billion yuan in 1987. Although it has recovered in recent years, the 1994 level was still 32% below that of 1978 Agricultural investment as a share of total agricultural gross domestic product was low in the beginning of the 1950s (Table 4.3.3). It increased from 5% in 1953 to more than 18% during 1976-78. However, this share began to decline sharply after 1978. In recent years, the share is back to the 5% level of the 1950s. Investment in agriculture as a percentage of investment for all sectors has also declined dramatically after 1979 from 20% in 1980 to only 2.4% in 1994.

Table 4.3.3 Agricultural Investment, AgGDP and National Total Investment

	Agricultural investment	Total National Investment	Aricultural GDP	Agricultural Investment as percentage of	
				Total investm.	Agricult. GDP
1953-79	17.80	n.a.	138.99	n.a.	11.56
1980-85	26.05	266.63	333.28	10.76	8.13
1986-90	24.41	492.51	454.42	5.05	5.42
1991-94	26.23	830.55	537.47	3.41	4.89

Source Fan 1995

Undoubtedly, the downward fluctuations of agricultural investments will make it difficult for the sector to achieve a more sustainable pattern of growth in the long-run. A most striking aspect of this recent trend was the decline in allocations of public investment for agricultural R&D, the resulting deterioration of the national agricultural research system, and the slowdown in the rate of release of new agricultural technologies. Another aspect is the decline of investments in the maintenance and repair of irrigation and drainage systems, rural roads, and soil improvement schemes, leading to a decline in the effective irrigated area and an increase incidence damages brought about by natural disasters like floods and drought.

Regional Differences in Agricultural Investment
There have been great regional variations in government agricultural investment. In 1981, Helongjiang and Guangdong provinces received more than 30% of the total government investment (Table 4.3.4), while the Northwest region (Xinjiang, Qinghai, Gansu, Ningxia and Tibet), which covers more than half of the total China's territory, received only 7%. In 1994, it is the northern China Plain (Henan, Shanxi, Hebei, and Shangdong) that received the government's top priority, while the northwest region stayed at the bottom of the government's priority list. Investment from private sources, i.e., from collectives and farmers, also indicated large regional differences in both the investment level and growth rates. It is the most developed regions of Shanghai, Jiangsu, Zhejiang, Guangdong, and other costal provinces where collectives and farmers have increased their investment the most rapidly in the rural sector since 1985.

Policy Issues
This section discusses the relevant policy issues related to investment in agriculture. Public investment greatly enhances agricultural productivity directly by making it possible to do research and extension and through the provision of irrigation services. If managed properly, however, investments can give rise to externalities that could sustain long-run growth at higher rates than they would otherwise be. How policymakers use government investment as a policy instrument to improve efficiency and equity of agricultural growth is currently under hot debate.

Reforms of public institutions
Like any other sectors of the economy, public institutions supported by government funding are undergoing rapid reforms. There are, however, no formal mechanisms nor guidelines for implementing these reforms, although the general trend is to make these institutions more financially independent

Table 4.3.4 Provincial investment in Chinese agricultural production.

Province	Public 1981	Public 1994	Farmers 1985	Farmers 1994	Collectives 1985	Collectives 1994	Annual Growth Rates Public	Annual Growth Rates Farmers	Annual Growth Rates Collectives	Public Investment as % of Production Value 1981	Public Investment as % of Production Value 1994
			Constant Million 1990 Yuan				%			%	
Beijing	1,062	1,441	850	1,434	729	2,611	2.57	5.98	15.24	4.18	1.47
Tianjin	512	956	665	1,396	206	2,513	5.33	8.60	32.05	3.02	1.49
Hebei	2,485	8,233	4,890	11,865	928	8,677	10.50	10.35	28.19	1.32	1.53
Shanxi	911	3,496	1,434	3,301	684	1,089	11.86	9.71	5.31	1.18	2.36
Inner Mongolia	1,157	1,360	1,434	3,214	77	328	1.35	9.38	17.54	1.54	0.65
Liaoning	1,650	4,632	2,455	4,698	514	2,872	8.98	7.48	21.07	1.18	1.14
Jilin	986	1,262	2,991	2,286	168	433	2.08	-2.94	11.11	0.98	0.46
Heilongjiang	7,399	4,168	2,384	3,964	221	444	-4.67	5.81	8.06	4.55	1.14
Shanghai	1,043	4,658	2,003	1,706	477	7,805	13.28	-1.77	36.43	2.71	4.91
Jiangsu	930	4,305	5,907	38,173	982	22,541	13.63	23.04	41.65	0.31	0.48
Zhejiang	1,271	3,810	4,694	21,175	1,441	21,511	9.58	18.22	35.04	0.70	0.80
Anhui	1,062	2,139	4,998	12,812	152	2,192	6.00	11.03	34.55	0.51	0.41
Fujian	1,650	1,472	1,119	10,892	435	2,646	-0.95	28.77	22.20	1.71	0.37
Jiangxi	1,043	1,810	2,003	5,838	233	1,245	4.70	12.62	20.47	0.74	0.51
Shandong	1,556	5,563	9,453	15,145	1,461	17,735	11.20	5.38	31.97	0.45	0.59
Henan	2,713	12,142	7,271	15,148	778	3,910	13.30	8.50	19.65	0.96	2.03
Hubei	2,239	3,464	3,277	8,498	708	2,201	3.71	11.17	13.42	1.06	0.65
Hunan	1,442	3,204	4,164	13,862	490	2,312	6.88	14.30	18.81	0.62	0.56
Guangdong	10,377	5,357	4,163	25,940	738	20,380	-5.36	22.54	44.59	4.93	0.69
Guangxi	1,783	1,860	2,246	9,418	70	1,993	0.35	17.26	45.14	1.35	0.51
Hainan	n.a.	3,410	n.a.	1,921	n.a.	363	n.a.	n.a.	n.a.	n.a.	2.96
Sichuan	1,954	4,019	3,867	15,811	1,133	4,765	6.19	16.94	17.31	0.51	0.48
Guizhou	531	673	1,539	2,488	81	655	1.99	5.48	26.23	0.70	0.36
Yunnan	2,068	3,233	1,080	6,370	242	1,460	3.79	21.80	22.08	1.96	1.34
Tibet	95	848	102	n.a.	3	n.a.	20.03	n.a.	n.a.	0.80	5.43
Shaanxi	1,138	1,629	2,191	5,946	200	911	3.03	11.74	18.33	1.27	0.80
Gansu	1,309	2,747	592	1,921	190	296	6.37	13.97	5.03	2.57	1.80
Qinghai	569	837	222	502	47	144	3.27	9.52	13.28	3.88	2.76
Ningxia	398	792	102	766	26	169	5.89	25.15	22.87	2.95	2.55
Xinjiang	1,783	3,942	407	2,793	78	447	6.83	23.86	21.43	2.83	1.90
Other	2,125	7,643	n.a.	n.a.	n.a.	n.a.	11.26	n.a.	n.a.	n.a.	n.a.
Total	55,413	105,104	78,503	249,284	13,490	134,646	5.48	13.70	29.13	1.39	0.99

Sources: China Statistical Yearbooks, 1982-94, and China Fixed Assets Statistics, 1950-85, 1986-87, and 1988-89.
Notes: 1. Fixed Assets Investment by collectives and farm households is total investment including non production investment.
Therefore, the figures are not consistent with those in table 2.
2. Hainan was part of Guangdong prior to 1988.

from direct government funding. The outcome of these institutional reforms will be critical to the future performance of these agencies in serving the agricultural sector. One set of issues concerns the decentralization versus centralization of management and financing responsibilities. Are there significant economies of scale and scope to be wrought from consolidating various public functions in a system that is often perceived as overly fragmented and uncoordinated? Is a more decentralized approach - where decisions regarding the financing and provision of public services and investment projects can be tied together more closely - justified? Or do different investment options require different degrees of coordination and consolidation? These questions relate to issues of institutional efficiencies. For example, the site-specific nature of agriculture calls for a more decentralized investment system, while scale economies and spillover effects mean that a more centralized and harmonized system could be more efficient.

Balancing these conflicting requirements is a critical problem in the design of an efficient organizational structure for public investment in agriculture. The overwhelming size and institutional disparity of public investment in Chinese agriculture requires more effective coordination of the institutions that are executing these investments. At present, however, institutions under various ministries operate somewhat independently. As a result, the linkages among these institutions are weak. Strengthening coordination and linkages among these institutions is particularly important for an effectively integrated public investment strategy. Aside from these more "macro" issues there are a host of "micro" issues pertaining to the so called institutional design that have a direct and often profound effect on institutional performance. For example, the reward structure facing personnel working in the public system affects the quality of services they provide At present, however, public employees work in accordance with standard civil-service regulations, which provide no explicit linkages between an individual's performance and the compensation they receive. Transferring to jobs either within or outside of the system remains difficult for a host of forinal and informal reasons. The major effect of such an institutional structure is the dampening of the incentives for personnel to work efficiently. A systematic, and where possible quantitative, characterization of these institutional constraints and an identification of their likely efficiency consequences could prove invaluable to policymakers seeking to improve the payoff to public undertakings in agriculture.

Public versus priate investment
While the private role in Chinese agriculture is growing, economic theory, bolstered by contemporary views of the sources of economic growth, points to an important role for public investment in the sector. The arguments rest largely on several notions of market failure, such as externalities; scale economies; failures in related markets like credit, insurance, and labor markets; non-excludability; information problems about benefits and costs; and the need to pursue equity or poverty alleviation objectives. As governments trim budgets, public investment is coining under closer scruny. There has been a naive belief that privatization is an easy and appropriate solution to this problem. There is rnuch that governments can do to enhance the private role in agricultural investment. To bring about this requires policy action. But even if such policy measure becomes successful, the available private action may not be enough Given the public goods nature of public investments, a market failure problem can arise wherein the private sector will invest less in agriculture than is socially desirable (investors may not be able to preclude others from taking advantage of the benefits from the investment, i.e., the flee-rider problem). Therefore, appropriate government intervention may be warranted to correct these market failures. This means agricultural investments,

especially in areas with relatively low private incentives and relatively high social payoffs.

Private investment in Chinese agriculture comes from primarily two sources one involves investments by farmers aimed at increasing their long-term production capacity, and the other involves private-sector investments in the input supply, processing, and marketing subsectors. The rural reforms have led to rapid income growth in rural areas, and consequently rapid growth in personal savings. This has created an enormous capacity for increases in agricultural investment within the rural sector. But due to the relative disadvantage of the farming sector and poor rural public services, this potential has never been realized, particularly in the poorer areas. Farmers' failure to assume their investment responsibilities appears to have been the product of various forces. One was no doubt a long-held belief that such matters fell within the scope of the government. Uncertainty over land use rights and a lack of confidence in the permanence of land policies, as well as a blurring of the nature of the contractual relationship between themselves and the government, were other more tangible factors.

Farmers' investment behavior can also be interpreted as a rational response (in terms of profit maximization) towards the opportunities facing them. The returns from investing in industrial manufacturing and commerce were much higher than these which agriculture offered. Moreover, within crop farming itself; the least profitable alternative was the cultivation of food grains. It is not surprising that a feature of evolving farmer investment patterns during the 1980s was a re-orientation of funds towards non-agricultural uses.

Private-sector involvement in the agro-industry and services (processing, marketing, and input supply) sectors is also embryonic in China There are several reasons why the private sector in these areas will, or should, play a bigger role in the future. As China's economy continues to grow rapidly, the demand for agriculturally related technologies will increasingly move off-farm. Further increases in the use of off-farm inputs such as fertilizers, pesticides, and machinery will stimulate increased demand for new technologies and know-how aimed at the input supply sector Rising per capita incomes are resulting in a rapid increase in the demand for processed agricultural products that will in turn stimulate the demand for post-harvest technologies related to the storage, processing, packaging, and marketing of agricultural produce. At present China's capacity in input supply particularly, post-harvest technology is minimal. The private sector can fill this gap, and these activities ought to attract private investments. The reasons why private sector investment continues to be small.

are complex. Public policies undoubtedly play a large role and we need to more fully understand the precise nature of those public policies and their incentive effects regarding private investments. And it is also the case that there has been no systematic evaluation of the use of direct public investments to mobilize private-sector investments in agriculture.

Future priorities for public investment
The overwhelming size and institutional disparity of public investments in Chinese agriculture raise an immediate set of policy concerns. Determining the level and optimal allocation of public investment in agriculture among different investment activities (e.g., R&D, irrigation, communication, roads, etc.), subsectors (crop, animal, fishery, post-harvest technology, etc.), and regions in order to achieve various policy objectives (e.g., food security, efficiency and equity) is a central point of debate among policymakers. Thus, it is important to incorporate effective review mechanisms to ensure these investment priorities and institutional performance will achieve the government policy objectives. This review process requires a systematic information support system that provides decision makers at all levels of government with readily comparable information on the amount and nature of public investments and a systematic assessment of the potential and actual impact of such public investments. To date, this information and analysis is lacking. Without such information it is difficult to assess whether past allocations have achieved government policy objectives and to provide a useful input into the specification of priority areas for future investment.

Conclusions
This paper attempts to analyze the trends of public investment and related policy issues in Chinese agriculture. The findings of this paper are clear. The evidence presented in various tables indicated that government investment in agriculture declined both in absolute and relative terms. The decline in government investment has been offset by a considerably increase in individual farmers' investment for production purposes. But in recent years, both public and private investments have stagnated or even declined. Due to rapid population and income growth in the future, demand for agricultural products is expected to continue to increase at an even higher rate. It is no doubt that government investment is the key to meet future food needs. Increased government investment will not only promote production directly through improved technologies and infrastructure, but also improve returns to private investments, resulting in more investment from other sources. Simply seeking more investment is not adequate. The financing, organization, and management of these investments are even more important in promoting future

production growth and further poverty reduction in rural areas. The key issues include ways by which government designs its policies to mobilize resources to support public investment, attract private involvement, and set investment priorities among different investment activities, geopolitical and agroecological regions, and different sectors. However, the mechanisms for increasing current research capacity and setting priorities are lacking. Without continuous access to information regarding various options of institutional change and priorities for public investment, it is difficult for government to make wise policy decisions.

References

Ash, Robert. 1994. The Peasant and the at State *China Quarterly*

Fan, Shenggen. 1991. "Effects of Technological Change and Institutional Reform on Production Growth in Chinese Agriculture. *American Journal of Agricultural Economics: 73:266-275*

Fan, Shenggen. 1996. Data Survey and Preliminary Assessment Agricultural Investment in China. A report to Food and Agriculture Organization, Rome, January 1996. Washington D.C. International Food Policy Research Institute.

Fan, Shenggen. 1997. Production and Productivity Growth in Chinese Agriculture: New Measurement and Evidence *Food Policy*, 22:213-228

Fan, Shenggen and Philiph G. Pardey 1992 *Agricultural Research in China:Its Institutional Development and Impact* The Hague: International Service for National Agricultural Research.

Fan, Shenggen, and Philiph G. Pardey 1997 Research, Productivity, and Output Growth in Chinese Agriculture. *Journal of Development Economics* 53: 115-137

Fan, Shenggen, and Mercidita Sombilla. 1997 "Why Do Projections on China's Food Supply and Demand Differ? *Australian Journal of Agriculture and Resource Economics* (forthcoming).

Huang, Jikun, Mark Rosengrat and Scott Roselle. 1995 Public Investment, Technological Change and Agicultural Growth in China. Paper Presented in the Final Conference on the Medium and long -Term Projections of World Rice Supply and Demand, Sponsored by the International Food Policy Research Institute and the International Rice Research Institute, Beijing, China, April 23-26, 1995.

Lin, Justi.nyitu. 1992. Rural Reforms and Agricultural Growth in China. *American Economic Review 82:34-51.*

Lin, Justin Yifu. 1991. Public Research Resource Allocation in Chinese Agriculture: A Test of Induced Technological Innovation Hypotheses. *Economic Development and Cultural Change* 40: 55-73.

McMillan, John, John Whalley, and Lijing Zhu. 1989. The Impact of China's Economic Reforms on Agricultural Productivity and Growth. *Journal of Political Economy* 97: 781-807.

Wen, Guanzhong J. 1993. Total Factor Productivity Change in China's Farming Sector: 1952-1989. *Economic Development and Cultural Change*, 42:1-41.

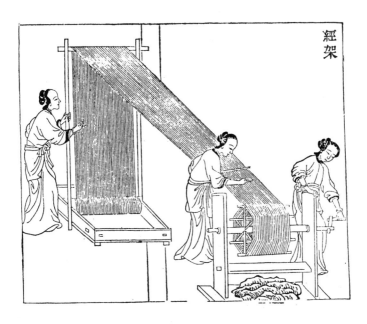

CHAPTER 4.4

Toward 2030: The Issue of Agricultural self-sufficiency in China

T. C. Tso
IDEALS Inc.

In projecting the food situation in the next century, there are many different views using various models. Some models are developed on a macro-scale that reflects the general broad situations; other models are of micro-scale and limited to isolated conditions. Among numerous variables considered in any model, the most important is the human intervention. This includes investment funding, natural resources management, preservation, and other policy issues such as trade, marketing, etc. (1). There are some other projections based on simple arithmetics of low quality data which cannot be considered seriously, although it may easily arose public attention in the short term.

In the past 20 years, China grew by four times in its economy. The World Bank predicted that China could grow by seven times in the next 20 years (2). China has become a major power mainly because of its economic strength. The agricultural sector in China, however, is still weak. The Chinese government in recent years, gradually realized the significance of agriculture in its economic future, particularly facing a population increase together with a decrease in arable land. However, so far, no really serious actions on national investment have been taken affecting agriculture. Under the current "family planning" situation, China's population is still growing slowly, it is expected to peak at Year 2030. The question is, can China maintain a certain degree of agricultural self-sufficiency in its food production by that time. China must act before it is too late. This paper addresses this broad question, based on current agricultural status and major issues of concern, and discusses key requirements in achieving self-sufficiency at the Year 2030.

SELF - SUFFICIENCY IN CHINA

Concepts of Agricultural self-sufficiency

Agricultural self-sufficiency, in its absolute sense, is not possible for any country, nor is it necessary for economical and ecological reasons. Agricultural self-sufficiency does not mean to produce everything in need, but to produce in full capacity that is most desirable and economical in harmony with the environment and ecosystem to satisfy the needs through commerce and trade.

There are various degrees of "sufficiency". The minimum is, of course, free from hunger (subsistence); followed by above-subsistence (below well to do); and then well to do. It is the goal of the Chinese government to achieve nationwide "well-to-do" status in the next stage of economic growth. The term "well-to-do" is used very loosely, it is desirable at this moment to clarify the minimum standard, as listed in below:

Ratio of population earning "minimum income" to total population	less than 35 percent
Food expenses in proportion to income:	less than 50 percent
(Daily protein intake at least 75 grams)	
Education	at least 8 years
Other factors, including health maintenance, housing, communication, electricity, etc., must be maintained	a respectable level.

Seventy percent of China's population is in the rural area and their lives are centered on agriculture. The balance between demand and supply of agricultural products, especially that of grain, is of great importance. Obviously, without certain degrees of agricultural self-sufficiency, to attain "well-to-do" level is not possible. In reporting the grain need, including both food and feed, China is using 400 kg per capita, which is its current base. This is much lower in comparison with approximately 800 kg per capita in the United States. It should be noted that the definition used for "grain" is different between United States and China; and also that the diet pattern in China will change as its economy continues to grow. Many factors need to be considered in addressing China's agricultural potential toward 2030. Whether China's agricultural production can achieve self-sufficiency, or to what level of sufficiency, depends on many variables. Most of those variables are manageable if giving adequate support and time.

AGRICULTURE IN CHINA 1949-2030

Areas of General Concerns:

In the rapid growth of the Chinese economy, agriculture falls well behind industry. The disparity becomes wider in time. Following are certain issues of immediate concerns as China moves toward the next century.

General social issues:

- There are 65 million people below the poverty line.
- Income disparity becomes wider between east and west, urban and rural society.
- As the population grows, competition for education and employment opportunity becomes more intensive. Current floating population of 50 million is already a concern for instability.

Rural-related issues:
- Government grain policy not stable.
- Farmers have low income and heavy tax burden
- Less opportunity for education.
- Surplus rural labor forces create high unemployment rate.

Public investment issues:
Agricultural investment is too low, resulted in shortage or deterioration of:
- Basic infrastructure
- Resources preservation and development
- Environmental protection
- Agricultural Science and technology development
- Rural Education
- Rural society development and improvement in the quality of life.

Marketing issues:
- Quota restriction
- Price restriction
- Commodity movement restriction
- Availability and affordability of farm input materials
- Mechanism for postharvest handling and processing

Land policy issues:
- Public ownership or private ownership

- Investment, management, and preservation
- Security and stability

Environmental issues:
- Waste and depletion of natural resources
- Serious environmental pollution (such as acid rain,water pollution, etc.), and serious ecological damage

Current Status

Agriculture in general has been growing during the last 20 years, although at a slower pace than industry with frequent fluctuations,. As shown in the 1996 report (Table 4.4.1) based on State Statistical Bureau (SSB) data of gross domestic products (GDP), agricultural development is well behind industry and service segments. In Table 4.4.2, it is of importance to note that grain production had reached 490 million tons. Significant increases also occurred in meat, aquatic products, as well as farm inputs such as fertilizers and pesticides. Whether this trend of grain production increase can continue, or how long it can it be continued is a debatable question. Farmers have already experienced problems such as lower purchasing prices resulted from over production, various taxes plus forced contributions, and higher prices for farm inputs.

The inflation in 1996 (Table 4.4.3) for input materials was 8.4 percent, while government purchase price for agricultural products only increased 4.2 percent. Farmers had to absorb the differences at difficult times. In addition, per capita income in rural population is much lower than urban population in 1 to 2:5 ratio (Table 4.4.4).

Table 4.4.1 Gross Domestic Product (GDP) 1996.

	1996 Billion RMB*	Compared with 1995 %
Total	6780	+ 9.7
of which: Agriculture	1355	+ 5.1
Industry	3315	+ 12.3
Service	2110	+ 8.0
*Approximately 1 USD = 8.3 RMB Yuan.		

Source: Peoples' Daily

Table 4.4.2. Agricultural Sector Status. (Source: Peoples' Daily)

	1996 (million tons)	Compared with 1995 (%)
Grain	490.0	+ 5.0
Meat	58.0	+10.3
Aquatic	28.0	+11.2
Fertilizer used (active)	38.0	+ 7.1
of which domestically produced	26.6	+ 4.4
Chemical pesticides (active)	0.4	+ 3.2

Table 4.4.3. Inflation 1996. (Source: Peoples'Daily)

Food	+ 7.6%
Grain	+ 6.5%
Vegetable	+19.7%
Agriculture production input materials	+ 8.4%
Agricultural product government purchase price	+ 4.2%

The major problem arises from low capital investment by government in the agricultural sector (Table 4.3.5). In fact, the investment in agriculture during the past 20 years decreased in relation to the GDP of each year. Region wise, developed areas '(east) have the lion's share of investment in comparison with underdeveloped areas, of central and western China.

Table 4.4.4. 1996 Key Statistics. (Source: Peoples' Daily)

Population net increase:	+1.04% to 1.22 billion (of which rural 70.6%)
Income (per capita)*:	
Urban	4839 RMB (+13%)
Rural	1926 RMB (+22%)
Employment:	
Nationwide	668.0 million
Rural enterprise	131.0 million
Urban registered unemployment**	5.5 million

* Wide variation among regions.** Rural unemployment not reported.

Major Factors Affecting Agricultural self-sufficiency:

In view of the immediate concerns and current status as mentioned above, there is much that needs to be done and much can be done to resolve the major issues in the areas of population, land, grain yield, water, and investment. Each item involves the adjustment of government policy, to place agriculture in top priority above other segments, an increase public investments. Agricultural improvements takes time, needs determination and efforts, and complete new concepts from government leadership. For the purpose of increasing a few tons of grain yield in a short period at the expense of permanent destruction of natural resources is not a wise approach.

Table 4.4.5. Selected 1996 Fixed Capital Investment.

	Amount Bil. RMB)	Percent of 1996 Total Investment
Agriculture, forestry, animal husbandry, fishery, water	33.6	1.9
Energy	293.8	16.9
Light industry	100.0	5.7
Transportation and communication	301.2	17.3
Regional Difference:		
East	1038	+17.4
Central	368	+16.6
West	211	+11.2

Population

Family-planning policy has demonstrated its effectiveness for population control in China. Reports on population dynamics in China (3, 4), based on the total fertility rate (TFR) or average childbearing per female showed very favorable results. Between 1970 to 1996, TFR values reduced from 5.9 to 1.8 (Table 4.4.6). These authors also identified the critical TFR values for population stability in the dynamic system, and project the population changes for the future For example, population will not reach the peak even at Year 2050 assuming TFR is 2.2; peaks at 2030 when TFR is 1.9; and peaks at 2020 when TFR is 1.6 (Table 4.4.7). Based on this formula, assuming TFR value is 1.9, the population in

Years 2000, 2030, and 2050 are respectively projected to be 1270, 1480, and 1420 million (Table 4.4.8).

It is, therefore, reasonable to assume that the population in China will not continuously and indefinitely increase, nor China's grain demand would starve the world. However, there may be fluctuations in the TFR values within a certain period of time for some reasons, such as in the years 1986, 1987, and 1988, the TFR value suddenly rose to 2.4, 2.7, and 3.3 respectively before it returned to 2.3 in 1989 (Table 4.4.6). Therefore, to be sure that our discussion is on solid ground, the population number of 1600 million in year 2030 is used in this paper, assuming TFR values between 2.2 and 1.9. Actually, based on current trends, the TFR value in 2030 should be much lower; thus a much low population.

Table 4.4.6 Population dynamics in China. Total fertility rate (TFR) 1970- 1996.

Year	TFR	Year	TFR	Year	TFR
1970	5.9	1979	2.8	1988	3.3
1971	5.4	1980	2.3	1989	2.3
1972	5.0	1981	2.7	1990	2.0
1973	4.5	1982	2.9	1991	2.0
1974	4.2	1983	2.4	1992	2.0
1975	3.8	1984	2.3	1993	1.9
1976	3.2	1985	2.2	1994	1.9
1977	2.9	1986	2.4	1995	1.8
1978	2.9	1987	2.7	1996	1.8

Source :Song 1997.

Arable Land Area:

Due to the rapid pace of economic growth, arable land lost to housing, construction, industrialization, etc., is a matter of great concern, even government policy had placed certain restrictions. China's State Statistical Bureau (SSB) data are widely disputed for its accuracy. There are claims that it might be understated by 25-44 percent, but so far, no reliable data from a Chinese official source can be cited. Unofficial estimate is between 124 and 132 million ha of arable land in China (5). Official SSB reports state the total agricultural land area, including arable, potentially reclaimable, grassland, and also aquaculture water surface are listed in Table 4.4.9 (4)

Table 4.4.7. Total fertility rate (TFR) and population.

Year	Projected Population (millions)		
	TFR: (2.2)	TFR: (1.9)	TFR: (1.6)
2000	1304	1272	1257
2010	1428	1365	1326
2020	1547	1443	1377*
2030	1635	1476*	1370
2040	1682	1468	1325
2050	1703	1421	1234

*Peak population projection   Source: Song and Yu, 1988

Table 4.4.8. Population in China.

Year	Population (million)
1953	583
1964	691
1982	1004
1990	1131
1995	1237
2000	1270*
2030	1480*
2050	1420*

*Projected population using TFR value 1.9. Source:Song 1997.

Here we need to mention that two-thirds of arable land are low and medium land; or only one-third is high yield land. In this particular aspect, one may question the 1996 data from SSB and Ministry of Agriculture (MOA) (Table 4.4.10) which reported the "changes" between 1994-1995 of arable land area based on simple mathematical differences between areas of arable land lost and areas of land newly reclaimed for replacement. The nature of productivity between normal yielding arable land and poor yielding newly reclaimed area is totally different, at best 5:1 ratio for yielding potential.

Taking into consideration that SSB data is understated, and that newly reclaimed land will be added, it is rather difficult to find a solid base to estimate exactly how much arable land would be available in China at Year 2030. For the purpose of projecting China's agricultural potential in an academic exercise, this author assumes that China would have 90 million hectares of arable land by Year 2030.

This assumption derived from the facts that (a) current SSB data is only 25 percent understated; (b) newly claimed land would be counted as 20 percent in production potential; and (c) multiple cropping index will increase in the future.

Table 4.4.9. Agricultural land (1997 estimate)*

	(1997 Estimate)
Arable land: of which two-thirds are low and medium yield)	90 million ha
Potentially reclaimable land:	20 million ha
Grassland:	250 million ha
Aquaculture potential:	
inland water surface	2 million ha
shallow sea area	300 million ha

*SSB data may be 25-44 percent understated. Source: Song, 1997

Table 4.4.10. Arable Land Area.

Year 1994-1995 reported changes*	
1994	95.1 million ha
1995	94.9 million ha
Changes during the year:	
Loss to housing, construction, industrialization:	708,630 ha
Gain from newly reclaimed loss productive land:	504,840 ha
Net loss:	180,790 ha
Year 1996-1997 reported arable land*	
Between	90-93 million ha
Year 2030 projected arable land area:	90 million ha**

*SSB, MOA data are considered understated 25-40 percent.
**Outside projection.

Grain Yield

During the past two decades, the world supply of food crops increased considerably, but the increase still fell behind the demand due to the world population increase and diet pattern changes. The balance was made up through expansion of additional arable land

This phenomena will not be possible in the future, as population will continuously increase, but undeveloped arable land will no longer be available. This situation is particularly true in China. Increase yield per unit area appears to be the only solution. In China, the currently reported yield per unit area by SSB is inflated considering the fact that it is calculated from understated total arable land area. Therefore, there is still room for yield increases in China to approach world yield standards. Many factors can contribute to yield increase, including the advancement of science and technology; improved and certified seeds; improvement of the quality and efficient use of fertilizer, pesticides, water; better field management and use of plastic film; and reduction of postharvest wastes. Take rice production for example, rice yield can definitely be elevated from five tons per hectare to six tons as in the United States, seven tons as in Korea, or even eight tons as in Australia. Investment for science and technology development plus farmers' incentive hold the key to this matter. Even under current conditions, some Chinese experts predicted the potentially attainable grain production in Year 2030 could reach 1026 million tons under most favorable conditions, and 534 million tons even under unfavorable condition (6).

As mentioned in previous sections, China's grain need would be 640 million tons by Year 2030, assuming per capita need of 400 kg for 1.6 billion population. If we consider the TFR value mentioned or even below the 1.9 level, the Year 2030-grain need would be 590 million tons or less. Even if TFR value reaches 2.2, the need would be 650 million tons of grain (Table 4.4.11). Those estimates may vary when considering the diet pattern change, such as increase production of fruits, vegetables, development of pastureland, and increase in production of aquatic products.

Water
Among all the factors affecting agriculture and human life, water is one of the most important. In fact, water is the most treasured element in the world. Tables 4.4.12 and 4.4.13 show, respectively, world water balance and Chinese water resources and distribution. Table 4.4.14 illustrates how much water is needed to produce major crops or meat (6). It takes 900 liters of water to produce one kg of wheat; 1912 liters of water to produce one kg of rice; 3500 liters of water to produce one kg of broiler chickens; and 100,000 liters of water for one kg of beef.

Table 4.4.11.	Grain harvest and project need (million tons)
1949	115
1980	330
1985	380
1990	440
1995	465
1996	480
2000	518*-520**
2010	550*-570**
2020	580*-620**
2030	590*-650**
2040	590*-670**
2050	570*-680**

* Calculated based on TFR at 1.9.
** Calculated based on TFR at 2.2.

Table 4.4.12. World Water Balance.	
Total available source of fresh water (from sea and land evaporation)	518,000 km^3
Net supply to land surface	46,000 km^3
Global effective supply	14,000 km^3
Controlled by human intervention	3,200 km^3

Table 4.4.13. Water resources and distribution in China.	
Total availability in normal year:	650 km^3
(per capita only 20% of world average)	
Estimated water requirement:	
1996	650 km^3
2000	700 km^3
Distribution:	
South 83% water, 36% cultivated land	
North 17% water, 64% cultivated land	

China's projected water need to produce 640 million tons of grain by Year 2030, considering 1500 liters water to produce one kg of grain, the 640 million tons grain would need 960 km^3 of water by Year 2030. China has an average of only 650 km^3 of water available in a normal year, and most water (83 percent) is distributed in south China where only 36 percent of cultivated land are located. In addition, we need to consider human and industrial water needs.

Table 4.4.14. Water need for production (liters/kg).

Crop:	potato	500
	alfalfa	900
	wheat	900
	sorghum	1110
	corn	1400
	rice	1912
	soybean	2000
Meat:	broiler chicken	3500
	beef	100,000

Source: Wen & Pimentel, 1997.

Obviously, the water situation is most critical in the future. Current government consideration to divert water from south to north would provide little but insignificant relief. To solve the water problem, China must improve its water-use efficiency as the first step. Table 4.4.15 illustrates the relative water-use efficiency on agriculture in China, Japan and Israel as 10 percent, 30 percent, and 80 percent, respectively (4). If China follows the Japanese pattern, it may save 84 billion M^3 water, or save 126 billion M^3 water if it follows Israel's pattern (Table 4.4.15). This conservation concept involves education, science and technology, facilities, and most important, the incentive of farmers themselves. The current "3-high" policy promoting high yield, high quality and high efficiency not only leads to reduction of soil productivity, damage soil quality, but also leads to waste of water, induces pollution of water, and gradual destruction of irrigation and drainage facilities for water system.

Table 4.4.15. Relative water use efficiency in agriculture.

China	10%
Japan	30% (save 84 billion M^3)
Israel	80% (save 126 billion M^3)

Source: Song, 1997

Water diversion and conservation cannot solve the shortage problem. New water resources must be found. With additional new resources, China can expand arable land area in the vast northwest region, which has 42 percent of the total China area with only 8 percent ofChina's population (Table 4.4.16). There are proposals to use water from Xinjiang in both surface and underground systems, divert water from

Ertrix He from north to south, and divert Tongtian He toward west and north. The total available water volume from those sources and covered area are not available (Figure 4.4.1).

Among the many thesis on increasing China's water resources, the division of water from Tibet to the northwest area is most challenging and workable (7). The Yarlung Zangbo (YLZB) River runs through the southern part of Tibet with an annual volume of 1.4 times that of Changjiang. Toward the southeast area, YLZB made a reversed "U" turn between Pie and Medong. The shortest points between this "U" turn is only 39 km. A water "tunnel" can be constructed which provides a sharp 2250 m drop.

Table 4.4.16. Northwest region development and water needs.

- 42% of total area, 8% of population
- Need at least 20 km^3 for agricultural development.
- Possible water resources from:
  Tibet - Yarlung-Zangbo-Jiang (to north
  Xinjiang - underground system, and Estrix He (to south)
  Qinghai - Tongtian He (to west)

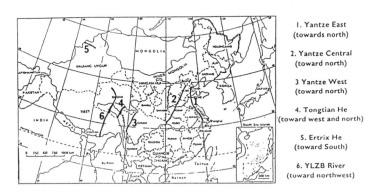

1. Yantze East
   (towards north)

2. Yantze Central
   (toward north)

3 Yantze West
   (toward north)

4. Tongtian He
   (toward west and north)

5. Ertrix He
   (toward South)

6. YLZB River
   (toward northwest)

Figure4.4.1. Six possible routes for water diversion.

286

The calculated hydroelectric power potential from this drop is twice that of the "Three George Dam" under construction. Such energy can provide sufficient power to transfer 26 km^3 of YLZB water to the northwest region of China for agricultural and industrial development. The diversion of water is only 19 percent of the total YLZB volume. In addition, this system can provide advantages to India and Bangladesh: the reservoir will improve microclimate and environment, reduce risk of drought or flood, plus share the use of inexpensive electricity.

To use YLZB water for the northwest, the capital investment will be high. However, in comparison with the "Three George Dam", the benefit is much, much higher. The vast northwest region of China will become promising land forever for agriculture, animal husbandry, mining, and other economic development. An intensive on location study is underway by scientists from Academia Sinica.

Public Investment and Policy

All those above-mentioned major issues cannot be resolved without government's determination to increase funding for public investment and make certain policy changes.

Despite the official statements that China places agriculture as its top priority, public investment in agriculture has not been increased. In fact, public investment decreased consistently relative to agricultural GDP during the past 20 years (8), even including operation and management costs. Since a separate article will deal with the investment issue (8), this author wishes to emphasize only the following items that are essential for promoting agriculture self-sufficiency.

Under the condition that no additional arable land can be developed in the future, China must increase production per unit area through development of new technology and efficiency. Postharvest handling, waste elimination, and increase utilization value are among the essential items. Agricultural processing, marketing and trade mechanism need to be established. In the use of water, it should improve irrigation and drainage facilities, increase water use efficiency and eliminate waste, plus, develop new water resources for agriculture.

Existing science and technology knowledge will not be sufficient to solve future problems. New frontiers have to be explored in basic science and technology to achieve breakthroughs for sustained agricultural growth. Obsolete agricultural research and development systems must be modernized, old facilities are updated to attract more talented scientists for agricultural research.

Agricultural education - should be promoted at all levels from research scientists to peasants. Fertilizer, pesticides, plastic films, and other important input materials are improved for quality, availability, and affordability. Soil management - conservation, preventing erosion and improve soil quality are essential. Pollution control - control acid rain, soil and water contamination, air pollution, particularly that generated from rapid rural enterprise development must be preserved. Basic infrastructure - electricity, roads, means of transportation, etc., must be developed.

In agricultural related policies, the main objectives should be aimed at preservation of natural resources, promoting farmers' incentive and welfare, increasing production efficiency and utilization efficiency. To achieve those goals the following policy changes should be seriously considered. Grant farmers the land ownership - this will encourage farmers' incentive for private investment in land improvement for higher productivity and quality to prevent soil and water erosion, to reduce or eliminate pollution. At the same time, it would maintain rural society stability.

Abolish price and quota restrictions, allow commodity free movement to stimulate higher and better productivity. Assure that all production input materials - fertilizer, pesticides, plastics, etc. - available and affordable, in price and in time. Provide facilities for drying, storing, processing and transporting. Reduce the burdens for peasants in various forms of taxes, fees, and contributions. Provide educational opportunities to all peasants, especially school-age children. Improve quality of life in rural society.

Possible Scenarios in Year 2030:
China has the potential and ability to achieve agricultural self-sufficiency, but China must act in time. New science and technology cannot be developed overnight, nor basic infrastructure. China will not be able to produce everything they need, no country can, and it is not necessary to do so. China will be an active player in the world grain market in Year

2030, and it is very likely that China will import more grains for feed as the diet pattern changes. They will also need to import food grains in coastal metropolitan areas for logistical and economical reasons. Import will be limited in quantity, that certainly would not starve the world in Year 2030.In summary, depending on government action, China's agriculture self-sufficiency can be either of the following two scenarios:

Scenario One: If China seriously considers the importance of current concern of factors that are affecting agriculture, and take immediate actions by increasing public investment and making policy changes, China's grain output can reach at least 675 million tons for 1.6 billion people by Year 2030, or 420 kg per capita, and move toward the well-to-do level.

Scenario Two: Facing the arable land loss, and without significant increases in public funding, China can expect only minor progress on science and technology, together with little infrastructure development. China might still be able to produce 450 million tons grain in Year 2030, which is near the current level, but with a much higher population to feed. The per capita grains available would be 280 kg, or at a subsistence level. The need for imports would be high and would cause considerable stress to the world food market.

References

1. Papers presented at the Post Conference Workshop of the 1997 American Agricultural Economical Association Annual Meeting on China's Food Economy in the 21st Century, Toronto, Canada, July 31, 1997.

2. The Washington Post editorial, September 21, 1997.

3. J. Song and J. Y. Yu. Population system control, Springer, 1988.

4. J. Song. No impasse for China's development. Speech delivered at Forum Engelberg, March 20, 1997. (In this monograph, Chapter 3.3.

5. Alexander von der Osten. See citation in "Food grains - the potentials for yield " (In this monograph, Chapter 4,22).

6. Y. C. Shi. Prospective analysis of potentials for the grain production in China. Science and Technology Daily, October 21, 1996.

7. C. Y. Chen. A tentative plan for transforming northwest China. Science and Technology Review, Beijing, Volume 10: 60-61, 1993.

8. Shenggen Fan. Agricultural investment in China: Measurement and policy issues. (In this monograph).

9. Wen Dazhong and D. Pimentel. 1998. Agriculture in China: Water and energy resources. (In this monograph.)

CHAPTER 4.5

Some Strategic Aspects of
Chinese Agricultural Readjustment: 2000-2030

Samuel Pohoryles
Agrindus International Ltd
Tel Aviv, Israel

Our common world ecosystem intrinsically demands the highest possible degree of cooperation among us who inhabit the area. Our comprehensive application of human inventiveness, our liberal exchange of know-how, and our joint efforts in extensive development projects can convert even desert into green, fertile fields. It is an ambitious goal, but it can be achieved by mobilizing the available scientific, technological, educational and organizational resources. According to FAO experts, a desert area such as the Middle East requires an entire generation - 25 years - just to double its agricultural output. However sophisticated technologies made possible the intensive irrigated farming that increased Israel's agricultural production twelve-fold during a single generation ... thereby poking holes in the classic Malthusian theory.

To achieve results such as these in a land burdened with a chronic water deficit, it is crucial that a mix of technology, genetic engineering, agrobiology and management be substituted as much as possible for water. In this readjustment, the dominant role in agricultural development shifts from the natural production factors - land and water - to the so-called "unexplained productivity" of know-how and technology. In the last decade, "unexplained productivity" contributed to 96% of Israel's agricultural output, while natural production factors, land and water contributed only 4%. The new "factor mix" should act simultaneously to the increase of water use efficiency.

Two axioms of agricultural capability are crucial to the developmental strategy: First, agricultural underdevelopment is not a natural disaster. A food shortage is not even a problem of agricultural capacity. In the short term, it is a problem for politicians and policymakers, a problem of agricultural philosophy. In the long term, it is a matter of transferring technology, progressing in agrobiology, improving the genetic pool, increasing the efficiency of planning and organization -- in short, the human factor. Secondly, it is possible to attain significant agricultural emancipation and technological independence even in less-favored areas such as the desert. Research into a new mix of agricultural inputs will increase the share of intellectual investment as a dominant production factor.

This makes it possible an alternative approach to self-sufficiency: not physical autarky, but economic self-sufficiency. Research can develop relative advantages for some of our crops and livestock products on the world market - and their export can compensate for whatever food imports are necessary.

Therefore, it is time to discard the traditional view of agriculture as nature's stepdaughter and a favored child of governments. Technologically oriented agriculture copes fairly well with nature's hardships, and the benevolence of today's governments cannot be taken for granted.

It is important to define the general philosophy of agricultural development in more specific and tangible terms. We can identify at least five basic modalities for international cooperation.

1.   Exploration and preparation of joint projects of research and developmental interchange---The first phase will be an examination of the problems and opportunities.

2.   Operation of pilot projects---The pre-development phase of operational projects will lead later to regional networks of pilot projects as an accelerating factor of agricultural transformation.

3.   Cooperative research programs--- This modality will enable to conduct collaborative studies, hold joint seminars and symposia, exchange scientists and share research date, including genetic, structural and logistic issues.

4.   Extension services and transfer of know-how--- To ensure that the technology reaches the farmers, the authorities should coordinate the transferring of know-how through extension services.

5.   Exchange programs and training fellowships--- The cooperation will be focused on training centers and the associated international agencies for fellowships.

In order to establish an adequate scientific technological infrastructure for technology interchange the Chinese agriculture has to benefit the existing already tools of agricultural progress. A clear linkage to the 21st Century capacities should be established.

Agricultural Technologies of the 21st Century

Today, at least a dozen advanced technologies of the 21st Century are having considerable impact on productivity in agricultural development. These are:

• Enhancement of photosynthetic efficiency, including the following: improvement in carbohydrate formation through genetic selection, physical modification and chemical modification; enhanced nitro-absorbing capacity for protein synthesis; enhanced growth rates through elevated levels of carbon dioxide.

• Water and fertilizer management; pest control; Greenhouses (controlled environment); multiple and intensive cropping and reduced tillage.

• Bioregulators (natural and synthetic compounds which regulate the ripening and aging of plants; bioprocessing; antitranspirants. Plants resistant to drought; new crops; and multiple births in beef cattle

These technologies are changing the face of agriculture in less favored ecosystems through comprehensive, integrated development of rural areas. As I mentioned before, technology has been by far the greatest influence on agricultural development in Israel. It is the most effective substitute input for water, and the knowhow is easily transferable. Nevertheless, two other major elements of agricultural development, redistribution of water, and a rational and responsive administrative system, are also important considerations that should be considered. Moreover, although advanced industrialization is not an immediate problem, the social and ecological implications of industrialization should be kept in mind during the readjustment planning.

The development of these advanced technologies concentrates mainly in developed countries. It is reasonable to assume that as far as the developing countries are concerned, the adaptation process of sophisticated production methods will take place now. Towards the end of the century there will be various developments regarding the advancement of agricultural production in both developed countries and developing countries. Here the Israeli experience can contribute particularly for adaptation of existing modern technology to the specific conditions prevailing in developing countries, including unique socioeconomic modalities of private and cooperative frameworks.

CHINESE AGRICULTURAL READJUSTMENT

The Interaction between Land and Water and Technology

It was agricultural technology, in its widest sense that exerted the greatest influence on agricultural development in Israel, largely off-setting the problems of water shortage.

Reference here is to a combination of agrotechnology, research, extension services and mechanization. All these raised the share of productivity not explained by production factors in Israel's agriculture, i.e. the share of what Hicks calls "*neutral technological changes of productivity*". This share in the first 20 years stood as mentioned before at around 50 percent of total productivity. Throughout the recent decade the intangible "factors of accumulated experience and research benefit" contributed 96 percent of total productivity rate, whereas the "tangible" factors contributed 4 percent only. This is an expression of high level of technology based on "human capital".

A sort of substitution was thus created between water, a tangible factor and the "*intangible*" - productivity, which largely counteracted the basic constraints on land and water. The technology of water use I undoubtedly an integral component of any agricultural technology, and it encompasses a broad range of irrigation innovations (such as drip irrigation, irrigation equipment, technical adjustment of irrigation systems to particular crops). The water quantity available for agriculture has been static or decreasing throughout the recent 18 years 1.0 - 1.2 billion cubic meters. Nevertheless, agricultural production value, in real terms, has gone up during this period 2.9 times, as a result of improved resources utilization and technological progress.

We found in the Negev desert very interesting phenomena of disconnection between physical quantities of water and the level of farmer's income: 10% reduction of water lead to 5.7% only reduction in net income; this result is a ceteris paribus result - it means a result in condition of static technology.

However, if we suppose a more realistic approach of dynamic technology (improvement of mechanization, fertilization, more efficient organization, more adequate adaptation of crops to the relative advantage of the region) the income drop will be 1% only. This means a significant water-income disconnection. This is of course a very tangible expression of agricultural emancipation.

An exogenous form of adjustment can be observed in the changes taking place in the village community that are not agriculture oriented. Reference here is to the fact that the Israeli village has undergone what may be called a typical stage of heterogenizing its economic activities, resulting in the fact that only 20 percent of employed village residents are actually engaged in agricultural production; the rest work in other sectors such as industry, services, construction, trade and so forth. The ratio in Moshavim is even higher - 33 percent; whereas in Kibbutzim it is much lower - only 18% (percent).

The Israeli growth rate, however, was made possible due to high-level technology, by which the natural constrains were successfully overcome, and also owing to development patterns based on national ownership of land and water.

A significant contribution to this progress is due to the cooperative structure of our agricultural sector: 80 percent of the total agricultural output in Israel originates in cooperative enterprises.

Israel's agriculture forms a relatively small segment of the national economy; it constitutes about 2.9% (percent) of the GNP, 3% of the national income, 3.0% of the national labor force and 3% of investment. The manpower in agriculture - in terms of full employment - stands at only 60 thousand. Agricultural land occupies 400 thousand ha. Thereof half under irrigation. Moreover, the socio-economic contribution of our agriculture is significantly greater than is share in the national economy. Israeli agriculture's contribution to the national economy is expressed not merely by a relatively high food sufficiency level and a sense of "food security". It also has a highly significant multiplier effect, as one employee in agriculture creates derived employment for 1.3 employees in non-agricultural sectors.

An important indicator which demonstrates the quality of Israeli agriculture is its technological level. Here, it should be pointed out that one employee in agriculture produces a quantity of food sufficient for 98 people (compared to the ratio of 1:17 prevailing 30 years ago). Israel's agriculture provides today some 97% of local demand for food, whereas its export covers the supplementary import. Agricultural import consists mainly of grains, sugar and meat of which domestic production is still limited by scarcity of water.

CHINESE AGRICULTURAL READJUSTMENT

The Socio-economic Philosophical Dilemma

In reality, the above problem is far from being a simple one; its solution calls for a socio-economic system organization encompassing the following:

Economic efficiency; Moral decency; Social justice; and Democratic management principles.

The inclusion of these four elements in the system complex guarantees an equilibrium between the economic and social aspects of the development process, thereby preventing economic predominance by means of compulsory equalitarianism.

Integrated rural planning, conceived as a central policy system, constituted an uppermost theoretical issue.

There is no doubt that the entire subject can be very controversial: it implies an expression of economic-development strategy; it is focused in the center of a philosophical dilemma arising from development in the wake of the market mechanism of development by planning.

The solution of the dilemma must be worked out by analysis of two diametrically opposed theoretical approaches:

The first: were we to see in agriculture a field of activity in which preservation of health was the main goal, our only consideration in the performance of any action, connected with agricultural production, would be whether that action was necessary for the preservation of human health. Economic considerations would be of no consequence.

The second approach represents an abstract model of the flow of resources; it takes into consideration only marginal product and the dynamics of market mechanisms. In its extreme form, this approach is an expression of disregard for social and political considerations. Absence of economic considerations such as strict profitability ultimately undermines social targets.Thus, arriving at an agricultural development policy that takes into account both economic and social constraints is much like solving a complicated crossword puzzle, where the vertical and horizontal line represent economic and social variables, and furthermore, the problem is dynamic, requiring periodical revision.

Planning is a basic condition for collective participation in decision making and has to be a flexible democratic expression of targets, criteria and strategies. These require constant reformulating and adaptation to changing economic, social and environmental conditions.

Agricultural strategy in the third development decade must take into account the fact that in 10-15 years agriculture will be totally different from what it is today - and the main problem is the need to find such technological economic, social and institutional means that will enable agricultural policy to shape the new form. This means comprehensive integrated development - heterogeneous development of rural areas.

The main objective of this statement is to demonstrate that even less favored areas can support a high level of modern agriculture, not lower, and sometimes even higher that than of countries with temperate climates.

Even here, a significant agricultural emancipation and a relative agrobiologic and technological independence could be reached, which permits a high degree of economic self sufficiency in food supply.

The indirect conclusion here is that the less favored areas, which seems to be the most complicated type of LFA, could be transformed - through technological substitution of natural constrains, the total world reservoir of land and water creates a real antimalthusian infrastructure for the right to food - for production of food.

One of the typical features of semi-arid zones is the lack of geographical proximity between agriculture's two natural factors of production, land and water. Another well-known and related feature is the shortage of the human production factor - i.e., labor - which is not attracted to such demanding regions.

The concrete solution to the problem of geographical distances between land and water sources lies in solving the problem of inter-regional water transportation. In Israel, this means transporting water from north to south, but more than a mere question of haulage, it entails a complex system of technological, sociological, judicial, organizational and administrative issues.

Simultaneously, we have to do with a consequent process of internal realignment of arid "factor mix" of the agricultural input basket; the share of land and water in the input composition decreases - the share of neutral technology steadily increases (new varieties, mechanization of adequate technology, genetic engineering) and will be more and more the dominant factor.

No less important is the process of optimization of the crop structure. Compensation in this case may be attained by finding a basket of crops hat, on the one hand, meets domestic and export demand, on the other hand makes the best use of available water. Reference here is to increase in the marginal output per unit of water used.

Questions and Modifications of Chinese Agricultural Readjustment

China has tremendous achievements in balancing a food supply for 1.2 billion population, assuring for many years the food demand supply equilibrium. The grain harvest has doubled over the past 30 years.The main strategy was linked to Deng Xiaoping's philosophy that *"without agriculture there is no stability, without grain there is chaos"*.

However, in recent years China has been facing crucial problems of agricultural and food strategy with strong and dangerous socio-economic impact.

Problems and Options of Sustainable Agriculture

The State of the Art: According to international sources, the forecasted demand for grains in China will be 600 million tons by the end of the century. China's grain imports are expected to surge above 10 million tons this year, with more than 1 billion US dollars worth of grains and edible oil from the United States. This, in addition to hundreds of millions of dollars in processed food via Hong Kong.

China will more than triple annual wheat imports to 15 million tons by the decades' end, due to the demographic growth of 17 million annually until 2044. The gap between this demand and the predicted current supply is 50% of the world grain export.

The Reasons for the State of the Art: (1)China's biggest obstacle to higher output is rural over-population and area atomization: for 400 million farmers the average cultivated area is 4 mu. (2) Farmland is lost to erosion,: (3) Farmland is built over as factories and industry sprout. Urbanization and industry continuous pressure farmland. (4) Stagnation of rural investments. (5) Exodus of rural areas - 110 million farmers flocked to the cities. This is the equivalent of the Sechuan Province out of production. Rural income is 39% of the urban income. The retail prices for grains increased from 50-200% in 1993. (6) Free market for grains leads to inflation spiral, dangerous for urban population. (7) Significant gap between producers' grain prices and cash crops.

Options for Solutions. The Strategy. China's traditional view of physical autarky has to be changed to a system of diversified farming, based on relative advantage verification with economic self sufficiency, as the main objective.

China is expecting a new Green Revolution. The first tangible expression should be geographical shifting of its agricultural base to Gansu and Xinjiang, linking the Yangtze River with Beijing, 1,200 km. to the North. This would require an investment of 4.6 billion dollars for aqueduct construction and irrigation.

The Policy. The dilemma how to raise rural income while keeping prices stable is a difficult political dilemma. Price reform works in the long run, however it is not sufficient to solve current problems. Supplementary regulatory transitional measures are important.

Grain production is inferior in market competition with cash crops. Free cash crops and vegetable oil prices pushed the production of wheat to marginal regions. An immediate important step with accelerated impact could be the introduction of certified seeds into Chinese agriculture. Today, only 5% of seeds in China are of certified genetic value. This would increase the yields from 50-100%, with the same input of land capital and labor. The cost of such modernization is marginal.

Another important step is to include desert areas of Xinjiang and saline soil areas in the production targets, including the use of recycled water and effluents. A network of sophisticated high-tech demonstration farms should be established throughout the country, with strong regional demonstration effect. The accumulated experience of developed countries, integrated with the assistance of international organizations, could be of tangible significance in this area.

CHINESE AGRICULTURAL READJUSTMENT

The Production and Demographic Linkage

China should be able to meet its own needs, which means that domestic grain production trods ahead of population growth and dodges the Malthusian quasi-axioma. However, if production stagnates or worse collapses entirely, grain could become a scarce commodity in high demand.

The increase of domestic yields of grains is linked to the motivation of farmers own interest. If the only profit available to the farmers is from cash crops, livestock and conversion of farmland to industry, the solution will be very difficult. The policy requires a symbiosis of the free market mechanism with well verified governmental involvement.

Voluntary cooperatives to reach economics of scale should be considered. An adequate integration of scientific technological modernization and relevant socio-economic adjustment is a condition sine qua non for acceleration of Chinese agricultural and agroindustrial development.

Suggested Project Areas

In this context, two main aspects should be considered, evaluated and transformed into reality. They are as follows:

1. Strategic readjustment of Chinese rural economy, including patterns of farms and villages, integration of agriculture, post harvest, marketing and processing based on regional agroindustrial entities. This includes macroeconomic and regional planning, computerization, forecasting, feasibility evaluation and preparation of development projects for domestic and international financing, based on financial requirements. This includes also strategic policy decision i.e. introduction of certified seeds, which means multiplying yields of crops with the same investment of capital and labor.

2. Implementation of specific projects in defined project areas as suggested below:
Water Resources
Water prospecting (hydrology) and development (engineering).
Efficient use of water.
Use of saline, brackish and wastewater.
Wastewater treatment and desalination.

Crops and Genetic Material
Quality cultivars for the desert.
Introduction of certified seeds.
Development of new desert crops (domestication).
Crop protection.
Agromanagement in Desert Areas.
Survey of plant genetic resources.
Livestock and Aquaculture
Semi-intensive dairy production systems.
Semi-intensive small (sheep and goat) livestock production.
Coordinated surveillance of animal diseases.
Pond-based fish production.
Agricultural Technology
Irrigation systems from runoff harvesting to the most highly engineered.
Post-harvest handling.
Food processing technology.
Infrastructure and marketing.
Protected Agriculture
Plastic technology.
Methods for controlling environment.
Genetic dimension of protected agriculture.
Biotechnology
Molecular biology.
Plant physiology.
Breeding or salinity, drought and pest resistance.
Ecology and Environment
Agroecological conservation of desert systems.
Crop protection techniques.
Organization of Human Settlement and of the Argoeconomic Sector
Agricultural cooperative.
Public/Private Initiatives.
Policy.

3. Optional Projects: Opportunities abound for cooperative projects in the region. The following subjects are among the issues with the most potential impact on agricultural development.

(1)Developing high-yield , high quality, pest-resistant varieties of field crops adapted to the region. (2) Establishing a gene bank from fruit species originating in the region. (3) Adapting temperate-zone fruits to a warm climate. (4) Recycling saline and effluent water for irrigation. (5) Increasing the efficiency of water use. (6) Preventing fertilizers form polluting ground water. (7) Controlling the environment with permanent greenhouses and temporary plastic covers. (8) Controlling pests and diseases. (9) Preventing the infiltration of new diseases though quarantines. (10) Increasing the resistance of crops to post-harvest diseases. (11) Replacing toxic chemicals with food-grade preservatives. (12) Developing improved crops-pasture systems. (13) Increasing the yields of rain-fed farming with little or no irrigation, by reassessing cultivation practices (tillage and soil shaping, planting desities and geometrics, fertilization, crop rotation, etc.) (14) Coordinating the surveillance of infectious animal diseases. (15) Developing genetic stock and low-cost feed to intensify the production of sheep and goat products, for which there is a heavy and increasing demand. (16) Intensifying dairy production systems. (17) Expanding pond-based fish production. (18) Introducing aquaculture production. (19) Organizing institutional frameworks and policies, agricultural cooperatives, public and private initiatives. (20) Improving post-harvest functions: transportation, storage, processing and marketing.

Recommendations for Immediate Action

Introduction of certified seeds of grains in general and of wheat in particular, simultaneously to verification of quality of existing seeds. This undertaking will be of marginal cost and will have an immediate impact in yields with the same labor and capital input;
Establishment of a national network system of pilot demonstration farms and projects adapted to different regions of the Country. The Israeli Yongledian Pilot Project (1125 mu) in Beijing, the Canadian activities in Hubei the Netherlands, USA and other assistance of FAO, ESCAP and international organizations can form the first components of such a network.

Development of Desert and saline water areas of Xinjiang and Gansu Provinces through geographic shifting of grains production to these areas. Such realignment will compensate the lost areas of grains production as result of urbanization and industrialization pressure and cash crops competition.

Reconsideration of self sufficiency policy linked to relative advantages criteria and focusing the economic self sufficiency as the main goal through additional expansion of export - import relations;
Development of a voluntary cooperative system of inputs and credit

supply, marketing and production in order to overcome the obstacles of small farms and to create economies of scale effect (to overcome the 4 mu average);

Development of modern irrigation system, including drip irrigation; Adjustment of the mechanization and agricultural equipment production to the new socioeconomic structure of Chinese agriculture is an urgent challenge;

Standardization and unification of the statistics infrastructure in accordance to the generally accepted UN/FAO methodology in order to avoid controversial estimation and for international comparison.

Long term and gradual genetic upgrading or dairy herds through artificial insemination, embryo transplant and high genetic heifers should be considered.

The Chinese model of economic reforms and the readjustment of agriculture and rural economy expresses a successful modality with clear advantages in comparison with other countries with similar problems. Additional improvements of the system will accelerate the entire process of modernization of agriculture and the agroindustrial complex in China. An important factor in development of Chinese desert areas could be Chinese participation in the multinational Desert Development Program initiated by Shimon Peres of Israel.

Specific Case Studies

A Chinese Program for Sino-Israel Cooperation in Agricultural Sustainability in Desert Areas. The desertification-prone land involves more than 100 countries of six continents all over the world. Twenty percent of the global population is facing the risk of desertification. The total land of the world's arid area has reached 6.1 billion ha., accounting for 37% of the global land. The arable land and pasture affected by the desertification are about 3 billion ha.. The desertification-prone land is increasing at the rate of 3% from 1980's to 1990's, resulting in over 40 billion dollars of economic loss every year all over the world. The desertification is the first one of ten environmental problems faced by human being, which were documented on the Environment and Development Conference in 1990.
Desertification is one of the most acute environmental problems in China.

The land suffering from desertification is estimated to cover 1,533,000 km^2, 71% of which is steppe and rainfed arable land of arid and semi-arid area suffering from the most severe desertification; 24% of the total desertification land locates in the irrigated area of arid and severe-arid zone named as oasis. On the other hand, desertification is even occurring in humid and subhumid areas. The desertification-prone land increases by 1.33 million ha. every year in China. The desertification land of Qinghai province increased by 1.9 million ha. from 1959 to 1989 and the total desertification land was 1.5 times than that in 1959. The desertification land in the plateau of Inner Mongolia reached as estimated of 600,000 ha. in the 1980's; The desertification land increased from 0.19 million ha. in 1961 to 0.258 million ha. in 1983 in Ningxia Hui Autonomous Region; The desertification-prone land increased by 0.646 million ha. from 1949 to 1989 in Heilongjiang province; The desertified land enlarged by 0.288 million ha. from 1977 to 1986 in South-East Maowusu Desert. It is predicted that the desertification-prone land will increase at a rate of 1.32% every year unless the effective measures are taken.

Desertification and wind-erosion have severally affected the development of industry and agriculture and people's living in some areas of China by leading to land degradation, decrease of land productivity, bearing capability and worse ecological and living conditions. The cropland, pasture, industry and mining, railway and infrastructure suffer from sand movement. Based on the statistics, about 1.33 million ha. arable land, 1 million ha. pasture land, 800 km. railway and a thousand of highway are reported to be threatened by desertification. The direct economic loss caused by desertification was estimated to be 4.5 billion yuan every year. Over 60% of impoverished countries with 17 million people located in the area are affected by wind erosion and sand movement in China. Desertification is not only environmental and economic problems, but also a severe social problem.

China's population is growing, which will intensify the contradiction among resource, environment and population, while arable land is still decreasing due to urbanization and extension of infrastructure. In order to reach the goal the total grain production increases to 500 billion kg., the most important strategically measure to be taken is the reclamation of desertified land for the development of agriculture like oases. With the economic reform and development, and the adjustment of industry structure and national policy in the East coast and Western area of China, the focus of economic development will move to the middle West of China on one hand, large mine, coal, oil and gas field will be constructed in those areas on the other hand. All of those need to control desertification and ensure economic development at the same pace. At the present, 80 million people in 60% of impoverished countries are living in the desertification areas. China only has to work hard to control desertification, but also has to rely on the development of agriculture and animal husbandry in this area to feed 80 million people and improve their living quality.

Previous Study and Control of Desertification. In the past the researches on desertification focused on the process of desertification, monitoring and mode of desertification control in arid and semi-arid area in North of China. A survey of the desertification situations and the map of desertification in the main desertification areas at different stages have been made, and the type, size of area cause and developing trends of desertification was determined. Based on the above research results, the way, measure and strategy for desertification prevention and control was proposed. Especially the network of more than ten experimental and practical stations for the prevention and control of desertification was set up in order to test and demonstrate the performance of the techniques for desertification control and explore the scenario for the technological and economical development. There are several national institutions engaged in on-spot investigations and researches on the prevention and control of desertification.

The researches on the prevention and control of desertification have been started since the 1950's. The early researches laid a particular stress on desertification process, geographical base of desertification, classification of desertification-prone land and summary of the farmer's experience in aforestation for sand-fixing.

During the 1980's, the prevention and control of desertification concentrated on the studies on classification of vegetation establishment sites, primary selection of their suitable plant species (tree, shrub and herb) and relevant techniques of plant establishment for different site type, silvicultural techniques of wind-breaking and sand-fixing shelterbelts for protecting oasis, aforesting techniques for sand-fixing in steppe by airplane sowing, aforesting techniques in salinizes sandy land in arid areas, techniques of regenerating and rejuvenating natural vegetation et. al..

Since the past few years, the prevention and control of desertification has paid attention to the studies on the impacts of aforestation and shelterbelts on water balance and environment, monitoring of desertification dynamics by using remote sensing, strategy and techniques of natural resource exploration and environmental improvement for the rural development. The great achievements have been made in the fields of the aforesting techniques by airplane sowing and planting shelterbelts of sand-fixation for protecting railways and highways, the techniques of planting eco-economic forest et. al.. The successful experience and may techniques for wind-breaking and sand-fixing have been obtained. Many research output and practical results reach or surpass world's advanced level.

Overall Objective of Desertification Control in China. Desertification in China is the major part of the global environment problems and its control and prevention will make a great contribution for the global environmental improvement. The desertification control of China is to carry out commitments of the world's convention "the prevention and control of desertification".

Chinese government has brought the construction of comprehensive control engineering of desertification into the state economic and social development plan and approved "the overall objectives of desertification control during 1991 - 2000" as the following:
• 13 million ha. desertification-prone land will be controlled and can be explored for agricultural use.
• Twenty national key research projects will be carried out in North of China.
• Twenty countries with severe desertification will be greatly invested for desertification control.
• Nine experiment and demonstration districts will be set up in order to produce the techniques of desertification control for 33 technical extension districts.
The priority to the control of desertification has been given and documented in China's 21st Century agenda as "The economic development strategy and measures of desertification control should be studies and taken to improve environment toward sustainable land use and to strengthen the sustainable development ability of China, which will greatly contribute to the prevention and control of desertification in the world".

There were many international cooperation and exchange among China, more than seventy countries and over ten international organizations in the course of prevention and control desertification in the past. The training courses have been jointly organized in China by China's institutions and international organizations. The several international symposia on the utilization and exploration of desertified land have been held in China. The may bilateral or multilateral cooperative projects have been finished or is going on. According to the past cooperative research output and experience, the cooperative researches and practices on the control of desertification have to be set up in order to make joint efforts to the control of desertification all over the world and realize the sustainability of socio-economic development, which would be beneficial to both parts.

Israel has the advanced techniques and excellent expertise and experience to use the upland and desert in arid areas for the technique intensified agricultural production. · China is willing to cooperate with Israel to strengthen the researches on desertification control in arid and semi-arid areas of North of China towards sustainable land use.

Objectives of the Future Cooperative Project. There are quite abundant radiation and heat resources, proper water and biological resources than can be explored for utilization in desertified areas of North of China. This area also has plentiful minerals resources. The future economic development in the West and the middle part of China will provide favorable factors and market condition for agricultural development of this area. The development of agricultural production and regional economy in this area will certainly be promoted by the rational exploration and utilization of the natural resource in the desertification areas, which not only can eliminate poverty, but also is a major means for the prevention and control of desertification. Major objectives of this research project are as follows:

• To develop high efficiency agriculture production system and relevant technological system suitable for the desertified areas, such as water-saving irrigation techniques, the techniques for greenhouse production, etc. .

• To develop the optimum scenario for the integrated utilization of all natural resource in the scattered distribution area of pasture and cropland so as to fully enhance both crop and animal production and to basically control desertification.

• To establish the monitoring and information system of desertification dynamics.

• To formulate the management systems and policy option for the adjustment of farmer's activities so as both to develop economy and to control desertification in the desertified area.

• To develop the pattern of the rational use of the natural resource in the desertified area for enhancing employment, improving the living quality and quarantining the construction of Northwest industry and energy base.

Activities. Both desertification control and the sustainable development of socio-economy are urgent strategically tasks in the desertified area. It has great theoretical and practical significant for enlarging arable land, solving problems concerning population increase, food security, paucity of the agricultural resource and environmental degradation and realizing overall strategically objectives of China's socio-economic development. The desertification control, elimination of the regional poverty and the sustainable development of socio-economy have to be integrated and the theoretical research, basic research for application and the applied research have to be combined in this project.

The researches will focuses on the mechanism of desertification formation and its distribution patterns in different type of the decertified areas, including the cause of desertification in the scattered distribution area of cropland and pasture, internal factors of dynamic link between climate change and desertification, the law and pattern of interaction among all factors, prediction model of desertification dynamics and the adjustment and control system.

• The scenario of sustainable development of agriculture and mode of desertification control will be developed in the areas with the different type of desertification. The experiment and demonstration districts will be established in the areas with the different types of desertification to study on the scenario of sustainable development of agriculture and techniques and its effects of comprehensive control mode of desertification and sustainable agricultural development.

• The desertified areas cab be roughly divided into the following types: (1) In pasture land with desertification, the experimental districts will be set up for comprehensive control of desertification, where the researches will out -- The engineering and techniques for the comprehensive control of desertification; techniques for the vegetation recovering, optimum structure of eco-economic shelterbelts and aforesting techniques for the protection of farmland and pasture; the rational pattern of the natural resource exploration and alternative pattern of living style. (2) In the scattered distribution area of cropland and pasture, the researches at the experimental district will concentrate on studies on the integrated utilization of the natural resource and comprehensive control of desertification. The research will focus on studies on optimum integrated utilization of all natural resource and the way to adjustment of agricultural structure, improvement of animal production, high efficiency utilization of the natural resource for mixed animal and crop production, crop production systems by using biological resources, the techniques for the added-value procession of crop and animal products. (3)In the desertified area with most arable land, the researches in the experiment district will focus on high efficiency crop production which produce the following:

The optimum integrated utilization of the natural resource and the adjustment of agricultural production, techniques and management systems for greenhouse production, water-saving irrigation techniques for crop and fruit production, planting - animal-procession systems with high efficiency by the exploration of resource.

The key technique for the sustainability of the agricultural development in the desertified area will be paid great attention to being developed.

The practical key techniques and reserve techniques suitable to the adoption to the desertifeid areas in North of China will be studied, and includes the following:

- The engineering and techniques for "oasis" construction.
- High-efficiency and intensive farmland construction and its relevant integrated techniques.
- Integrated techniques for enhancing grassland production.
- Integrated techniques for Greenhouse production.
- Techniques for the exploration of water resource and water-save irrigation.
- The techniques for the exploration and utilization of wind energy and radiation energy in the desertified area.

The strategy of sustainability of agricultural development and bearing capacity of water and land in desertification area will be studies in a large scale, covering the following topics. The bearing capacity of land resource in desertification area and pattern of sustainable land use; water resource potential and the bearing capacity of water resource in the desertification area; strategically scenario for the sustainability of agricultural development in the desertification area.

The Priority Research Project Will Be Carried Out If The Fund Is Secured. The experimental districts with different type of desertification control and pattern of agriculture development will be set up such as 4-6 experimental districts for comprehensive control of desertification and the development of integrated agriculture in Xinjiang, Gansu and Inner Mongolia provinces. The techniques of water resource exploitation and water-save irrigation will be greatly invested to be developed, covering the different type of water resource exploitation, drip irrigation technique for agriculture and forest and fruit production. The techniques for greenhouse production of the desertified area will be evaluated and improved, covering greenhouse structure and building material; techniques to increase greenhouse production and management.

Research Sites. The different type of desertified area will be selected as the research and practice sites in Xinjiang, Gansu, Inner Mongolia province where the desertification is acute while at the same time it is most important area for future exploration and development in the middle western part of China, where the majority of the population live under the average living standard and there is an abundant resource. In the past decades the most desertification control practices and research supported by the Chinese government in this area so that the plentiful experience and data might be accumulated. The future exploration of the middle western economy and the construction of "European and Asia continental Bridge (Railway)" will bring a great challenge and chance in agricultural and regional evelopment. Concrete research and practical area will be decided after field investigation by both sides.

The Period of Project Operation and Fund

The project will be divided into several stages. Every stage cover five (5) years. Project funds will be supported by Israel and China. The profile will be discussed in detail by both sides.

Implementing Institution

The Bureau of Agriculture Resources and Management, Ministry of Agriculture, P.R.C.

Some Lessons and Conclusions on Agriculture Water Resources and Irrigation System in the Shiyang Basin of Gansu Province. Analysis of the climatic, hydrological and agricultural data leads to the conclusion that the paramount problem facing agriculture in this area can best be defined as : management and efficient use of water for irrigation under conditions of scarcity. The dominant factor influencing life in this region is the shortage of water.   Two aspects to this situation is (a) water determines the boundary of desert encroachment; and (b) water sets the limits to agricultural production.   It appears that there is now a delicate balance at the edge of the desert bordering the project area, that could be upset by utilizing additional water for increasing the irrigated area. Consequently,   any attempt at raising the income of the farming community should seek ways to increase the value of agricultural production, using the same quantity of water.   Concurrently, a thorough investigation should be undertaken to examine the possibility of exploiting additional water sources, particularly groundwater.

The Shiyang Basin is situated in a semi-arid to arid region, where crop production is dependent on irrigation.   Since all of the water resource potential is being exploited,   irrigated agriculture can be further developed only by using the irrigation water more efficiently. However, before pointing to specific solutions,   it must be mentioned that the principal constraints that has to be taken into consideration when proposing changes or improvements. The constraint is the   regional wheat quota that must be produced as directed by the national agricultural policy.   In this region the wheat crop yields on the average 4 tons/ha, therefore, in order to produce the prescribed quota,   wheat takes up to 70% to 75% of the irrigated farm land.   Consequently, as long as wheat yields remains at this level,   only 25% to 30% of the irrigated land is available for possible improvements in production and income per unit of water.   Clearly, it way can be found to increase the wheat yield per hectare,   then the area planted to wheat can be reduced and more land would become available for changes and improvements.

In practice,   improvements can be brought about by focusing on the following subjects: Throughout review and analysis of the water resources, followed   by well planned management and operation, may lead to the conclusion that additional quantities of water can be safely utilized,   which in turn would enable the irrigation of additional lands.

Improving the efficiency of water use for irrigation on lands retained for continued wheat production to be irrigated as before,   by the surface method of flooding.   Higher irrigation efficiency can be attained by between water management for irrigation through the adoption of technical, agrotechnical and administrative measures (as detailed in the following).

Replacing the traditional methods of irrigation (on land not planted to wheat) by modern, water conserving pressure irrigation systems: low volume irrigation such as drip, micro-jet or micro-sprinkler, combined with the introduction of high value crops. Savings in water together with higher crop yields will increase the farmers' income per unit of water used, and will enable the development of additional land for irrigated cropping.

Development of supporting services for agriculture, particularly post-harvest handling of produce, agroindustries and marketing facilities, enabling the shipment and sale of produce to distant markets, attaining higher prices and better income for the producer.

Improving agricultural research facilities to accompany and assist the process of modernization in cropping and irrigation methods. Equally important is the development of an efficient agricultural extension service to facilitate the transfer of new know-how from the laboratory and experimental plots to the farmer.

The Shiyang River Watershed is a closed inland basin that is fed by rains and snow falling on the surrounding mountains. The surface runoff and underground inflows from the mountains and find their natural outlets as follows: Free flows in streams and creeks towards the low-lying saline marshes. Outflow from springs. Exit from aquifers (where the water table is close to the surface) by evaporation or evapotranspiration.Similarly, evapotranspiration by natural vegetation, not necessarily in the salt marshes but by deep-rooted plants, whose roots penetrate down to the groundwater.

The natural flow regime has undergone considerable changes, as a consequence of the intensive exploitation of the groundwater by means of wells, and of surface water by dams and diversion structures. These changes have led to a marked reduction in the quantities of water reaching the above listed natural outlets, mainly because of the large-scale withdrawal of water by thousands of wells. The expansion of irrigated areas has created an additional potential groundwater source through the infiltration of irrigation return water, the content of which is a function of the scope and efficiency of the irrigation systems. Analysis of the available data leads to the conclusion that the central component of the regional water resource system is the aquifer. It is the acquider that has to be thoroughly studies - its structure, extent and borders, its division into sub-aquifers, the relations between its inflow and outflow areas, and its capability to serve as the principal operational reservoir in the Basin. The answers to these questions will have a crucial influence on the potential water resource and its future exploitation.

Available data indicates that the depth of the regional aquifer may reach hundreds of meters. It is composed of sandy layers, separated by clayey strata that divides this aquifer complex into several units and sub-units showing geological sections in different parts of the Basin. Most of the currently available information has been gained from the wells drilled in the central plain along the river beds and in the cultivated areas. Most of the thousands of wells sunk in these areas penetrated no deeper than 100-150 m, with the exception of a few deep drillings (300-400 m) that given some indication of the deeper aquifer sections.

Whilst the majority of wells pump from the upper acquifer layers, there is reason to believe that large quantities of water flow under high pressure directly from the mountains into the deeper, unexploited aquifer strata. Some of these deep flows emerge as springs in various parts of the valley and some drain into the more distant saline marshes bordering on the desert. The possible existence of such a sub-system of flows is indicated by the location of springs in places where the level of the groundwater table is lower (as evidenced by the boreholes) by several meters than that of the springs. According to locally available information, the discharge of springs has not changed significantly in spite of the massive withdrawals from the upper aquifer layers and the drop in the water table - confirming the possible existence of separate sub-systems.

There are some 11,000 boreholes in the Shiyang Basin, 8,000 of them operational and more than half of them concentrated in the Mingqin area, which constitutes a small part of the Basin. Boreholdes have been drilled in groups, concentrated in long, narrow strips along the banks of streams and in irrigated areas bordering on the desert dunes. Geological maps of the Basin indicates the existence of a number of transverse faults in a West-East direction, descending from the mountains towards the desert, possibly creating hydrogeologial boundaries between the sub-basins along the entire depth. The flow between these sub-basins is either blocked altogether or there may be narrow connecting passages. Therefore, one may assume that a well field pumping from a sub-aquifer may cause a significant drop in the water table, whilst barely affecting a neighbouring sub-aquifer. This heterogeneous character of the Basin has an important bearing on the distribution of production wells, on the efficient exploitation of the ground water potential and on the prevention of adverse environmental effects. Fluctuations in groundwater levels are confined to the withdrawal areas; whilst there is little information about the situation in the desert and its proximity, one would expect a much more moderate decline in water levels. It is of great importance, therefore, to adopt a properly controlled and managed groundwater policy, closely integrated with surface water utilization in order to attain maximum efficiency, taking into consideration the environmental aspects.

It is of great importance to verify the existence of deep sub-aquifers that drain into the saline marshes where the water evaporates and cannot be exploited. Assuming that their existence can be confirmed, the volume of additional exploitable water may reach 100 million cubic meters per year. This estimate is based on the following: (1) Originally, before the utilization of surface waters by dams and reservoirs, and before the start of intensive pumping of groundwater, the order or magnitude of natural groundwater flow in the upper aquifer layers was estimated at hundreds of millions of cu.m. (taking into consideration that much of the surface water flow estimated at more 1,000 MCM/year, infiltrated into the aquifer along the revived beds). This groundwater flow drained into the saline marshes and into shallow groundwater basins in the plain and evaporated under conditions of a closed inland basin. At present, there are no signs of drastic shrinkage of the saline marsh area, and the groundwater tables have also remained high, as shown by topographic and hydrogeological maps. Evaporation too has continued, although not at the former rate of hundreds of millions of cu.m. However, considering the extent of the areas (about which there is no accurate information) and the past flow rate, a reasonable estimate of the present outflow would be between tens of millions to hundred million cu.m.

(2) At present, it seems that groundwater levels in most of the Basin area are not more than a few meters below ground level and that the general flow gradient along the slope is not markedly different from its natural state in the past. Thus, the changes are not sufficiently great to justify a drastic reduction in the flow rate by hundreds of millions of cu.m/year.

It seems more probable that the halting or significant reduction of outflows would necessitate the lowering of the water table by tens of meters, reversing the flow direction and creating local depressions - none of which has occurred in the Basin.

(3) If indeed the deep aquifers feed the springs in the valley - and taking into consideration the actual spring discharge - it is then reasonable to assume that the groundwater flow downstream to spring outlets would be at least in the order of magnitude of tens of millions of cu.m. (No data has been received on hydraulic conductivity in the aquifer layers, nor about pumping test results and other relevant hydraulic characteristics). Since the geological sections show no sign of the existence of physical hydrogeological obstructions, it is assumed that the flow continues down the slope and either emerges in the desert area or remains close to ground surface.

To gain a deeper understanding of the subjects discussed above, it will be necessary to conduct comprehensive studies and surveys of the hydrogeological systems, the available potential water resource, the possibilities of utilizing additional water and the efficient operation of the resources. There is a reasonable prospect that on the basis of the study results the available quantities of water can be increased significantly, which would certainly justify the expenditures involved in the studies. However, even if the studies will not procure additional water, they will make a valuable contribution to the efficient management of the existing water resources, making possible substantial savings in irrigation water, in addition to the savings effected by the adoption of advanced irrigation methods.

It has been mentioned above that about 70% to 75% of the irrigated area will continue to be allotted to wheat. Clearly, there would be no point in adopting modern irrigation methods, like drip or sprinklers, for the wheat crop; it would be unwarranted for technical as well as economic reasons. The income from the wheat crop per unit area is very small compared to other crops, consequently, the traditional flood irrigation method will be retained.

Whilst this method of surface irrigation is practiced here comparatively efficiently, it should be possible to find ways to further improve the efficiency of wheat irrigation. Improving wheat irrigation efficiency even by a small percentage would result in savings of water in significant quantities - water that could be sued for the development of additional land for irrigated cropping. At present, water losses in main, secondary and third canals supplying irrigation projects, average about 27%. Some of the loss is through evaporation from the canals but most of the water is lost by infiltration through the sides and bottoms of the canals, in spite of the fact that all canals are lined. It is certainly possible to reduce these losses by proper care of routine maintenance words: removal of weeds, immediate repair or cracks, fractures or faults in the lining. The maintenance of these canals is the responsibility of the Prefecture, and it should not be difficult to lay down rules for the routine monitoring of susceptible sites and for the quick location and repair of flaws.

Fourth and fifth distribution canals, fed from the main supply network, are the responsibility of the villages. These are earth canals and the losses are mostly through infiltration. As the water is metered before passing into these distribution canals, these losses not only reduce the quantity of water reaching the village for irrigation, there is also a financial loss, for the village has also paid for the lost water.
The villages should therefore improve the maintenance of earth canals - according to the directives and under the supervision of the Prefecture - by setting up a system of continuous monitoring for the spotting of faults and immediate repair, and for the routine maintenance of the canals, also in the off-season, with particular attention to the removal of weeds and patching of the canals.

The villages receive their allotted water via division structures installed on the main conveyance system. The water is measured at the head of the fourth canal that conveys the water to the village. These divisions and measuring structures are controlled by the Prefecture. In order to ensure accurate distribution of water according to the quotas laid down by the Prefecture, it is essential to maintain the division structures, especially the gates, in good repaid; they should be impermeable so that no water is lost when there is no supply to the village and that they open and close accurately during supply. The measuring structures must be calibrated and properly maintained to enable exact measurement of the water supplied to the village.

It is important to check whether there are properly functioning division and measuring structures at all outlets from the conveyance system to the village distribution canals. Wherever they are missing or are in a deteriorated state, new structures should be installed and the worn ones replaced. The objective of efficient water management in irrigated agriculture is to provide timely and adequate water supply to the crops, for obtaining maximum crop yields with the economic use of water. This objective can be attained only by maintaining proper soil-water-plant relations by employing suitable technical and agrotechnical tools, based on reliable up-to-date information on crops, soil and climate.

It is necessary to develop more accurate tools - in the form of field trails, as well as basic research work - in order to establish the exact water requirements of crops. In practice, this means laying out many more experimental plots, distributed in the project areas, so that they accurately represent the typical features of the surrounding irrigated lands. A number of small meterorological stations will also have to be established in order to provide the climatic data needed for calculating crop water requirements. Based on climatic information collected in the irrigated areas, there are now reliable methods for accurately calculating crop water consumption. There are very few meteorological field stations in the Basin area, in fact, records are being kept only in two centres: Wuwei and Mingqin. Once the required number of additional field stations are established, the recorded data will be used in conjunction with the information gained from the experimental plots.'

More soil tests will have to be carried out, taking more samples in the irrigated areas. At present, tests are sporadic, samples are few and far between providing little information on structure and moisture holding capacity of the soil. Information on soil moisture holding capacity is most important; this and the water requirements of the crops are the two principal parameters for determining the quantities of water applied in each irrigation and the intervals between irrigation's. These in turn are the precondition for efficient water management in the irrigated area, making it possible to provide timely and adequate water applications to the crops. When information on soil-water-plant relations is lacking or is unreliable, excessive quantities of water are often applied at the wrong intervals, resulting in losses of water by percolation into the subsoil, carrying with in nutrients from the root zone. An additional measure - improving the physical structure of the irrigated basins - is also important. The land should be level, without slopes, depressions and bumps; only then can water be distributed uniformly to all parts of the I irrigated plot.

People, particularly farmers, living in arid and semi-arid regions generally realize the importance of saving water; nevertheless, the Prefecture should take steps to enhance awareness of this subject. A comprehensive information campaign should be set in motion, using radio and television broadcasts, as well as youth and adult education, starting from elementary schools to farmers' organizations and to the public in general. The campaign will stress the fact that water is the most important natural resource in the region and that it is in short supply. Its wasteful use will halt development in the region and reduce the chances of improving the standard of living of the population.

In view of the fact that 70% to 75% of the irrigated area will continue to be planted to wheat and be irrigated by the surface method of flooding, 25% to 30% of the area remains for the introduction of modern methods of irrigation, combined with high value crops and advanced cropping techniques, in order to achieve maximum income per unit of water used. It is proposed to replace gravity irrigation by low volume pressure irrigation, such as drip or micro-sprinklers, as the primary means of improving irrigation efficiency.As these irrigation methods involve relatively high capital investment, their suitability for the various crops will have to be assessed by economic analysis. From the technical viewpoint there are no obstacles to the adoption of low volume irrigation (LVI); farmers in the Prefecture possess the technical skills needed for the efficient operation and maintenance of LVI systems. Also there are crops suitable for this type of irrigation, particularly apple and pear varieties, that will probably justify economically the heavy investments in the irrigation installations.

Having said that much about the favourable prospects, it should be noted that the introduction of modern irrigation systems requires major improvements in the collection of basic data through observations, field trails and fundamental research. The advantage of advanced irrigation systems lies in the precision they afford in controlling water applications, quantities and timing, as well as their capability of supplying the required plant's nutrients through the irrigation water. Hence the need for improved data collection facilities in the region, that can provide all the relevant information on climate, soil, water and plant.

In selecting the preferred source of water for LVI, two main factors must be taken into consideration. They are as follows: (1) The water must be supplied under pressure via closed pipelines, which implies that the water source should be near the irrigated land in order to reduce the cost of piped conveyance. (2) The water should be clean, with minimum sediments, suspended solids and organic impurities in order to reduce the risk of clogging in the emitters.

For these reasons as mentioned above, preference should be given to existing wells for supplying neighbouring lands through newly installed, modern irrigation networks. Well water is generally pure and even if it contains some sand, that can be removed by mechanical means before the water enter the pipeline. Water from wells can be supplied through the year (which is not the case with gravity irrigation systems). This is important because some of the crops proposed under this plan are fruit-trees that have to be irrigated in all seasons and others are seasonal cash crops under intensive cultivation of at least two crops a year, requiring regular virtually year

As mentioned above, economic analyses will have to be undertaken to ascertain which of the proposed crops can be produced profitably with the LVI system. Form the agrotechnical point of view, the system is suited to orchards and to a wide variety of row crops - vegetables and field crops, like melons, sugar, beets, cotton, maize, oilseed crops and others Due to the precise control afforded by the LVI system over water and nutrient applications, the planning and installation of the system must also be carried out with extreme care and accuracy, paying close attention to growing conditions related to climate, topography, soil, plant and water quality. Therefore, the planning and implementation of LVI must be preceded by data collection as follows: (1) Topographical maps of each irrigation plot, to a scale of 1:1,000 with contour intervals of 1 m and indicating the planned trees and crop rows. (2)Soil survey, including physical and chemical laboratory analysis, with particular attention to soil structure, moisture holding capacity and chemical composition. (3) Water quality tests of physical characteristics (sediments, suspended solids, organic matter) and chemical composition. (4) Detailed, accurate date on extreme climatic events such as exceptionally high and low temperatures, including frosts. (5) Characteristics of the crops required for LVI design must consider: (1) crop species and varieties; (2) times of sowing/planting, the growing of harvesting seasons; (3) exact plant water requirements throughout the growing season; (4) and sowing/planting distances within the row and between the rows.
The following three LVI methods are considered to be suitable for the proposed crops drip irrigation, micro-sprinklers and micro-jets. They all have the same advantages: Saving water; Accurate control of water and

fertilizer applications and irrigation and fertilization schedules; High quality and more uniform crops; Higher crop yields; Adaptability to marginal soils; Suitability to poor quality (saline) water.

All LVI systems comprise identical components - the difference lies in the emitters and in the wetting pattern. The main components are:(1) Control head. It is installed at the head of the plot, on the main distribution line supplying the plot. Its components are: filter, fertilizer application system and water application control system. (2) Distribution System, consisting of plastic pipes, fitted with valves and pressure regulating devices. (3) Laterals, consisting of small diameter plastic pipes, fitted with emitters - drippers, micro-sprinklers or micro-jets.

Regarding the layout, planning and operation. there are no significant differences between the various types of LVI system. The difference as mentioned earlier is in the emitters and in the shape and size of the wetted area. The irrigation method and the type of emitter will be selected in the detailed design stage of the irrigation system, according to the crops, the soil types and the local climate conditions. The introduction of the LVI system will be preceded by a more detailed study of agricultural development in the region, including the scope of irrigated agriculture as imposed by the potential water source, soil suitability and the profitability of the planned crops. It is proposed within the framework of this presentation, to establish in the initial stage of the project a number of demonstration and experimental plots, with the objective of demonstrating the operation of the LVI system and to prove its suitability to local conditions, and also to investigate plant, water and soil relationships.

Also see : WU WEI Readjustment of Water and Irrigation System,
Evaluation Report,
SDSUF TAI TUNG Delegation for Technological Cooperation
1988.

CHAPTER 4.6

TOWNSHIP and VILLAGE ENTERPRISES IN CHINA[1]

He Kang
World Food Prize Laureate
Former Minister of Agriculture, PRC.

The phenomenal growth rates of township-and-village enterprises (TVEs), dubbed China's secret weapon, have attracted worldwide attention since nearly 20 years ago. From their humble origins as cottage industries and Commune workshops, these smallscale rural enterprises languished through the catastrophic years of the "Cultural Revolution (1966--1976)." But all that has rapidly changed since China opened up to the outside world. Recently, China has been able to maintain double-digit economic growth rates. This has been the trend since 1993 when rural areas contributed more to the Gross Domestic Product (GDP) than urban areas.

Growth Rates of GDP: 1992 --- 14.2%; 1993 --- 13.5%; 1994 --- 11.8%; 1995 --- 10.2%; 1996 --- 9.7%.

In 1995, the growth rate of our GDP was 10.2%. Rural areas contributed 72% (=7.35% of 10.2%) and urban areas 28% (=2.85% of 10.2%) to this growth. In 1996 the growth rate of the national GDP was a bit lower, reaching 9.7%, but still, 86% of 1996 GDP growth rate was achieved by rural areas.

Township and Village Enterprises (TVEs) and the rural economy they support, have evolved to the point when in 1996, rural areas contributed 58.3%, whereas urban areas contributed 41.7 to China GDP.

It is clear now that non-farming sectors of TVEs have been a vital force in the national economy. The 1996 rural gross output value is revealing:

---

[1] The editors wish to acknowledge with thanks to Madame Yu Junmin for her contribution and translation of this manuscript. Received Sept. 2, 1997.

TOWNSHIP AND VILLAGE ENTERPRISES

Rural Gross Output Value, 1996

Agriculture ------------- 25.1%
Rural industry ---------- 63.5% (up 1% from 1995)
Rural tertiary sector--- 1 1.4% (up 0.4% from 1995)

The rural industry greatly increased. In 1996, TVE gross output value reached 1,766 billion yuan ($213 billion), a growth rate of 21% at fixed price. This provided 56% of the total 1996 rural GDP, or 26% of the national GDP. The value of TVE rural industries increased to 1,263 billion yuan ($152 billion), up 16.9% from 1995. The increased value of TVE rural construction, transportation, commerce and other service industries was 326 billion yuan ($39 billion), up 45.8% fi7om 1995. Thus the rural industries reached a share of 47% of the 1996 national industrial gross output value. The strength of TVEs increased, their fixed assets reached 1354 billion yuan ($163 billion) in 1996. Their profits amounted to 435 billion yuan ($52 billion), up 17.7% from 1995.

ORIGINS

As one popular saying goes, "hotcakes do not drop from heaven." The origins of China's TVEs emerged, paradoxically, in the 1950s, a time when central planning, based largely on Soviet models, had its tight grip on the countryside. Those were the days of the People's Communes, the inevitable outcome of a whole process of collectivization known as socialist transformation, hastily done in a short span of six years.

Historically in developed countries, handicrafts were separated from agriculture at some stage and later developed into independent, strong industries as capital was accumulated and peasants went bankrupt to become workers. However, in old China, it was a different story. It was impossible for national industries to grow vigorously in a semi-colonial and semi-feudal society. Therefore, when New China was founded in 1949, there were not many modem factories to start with. Handicrafts were then largely sideline occupations of farming households and an important source of cash income for peasants.

Take the example of coastal Jiangsu Province, a comparatively more developed area before 1949, and currently one of the most developed areas of TVEs in China. Jiangsu was then known for its cottage industries, producing cotton and silk, pottery, bricks, tiles, iron- wares, rice, wine and liquor, and more than 100 kinds of local crafts using native raw materials. Handicrafts also thrived in some small towns where a much greater proportion of specialized products were sold as commodities.

Statistics in 1949 showed there were roughly 150,000 peasants engaged in sideline handicrafts, and full-time handicraft workers amounted to 730,000 about 11.3% of the national totals. Even so, the gross output value of Jiangsu's handicrafts was 589 million yuan ($70 million), or 13.6% of the gross output value of both the province's farming and industrial enterprises, and 38.9% of the gross industrial output value.

After New China was founded, Jiangsu's individual handicraft workers and businessmen were organized into productive coops, and moved into small towns. Later many of these were turned into State-owned enterprises. Structurally this intensified the separation of rural cottage industries from farming. And what was left in the villages was consequently a very small proportion of minor crafts. Although these were not the only basis from which the present TVEs directly emerged, they were nonetheless meaningful to rural China in the days to come.

REAL PREDECESSORS

Beginning in 1958, to answer the call of the Communist Party to set up industries for developing agriculture, the People's Communes did set up various rural workshops to repair and make farm tools, make bricks, tiles, cement and process farm products, etc. Such small-scale factories relied mainly on handicraft skills. These "Commune and Brigade Enterprises" were so called because they were run, **at** different stages, by the three tiers of Commune administrations: (from bottom to top) the production team, brigade, and Commune.

Peasants in China were then organized in these People's Communes, which assumed both the role of grass root governments and of economic entities engaged in farming. They were unsuccessful experiments born of ultra-leftist thinking and a misunderstanding of the principles of socialism. Consequently, the Commune and brigade enterprises faltered in the two decades that followed. Political movements, wrong policies, and misguided economic systems... all resulted in their slow growth rates. In 1978, the gross output value was only 49,300 million Yuan ($5,939 million), a mere 7 % of the national GDP. Their employees numbered 28 million or 7% of the national labor force. The tax they paid was only 2% of the national revenue.

DEVELOPMENT

By the end of 1976, the notorious Gang of Four were imprisoned after Mao Zedong died. Thus the ultra-leftist "cultural revolution" was ended. The Third Plenum of the 11th Central Committee of the Communist Party convened in 1978, then called for a "great development" of the "Commune and Brigade Enterprises." It stressed that State-owned factories in cities should help Commune enterprises to process farm products or make machine spare parts. Soon the People's Communes themselves were dismantled. Supported by the peasants, who knew best what was better for them a new system of farming called "household contracted responsibility system " was officially endorsed in late 1981 and still predominates today.

Now land is still owned collectively by the villagers and by contract, peasant households till the land as long as they pay taxes and certain fees to support the village and *Xiang* (lowest level of government[2]). Also, they now finally get the right to plant what they choose. In the Commune days, peasants were laborers only, deprived of all decision making. Often they did not get proper rewards for the work they have done. At one time, they were even forbidden to take up sideline occupations, labeled then as "capitalist tails."

For decades, peasants had to stick to their native villages unless they were drafted. They could not live or work in cities because of the rigid nationwide household registration system and grain rationing in urban areas. Today, after fulfilling production quotas to sell to the State a fixed amount of grains at State procurement prices, they have freedom to make money in many ways: some sideline occupations including planting cash

---

[2] Towns are often the seats of *Xiang*. Unlike the *Xiang* leaders who are appointed by county governments, heads of villages should be democratically elected by villagers. Villages are autonomous units.

crops, fishing, raising poultry, transportation, commerce etc. They can now sell surplus grain and other products at free markets, or sell more grain to the State at higher, negotiated prices. They can work part or full time at nearby TVE factories, engage in service industries, or even, join the 60-70 million strong "floating population," and travel to far-away cities to find work. All these would be fantastic tales in pre-reform years. When leftist policies and inefficient central planning joined forces to stifle the peasants' initiative, agricultural productivity was doomed to drop. It dropped to the lowest level in the 20 years between 1958 and 1978. Now, when the shackles are gone, productivity soars. TVE emerged, and was market-oriented from its inception. The peasants themselves learned from bitter experience that without the markets, they would never grow richer. Gone forever is the leftist notion that markets are linked solely with capitalism. Now when incomes of both the peasants and township and village administrations rose, investment into Commune and Brigade enterprises also increased.

By 1983, the gross output value of Commune and Brigade enterprises reached 101.7 billion Yuan ($12 billion), or 9% of the national GDP. Industrial output accounted for 12% of the gross national output value, amounting to 76 billion ($9 billion). The first document issued by the Communist Party after 1984 New Year's Day said not only collectively owned enterprises that were run by grass-root level governments or village committees should be promoted, but those set up by individual peasants should also be encouraged.

Before the document was issued, the Ministry of Agriculture suggested that since "People's Communes" had already been dissolved for several years, the question of proper nomenclature should be considered. To facilitate building up infrastructure and administration over the widely scattered rural enterprises, they should be encouraged to move to or be built in and around small towns. Thus, the document formally changed the term "Commune and Brigade enterprises" into "township and village enterprises." Later not only were agriculture, industries, but also commerce, transportation, and service and building industries were promoted as TVEs. In this congenial climate, newly granted loans in 1984 by banks to TVEs amounted to 65 billion yuan ($8 billion). The gross output value in 1988 amounted to 702 billion yuan ($85 billion), a 24% share of the national GDP. TVEs numbered 19 million, an increase of 13 fold over that in 1983. In Jiangsu Province, for the first time in history, the gross output value of TVE surpassed that of farming and sideline production.

In 1987, Deng Xiaoping said to foreign visitors: *"It is totally unexpected to us that the greatest achievement in the rural reform has been the development of the TVEs, a new force suddenly coming to the fore. "*

But owing to the rampant growth of TVEs, management, low labor force quality, pollution and other problems became evident. From 1989 to 1991, the central government adopted a readjustment and streamlining policy towards the national economy. In 1990, no new loans were granted to TVES. Many small scale and badly run enterprises closed down, were merged, or went bankrupt. Several million peasant-workers returned to the fields.

Nevertheless, after the streamlining, stronger and more efficiently run businesses emerged. They imported advanced equipment and technology, learned management science, and won investments from abroad. Export-oriented TVEs quickly developed. In 1991, export products delivered were valued at 67 billion Yuan ($8 bhlion), a 1.5 fold increase over that of 1988, or, one-fourth of the total amount of national export value. TVEs .then numbered 19 million, a 10% increase over that of 1988. Their gross output value was 1,162 billion Yuan ($140 billion), an increase of 65% over that in 1988, or 26% of the national GDP.

Since 1978, TVEs have contributed much to the national economy. But there has always been skepticism and unwieldy restrictions against TVEs from certain government departments. Those few who still believe in a centrally planned economy worry about the role of State-owned enterprises, if TVEs grow too strong. Basically this kind of skepticism originates from the already mentioned misunderstanding of the nature of socialism. This indicates that theoretical studies on the principles of socialism should be developed, to ensure smooth sailing into the socialist market economy.

Nevertheless, 1992 brought good news for TVEs. When the retired senior leader Deng Xiaoping visited southern China, he was deeply impressed by the changes in the Special Economic Zones and other coastal areas. He commented that TVEs were just one indication of the "superiority" of "'socialism with Chinese characteristics." Secondly, it was in 1992 that the Conunutfist Party officially pronounced China was building a "socialist market economy." This is a system that should uphold both social fairness and market efficiency.

As was predictable, the year of 1992 saw a rocketing GDP growth, and then the unsaddled investment tide led to an economic overheat, later, a high inflation rate of 21.7% in 1994. Since 1993, the Central Govenunent has enforced macro-control by making successful, flexible use of market-oriented economic leverage such as taxes and price controls, and credit and interest rate adjustments to guarantee both high growth and low inflation.

Many TVEs in this period of high growth sustained restructuring that led to improved management and better economic scale. They formed groups of firms, companies or corporations, reformed ownership systems, and made more rational and economic use of resources. Some of the most successful ones have not only expanded export trade but made their debut in international securities markets. High-tech TVE industries have emerged.

Take the example of the better developed Jiangsu Province. During the five-year period of 1991-1995, per year growth rate of technology and capital intensive enterprises reached 55%, and one fourth of these have set up research and development institutes. More than 100 large and medium scale Jiangsu TVEs have acquired export licenses. There are now more that 200 Jiangsu TVEs which are operated overseas.

Today, the term TVEs includes not only those classic ones originally set up and run by village or township administration, it also includes those collectively or individually owned enterprises invested in mainly by peasants as well as other sources of investment, including foreign capital.

By the end of 1996, TVEs numbered 23 million, an increase of 6% over 1995. Among those collectively owned amounted to 1.6 million, down 4% from that in 1995, and a share of 7% of the TVE total. Those owned by individuals and by several households numbered 22 million, up 7% over that of 1995, and were a lion's share of 93% of the TVE total. There were 1.3 million TVEs with eight or more people, coming to 6% of the total of those owned by individuals and by several households.

Some experts have their own way of classification. Besides ownership and sources of investment, they also take into account historical background, kinds of commodities produced, way of production, management or circulation etc. They use the term "modes of TVEs." The well-known ones include Southern Jiangsu Mode, Wenzhou Mode, or Jingjiang Mode. The diversification will require an in-depth study to explain.

The Law of Township and Village Enterprises passed in October 1996 stipulates that all TVEs have the duty to support agriculture financially.

BENEFITS

As mentioned above, old China had no strong industrial base. For new China, although priority had been given to building up heavy industry, it was evident that without striking a proper balance between agriculture and industry, there was no way to have a sound, sustainable fast growing economy. Furthermore, family planning was ignored until after the " Cultural Revolution". It would be too arduous a task to industrialize a whole developing country with a huge rural population (now: roughly 900 million in a total of 1.2 billion) by relying solely on State-owned industries. Our peasants finally stroke a chord: the role of TVEs has hastened the tempo of industrialization. TVE is thus a purely Chinese rural phenomenon. Tangible advantages include:

1. Strengthening agriculture. TVEs have easy access to land to build workshops, since land belongs to villagers collectively. Peasants can then work part or full time throughout the year and need not to swarm to big cities to find work. When out of work, they can still till the land. Hence, growing out of the rural community, TVEs logically win approval to help support the agricultural economy. China's per farm household's 0.42 ha land is too small for peasants to earn a high return before agriculture is fully modernized and switches from extensive farming to intensive. From 1978 to 1995, TVEs contributed 100 billion yuan ($12 bilfion) to boost agriculture, a sum that equals 80% of State fiscal expenditure on agriculture in the same period. TVEs will be a force that will help to modernize

2. Increasitig effective supplies to society. TVEs not only supply daily necessities but also many kinds of means of production.

The proportional contribution of some TVE products to the domestic economy include:

Electronic & Communications Equipment	17%
Machinery	26%
Coal	40%
Cement	40%
Food & Beverages	43%
Garments	80%
Medium-sized & Small Farm Tools	95%
Bricks & Tiles	95%

3. The main channel to transform surplus rural labor. In 1996, 6.5 million people newly joined TVES, which then had a total force of 135 million(up 5% from that in 1995). This surpassed the total employment by State-owned enterprises.

4. A new force for industrialization and for building small towns. The increased gross output value of rural industries in 1996 was 1,263 billion yuan ($152 billion), or 47% of the national total. In coastal Zhejiang Province, the gross 1996 industrial output value of TVEs was 76% of the provincial total. The following figures for this year are also convincing:

Output sales value from January to June 1997
State-owned enterprises: Industrial output sales value—1,390 billion ($167 billion) (up 6% from that in Jan--June 1996)
TVEs: Industrial output sales value—1,069 billion ($128 billion) (up 16% from that in Jan--June 1996)

TVEs have been directed to move to small towns or newly exploited areas called small non-industrial parks which can then grow into small towns. Currently there are 6,209 small industrial parks. Six-thousand small towns boast gross annual output value of more than 100 million Yuan' ($12 million) each. Small towns facilitate the growth of service industries which are badly needed in rural areas, and have become political, economic and cultural centers in the rural communities. It would have cost the governments' coffers an astronomical sum of money to build industrial towns without TVEs shouldering the bulk of the burden

5. Improving peasants' livelihood and public facilities. From 1991 to 1995, 30% of the net increase of peasants' income came from TVEs. For the better developed areas, it was over 50%. In 1996, TVEs paid total wages mounting to more than 534 billion Yuan ($64 billion). This means the per capita average annual income for TVEs rose from 1,380 yuan ($166) in 1990 to 4195 Yuan ($505) in 1996.

In some villages of Jiangsu Province, farming households' income from TVEs often constitute more than 90% of the total. Collectively owned TVEs run by *Xiang* or town governments and village committees contribute funds to improve or set up schools, retirement homes for the aged, and other educational, cultural or medical establishments. New roads and communication and irrigation facilities are often wholly or partly funded by TVEs.

6. Contribution to national tax revenue and export. In 1996, TVEs paid 144 billion yuan ($17 billion) tax to the State, up 12% from that in 1995. It came to 25% of the 1996 total national revenue. In 1996, TVE exports delivered was 601 billion yuan ($72 billion), up. 111% from that in 1995, a share of 3 5.7% of the national total.

Foreign investments into TVEs amounted to $8 billion in 1996. The total foreign investment up to last year then reached $31 billion. Most of the products of joint ventures with foreign capital are for export. Altogether 150,000 TVEs enterprises furnished products in 1996 for export, 598 of which have export license, while the rest exported through State-owned corporations.

7. Furnishing experience for setting up a new socialist market economy. Since TVEs were market oriented from the beginning, they've furnished experience in marketing and market research and management. Especially valuable are their experience in searching for better ownership systems, which will be explained in the following pages under the subtitle "Problems."

8. TVEs, good schools for peasants. Before 1978, most peasants were confined to the small world of their native villages. The per farm household's 0.42 ha land was their biggest concern. Attending seasonal peasant market in nearby villages or local fairs was their greatest delight in life.

China has spent too small a portion of her budget on education, smaller than some other developing countries. The majority of the huge rural population have had only 6--9 years schooling and illiteracy or semi-illiteracy is not uncommon. Now TV]Es have encouraged peasants to end their isolation. The rural population is now increasingly aware that money can be made from a variety of sources. They are aware that to run or to work in factories require knowledge and skills. "Technology," "science," "management" and "modernization" are becoming common words in their vocabulary. Peasant entrepreneurs are increasingly common. More and more peasants have begun correspondence courses or joined training classes. Many go to classes and pass examinations to earn " Green Certificates" in recognition of their farming skills, and professional qualification certificates for non farming jobs.

DISPARITIES AMONG EASTERN, CENTRAL, AND WESTERN REGIONS

The gap in economic development between the more developed eastern part of China and the less and least developed central and western parts is great.

Eastern part---- 12 coastal and neighboring provinces
Central part ---- 9 inland provinces and regions
Western part ---- 9 northwestern and southwestern provinces and regions

Most of the poverty-stricken, mountainous and karst (limestone) areas are in the western part. Among the 592 poverty stricken counties classified and specially aided by the State, 307 are in the western part, 180 in the central part and 105 in the relatively under-developed mountainous areas in the eastern part.

Aside from the harsh, disadvantageous natural and ecological conditions that have hampered economic progress in the central and western parts, another determining factor that has caused the wide gap is lack of strong TVEs. Industrial growth rates in the central and western parts have always lagged behind. Besides the poverty-elimination program that began in 1993 and has helped more than 22 million people to get rid of poverty, the Central Government has recently approved new preferential loans and other poverty-relief projects to help inland China.

TOWNSHIP AND VILLAGE ENTERPRISES

Percentages in 1996 National Totals

	Eastern part	Central part	Western part
Population	41	35	24
Cultivated Land	32	44	24
Horsepower of Farming Machine	50	35	15
% Agricultural in nation GDP	64	24	12
Gross output	48	34	18

However, natural resources, if wisely exploited, will compensate in the coming years for the poverty of the western part:

Western Part Percentages in the National Totals

grassland	94%
forest reserve	51%
coal	50%
non-ferrous metals	90%
land reserve for cultivation	70%

We now still have 58 million poverty-stricken people. For 1997, the yearly 108 million yuan ($13 million) fund will be given as usual. An extra fund of 150 million yuan ($18 million) for relief and 300 million ($36 million) poverty-relief loans has also been issued by the Central Government. What is more, the State has added 10 billion yuan ($1.2 billion) loans for building up TVEs.

Even so, aid from the State will have to remain very limited. The bright hope for central and western China will have to be pinned on tapping the great potential of agriculture and TVEs. If TVEs can be involved in not only promoting farming, animal husbandry etc., but also processing and circulating their products by forming conglomerates, that might be one break through for the people there.

The eastern part has already begun to answer the call of the State and the China Township Enterprises, Association to help develop TVEs in the two fraternal parts and set up joint development programs. But in the long run, *"help" or "benefits"* should be two-way traffic for sustainable cooperation.

Economists say that the outstanding feature of TVE development in 1996 was that the pace in central and western parts had obviously accelerated. The growth rates of value increased in rural industries were 18.7% and 30.1% respectively for central and western parts, while that for the eastern part was only 13.6% in 1996. Infrastructure in both central and western parts is improving. Several trunk-line railways are under construction. They will link inland, northwestern and southwestern parts of China with Northeastern Asia and the Euro-Asian continent, a prerequisite for starting direct foreign trade. However, all in all, the efforts of the local people will be basic for success.

PROBLEMS

1. *Reform ownership systems.* To solve the problem of ownership system, TVEs, always pioneers in reform, have searched for solutions since the early 1980s. Now two systems are in vogue. Both large scale State-owned enterprises and TVEs have adopted the "share-holding-system." When enterprises merge or new ones formed, they can form corporations, according to the Law of Corporations passed in 1993. "Shares" which were non-existent in pre-reform years, are now well accepted.

The second system "share-holding-cooperative system," has caused controversies, but is welcomed by peasants. An August 26, 1997 report in the Party organ People's Daily by Yia Jun says that by the end of 1996, there were more than 3 million share-holding cooperative enterprises in rural areas that had various ways of issuing shares and cooperation. Conventionally it is assumed in China that the basis for share-holding-system in the West is joining "capital" together, and the basis for cooperative system in a socialist society is cooperation in "work." Now how can one join "capital" with "work?" Some theorists said "Never heard of it!" But since it really works, people jokingly admit that these two different things have got "hybrid vigor!"

Take the example of collectively owned enterprises run by township or village leaders. They are in charge of both township or village administration and the enterprises. In actual work, things are hopelessly commingled: money of an enterprise is spent for drafting people into the army, which should be paid by town administration; a development program of an enterprise might be okayed by one or two cadres' work who know nothing or very little about the business.

Furthermore, the villagers or township citizens, the collective owners of the enterprises, have no say in and are not involved at all with the enterprises. There's often no effective way to keep a check on the cadres' work. Although lots of cadres are honest and hardworking, corruption and fraud are not uncommon either.

Although, individual owners of small business know there are policies to guarantee growth of their businesses, yet they feel that psychologically there's often something uncertain. Facing the challenges of competition, they like to join hands and set up stronger, larger scale businesses. The two new systems clearly define ownership and authorize shares to those concerned or who invest in the businesses. there are various ways to realize collective ownership. Those who work in the enterprises now care about not only their wages, but bonuses for their shares and business management. Village and township administration can be completely separated from the enterprises. Corrupt cadres will find no loopholes to meddle in business management and honest cadres have a lightened work load to improve administrative work.

In the past five years, peasants in Fujian Province bought shares exceeding 30 billion Yuan ($3.6 billion), about half the newly invested capital. The four cities of Suzhou, Wuxi, Changzhou and Nantong in Jiangsu Province collected in 1996 capital investment from non-collective sources amounting to 4.7 billion yuan ($0.56 billion). The Wenzhou Mode TVEs in Zhejiang Province has been famous for their prolific small individual businesses. Now the city of Wenzhou has more than 400 share-holding-cooperative system enterprises with annual output value over 10 million Yuan ($1.2 million) each. This is a five-fold increase from 1992.

In feudal China, the patriarchal system was deeply rooted in many spheres of life, rural life being no exception. Today in some TVES, people from one family share the most or more important positions. This is not in conformity with modem corporation structure. But still, it'll take some time to completely remove this vestige of antiquity.

2. Improve Management and Raise Efficiency. First of all, accounting and auditing should be standardized, and a series of systems should be set up to verify, appraise and register ownership, authorized stocks, and other properties or capital. Managerial personnel should learn modem management science. While labor-intensive TVEs should be there to draw surplus rural labor, more and more technology and capital intensive TVEs are emerging. Therefore, to ensure product quality and working efficiency, the best TVEs have already adopted systems from abroad like ISO-9000. Advanced managerial science and working methods should be more widely and quickly popularized.

3. Set up more agriculture-related TVEs. In a recent survey of 14 industries in both rural and urban areas, it was found that 70% of the categories surveyed were found in both urban and rural areas. Yet agriculture-related processing industries are very much underdeveloped. TVEs should stress categories that directly support agriculture.

4. Save more land and have less pollution. China has a huge 1.2 billion population to feed, while the per capita cultivated land is much lower than most developed countries. Industrialization and building up small towns, building expressways, railways, air and sea ports are quickly "eating up" the best farmland. This alarming loss of arable land has aroused the attention of economists abroad.

From 1986 to 1994, China repeatedly passed laws and regulations to strengthen management of land, In 1994, the Regulations for Protecting Basic Farm Land was passed by the State Council. It stipulates that while farmland must be used for non-farming purposes, the same acreage of land must be reclaimed or cultivated somewhere else to compensate for the loss. But in practice, violations occur where supervision and inspection were weak. A bureau was set up in the 1980s in the State Council to strengthen management of land. Recently more strict regulations have been issued. Implementing the rules must be strengthened,

TVEs too, have used large tracts of land and they are still too widely scattered, albeit small towns and industrial parks are being built.

Pollution is caused by both State-owned industries and TVEs. Since the latter are much widely scattered, they should pay more attention to environmental protection.

5. Upgrade professional standard of TVE personnel. To meet the competition of the market economy, and the challenge of a fast changing world, China's TVEs must give priority to personnel quality.

Of course there are TVE that have hired specialists with master or doctor degrees, trained either at home or abroad, or having had working experience abroad, but they form too small a portion to mention. 1996 survey shows the following results:

Distribution of the 59.53 million TVE, personnel

Categories	Number (million)	% of total
Work categories		
Engineers & technicians	3.7	6.2
Workers & Apprentices	46.22	77.6
Managerial Personnel	14.99	8.3
Service Staff	4.61	7.7
Education levels		
College & University graduates	760 thousand	1.3
Secondary school graduates	1.73 million	2.9
Technical workers, Senior & junior high school	57.05 million	
Primary school graduates		95.8

Toward Year 2030: Challenges and Opportunities.

Musing over the TVE situation has prompted this writer's thoughts to move to some related areas, such as challenges and opportunities toward Year 2030, when China's population is expected to reach the peak. TVEs must move forward with time, in technology, and in its role in rural society and national economy.

1. Balance between industry and agriculture. TVEs and agriculture are closely related. Both have contributed much to the national economy. On the other hand, State-owned enterprises have been the key link in the economic framework. For a sustainable fast growth, it is necessary to strike a balance between industry and agriculture. However, the following table shows a disturbing long term tendency:

On the other hand, contributions of rural areas to the State have increased in the past 20 years. For instance, in the 1991-1995 period, State expenditure on agriculture was $28.2 billion, while the taxes paid to the State by peasants and TVEs were $71.6 billion. The net contribution by rural areas came to $43.3 billion.

State investment in agriculture as proportion of State investment in fixed assets and total State expenditure

Five -year periods	Proportions that agriculture gets from the total State investment in fixed assets(%)	Proportions that agriculture gets from the total State expenditure (%)
1976-1980	11.4	N.A.
1981-1985	6.3	9.5
1986-1990	3.6	8.4
1991-1995	3.1	9.0

The credit-loan situation during the same 1991-1995 period also shows the same tendency. Banks in this period kept a net reserve of $65.6 billion from the difference of savings put into them by rural areas and the State loans granted to them.

Furthermore, the peasants must bear the difference between State procurement prices for grains and cotton, etc. and the market prices, and, in the "scissors gap" between prices of industrial and agricultural products. They also pay various surcharges which are not levied on urban residents such as education or army drafting fees. This is why that in recent years, the Central Government has repeatedly called to lighten the burden on peasants.

Now the country is reforming State-owned enterprises. The focus now is to reform or define ownership systems. It is hoped that as this reform goes on, and production efficiencies of State-owned enterprises are raised, the State can invest more and more in agriculture and effectively raise agricultural productivity--still low, judging by world standards--and the peasants' income. On the other hand, besides the taxes the TVEs pay to the State, they're also burdened by various extraneous "fees" and "donations" imposed upon them. These unreasonable financial burdens must be more strictly regulated. The Law of Township and Village Enterprises clearly defines both their duties and rights. Enforcement of the Law should not lax.

2. *Education.* The future of TVEs depends in large part on their attracting better educated and professional personnel. Although, many TVEs, spend large funds on inhouse education, the Central Government should still cut a bigger portion of the pie for the nation's education. Our

State expenditure on science and education and on cultural and health undertakings has always lagged behind others. The quicker the situation changes, the better and more balanced our national budget will be.

3. State-owned enterprises and the TVEs. As was specified above, to build a sound socialist market economy so as to speed up the tempo of industrialization in a developing country with a huge rural population like ours, one wise way is to foster the growth of TVEs. State-owned enterprises should still form the key link, the vital or lifeline part that mainly control infrastructure areas like energy and communications, national defense and heavy industries. Basically light industries and part of the heavy industries can be taken care of by TVEs. This will greatly cut the State expenditures and funds can be focused on more efficient State-owned enterprises. It will also streamline government structure and reduce bureaucracy. Greater funds can also be spent for education, infrastructure, environmental protection, social welfare etc. Thus, the national economy will be dwelt on a more rational and effective basis.

REFERENCES:
1. Guo Shutian: Promote Soft Science Research for Accelerating Rural Reform, China Soft Sciences Monthly, No. I 1, 1996 22
2. Guo Shutian: Narrow the Gap between the Eastern, Central, and Western Parts. Information, No.4, 1997, published by the China Management Science Academy
3. Guo Shutian: General Situation in the Rural Economy. Information, No.6, 1997, published by the China Management Science Academy
4. Qi Jingfa. Vitality, Vigor and Hope. China Market magazine, October 1996, Hong Kong
5. Yia Jun. New Creative Ownership Systems of the Township and Village enterprises. People's Daily, page 2, August 26, 1997
6. Pan Yue. Share-holding: An Important Way to Manage State-owned Properties. People's Daily, page 9, August 26, 1997
7. Development of the Rural Industries in Jiangsu Province, edited by Mo Yuanren, publication of the Nanjing Engineering College, 1987
8. New Landmarks of the Township and Village Enterprises, proceedings of the National TVE Working Conference, edited by National Rural Entrepreneurs Training Centre, September 1996
9. Basic Information and Economic Analyses on Rural Enterprises--1996, 194 pp. Edited by Township and Village Enterprises Bureau, the Ministry of Agriculture.

10. Economic Green Book --- 1996 Annual Report on the Economic
Development in Rural China & 1997 Development Prospects.
201 pp.. edited by the Rural Development Research Institute of
the China Social Sciences Academy & Rural Socio-econornic
Survey Team of the State Statistics Bureau.

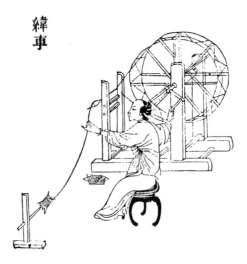

CHAPTER 4.7

Resisting Neo-Malthusian Pessimism
A New Century for Rural China

Klaus Lampe
Former Director General
International Rice Research Institute

The least costly experiences are those made by others

No doubt there are enough people around the world providing unsolicited advise what China should and could do to cope with one of the most challenging tasks: how to provide employment and food for its population in the next century. The rigid population control-program 20 years ago harshly criticised in the western world, is no longer an issue for the so called international expert community. After the fall of the political system in the Soviet Union, democracy in its western definition was seen as the only acceptable form of governance. Today many observers including the critical ones of yesterday comparing the Newly Independent States including Russia with the PRC of today have learned that a system shift is not a question of days but decades. There is also a growing understanding that market systems have to develop over time. Only too often shortcuts on the road to a system shift end up in detours.

German unification has demonstrated and is showing until today on a microlevel - in comparison with China - the complexity of adaptation processes. Even within one ethnic group separated for only 50 years and despite enormous capital transfers the problems of building a generally accepted market society are painstaking. The circumstances in Germany 1990, the political and market forces did not allow a slower step by step approach, but the price is obvious: high unemployment, unfulfilled expectations and a too often still existing invisible wall in the minds of people. The privatisation process pursued at high speed, has confronted people born and grown up in a socialistic system, with a free market society of today in which the social component in getting less visible and less influential. The internationalisation of production, markets and capital under the agreement leading to the World Trade Organisation (WTO) has left even less room for a public control over the private sector as originally envisaged and planned in Europe after World War II by "Neo-liberal Socio-Economists" in their vision of a market society with a human, a social - not a socialistic - face.

With this background in mind the value of outsider opinions related to China's food security in the future must rightfully be questioned. The results of western technological knowledge combined with political and economic value systems introduced over the last 40 years in economically less developed countries and the experiences made in Russia as well as in the Newly Independent States of the former Soviet Union call first of all for modernisation. But does that mean that there are no experiences outside the China made since World War II which could be used in the transformation process China is going through from a centrally planned to a more market oriented society? On the contrary China could save time and enormous resources; human and financial, simply by profiting from mistakes made elsewhere.

In the western world numerous research institutions concentrate on Asian studies with strong emphasis on China. The results are intensively used in public and private sector decision making. Japan's economic strength of today is among other reasons the result of a systematic observation, evaluation and utilisation of western innovations, systems, technologies, of new knowledge in general. This leads to a series of questions I want to raise in this paper that has no solutions to offer.

Question 1: If western countries invest remarkable human and financial resources in studies related to Asia and specifically to China, would not it be economical for China to observe and analyse economic, social and political developments in the western hemisphere with special emphasis on issues crucial to the economic and social development of China in the coming century. Would not it be worthwhile to invest in existing institutions - or even new ones within or outside China to tap western sources of knowledge? Would such a search for knowledge already existing save time and resources and enable the country to develop the needed food production systems? Could it help to develop a "Rural Habitat Agenda" for China?

Doomsday scenario-prophets painting a dark picture are getting public attention once the basic question is raised: Can the world of 10 billion people in the next century really feed itself? And in case we come in principle to a positive answer the second even more difficult is: how and for how long. Unfortunately there is an easy escape by simply quoting those who already predicted centuries ago that population increases would outgrow the potential for food production. The Neo-Malthusians of today are again predicting the same scenario for the coming generation. China with its population of 1.24 billion people is often taken as an example to prove that life in the next century will be linked to starvation, to conflicts over food- and water-supply. With the data on available land- and water-resources, present yield levels and technology standards such prediction are not totally unrealistic. And let us not forget hunger and malnutrition is affecting already today almost a billion people world-wide. About 1.000 children under the age of 5 years only are passing away every hour in this world - due to lack of food and related diseases; even more so due to a lack of politically responsible leadership. Poverty - as we know - does not stop at the boundaries of the People Republic of China. And at least today malnutrition in most cases is the direct result of poverty and not so much the absence of adequate food supply. This scenario might change with the expected doubling of the worlds population in less than 50 years from now, at a time the real population figure for China might come close to 1.8 billion people, taking the least optimistic scenario.

Until today nobody has the exact data needed for reliable forecasts be it population growth, arable land- or water resources in China. What we know today however is enough for fundamental policy decisions to avoid food shortages of critical dimensions. Grandfather technologies alone will not allow us to cope with the problem of tomorrow. The investment in fertiliser use may serve as one example only. The country has doubled its per capita use of mineral fertilisers from 1982 to 1992 by about 100 %. Its present use is about three times higher than that of the United States and comparable with the one of Japan and several European countries. An other figure however is indicating a shift in priority setting that must disturb not only everybody interested in agricultural but in stability. The share of Agriculture of China's gross national product (GNP) dropped from 37 % in 1983 to only 19 % in 1993. In 1984 about 40 % of Chinas export was agricultural based. Within 10 years only it dropped to 12 %. In 1986 almost 75 % of the working population was employed in Agriculture. In 1993 that figure dropped to about 60 %. In fact it means that 15 % of Chinas work force shifted within 10 years only from food producers to market depending consumers.

Question 2: Is it seen as a political priority to modernise agriculture and increase productivity of the rural sector to insure national self-sufficiency at least for the staple food supply of the country including vegetables and animal protein?

China's resource base

It could long be argued if self sufficiency should still be a policy to follow instead of ensuring food security through import and dependency on world market prices. This holds specifically true for rice a commodity with an almost negligible international market. It is important to note that 90 % of world rice is produced in Asia predominantly, by small farmers whose families are consuming about 50 % of the harvest themselves. Local and national markets are serving the landless population. The international export market of milled rice amounts to about 16 million t only. In 1994 China alone produced 118 million t. An import of 15 % of its annual production would absorb the whole international supply, not to speak about the price implications and the political vulnerability.

China's land base

Given the worlds land reserves, the economic potential for production increases and the expected demand, prices for internationally traded staple food will sharply increase also due to the agreements related to the World Trade Organisation (WTO). Subsidised production and government financed export like the one of the European Union will in future not influence market prices like in the past. At the same time imports from developing countries will increase sharply. They are expected to grow from 1988 to the year 2000 from US $ 40 billion to almost 65 billion, an increase of 25 billion. Asia alone is expected to import in the year 2000 Food the equivalent to US $ 42 billion.

Question 3: World food prices are fluctuating due to several unpredictable facts. Chinas export market at the same time is depending on the economic climate in Asia and the rest of the world. Given these interdependencies should China in future count on the world market to fill eventually occurring unavoidable supply gaps, but at the same time make all efforts to avoid importation of staple food? As mentioned the existing statistics related to agricultural land, water etc. are allowing at best guesstimates, leading to optimistic or less favourable scenarios. Under western-style democratic systems foreseeable problems of the more distant future remain only too often unsolved since unpopular decisions are seldom accepted by a given electorate. By political leaders interested in re-election they are therefore simply left open.

Political systems with leadership continuity less dependant on election processes provide the opportunity to identify major challenges ahead of time and take decisions which can lead to problem solving or at least minimising consequences. Family planing is such an example. China has made over the last 15 years enormous progress in stimulating private initiative and develop a climate conductive for entrepreneurial activities on the national and international level. To a distinct extend that holds true also for the agricultural sector. However the substantial growth in agricultural production should not detract from the fact that the price paid might be higher than the country can afford on a long-term basis: soil fertility and environmental balance. In Korea, to take another example, agriculture has been the root of economic growth, of industrialisation, of wealth. The present economic problems the country is struggling with are not due to Korea's agricultural sector. The "secret" of Korea's success is well known and well documented: land reform, stimulation of family farm-based production, effective agricultural research, need-based market-oriented extension services, rural support industries and infrastructure development in rural areas.

Hundreds of generations tilled Chinese soil. Rice production dates back about 10,000 years. However the present productive systems are closer to soil mining than soil conservation and the thinking horizon has been too often reduced from a generation or two of people living from that land to a single harvest or two. The major reason can be reduced to a simple factor: ownership. Anonymous ownership, without the mechanism of individual interests, long term responsibility and the social control of a village community, is not only leading to the lowest possible output but the absence of any conservation efforts. The agricultural production system in East Germany after World War II for example was not only uneconomically managed but has led to high pollution and losses in productivity. With a change in ownership not only private investment stimulated economic production systems.

Since the well being of present and future generations are directly linked to the degree of ecologically sound production systems used, farmers do not need to be told what to do. What they need is the freedom of decision making access to agricultural inputs including credit and a conducive market system that provides adequate prices to stimulate production.

When public land ownership for a long time has been introduced for reasons what so ever a radical change to privatisation could become even harmful. Again the East European developments over the last almost 20 years provides many lessons already learned. Privatisation of land can be implemented in steps once the principle decisions at the political level are taken and confidence that these decisions will be implemented has been developed within the constituency. The most important steps towards individual land-use responsibility are:

• land-reform allowing only lease but no sale of public land within a first phase
• holding sizes visibly above subsistence level to allow market oriented production,
• terms of lease long enough to stimulate a "next-generation-oriented" land use
• rental value and rental terms attractive specifically for young families,
• credit schemes to finance the use of modern technologies and inputs,
• insurance schemes against natural hazards,
• promotion of cooperation for the joint purchase of inputs and sale of agricultural produce, knowledge exchange and transfer-systems introducing innovations without long delays.

Question 4: Is it correct that the Chinese farmer will be the most efficient and effective user of agricultural land and promoter of soil fertility once he is granted with a generation based right and responsibility over the land he is tilling. If so would it be thinkable to introduce stepwise a land-tenure system with a land lease of not less than 50 years in the 1. Phase.

China's water base
There are only a few countries in the world equally dependant on irrigation water like China. T.C.Tso has documented in this publication the increased water demand very clearly. Therefore, it

does not require extensive discussion. In future the agricultural sector has to compete for efficient water use with at least three important sectors.

Energy supply for China is growing, the increase in population will lead to the development of hydroelectric power plants wherever feasible. Experience elsewhere has shown that it becomes very difficult to harmonise the needs of agriculture for irrigation water with the needs for energy supply. In this power struggle agriculture has been in the past mostly - and world-wide: the second winner.
1. The dynamic growth of the industrial sector in China has led to an exponential growth of water use. This development will continue in the years to come with growing speed. The influence, the importance, the dominance of the industrial sector will force agriculture to economise water use and compete with the industrial sector in water use efficiency, in very much the same was it is the case in Israel since decades already. Urbanisation finally even if drastic measures are taken is diverting a growing portion of available water resources to cities, again on the account of agriculture. Today 30 % of all Chinese live already officially in cities. The real figure might be much higher. And there is no indication today that this movement from rural to urban China is being stopped or at least slowed down (Figure 4.7.1).

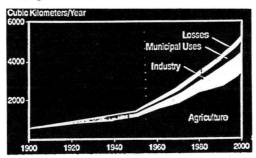

Figure 4.7.1 Estimated world water use total and by sector. 1900 to 2000.
(Source: Last Oasis, 1992)

Question 5: Conflicts over water rights and water use are not new to China. They are part of human history back to the early days. With growing demand per capita, industrialisation and demand for "clean" energy, water is becoming scarce in tomorrow's China. Would it be possible to develop a "master plan 2030" for using old and new water resources in a way that satisfies real needs, avoids misuse and serves as a cornerstone for human settlement planning in China? Could such an endeavour become a joint effort of the public and the private, the urban and the rural sector including all relevant sciences and user groups?

China's human resource base

The world oldest documented knowledge about agriculture has its roots in China. The skills, the stamina and the knowledge of the Chinese farming community has prevented not only hunger but also has laid the base for the industrial development. Agriculture has provided under difficult economic, political and social circumstances not wealth but a living base for the overwhelming majority of Chinese people. Factor productivity decline has been observed in continuous rice growing as well as in rice-wheat production systems in several parts of Asia. Even modern varieties do not respond adequately to even largely increased fertiliser applications. There are several theories developed so far but an answer to the question of this "soil fatigues" has still to be found. Rural China until recently was the shelter of more than 200 million unemployed or predominately/temporary unemployed men and women. With the privatisation of the industrial sector a movement of people from rural to urban of unknown dimension in modern China took place and is still ongoing. Unfortunately this development is often compared with the mobilisation of rural surplus labor during the first. phase of industrialisation in Europe in contrary to China today. At that time urban Europe developed a labour-demand not on a temporary but a much more permanent industrial sector driven base. Urban China today is mostly in need of temporary employed construction workers. The result is a growing number of migrant labourers. The absence of income generating employment in the villages is stimulating the move to the richer coastal area, leading to an oversupply of labour.

Many observers today question Chinas ability to feed itself in the next century. The much more serious and critical question however is: Can China generate enough employment or at least keep unemployment rates within a tolerable and controllable level. The conflict between this macroeconomics goal and private sector interests is obvious. The globalisation of markets and Chinas intend to play a key role in this development is influencing the national labour-market in an unprecedented manner High-Tech-

Mechanisation, Automation and the introduction of robots in Chinese industry is only a question of time. The pressure towards reduced costs, high and uniform quality standards and price-competition on a world-wide level dictates productive systems in which the human factor at less skilled levels will be marginalised. At the same time the cost of living in urban agglomerations will rise sharply in the next century. The investments for public services like water supply and water recycling, transportation and energy, housing and waste disposal will strain the public budgets leading to a high urban taxation or the deterioration of services. The slums of many cities in Asia and other parts of the world are strong signals of what is seen by some as one of the nightmares of the 21st Century.

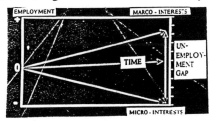

Figure 4.7.2 Employment, the conflict of interests.

The interest of the public sector (macro interest) maximizing employment opportunities and the private sector (micro interest) minimizing costs are in conflict with each other. The gap widens with time.

Question 6: China today has still one of the lowest urbanisation rates world-wide. The growing rate of visible unemployment in is undebatable. Would it be possible to develop a national program transforming the existing township- and village enterprises of smaller urban settlements in viable private entities in order to:
• provide low cost agricultural inputs
• process and market agricultural products
• reduce post-harvest losses
• stimulate the development of small and medium sized townships
• prevent or reduce the migration of larger number of unemployed people.
• provide employment produce all sorts of goods for the local and regional market.
There are no examples in Asia of today how best such a policy could be implemented. However there are enough mega-urbanisation exist

which could serve as a warning signal for the future. Finally, European small and medium scale industry, its historic roots, its rural rooting and the present importance of this sector might be worthwhile to study. What was once a small watchmakers shop more than 100 years ago in the most backward part of South Germany is now a regional centre for electronic instruments. Grandfathers carpentry backyard-shop is today producing prefabricated houses. Thousands of such factories today not only provide jobs. The location in the countryside has helped to avoid the development of urban megaagglomerations of unmanageable dimension. China has the governmental structure to establish the ground rules needed within a reasonable short period of time to avoid what seems in other parts of the world preparing for a disaster. With the existing township- and village enterprises there is a base which needs adaptation to the new environment of privatisation and a new market oriented economy. There is still time to act, will it be used wisely?

Chinas Agriculture 2030 - or Rural versus Urban China

The degree of self-sufficiency in food China can and will achieve until 2030 depends only practically upon new knowledge, new technology, farmers practice and the market related services. First and foremost the productivity of China's agricultural sector will be influenced by the attraction of the living standard in "Rural" compared to "Urban China". In many parts of Asia "rural" is synonymous for poverty, backwardness and neglect. The only technology that entered almost every village is the Television- and Video-Service. It provides entertainment but more often an idealised vision of life in urban effluence. In contrast, rural life is determined by the absence of adequate health- and education-services; modern drinking water and energy supply systems and employment conditions at the edge of subsistence. For the landless the unit prices for basic goods are in rural areas higher than in cities and labour cost often a fraction of what an urban laborer can earn. A study undertaken in India showed, that a labourer has to work 4 - 5 times longer in rural than in urban regions for basic commodities like rice, tea or soap. "Rural" in Asia including China is synonymous for old, backward, poor.

Those who decide to leave the land, the village in search of employment elsewhere belong to the most active, most mobile and most skilled part of the rural population. This selection process will not only affect modernisation of agriculture in China, it may lead to a continuous drain of the best, the innovative dynamic part of China's farming community. They however would be urgently needed to transform Chinas agriculture into a modern, productive, private initiative driven sector serving the food market needs for not less then 1.6 Billion people: China in 2030 and beyond.

Question 7: How can Chinas leadership be convinced of the need to give much higher priority to the development of the countries rural regions in general, the development of middle sized townships and provide these regions with modern services, with opportunities for work and prospects for the future? How can the agricultural sector (re)gain the political priority needed to boost long-term production increases? How can the private sector get interested in rural China and not restrict itself to Shanghai only?

From agricultural labourer to rural entrepreneur
Chinese farmers despite the political system that did not allow them to make independent decisions have never lost the knowledge and experience to produce efficiently. Once the constrains of a planned centralised economy are replaced by a framework of laws which promote ownership, stimulate production and ensure that markets forces can regulate prices and supply, there might be no real reason any longer for pessimism in view of the agricultural potential to feed China. We know that there is much more land available than official statistics show. We know as well that yields per land unit are lower than anticipated. Nevertheless annual increases of about two percent from a level of about 4.0 t/ha grain production cannot be taken for granted. More food has to be produced on less land with less, much less water and a much more economical use of agrochemicals not only for ecological but for economic reason. New knowledge and new technologies have to be developed and introduced. Since more than 50 % of China's cultivated land (43 million ha) is irrigated the productivity of half of its agricultural land depends on sufficient water availability at the right time. These land resources belong to the most fertile ones and therefore by far the largest productivity increases must be generated from these soils. Under the assumption that 50 % of this land can be double cropped about 65 million ha can be used for staple food (grain) production. T.C. Tso is assuming in his paper that about 650 million t of grain will be needed in 2030. Since about 70 % of this grand total or 455 million t has to be generated from irrigated land, China must harvest not less than 7.0

t/ha on irrigated land per crop. In 1988 about 400 million t of grain have been harvested in China. Taking the same ratio of 70 % for irrigated farmland it is estimated that only 280 million t or 4.3 t/ha have been harvested 10 years ago. The respective figure for 1996 comes up to 5.2 t/ha or an increase of roughly 2 % per annum or 0.9 t in 10 years. For the 30 years ahead additional 1.8 to/ha have to be harvested. Given the hidden land resources not yet included in the official resource data system such targets seem to be achievable, however not without special efforts.

The agricultural research agenda
China has unlike other countries in the region a large number of research institution in all eco-regions and at various levels from adaptive R+D to strategic research at the cutting edge of science. The latest available figures of about 200,000 researchers and 10,000 institutes do not give any indication about the potential of Chinese agricultural research capable to develop the base for the increase needed in the decades to come. However the challenge as mentioned earlier is at least threefold. The factor of productivity must be improved visibly. Better utilisation of sunlight, nutrients and water is urgently needed. But research in India, Pakistan, Nepal and Bangladesh as well as studies undertaken by CIMMYT and IRRI has clearly shown a factor productivity decline over the last years at least for nitrogen fertiliser application. The well known law of diminishing return is not the only reason, all relevant research results on this topic have clearly proven the need for a wider, interdisciplinary research agenda related to yield- and productivity increases without compromising neither long-term soil-fertility nor product quality. Also the experiences with hybrid rice in China have shown that with growing income and before a shift to more animal protein based diets consumer demand for higher quality staples like rice and vegetables must be met. The necessary research agenda for the next 10 - 20 years will therefore be very complex.

Question 8: Is it feasible to interest leaders in policy making and science to commission a master plan for agricultural research comprising of at least 4 major segments which need close inter-linking? The following areas could belong to a new research - and development - agenda.

Plant soil-water-interrelations

The objective: Developed plant-production systems that meet the food requirements of China using all resources wisely in view of the generations to come.
Key research areas:
• Exploit hybrid vigour for all staples with special emphasis on mobilising apomixis to avoid the purchase of new seed for every harvest
• Increase yield potential through the development of new plant types for all major food crops.
• Improve photosynthesis efficiency through traditional and non-traditional research approaches
Enhance nutrient efficiency uptake through strategic root zone research and improved root systems for all crops. In irrigated rice systems e.g. 50 % of all nitrogen applied is getting lost into the groundwater or through volatilisation.
• Strengthen resistance and tolerance against biotic and abiotic stresses (pests, diseases, frost, heath, drought etc.) using modern biological tools including genetic engineering.
• Control pre- and post-harvest losses using traditional and nontraditional research tools and technologies (Herbicide tolerance alellopathy, loss reducing harvest- and post-harvest technologies and management systems). It is obvious that private sector grain storage losses are only a fraction of those experienced by public institutions.
• Reduce risks and increase productivity of plant production in marginal areas through e.g. water harvesting technologies for arid/semiarid ecosystems, plant types with higher water and nutrient uptake capacity and water holding capacity.
• Develop agricultural practices and systems combining increased production with increased factor productivity. For 1 kg of rice in USA only about 2000 L of water are needed. In most parts of Asia for the same amount up to 5000 L are consumed in irrigated ecosystems. This is due to water wasting land preparation- and transplanting methods instead of dryland ploughing and dryland seeding. Research is needed to develop new plant types for dry seeding as well as land preparation on small plots without flooding.
• Improve Vegetable cropping and production methods specifically for vegetable growers near the perimeter of urban areas. These systems are providing best opportunities to absorb labour, generate income and serve the growing urban needs for high quality fruits and vegetables and even ornamentals.

Animal production

The objective: Developed animal production systems specifically for small and medium scale producers that will meet the growing demand for animal protein predominately based on pig, poultry and small ruminants.
• Developing cost efficient and environmentally sound models for small and medium sized egg,-broiler- and pig production
• Replacing traditional maize and wheat based animal feed with non food crops and waste products
• Modifying waste wood and other non digestible products through e.g. genetically modified micro-organism in animal feed.
• Improve the efficiency of poultry, small ruminants and pigs specifically suitable for less favourable conditions and adverse climates.

Processing and marketing

The objective: Enhanced marketable agricultural products meeting in quality and price needs and growing standards of different consumer groups.
• Developing small and medium scale processing and storage systems specifically for vegetables and meat.
• Enhancing transport and marketing facilities
• Reducing losses to the economically viable minimum

Social sciences

The objective: Rural regions of China are in social and economic terms an attractive alternative to urban life for present and future generations.
• Formulating proposals for policy decisions related to land tenure and holding sizes and land ownership.
• Developing policies to promote production, marketing and crediting of the rural sector.
• Establish price information systems for agricultural products and inputs.
• Developing self help structures for purchase of inputs and sale of agricultural produce.
No doubts these areas of possible intervention can only serve as examples demonstrating the character and potential effect of a new comprehensive research and development agenda. What should not

be underestimated is the expected rate of return of such investments. It is proven that they often exceed those of the most profitable investments in industrial production.

Dreaming of a new R+D structure
China's best scientists in agriculture belong to the best in the world. Provided with the freedom researchers need to be productive and the resources for modern equipment as well as project development and implementation, advances in agricultural sciences of the expected and needed dimension are most likely achievable in time. To make that happen a refraining restructuring and reorganisation of the agricultural R+D system in unavoidable. The successful privatisation of agriculture is depending on research outputs that are conducive to privatised farming practices. Thinking of the more than 150.000 scientists, researchers and technician estimated to be engaged in agriculture related research the idea of a fundamental reform seems only unrealistic for those who underestimate the possibilities of a centralised system. Once the government can be convinced of the urgency of China's food- and employment-problem and the potential of research to find solutions agriculture, rural habitat in China will get a priority equal to family planing. It is hoped that these decisions to ensure food-security and a maximum of self-sufficiency will be taken soon enough to reorient China's agricultural R+D to new needs. Adaptation at least at three fronts are needed: the structure, the human resources and finance.

Question 9: Is it correct that China's agriculture and related industries must become attractive enough to absorb as much surplus labour as possible and produce food for 1.6 Billion people at acceptable prices and quality standards? Today research in China has to serve a new clientele of farmers especially in view of the shift from large sized public owned holdings to family managed enterprises. The needs of Chinas private farmers are different today and the science community in China until recently hasn't had the chance to produce research outputs for a farmer driven market. At the same time the different levels

of research from strategic to adaptive have their own individual agenda the same holds true for the different existing structures like the Chinese Academy, the Agricultural Academy, Provincial and Prefectural research organisations, not to speak of their different and independent R+D-programs. What is needed instead is an ecosystem based research strategy as well as a market need (farmer need) based work plan that stimulates the respective research mostly at an interdisciplinary and inter-institutional level. This might not be possible without input from outside or better "insiders from abroad". Only China can claim to have a vast intellectual potential of highest quality - living and working abroad: the Overseas Chinese Science Community.

Question 10: Is it possible to establish a Chinese Agricultural Research Council consisting of scientist, planners and private sector research managers provided at a first phase with the task to help develop a new Agricultural Research System for Chinas next century? The terms of reference of such a council - besides serving as an Advisory Board to the Government - could among others include:

- The identification of key research areas of highest priority
- The creation of a masterplan for a new R+D-System for China
- The nomination and review of research centres with the highest degree of excellence and relevance
- The identification of the most qualified scientists in major fields of agricultural sciences The development for a new organisational structure for „rural research" combining the need for excellence, relevance, timeliness, freedom and accountability. Such a group representing not only different sectors and disciplines should comprise of different age groups as well, profiting not only from experience but new knowledge, new science. To be successful it must be provided with the freedom and full support from the highest levels within the political hierarchy.

Indonesia after having gained independence had practically no national cadre of agricultural or related scientists. Many of those trained abroad in the following years hesitated to return to their country for many, including political reasons. Today, 25 years later, Indonesia can rely on an agricultural research network that has been instrumental in developing an agricultural system that feeds a growing country of more than 200 Million people. China's scientific base today compared to that is enormous. Once research programs between the different independent institutions are coordinated and research priorities set, Chinas researchers have the capacity to find solutions to pending and new problems. But will that be the case in future as well given the limited attractiveness of Agriculture for the younger generation? Again political priority setting and the public interest in a specific sector strongly influences individual decisions related e.g. to professional options and career planning. If food production is given the needed national priority highly qualified young Chinese must get attracted to study agriculture and get involved in the modernisation of all rural sectors in the broadest sense. Perhaps this is the most important, most crucial and most difficult task for China in its effort to retain political independence through self-sufficiency in food. The free and price wise acceptable import of food for a country of the size of China cannot be taken for granted in the next century. Almost no Government in the world can dictate its upcoming generation what to learn, what to study, what profession to choose. But it can make a sector, a region, a programme attractive for "the young and the bright". China's food-security will depend on its ability to stimulate the interest of a larger part of the countries young, educated bright generation in modernising rural China, through new knowledge, modern infrastructure and a balanced development of the private sector driven industrialisation in a rural setting. The key however is a modern, science driven, new knowledge based agriculture of sufficiently high labour productivity that can attract a large proportion of the coming generation - to be farmers.

Research alone can at best provide the tools. China's future is in the hands of those who take political decisions today. Thinking 30 years ahead this sounds not very realistic. But it must be noted: the research results needed for living peacefully in the year 2030 and beyond take time to develop, they are costly and the risks of failure have to be taken into account. 20 years to develop new yield plateau and production system is a rather short period of time. 10 years for their introduction in farmers fields even shorter. What is therefore needed most are future-oriented, even unpopular decision based on a vision: A China without poverty, without hunger, without environmental degradation. - A China with 1.6 billion people whose basic needs and more than that are met: grounded on individual freedom and decision making and supported by a government committed to social freedom.

The investment needed might be high, but the decision is easy, since there are only two options to invest in research, in mobilising the intellectual potential of China and bring the results to farmers fields or - spend foreign currency to import food - if it is available on the market.

Conclusions

World-wide agriculture is more and more seen as a backward, uneconomic, inefficient, ineffective and therefore unattractive sector of the economy. The price for this misjudgement, short-sightedness and ignorance given the importance of food production and environment conscious natural resource development practices will be very high; more so it is irresponsible in view of the next generation. China, the nation with the worlds largest population - thanks to a rigid family planning program and a very hard working rural population - was able to avoid major food shortages during the last decades. As it stands today the country will have most probably enough agricultural land available to feed its population in 2030 provided the annual loss of arable land will be drastically reduced and the factor of productivity increased. Other measures needed are the reduction of pre- and post-harvest losses and the full mobilisation of market forces to stimulate production. Again: first and foremost target-oriented output and impact driven agricultural research including forestry, animal husbandry and fisheries. The needed Chinese scientific knowledge within China and overseas is theoretically available. It needs mobilisation as well as reorientation towards new goals and objectives based on a market oriented production system. To be productive it needs political support, long-term commitment and financial resources. Experiences relevant for agricultural change in China including those related to the transformation of centrally planned economies are existing in other parts of the world. It would be highly economic to review, screen and eventually adapt and adjust relevant overseas knowledge related to the promotion of an effective and efficient private agricultural sector and the development of rural areas in general. This approach driven by Chinese experts is by far superior to the import of knowhow through predominately non-Chinese.

As critical as land- and human-resources for Chinas food self-sufficiency at 2030 and beyond will be the supply of water and plant nutrients. Research to increase the productivity of these inputs must be given high priority. The major part of increase in food production has to come from the highly productive irrigated production systems. Nevertheless infrastructure development, research and development related to the less favourable continental, arid and semiarid ecozones cannot be neglected. Unconventional research tools of e.g. modern biotechnology must be applied in plant- and animal breeding to increase the tolerance levels towards biological and physical stresses.

New plant types for the major ecosystems including fodder crops and new animal feed species will allow the production of animal protein and simultaneously reduce the amount of grain used as animal feed. The changes within China over the last 20 years exceed by far the most optimistic expectations. The speed of economic growth that might effect the economic and social environment needs careful monitoring to avoid a rural-urban imbalance of unacceptable dimension. The recent economic development in several Asian countries might not leave the Chinese subcontinent unaffected.

The most serious factor China will have to cope with is labor. The high speed with which the industrial production is getting robotised world-wide is nothing less than alarming. Large numbers of unskilled or on the job trained labour of today will be replaced by robot controlled automation.

Goods will be produced in the next century not where cheap labour is available but close to the markets reducing transport cost and time. It will furthermore allow timely adjustments to market needs.

Producing food for people without purchasing power is even more dangerous than famine due to national disaster. For China's 1.6 billion people in the next century it is decisive which political, macroeconomic and socio-economic road its political leaders are choosing today. Decentralisation, liberalisation of land tenure policies, viable urban-rural inter-relationships build on small/medium sized townships, promotion of a private initiative driven service sector are only a few of many tools serving the same goal: reducing unemployment to the absolute minimum.

Modernisation of agriculture at large through a forward looking R+D policy, modernising "Rural China", might serve this objective best.

Political leadership in China can plan, project and implement decisions based on a long term strategy. This opportunity is missing in most other Asian countries with fast changes in government leadership. This advantage must be exploited; through strategic planning involving the best thinkers and planers available. The output of that thinking process for the next 3 decades if well prepared and monitored could serve as a master plan for Chinas government committed to responsible, future-oriented leadership.

Literature cited:

UNDP Human Development Report 1994 and 1995 New York
Oxford University Press ISBN 0-19-509170-1
N.N. China Agriculture Yearbook 1991. Agricultural Publishing
House
BULATAO R.A., Bos E, Vu M.T. World Population Projections
1989/90 Edition, John Hopkins University Press, ISBN 0-
8018-4094-5
PARDEY P. Roseboom, J. Agricultural Research Indicator Series
1989 Cambridge Univ. Press, ISBN 0-521-37368-9
TSO T.C. Agricultural Reform and Development in China. Ideals
Inc. 1990.ISBN 1-878670-00-x
BARATTA, N v. Fischer Weltalmanach 1987-1996. Fischer
Taschenbuchverlag, ISBN 3-596-19096-7
LONGWORTH J.W. China's Rural Development Miracle University
of Queensland Press 1987 ISBN 0-7022-2264-x Pop. Ref.
Bureau World Pop. Data Sheet.Washington DC 1997

CHAPTER 4.8

Developing China's Agriculture in Intensive Mode Making Full Use of International Resources and

On the Future International Technology Cooperation
Regarding China's Agriculture

Bai Ji-Xun,
Gong Yan-Ming,
Wang Hal-Yang,
and Ma Jun-Ru

State Bureau of Foreign Experts,
Beijing,China

Generally speaking, resources in the world can be divided into two kinds: natural resources and human resources. For any country in the world, both natural resources and human resources are limited. Very few countries in the world have all the necessary resources. The shortage of resources is more severe in some countries than others. Therefore, it is very important for each country to adopt proper strategies for economic development according to their respective resources. Being primarily agricultural with the largest population in the world, China's per capita natural resource is very limited. From the end of this century to the year 2030, China will be facing more and more pressure from its burgeoning population. The per capita arable land will become less and less. Therefore, maintaining an ample supply of grain and other agricultural products for the long haul will directly influence China's success in modernizing as well as its overall prosperity and stability. Presently, it seems that in order to sustain China's agricultural development with high quantity, quality and efficiency, we should seek to apply the state of the art scientific and technological advances, although there are many other factors involved. Thus, the only way to successfully ensure China's agricultural and rural economic development in the future is to implement all-round international cooperation in economy and technology. It is also necessary to learn and absorb all advanced achievements in these areas and to gain experience in management so as to develop an export-oriented rural economy.

The necessity of developing China's agriculture by way of expanding international cooperation requires to examine:

The current status of agriculture in China:

The population of China is immense with very limited arable land and per capita natural resources. It is estimated that the population will reach 1.3 billion at the end of this century and will peak to 1.6 billion by the year 2030. According to some statistics concerned, currently the per capita arable land in China is only 0.086 ha, one fourth of the world average level. The distribution of the arable land is very inequitable between regions. The developed area with more demand for agricultural products has less per capita arable land, with more pressure for production. The water resource in China is also very limited. The per capita annual rate of water flow is 2474 cubic meters, one fourth of world average level, which ranks 88th in the world. The distribution of water resources in terms of seasonal rainfall and regions as well as per capita water resources is also unbalanced. Moreover, there also exist problems such as soil erosion, desolation, soil alkalization, etc. At present in China, among the wasteland resources, 0.287 billion ha. Wasteland can be reclaimed as Pasture, 66.67 million ha. Barren hills can be reforested. The reclaimed land available for agricultural use is 13.33 million ha. The inland waters cover 16.67 million ha., 5 million ha. of which are suitable for agriculture, several million ha. are coastal plains with sandy soil conditions. Reclaiming and making use of these wastelands can provide China with a lot of food and other necessities for life. These offer great prospects for the future. However, the investment for all this will be very big, so high productivity can not be achieved in the short run.

The rural economic development is unbalanced varying from region to region. The developed areas, such as the Pear River Delta, Yangtze River Delta, Jiaodong Peninsula, Liaodong Peninsula have provided excellent environments for economic development, and acquiring funds and materials which are vital to sustaining agricultural development. However, quite a few underdeveloped economies only have a single-product economy and are slow in improving the conditions for agricultural production. People in some poorer areas even lack adequate food and clothing with 65 million people still living in poverty. In these areas, the agriculture development is in extensive mode with low efficiency.

Eighty percent of China's population live in rural areas. In recent years, there are more and more unemployed laborers due to increased population without reciprocal increase in jobs. In fact, they account for one third of the total number of laborers in rural areas. Finding ways to employ these surplus laborers or transfer them to other jobs has become a very serious problem and has constituted a heavy burden hindering the development of China's agriculture and rural economy.

Relieving the pressures is necessary for cooperation.

The opening to the outside world and implementing international cooperation in terms of economy -and technology is helpful for relieving pressure from the restraints and limitations of natural resources. According to the theory of B. Ohlin (a Swedish expert on economics) concerning the rate of distribution of resources, a country with different resources reserve should make full use of its own rich resources of the country for its economic development. Also through international exchange, participating in and contributing to the world's economy, the country can gain more profits for and maximize the interests of its people despite its limited resources. The success of economic development in some countries has proven this theory. For instance, Japan is poor in natural resources and has to import most of its energy and mineral resources except coal and copper. Japan does not have enough arable land. In 1980 it had 5.8 million ha. arable land. The average per capita arable land was only 0.047 ha. During the World War 11, Japan was defeated, and its economy was destroyed. There was a big shortage of capital and technology. But after the war, the economy in Japan got off to a flying start. In fact, from 1951 to 1970, less than 20 years time, Japan jumped to the second strongest country in terms of economy in the world, creating a miracle in world economic development. There are many reasons for this phenomenon, but one of the most important reasons was that Japan continued to implement the open door policy started from Meizhi Reform while developing an export-oriented economy. In the different periods, Japan made full use of the advantage of its own technology and personnel to produce export-oriented products in exchange for all kinds of raw materials, which were deficient in Japan. At the same time, Japan also introduced new technology and attracted foreign investment. With the increased quantity of exchanged raw materials from foreign countries, the economic scale and speed were expanded and sped up resulting in a rapid improvement in the living standard of its people.

South Korea is also a country with very limited natural resources. It imports most of its raw materials and energy. In 1982, the total arable land in South Korea was 2.18 million ha., the whole population was 43.1 million, the per capita arable land was only 0.051 ha., which was almost the same as that in Japan. However, in a short time, South Korea became a newly industrialized country. South Korea also advanced by developing an export-oriented economy. Israel is even more confined by the lack of natural resources. Although the Middle East is rich in oil, there are few natural resources in Israel. It has neither oil nor mineral resources. The territory is small with very adverse circumstances. However, Israel has created a farmland in the desert. In the beginning of 1980s, there were 200,000 ha. Irrigated farmland in Israel. Compared with its 4.8 million population, the average per capita arable land was only 0.042 ha. However, even under such conditions, Israel not only produces enough grain for its people, it also exports a lot of its agricultural products to some Western European countries despite the intense competition in the world grain market. By doing so, the economy in Israel took off in short time and has entered the ranks of the developed countries in the world. As early as 1976, the per capita GNP in Israel was $3400, which was higher than that in Italy. Israel's success illustrates the advantage of engaging in export-oriented economy. In its development, Israel has absorbed state-of-the-art science and technology from many countries of the world while also attracting a lot of foreign investment. The combination of significant foreign investment in Israel and its capable personnel has created the economic miracle in Israel.

Concerns about the integrated world's economy.

The development of the world's economy has become more and more internationalized and integrated. Truly, the world has become more and more like a global village. So in this kind of environment, if a country adopts a closed-door policy, it will be eliminated through world competition. The international cooperation in the fields of economics and technology is a new practice started from the 1960s in the relationship of international economy. It is a kind of international econon-fic phenomenon, which comes out naturally when the social productivity has developed to a certain stage. Currently, within the development of the world's economy, the economy in one country can not exist separately from other countries. No matter how developed the economy of the country may be, how rich its resources, how advanced its technology, or how much capital it possesses, it can't develop its economy in isolation on its own without cooperating with other countries. People will meet each other's needs through international exchange, learn from one another to make up their deficiency through international cooperation in terms of economics and technology, and gain international comparative interests by participating in the global economy. Thus, opening to the outside world and international economic integration has become current world trends for economic development.

Moreover, in the modem civilized world, information and communication have developed rapidly. Large amounts of information can be transmitted rapidly from one place in the world to another in one second. The travel time needed for people to go from one place to another has become shorter and shorter. World travel is just like visiting your relatives and neighbors next door. It seems that the earth which we inhabit has become smaller and smaller. Indeed, we live in a global village. In this modern milieu, a country that is not willing to join in the international family by continuing to adopt a closed door policy will be abandoned by history.

The role-played by implementing international cooperation regarding agricultural technological development in China

Long ago, China began its international agricultural cooperation. In fact, its advancements in agriculture made great contributions to the development of the world's agriculture including the invention and improvement of many farm tools, the improvement of cultivation systems and crop strains, etc. Soybeans were introduced to America from China. On the other hand, China also introduced significant agricultural technology and good crop and vegetable strains from foreign countries, such as tomatoes and cabbage, etc. In 1898, an experimental silkworm farm was established and they invited a Japanese expert to help them raise the silkworms. The farm used a new method introduced from Japan to raise silkworms and improve the breed of silkworms. This successful experiment served to promote the development of silkworm farming in China.

However, as a result of the "Big Leap Forward" movement, the People's Commune and the Cultural Revolution which started in 1966 and lasted for 10 years, the agricultural development in China was misdirected. The entire country's agriculture and rural economy suffered a tremendous loss. China paid a tremendous price for violating the scientific law and not learning from other countries and absorbing their advanced experience.

After implementing reform and the open door policy, the international cooperation in agriculture with China has developed rapidly and achieved significant results. According to the China Agriculture Development Report '96, China initiated over 400 agricultural projects in cooperation with foreign countries from 1979 to 1995 with the total capital investment of $ 3.4 billion. The overall foreign capital investment expended on agricultural development was $10.1 billion. China has established science and technology exchanges and cooperative economic relationships with over 140 countries as well as with large international organizations such as the World Bank, the Asian Development Bank, the Food and Agricultural Organization of the United Nations, the United Nations Development Program and the International Fund for Agriculture Development. Furthermore, China has set up bilateral joint comnfittees or joint working groups in agriculture with 30 nations to regularly discuss bilateral exchange and cooperation projects. The country also put the development of agriculture on the top agenda in terms of introducing international expertise. According to the incomplete statistics, China has invited over 7100 foreign experts in the fields of agriculture, forestry, animal husbandry, sideline production and fishery by using the state's special fund for inviting foreign experts. At same time, 15 thousand agriculture technicians, engineers and management personnel were dispatched to foreign countries to learn advanced technology in planting, agriculture and management in many different settings. The country also introduced several thousand of advanced practical technology and over one thousand excellent breeds. All this has enhanced the level of science and technology in agriculture in China. The income of the farmers has increased. And some poor areas in China have shaken off poverty and built up a fortune. Thus the whole process of agriculture in China has been accelerated.

Inviting foreign experts.

By inviting foreign experts to work in China and sending people abroad , for training a lot of advanced and practical cultivation and excellent breeding techniques have been introduced to China. The result of such policies has yielded a great increase in the quantity and quality of grain, cotton and edible oil. For instance, the technique of raising rice seedlings in dry lands was introduced to China by the Japanese expert, Shoichi Hara. In fact he has visited China more than 40 times. At present, this technique can be used in different soils in closely with the aid of Chinese technicians and engineers. The technique of raising rice seedlings in dry land, which has the advantage of saving seeds, fertilizer, water and labor, has been used in the vast rice growing area in China. This technique has increased rice production more than 10 to 20% per ha. over the traditional technique and. reduced the cost by 700 yuan. In 1996 this technique was used in 9.33 million ha. rice land, as a result, there was an increase of 7 billion kilos. of rice. With a technique learned from the Canadian experts, the average yield of spring wheat per mu (one ha. is equal to 15 mu) is 300 kilos in the black earth area, 250 kilos per mu in white muddy earth area and increased output by 28% as compared with the farmland not using the technique. This technique has been widely used in Gansu Province and Xinjiang Urgur Autonomous region. Through cooperation with the foreign experts, 10-odd new cotton breeds have been introduced to China and have been cultivated and used in other areas in China. In some counties and towns the ginned cotton output per ha. is 1,500 kilos, for some high yield cotton field, the cotton production is 2250 kilos per ha. Here, special mention to be made is of the help of the Japanese expert Ishimoto Masaicm. At the end of 1979 he introduced a new technique using plastic film as a plant covering which brought a revolution in China's agriculture. This technique has been widely used in more than 40 kinds of crops in all the provinces and municipalities in the country. From 1984 to 1995, the accumulative total agriculture output value is 60 billion yuan.

Results of foreign experts activity.

By introducing foreign expertise, a diversified economy has been established enriching the food supply for both urban and rural people while increasing the income for farmers. For instance, the coastal provinces of Shandong, Guangdong, Zhejiang and Fujian invited foreign experts to help them with shrimp farming in sea water and achieved good results. The Nanyang area in Henan Province cooperated with Italian experts in an experiment using frozen semen and embryonic implantation to improve the breed of local cattle. After 18 months, the improved breed of cattle grew to 500 kilos each, and the net meat produced was 53%, yielding a 15% increase. Now every new breed of cattle can produce 72 kilos more meat. Since 1987, Yantai in Shandong Province has invited several dozen foreign experts from Japan, the Netherlands, Australia and Canada to supervise the whole production process. At same time, Chinese agriculture technicians have traveled to Japan and New Zealand many times to study fruit tree cultivation techniques. By doing so the technique for fruit tree production has been improved a great deal. The average weight of an apple has increased from 100 grams to 225 grams. The output of per mu apple has increased from 1000 kilos to over 2000 kilos. Many provinces like Hebei, Jiangsu, Shaanxi, Anhui and Liaoning have achieved excellent results in planting a new breed of Fuji apple and promoting the technology in some other areas in China. In 1995, the Fuji apple was grown in 0.773 million ha. land with the output of 3.5 million. Now the Fuji apple has become the main commodity in the apple market in China and is very popular among Chinese customers.

Effect of foreign experts on local economy.

By introducing foreign expertise, the local economy in agriculture has developed rapidly and farmers have more ways to gain income. Now farmers have diverse avenues for increasing their income.

Yu Xi Mountain area in Henan Province is rich in oak trees. In the past, the people in this area utilized this wood for mine timber and firewood. The value of every cubic meter of wood was less than 200 yuan. Later on, they introduced a good quality of Japanese mushroom fungus and invited experts from Japan and South Korea to share their know-how in growing this kind of mushroom. Now the farmers can earn 2000 to 3000 yuan for every cubic meters of oak by growing mushroom on it, which is more than 10 times the value than the original wood. This initiative has enabled the farmers in this remote area to significantly improve their living standard.

Since the end of 1980s, Zhanjiang city in Guangdong Province began to consult with experts from America, Canada, Thailand, Malaysia, Hongkong and Taiwan. With their help, farmers solved some of their difficulties in breeding, raising, and monitoring prawns as well as learned new disease prevention techniques. The prawn farming developed quickly as a result. The prawns produced are not only sold in large quantity in the domestic market but are also exported. Now many households in poor and remote areas now make an annual income of 10 thousand to 100 thousand Yuan contrasted with a per capita average annual income of less than 200 Yuan in the past. The development of prawn farming has also stimulated growth in the processing industry with a corresponding rise in the employment as well.

Effect of foreign experts on export.

The introduction of foreign expertise has provided an effective way for developing an export-oriented agriculture in China. In recent years, although export-oriented agriculture has developed rapidly in quite a few places, it is still restrained due to some unsolved problems in techniques. For instance, Shangshiling pig farm in Pengze County, Jiangxi Province raised pigs for export, but owing to some problems in technique and management, the quantity of pork for export has fluctuated for many years. In 1993, the province invited experts from France to work at the pig farm. Within 40 days, the experts helped to solve over 100 problems, such as pig breeding, treatment and prevention of pig disease, etc. The survival rate of pigs has increased by 12%.

In another situation, Mr. Zhao Hongen, the director of Dalian Research Academy of Aquatic Products was sent to study the technique of abalone fishing in Japan. After he came back from Japan, Mr. Zhao continued to concentrate on studying and adapting the technique for use in China. Later he even invented the technique of fishing abalone fries in great density. He has made a tremendous contribution to the industry of artificially raising abalone fries . This technique has achieved an advanced international standard and has been used in the provinces, like Liao Ning, Shandong, Jiangsu, Zhejiang and achieved significant results.

Effect of foreign experts on animal husbandry.

Through the introduction of foreign expertise, the animal husbandry economy has grown in China and the construction of electrification has been accelerated in China's rural areas. Provinces like Yunan, Xingjiang Autonomous Region, Inner Mongolia, Gansu, Qinghai Zhuang Autonomous Region, Guizhou, Hainan and etc. sent their people to Australia and New Zealand for in-service training and introduced herbage of good quality to China. They also gained experience in husbandry management and changed the traditional nomadic life to modem settlement life by enclosing the herding sheds. The China Academy on Tropical Agriculture in Hainan Province sent their people twice to foreign countries for training as well as invited experts from France to share their knowledge. The cooperative exchange resulted in the cultivation of several excellent breeds of herbage to be used in large pasture areas greatly enhancing the development of animal husbandry in each respective region.

In order to promote the development of electrification in China's rural areas, Jiangsu Province worked closely with the American National Association for Agriculture Electricity Cooperation from 1990 to 1995. Methods of management were learned and experience gained in accord with the international construction standard. They joined hands to establish a model county project using rural electrification. Qingzhou City in Shandong Province implemented all kinds of effective technical procedures to meet the standard used by the association and rebuilt the rural electrical network. The actual cost of power transmission for the network was lowered by 3.3 % than the originally planned cost of power. Within the year of 1993 alone, over 12 million kilowatts of power was saved. The stability of voltage transmission was also improved. The Bureau of Agriculture Electricity in Shaanxi Province achieved great results by reforming the rural electricity management system according to the American agriculture cooperative model.

Effect of foreign experts on basic research.

Through the aid of foreign expertise, China's basic research level in agriculture has been strengthened and expanded. Such cooperation has enabled the country's comprehensive agricultural development to be intensified. In addition, a large group of personnel have received training.

According to the statistics, from 1991 to 1995, the institutions of higher learning attached to the Ministry of Agriculture and Ministry of Forestry invited altogether 2,124 foreign experts and teachers. They helped the country to improve 77 key disciplines, establish 44 new disciplines, better equip 76 key laboratories and set-up 33 new laboratories. They also helped train 5,254 personnel of different levels and improved greatly the fundamental research level of agriculture in China. Also, a number of technological achievements at an international standard have been achieved. Some disciplines and research projects have ranked into the world's advanced domain. Thus, China's potential for agricultural development and great achievement has begun to be realized. For instance, the Agriculture University of Central China joined hands with experts from Britain in doing research on genetic engineering relating to the heredity of streptococcus molecules and antibiotics. The Agriculture University of Shenyang cooperated with experts from America to do research on soybean cysticercus. The Agriculture University of Northwest China along with experts from Germany jointly researched the breaking up and transplanting of dairy cow's embryos. All these projects represent the world's advanced level in research.

Under the supervision of German experts, the Agriculture University of China designed a computer consulting system for use with chemical fertilizer. This system has been used in 1 million ha. farmland in Beijing, Hebei, Inner Mongolia, etc. By using the new system, about 4 hundred million kilos more grain has been produced than through use of the traditional method while at the same time saving a large quantity of chemical fertilizer. In addition, the following key state projects have achieved great results with the help of foreign experts: the research on the ecological effect on shelter-forest in the upper and middle reaches of Yangtze River, the key open laboratory of water and soil conservation of Ministry of Forestry, the research on structure and economic efficiency of shelter-forest in the sandy conditions, the key open laboratory of forest plant ecology of the State Education Commission, the technological training center for transforming mountains in China's loess plateau, and the China training center for prevention of soil erosion. All of these projects have played a very important role in establishing the economic shelter-forest system in sandy soil environments, the mastery of erosion prevention and control technology, environmental protection, the development of models for exploration of resources, water and soil conservation and refomfing of loess plateaus.

Therefore, the positive and negative experiences of agricultural development both in China and in the world demonstrate that if China wants to have the stable supply of grain and agricultural products, to develop both in agriculture and rural economy, an export-oriented agriculture strategy must be developed and adopted.

Use of international resources for developing agriculture in a sustainable way.

How to develop China's agriculture in the future is the question at hand. The answer is very clear. China must rely on improving its technology and the ability of personnel while continuing international exchanges and cooperation in economics and agricultural technology. Developing China's agriculture in an intensive and sustainable way can best be achieved by absorbing all aspects of agricultural civilization created by human beings.

Historically, there has been the tendency to develop agriculture and bring about a prosperous rural economy by using the knowledge of science and technology and by improving the capability of laborers. Among all the factors impacting the world's future agricultural development, science and technology will play an increasingly more important role. According to the prediction of the Food and Agriculture Organization of the United Nations, in the coming 20 years, the grain output needs to increase by 1. 8% per year to meet the world population's demand. 1. 5% of the increased grain production will be realized through the input of science and technology. However, at present the ratio of scientific and technological contributions to the development of rural economy is about 35% to40% in China where as in the developed countries the ratio is 60% to 80%. China lags far behind in this regard, but has great potential to lessen the gap. Much work is yet to be done.

In the coming 20 to 30 years, the international cooperation in agriculture in China, especially the introduction of foreign expertise, broadens even further the prospects for development in the following ways:

Need for international cooperation for intensified sustainable agriculture.

China will carry out international cooperation with a view toward developing sustainable agriculture in an intensive mode and surmounting scientific and technical difficulties. For instance, we should make full use of the biological heritage potential in order to select good quality breeds of seeds, to maintain and improve the fertility of the soil and the physical, chemical, and biological quality of the soil, Such knowledge will enable agriculturists to try to create the best conditions for crops to grow, to protect and make good use of water resources, to improve the water

370

conservation and irrigation technologies, to try to improve the crop planting and aquaculture techniques as well as techniques for storage, transportation, and processing. International cooperation will aid China in its comprehensive utilization of agricultural products, the safe monitoring of food and the development of early warning systems technology. The mechanization and industrialization of agriculture will also be improved. Furthermore, cooperative efforts will enable the comprehensive harnessing of large sections of China's great rivers, lakes and deserts as well as the transforming of saline, alkaline, and waste lands into usable agricultural resources. Joint research will lead to greater knowledge concerning the ecological effects on the shelter-forest system, to better ways of water and soil conservation, and lead to the application of new and advanced technology in all areas of agriculture. At the same time, attention will be paid to introducing advanced practical technology used internationally. These technologies would include those used in crop cultivation, aquiculture, the development of good quality breeding of seeds, the prevention of plant diseases and insect pests, the thorough processing of agricultural products and the development of advanced managerial expertise. Gradually, we will seek to set up an ecologically sustainable agricultural environment and also make use of natural resources in a sustainable way. In time, we hope to establish a comprehensive agricultural system with state-of-the-art industrial equipment while applying modem science and technology and advanced business management to every aspect of our agricultural development.

International cooperation is needed for the industrialization of townships.

China seeks also to implement international economic cooperation with the view to realizing the urbanization of counties, the industrialization of townships and their agricultural potential. At the core of modernizing agriculture is the industrialization of agriculture. And agricultural industrialization may be realized through the urbanization of countryside. Our country's policy is to exercise tight control over the expansion of big cities, develop middle-sized cities in a moderate way and try to develop small cities rapidly. The process of agricultural industrialization is in fact a process of urbanizing counties. This involves in no small part, finding ways to employ a host of surplus laborers flowing to cities as a result of massive changes in the surrounding countryside. This influx of laborers to cities is a major issue which promises to impact the whole of China in the future.

The question still remains as to how to effectively achieve the industrialization and urbanization of China's countryside? The key issue is to take measures suitable to local conditions and develop industry in line with the advantages of local agricultural resources. For instance, in the future the vast counties and townships can develop their deep processing of agricultural products on a large scale. Numerous local farmers can work in these industries instead of looking for jobs elsewhere. The processing of agricultural products in many countries around the world has a significant place in their respective national economies. Take America as an example, the people engaged in industries which process agricultural products has accounted for 43% of the total number engaged in the comprehensive industry of agriculture. In 1981, the number of people engaged in product processing industries occupied 11% of the total workforce in manufacturing industries. The ratio between agriculture output value and the output value of agriculture product processing was 1 to 3. While in the middle of 1980s, the respective ratio for China's agriculture output value and the output value of agriculture product processing industry was 1 to 0.76. Additionally, the ratio between China's agriculture and the food processing industry was even lower at 1 to 0.26. In 1994, the people engaged in food manufacturing industry and the overall processing industry was 2.43 million persons yielding a product output value of 333.81 billion yuan. Among them, 0.7368 million people worked in township enterprises which accounted for 33.3% of the total number of people engaged in the food processing industry. The financial output value was 73.3 billion yuan which occupied for 22% of the total output value of food industry. In China, processed food only occupies 25% to 30% of the total food produced, whereas in developed countries it is 80% to 90%. China also has some undeveloped processing industries. For example, insufficient amounts of fruits and vegetables are processed in relation to the market demand. There are not enough high quality foods in the market. At this point the use of high-tech in the agricultural product processing industry is minimal to none. Therefore, international cooperation could be of great help in developing these areas of expertise.

Exchanges of world market resources improves agricultural efficiency.

Exchange of resources in the world market improves comprehensive agricultural efficiency. It is easier to regulate the fluctuation of agriculture production in the big international market than the small and closed domestic market. In this respect, firstly, we should take the advantage of inexpensive price and low cost of production of our agriculture products. According to the needs and demands of the international market, we should make efforts to export our agricultural products and processed agricultural products and develop our export-oriented agriculture. Especially we should give priority to exporting famous, special, top quality, and rare agriculture products and green food to international markets. Additionally, China will seek to exchange the species resources and other agriculture products which are badly needed in the domestic market. Secondly, we should take advantage of rich labor resources and try to export labor services on a large scale and earn foreign exchange for the country. Thirdly, according to the request of the Uruguayan Round Talk, we should selectively open the domestic agriculture and agricultural products market and try to absorb foreign investment to develop the joint-venture agriculture enterprises and agriculture product processing industries. Hence, the scale of using foreign capital should be expanded. At same time, we should try to develop agriculture and industries of agriculture product processing overseas and encourage the Chinese enterprises with capabilities to run farms and factories and use the foreign investment directly in foreign countries. In this way, they can learn the foreign technology and open and do business in the international market.

Learning from collective international experience helps to perfect the agricultural service system.

Learning from the collective international experience and exploring the new system of developing agriculture in China certainly helps to perfect the agricultural service system. The proposal put forward at the seminar on cooperatives in 1990s and strategies on cooperatives developing jointly with governments is "no matter in the past or in the future, cooperative will be an important way for social economic development among Asian countries". This conclusion is also true for the current development of agriculture and rural economy in China. After the reform was implemented in China's countryside, there appeared 160,000 different kinds of rural professional technology associations. Some were in the first stage for technical exchange, some were in the second stage for technical services and a few were in the advanced stage for technology and economic entities. The existence and development of these associations energized the development of agriculture and rural economy in China. They have played a very important role in spreading the knowledge of agricultural technology, in training agricultural personnel and in exchanging agricultural information. However, the development of these associations were restrained by their complex structures and relationships. For instance, properly balancing the governmental support and control so that the cooperative can run effectively while also allowing for local managing and development presents some challenges to be solved. In t s regard, the goo experience of developed countries should be learned so as to strengthen the management of structure, personnel, finance and promote marketing by the agriculture professional technology associations. We should offer support in terms of law, policy and finance so that these associations can help with the modernization, industrialization and socialization of China's rural economy.

Training of personnel needed in modernization.

Making efforts to train all kinds personnel needed by the modernization of the countryside in China. We will continue to be engaged in the introduction of foreign expertise. We will invite more foreign experts to China and send more Chinese professionals abroad for training. We will try to train a larger number of technicians and managers in terms of agriculture, forestry, husbandry, sideline production and fishery, who will be dealing with the international market. We will also improve the ability of all farmers to develop agriculture and the rural economy steadily.

Conclusions:

The experience of reform and opening to the outside world for about 20 years in China has already shown that there is great potential for the development of agriculture in China. There is great enthusiasm and creativity existing among the great number of farmers in China. So long as we continue to open to the outside and adhere to the ideology that science and technology is the first priority while also strengthening the international cooperation in agriculture and making more efforts to develop science and technology we will develop our agriculture in an intensive mode and a sustainable way. Truly, there is a great prospect for the development of agriculture in China. China can support itself and will not become a threat to any other countries. We can not only support ourselves, but we can also improve our standard of living year by year. China's agriculture is full of life and vigor will become even more splendid and magnificent.

References:

1. Lu Liangshu " Strategies and Prospects for the Development of Modern Agriculture" In: The Development of Modem Agriculture in China. Collected works from the Seminar on the Development of Modern Agriculture across the Straits. China Publishing House of Agriculture, July 1996.
2. China Agriculture Development Report '96, Ministry of Agriculture, People's Republic of China, August 1996.
3. Economic Planning Academy of South Korea: Main Statistics of South Korea Economy 1988.
4. Lu Wen, Pei Changhong, and Wei Wei. Rural Areas in China: Opening up and Development (Project of State Social Science Foundation) China Publishing House of Agriculture, March 1997.
5. Jiang Chunyun. Attention Should Be Paid for the Change of Mode of Production for the Development of Agriculture. Seeking Truth No. 1 Issue, 1996
6. China Year Book of Statistics. 1995. China Publishing House of Statistics, April 1996.

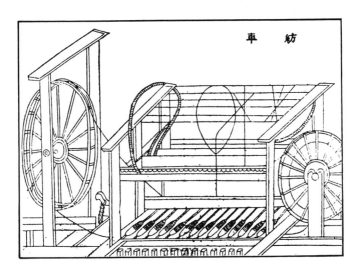

CHAPTER 4.9

INTERNATIONAL COOPERATION

Alexander von der Osten

Consultative Group on International Agricultural Research (CGIAR) The World Bank

China's huge agricultural research system cannot be expected to deliver a stream of innovative technologies to farmers if it is fragmented, and unable to fully participate in global research collaboration. Conversely, outside scientific institutions need to develop better knowledge of China's needs and potential if they want to find the right research partners.

The agricultural malaise

Agriculture has become a weak link in national economic development and is since about 1990 again considered one of the "four difficult problems." As a result, rural economy and agriculture have become one of the four 'Key Points' (the others being vitalizing state-owned enterprises; increasing state revenues; and protecting the environment).

While official sources profess that "the investment made by the Chinese government in agricultural production and agricultural science and technology as well as education is increasing year by year,"[1] independent observers see an opposite trend. Over the last ten years, investments in support for agricultural research from all sources -- at state, province and county levels -- have declined in real terms and research intensity has slowed down. Private sector support for research is very limited and relates only to a few applied areas.

China's agricultural research system

China's national agricultural research system is complex and not well understood outside China. It is the largest publicly funded and administered research system in the world.

Almost 60,000 researchers and technicians' work in about 400 research institutions and 70 agricultural universities throughout China. International research centers have be involved over the past twenty years in training Chinese nationals to develop strengths in key disciplines related to crop production, genetic resources, and fertilizers.

The Chinese Academy of Agricultural Sciences (CAAS) is the national agricultural research organization in China, directly affiliated to the Ministry of Agriculture (MOA). The Academy has about 10,000 employees working in 38 research institutes located in Beijing and 14 different provinces. These institutes play a leading and coordinating role with close collaboration with research institutes at provincial and local levels. The provincial academies of agricultural sciences are larger than many national agricultural research systems in sub-Saharan Africa. Among the academies, 20 are, for instance, heavily active in rice research. As a result, there is a good deal of duplication of efforts and, overall, heavy spending on rice research at the expense of other commodities, for instance sorghum, millet and sweetpotato, and especially livestock. The livestock sector receives less than 10 percent of total funding, but is essential for food security and contributes about one third to total agricultural GDP.[2]

In spite of the improved capacity of China's agricultural universities and key research institutes and their improved ties with the global research communities, the Chinese research and educational institutions remain constrained by several factors. These include inadequate financial support from the government, outdated equipment and facilities, duplication of work at federal, province, and country level institutions; lack of operational funds, public sector dominance and inadequate links between agricultural research, education and extension.

A new beginning

Although China has made remarkable progress in technology generation and its dissemination to farmers, contributing to the phenomenal increase in China's agricultural productivity and rural economic development. Nevertheless, politicians as well as the Ministry of Agriculture and State Science and Technology Commission officials have realized that further progress in agricultural and rural development will depend on the availability and application of new technologies.

Attention paid to the agriculture sector has increased substantially since about 1995, especially because of concern about increasing pressures on water and soil resources. China wants to develop a mid- to long-term plan to ensure availability of relevant technologies and its dissemination to farmers. The Chinese government pursues a policy of 'vitalizing agriculture' by using science and technology. Sustainable agriculture is a major element of the government's "21st Century Agenda."

The White Paper says in this respect: "China will improve its agricultural scientific research structure and set up a new scientific research system featuring advanced disciplines and the close integration of scientific research with production. The state and local governments will selectively support leading agricultural scientific research and education institutions to foster a capable contingent of scientists for important basic practical technology and new- and hi-tech research. Some people engaged in scientific research and education will be encouraged to go deep into the countryside to spread technology and train farmers. The government's unified plan for agricultural science and technology and relevant education should be strengthened to speed up the integration of agricultural scientific research and education, and popularization of agrotechnologies to improve grain production."[3]

Agricultural research has moved up on the list of priorities, and China intends to assume a leadership role among developing countries in their quest for food security. Resource flows for science and technology development have not yet reached adequate levels but China is opening up more rapidly to international scientific collaboration.

The institutional climate is changing. A broad range of institutions involved in agricultural research are moving from a position of fragmentation and competition toward shared goals and vision, coordinated efforts, and collaboration. They have discovered the benefits of compliment and synergies. The ongoing process is aided by the need to diversify sources of funding and to actively mobilize resources under increasing influence of market forces.

The research programs are increasingly client-driven. Producers' needs for technology are now complementing and supplanting development goals as defined by national and provincial governments. Research organizations take greater care in ensuring that their research products reach the farmers and are applied.

Cooperation in agricultural research

The Chinese government is generally interested in expanding contacts and strengthen communication with actual or potential agricultural research partners worldwide. Overarching specific research and program areas for future cooperation is the desire to receive assistance in improving national agricultural research and attaining sustainable development in agriculture.

However, cooperation with the private sector is hampered by the fact that China, as a rule, does not recognize international protection of Intellectual Property Rights (IPR). This limitation also affects cooperation with advanced research institutions (ARI) in developed countries because the latter increasingly establish IPR for their research products. Consequently, China is mainly bound to cooperate with partners that do not claim IPR, especially the international agricultural research centers (IARCs) for developing country agriculture, most of which are supported by the Consultative Group on International Agricultural Research (CGIAR).

Recently, a mechanism of potential importance to China has been created to facilitate and enhance international agricultural research cooperation on a more systematic basis. In October 1996, a Global Forum on Agricultural Research met for the first time at CGIAR in Washington, DC, in which major actors from among national agricultural research services (NARS) participated, together with international centers, advanced research institutions in North and South, the private sector and NGOs. The Global Forum built on a series of preceding regional and subregional fora, mainly organized by NARS and their subregional and regional groupings.

The Global Forum serves to integrate the various components of the global agricultural research system, and provides the platform for greater interaction among regional and sub-regional fora, and other actors in the global system. The operationalization of the framework, however, will take place at the regional, sub-regional and national levels. A Global Forum Steering Committee serves as the mechanism through which the objectives of the Global Forum are pursued; its primary role is to facilitate and coordinate implementation of the Global Forum Action Plan, adopted in October 1996.

Given the rapidly expanding national, regional and international research agendas and the decline in public sector funding, increased global and regional collaboration responds to the need for securing synergies and scale economies in a bottom-up, demand driven approach. For China, more systematically organized partnerships with other stakeholders in agricultural research could thus help optimize the use of resources, the exchange of knowledge, and mutually strengthen institutions, as recommended by Premier Li Peng[4].

China and International Agricultural Research

Important partners of China's international cooperation are the international agricultural research centers of the CGIAR and other centers not affiliated with the CGIAR (e.g. IBSRAM, AVRDC). Informal cooperation with the CGIAR centers started in the 1960s and 1970s, long before China formally joined CGIAR.

In 1984 China became a donor member of the Consultative Group on International Agricultural Research (CGIAR) through its Academy of Agricultural Sciences (CAAS), and that year funded IRRI, CIMMYT, the International Center for Tropical Agriculture (CIAT), the International Potato Center (CIP), the International Board for Plant Genetic Resources (IBPGR -- now International Plant Genetic Resources Institute IPGRI), the International Center for Agricultural Research in the Dry Areas (ICARDA), the International Crops Research Institute for the Semi-Arid Tropics (ICRISAT), the International Food Policy Research Institute (IFPRI), and the International Livestock Centre for Africa (ILCA -- now International Livestock Research Institute ILRI). In 1986, China began providing support to IITA, and to the International Service for National

Agricultural Research (ISNAR) in 1989.

For China, the CGIAR is one of the few available sources of property rights-free access to cutting edge technology. Cooperation is strong despite the fact that the CGIAR is focused on tropical and subtropical agriculture whereas China's climate is mainly temperate and subtropical.

The State Science and Technology Commission (SSTC) is currently expanding its information base on international agricultural research. Despite twenty years of China's membership in CGIAR involving large scale seed exchange activity, as well as many collaborative projects with CGIAR centers, and despite the training of numerous Chinese national scientists at CGIAR centers, knowledge of the CGIAR system, its role and potential, is still very limited among research institutions in China. Only few of China's agricultural agencies and institutes have adequate access to the new technologies developed by the CGIAR centers.

Germplasm Transfer and Utilization

China and the international centers have a tradition of extensive germplasm exchange. In the case of barley and rice, the centers collaborated with Chinese crop scientists, conserving samples of China's wild diversity of germplasm. Through the international centers, China has released varieties of rice, wheat, maize, barley, potato, and pulses.

In the 1950s, China's Ministry of Agriculture organized several nationwide and supplementary collections of crop germplasm resources. Some 50 species of field crops with nearly 200,000 accessions, including some duplicates, were collected. However, some of these collections were lost during the Cultural Revolution. In 1979, a nationwide, supplementary collection was again made, from which about 60 species of crops with nearly 100,000 accessions, including some duplicates, were collected. Since that time, more strategic collecting has been conducted, often in association with IBPGR/IPGRI.

In 1978 the Institute of Crop Germplasm Resources (ICGR) of CAAS was established. A national genebank was built with support from the Rockefeller Foundation, technical advice from IRRI, and cooperation from IPGRI.

Crop improvement

The International Rice Research Institute (IRRI) has long benefited from and contributed to rice research in China. One of the parents (De Ge Wu Gen) of the first of the so-called miracle rices, IR-8, came originally from China. Other such short-statured rice varieties have moved from China through IRRI's germplasm bank to rice researchers around the world. In turn, rice lines and varieties coming indirectly from IRRI were being used by Chinese researchers in their rice improvement programs long before formal working relationships were established in the mid-1970s.

China has been one of the world leaders in the creation and dissemination of high yielding short-statured rices. In the 1960s, high yielding rice varieties were released, following the development of semi-dwarf varieties in the mid-1950s. By 1970, semi-dwarf rice was grown extensively in most rice-producing areas. By 1985, more than 95% of the total rice area was cultivating modern high-yield varieties. The IRRI variety IR8 had been imported in 1967 to Guangdong, and in 1971 to Shansi. IRRI has supplied China with additional parent material, especially for hybrid rice. IRRI and China have a long common history of cooperative exchange of rice plant stock.

IRRI and China cooperated extensively in enhancing the creation and the use of hybrid rice. Such hybrids were first developed in China in the early 1970s. Appropriate genetic materials for hybrid rice research were made available to IRRI from whence they were provided to rice researchers in other countries. To enhance the development of improved hybrids in other countries, a jointly sponsored hybrid rice training course was established in 1979, with part of the training being offered at IRRI (in the Philippines) and part at the Hunan Academy of Sciences in China.

At the request of the leaders of the Chinese Academy of Agricultural Sciences (CAAS) in Beijing, IRRI scientists and administrators joined their Chinese counterparts in selecting a site for the Chinese National Rice Research Institute (CNRRI) located near Hangzhou. This cooperation was expanded as the CNRRI was organized and staffed. Similar interactions occurred as China planned and established its National Germplasm Bank (NGB) in Beijing.

Wheat materials from the International Maize and Wheat Improvement Center (CIMMYT) were introduced into China during 1968/69, followed by more formal collaboration dating from 1973. Since then, CIMMYT materials have been used in breeding programs in China.

Maize development was based on hybrids in the 1950s and 1960s from U.S. corn-belt lines and Chinese flints. From 1974, interest in quality protein maize led to the use of CIMMYT breeding materials. CIMMYT populations have been used for some open pollinated varieties and to develop inbred lines. Inbred lines from the International Institute of Tropical Agriculture (IITA) were introduced into China in 1986.

A scab-resistant ICARDA barley variety, identified by the Shanghai Academy of Agriculture in collaboration with the International Center for Agricultural Research in the Dry Areas (ICARDA), was introduced in 1995 in Shanghai and three neighboring provinces, with yield increases between 20 to 25 percent. ICARDA also exchanged faba bean breeding material with China, which is the largest single producer, accounting for over 50 percent of world output.

The International Crops Research Institute for the Semi-Arid Tropics (ICRISAT) is collaborating with China in sorghum and groundnut improvement. In 1995, ICRISAT, CAAS and the Shandong Academy of Agricultural Sciences cosponsored an international workshop to study management practices in groundnut production that result in yields of over 5 tons/hectare.

The International Potato Center (CIP) provided CAAS in 1978 with CIP 24, a potato variety that has rapidly spread in China because of its hardiness, high yield, and market appeal. Originally distributed to many counties in the northern provinces, the variety was also adopted by farmers in the southern provinces, in the late 1980s. Potato production in China doubled between 1978 and 1996. A CIP-coordinated UPWARD project in Zheijiang province showed that potatoes grown in paddies before rice improve yields of both crops considerably.

While China produces about 80 percent of the world sweet potato harvest, cassava has received little attention; the cultivation in China of both crops has been studied by CIP and the International Center for Tropical Agriculture (CIAT), respectively.

Partnerships and collaborative agreements

The International Plant Genetic Resources Institute's (IPGRI) predecessor, the International Board for Plant Genetic Resources (IBPGR) and CIP developed research partnerships with CAAS and in the late 1980s opened offices in China. Other international agricultural research centers established a number of collaborative agreements, for instance ICARDA in 1987 and ICRISAT in 1988. Ten centers have been involved in training with China's scientists.

The International Center for Living Aquatic Resources Management (ICLARM) established research contacts in 1984 in recognition of China's long experience in fish farming. 'The International Irrigation Management Institute (IIMI) developed links in the early 1990s because more than a decade of progressive Government decentralization of irrigation responsibilities in China was found to increase the productivity and profitability of irrigation systems. The China/IIMI collaboration continues to develop.

Early collaborative research between China and the International Food Policy Research Institute (IFPRI) focused on the organization of national and provincial data on food production and inputs to agriculture. Data collection methodology was emphasized. Systematization and analysis of food production data from local levels and their integration into national macro-trend research with macro-policy were major objectives. A wide range of micro- and macro-level policy issues studied were related to water, fertilizers, and high-yield crop varieties.

Irrigation and fertilizer-policy studies by IFPRI involved water management, including salinization control of the Yangtze diversion, and fertilizer production. China has an increasingly significant effect on international fertilizer markets. Demand for fertilizers at the local level allowed the extrapolation of household survey data for the systemic and policy adjustments in the administration of fertilizer production, marketing, allocation, and distribution systems.

Food policy research in China involved a range of studies with CIMMYT, CIP, and CIAT on crop production within different agroecological zones, and analyzing the variability in area production, utilization patterns, and crop development. The China/IRRI partnership conducted farm-household surveys on rice adoption, especially in environmentally advantaged and less advantaged areas in China. These surveys revealed a distinct association between farmers' choices of investment and production techniques, and improved crop productivity.

Rural employment promotion strategies and shifts in employment patterns emphasized the effects of rural infrastructure. IFPRI helped to analyze China's experience with commune systems and shared the findings with other developing countries, as well as with more developed nations.

At the national level, collaborative work with IFPRI on staple food crops helped the Chinese Academy of Agricultural Sciences plan its Center for Agrotechnical Economics. Changing demands for livestock products led IFPRI to conduct a study on the technological and policy options for livestock development.

In 1996, two Memoranda of Understanding were signed by IFPRI and CAAS. An IFPRI office was established at CAAS, and two mutual projects launched, one dealing with public investment policy issues and priorities, and another to study the national agricultural research system with a view to helping improve the priority setting exercise and the resource allocation mechanism.

Agricultural Research Management

China is aware that some segments of its agricultural research system require revision. Improved knowledge of the national agricultural research system enables Chinese leaders to strengthen the performance of the farm sector. A better understanding of the contributions of the national agricultural research system to agricultural sustainability and intensification can be of value on a global scale.

With these factors in mind, during the early 1990s, China and ISNAR undertook a thorough study of China's agricultural research systems in order to:
•   describe the Chinese agricultural research system, including its institutional development and the growth of its research capacity;
•   estimate the effects of agricultural research on production growth at national and regional levels; and
•   conduct a preliminary analysis of key policy issues affecting resource allocation to research on crop production, forestry, livestock, and fisheries

in different regions.

Biotechnology

Biotechnology is one of the sectors in which China is a leader and can contribute much to global progress through research collaboration while benefiting from joint efforts. Research alliances on biotechnology with CGIAR centers and other partners -- above all the private sector -- are important possibilities for broadening and updating biotechnology skills. Collaboration with CGIAR centers -- which are set to increase their biotechnology capacity -- will generate products and processes which are international public goods and can be freely accessed by developing countries for more rapid and sustainable agricultural growth. For China, the world's largest agricultural economy, it will be crucial to remain at the cutting edge of biotechnology if the food security and poverty alleviation challenges of the coming century are to be met.

Conclusion

A major obstacle to expanded international cooperation is the world's limited knowledge of what is going on in China and China's limited knowledge of what is going on in the world. Who are potential research partners, who are doing what, who can offer what, who needs funding, who can provide funding? Very elementary spade work still needs to be done on both sides to provide Chinese institutions with knowledge about international potential and activities, and to help outside institutions gain more insight into China's agriculture and agricultural research. Expanded cooperation in agricultural technology hence requires, as a first step, greatly enhanced cooperation in information retrieval and exchange.

Building such bridges means for research institutions outside China that personal contacts among scientists, occasional visits to China and attendance at bilateral meetings are not sufficient. Serious investment in exploring China's agricultural research system is needed as a *sine qua non* for future productive cooperation ventures.

For China, bridge building requires a conscious effort to explore possible international solutions to national problems by encouraging and enabling scientists, including provincial and junior research staff, to find and use existing gateways to outside expertise, and engage personally in international contacts.

---

[1] State Science and Technology Commission, informal discussion paper (1996)
[2] Shenggen Fan IFPRI, pers. comm.
[3] Information Office (1996): White Paper--Grain Issue in China, 11
[4] Science 273, 7/96, 13

CHAPTER 4.10

Agricultural Biotechnology

Ray Wu,

Section of Biochemistry, Molecular & Cell Biology,
Cornell University, Ithaca, NY 14853 U.S.A.

China has limited per capita agricultural lands, shortage of water and other resources, biotic and abiotic stresses on plants, and a very large and growing population. In terms of grain production, there is a need to increase yield by at least 30%, so that by the year 2030 it will be possible to produce 640 million tons of grain annually. Thus, a sustained increase in agricultural productivity is an urgent and number one priority for the next 30 to 40 years.

Several factors affect the yield of grain per hectare of land. Science and biotechnology can be helpful in the following areas towards increasing productivity: planting improved cultivars, improving post-harvest handling, and minimizing losses due to drought, diseases and insect pests. Between 1960 and 1985, significantly improved cultivars have been developed mainly through the efforts of plant breeders. However, during the last ten years, progress has been much slower. This is because utilization of available germplasm is limited, and plant breeders' efforts to exploit wide crosses are often set back by problems of incompatible species barriers. It seems that efforts in conventional plant breeding alone will not be able to make sufficient improvements in cultivars. On the other hand, agricultural biotechnology has recently been developed to a point where substantial improvements in cultivars can be anticipated in the next 30 years.

Of all the food consumed in China, over 70% (in terms of calories) comes from cereal grains (rice, wheat, corn, barley, oat, sorghum, rye and millet). Therefore, in this article, possibilities of increasing yield of cereal grains in the following ways will be briefly summarized.

In 1996, the total grain production in China was estimated to be 490 million tons. Out of this, rice represents a major component, which amounts to close to 160 million tons. An estimate on the increase of rice yield in the world was made recently by Evenson (1996), among others.

He gives estimates showing the amount of yield increase which can be achieved by around the year 2020 with today's knowledge, as compared to the yield in 1994 (shown in Table 4.10.1).

Estimate made by this writer based on the amount of salt-affected land which gives either low yield or is non-arable for rice (i.e., salt content higher than 90 mm).

Table 4.10.1. Estimated Yield Increase of Rice in the World.*

Yield Increase Due To Value	Percent Yield Increase Relative to the 1994
Introducing beneficial genes to increase yield	24
Growing transgenic plants with resistance to various insect pests**	8
Growing transgenic plants with resistance to various diseases**	6
Introducing beneficial genes to withstand drought stress or salt stress	10
Utilizing presently low-yield or non-arable land by using salt-tolerant plantst	8

*The data collected by Dr. Evenson was based mainly on information known to him before May 1994, and developments that can be readily predicted. However, unexpected new discoveries and developments in agricultural biotechnology have already been made in 199~1997, and many more discoveries are very likely to be made during the next 1~20 years. Thus, the actual percent increase in crop yield can be a lot higher than the value shown in this table.
* *Increasing yield by decreasing losses.

From this table, an estimated worldwide yield increase of rice can be as high as 56% by 2020. If China can achieve two-thirds of the estimated value for rice, i.e., 37%, and if the yield of other major cereal plants can also be increased to the same extent, it would be adequate to provide sufficient food for China without an increase in the use of fertilizers and pesticides. In fact, the use of chemical pesticides is expected to decrease when insect- and disease-resistant transgenic plants become available. Thus, the application of biotechnology is also beneficial for the environment. It should be pointed out that in making this estimate, Dr. Evenson assumes that substantial financial support for research and development in rice biotechnology will continue to be made in the next 25 years, as has been done for the last eight years or so. In addition, improvements in infrastructure, management, etc. are also to be made at the same time.

Potentially useful genes and approaches for rice genetic engineering have been summarized by Toenniessen (1991) and by Wu (1994). The potentially useful factors and genes reviewed by Toenniessen (1991) include the use of different cytoplasmic male sterility systems to improve hybrid rice, use of genes that encode insecticidal proteins (such as *B.t.* crystalline protein and protease inhibitors) to provide resistance to insect pests, use of genes to provide resistance to diseases caused by fungi, bacteria and viruses, use of genes to improve tolerance to drought and salinity, etc. Some of these potentially useful genes showed promising results after introducing them into transgenic rice plants; these results were reviewed by Wu (1994). They include introducing genes that encode insecticidal proteins (such as *B.t.* crystalline, protease inhibitors, snowdrop lectin, and ribosome-inactivation proteins), and those which encode proteins to confer disease resistance (such as chitinases, -glucanases, and antifungal peptides). Some of the more recent developments are summarized below.

Methods to Increase Yield of Grain or Seeds

Use of hybrid rice - By making use of the mal~sterility trait, hybrid rice was developed twenty years ago in China by Long-Ping Yuan. Hybrid rice has been grown in increasingly larger. areas of land in China, and the yield has been approximately 20% higher than traditional rice varieties. However, there are very few varieties of hybrid rice available for planting in the field; thus, more varieties need to be developed. In addition, other useful traits, such as insect and disease resistance, and drought, salinity and low temperature tolerance, need to be introduced into hybrid rice by biotechnology techniques.

Mapping quantitative trait loci (QTL) for high yield - By using the molecular biology techniques, restriction fragment length polymorphism (RFLP) and simple sequence length polymorphism (SSLP), one can identi~ genetic regions on the chromosome that co-segregate with high yield. One example is the mapping of two QTL loci from a wild rice species that can result in a 20% higher yield of rice (xiao et al., 1996). Once such QTL loci are identified, one can then use marker-assisted breeding to introduce the relevant chromosomal regions into elite rice cultivars much faster than using traditional plant breeding techniques. In the future, it will be possible to use a map-based approach to clone the two genes at the QTL loci. Finally, it may be possible to introduce these high-yield-associated genes into rice or other cereal plants by genetic transformation techniques, followed by regeneration of fertile plants.

Introducing a foreign gene into plants to delay senescence and to increase yield - During senescence, leaf cells undergo coordinated changes in metabolism and gene expression, resulting in a sharp decline in photosynthetic capacity External application of cytokinin was found to delay senescence. The work of Gan and Amasino (1995) showed that by introducing a cDNA that encodes for an enzyme responsible for cytokinin biosynthesis into tobacco, when this cDNA expression is driven by an autoregulatable promoter, the transgenic tobacco plant produced more seeds and biomass as compared to untransformed plants. If this beneficial gene also produces the same effect in transgenic rice and other cereal plants, a higher yield of grain can be expected.

Introducing Disease-Resistant Genes to Minimize Yield Loss Due to Diseases

Cloning and then introduction of a rice gene to confer resistance to bacterial blight disease -- Song et al. (1995) have cloned from rice the *Xa21* disease resistance gene by using a positional cloning method. The *Xa21* gene encodes a receptor kinase-like protein. Mter transforming a rice variety that is sensitive to bacteria blight with this gene, the transgenic plants are highly resistant to many isolates of the pathogen *Xanthomonas oryzae.*

Introducing cDNAs to control rice tungro disease -- An important disease in rice is known to be caused by infection with the rice tungro virus. There are two types of tungro viruses, the RTSV and the RTBV. Genes that encode the RTSV coat protein and the RTBV coat protein have been introduced into rice by the biolistic method. Third-generation transgenic plants have been tested for resistance, and the preliminary results showed that some of these transgenic plants are resistant to the rice tungro virus (Fauquet et al., 1997). These authors also planned to introduce several different viral genes into rice for full control of the tungro disease.

Introducing cDNA to control rice yellow stunt rhabdovirus - To develop a strategy for genetically-engineered resistance to rice yellow stunt rhabdovirus (RYSV), transgenic rice plants have been generated which express the RYSV nucleocapsid protein (N). Virus-resistance assays have been performed on Ti progeny of 23 transgenic plant lines. In 8 plant lines, over 65% of the progeny plants showed resistance (Fang et al., 1997).

Introducing a chitinase cDNA to control fungal disease -- A chitinase gene under the control of the CaMV 35S promoter was transferred into rice by transformation. Several of the second-generation transgenic rice plants were shown to be resistant to the sheath blight fungus, *Rhizoctonia solani,* which results in sheath blight disease in rice (Lin et al., 1995).

Introducing Insect Resistant Genes to Minimize Yield Loss Due to Insect Pests

Introducing crystalline protein genes into transgenic rice to control insects -- A synthetic *crylA(b)* gene from *Bacillus thuringiensis* has been synthesized and introduced into rice by transformation. The second- and third-generation transgenic plants showed a significant insecticidal effect on several lepidopterous insect pests (Wunn et al., 1996). Thus, introducing stem borer resistance into germplasm of rice breeding lines would make this agronomically important trait available for conventional rice breeding programs.

Introducing a protease inhibitor gene into transgenic rice to control insects - A potato proteinase inhibitor II (PINII) gene *(pin2)* was introduced into rice. The introduced *pin2* gene was stably inherited in the second, third, fourth and fifth generations, as shown by DNA blot analysis. Bioassay for insect resistance with the fifth-generation transgenic rice plants showed that these plants had increased resistance to a major rice insect pest, pink stem borer (Duan et al., 1996).

Introducing two insecticidal genes into transgenic rice - A modified delta~ndotoxin gene, *crylA (b)*, and a cowpea trypsin inhibitor gene, *CpTi*, under the control of rice actin 1 promoter, were separately introduced into rice. Several transgenic rice plants showed strong insecticidal activity against yellow stem borer (Wu et al., 1997).

Introducing a Foreign cDNA Into Rice for Drought and Salt Tolerance

LEA gene -- A late embryogenesis abundant (LEA) protein gene, *Hval*, from barley was introduced into rice suspension cells using the biolistic-mediated transformation method. Second-generation transgenic plants showed significantly increased tolerance to water deficit and high salinity (Xu et al., 1996). By planting rice (or other plants) that harbors genes to provide drought or salt tolerance, the yield loss can be minimized under conditions of water deficiency, or to give reasonable yield in soil that contains high levels of salt.

Other genes - Other useful genes to confer tolerance to water deficit and high salinity include those that encode for the biosynthesis of mannitol (Shen et al., 1997), proline (Kishor et al., 1995) and polyamines (Masgrau et al., 1997), which have been shown to increase tolerance of transgenic tobacco plants toward salt or drought stress. Additional useful genes, such as those that encode for the synthesis of glutathione S-transferase/glutathione peroxidase (Roxas et al., 1997) and Fe-superoxide dismutase (Camp et al., 1996), have also been shown to increase stress tolerance of transgenic tobacco plants. It is likely that several, or most, of these genes, when joined to a suitable regulatable monocot promoter and introduced into rice or other cereal cells, will produce transgenic plants that are tolerant to water deficit and high salt.

It should be mentioned that introducing one beneficial cDNA or gene is not sufficient to provide a high degree of drought and salt tolerance. One needs to introduce several beneficial cDNAs or genes into the same transgenic rice plant to provide a high degree of stress tolerance.

Introducing a Foreign cDNA Into Rice for Low-Temperature Tolerance

An *Arabidopsis* cDNA, known as C0R47 (Gilmour et al., 1992), was joined to the rice actin 1 gene promoter and introduced into rice cells, and a number of transgenic plants were regenerated (Cheng et al., 1998). After growing 60-day-ld second-generation plants at 4^0C for 5 days, most leaves of non-transformed control plants wilted and turned yellow, whereas leaves from two lines of transgenic rice plants remained green and relatively healthy looking, as compared to plants grown at 25^0C. The transgenic plants also grew much better in 20U mM salt, as compared to control plants. Thus, C0R47 is shown to be useful to confer low-temperature tolerance and high-salt transgenic rice plants (Cheng et al., 1998).

By planting rice (or other plants) that harbors genes to provide cold tolerance, the growing season can be extended in places of cold climate and thereby increase yield by allowing more time for grain filling. It should be pointed out that for a high degree of low-temperature tolerance, one needs to introduce several beneficial cDNAs or genes

Use of Other Genes

There are other potentially beneficial genes that can be used. These genes encode proteins that can make the metabolic pathway responsible for the synthesis of starch, lipids or protein more efficient. Some of the possibilities are long-term goals toward which scientists have started to carry out more research work. These include the modification of the photosynthesis system to make it more efficient and thereby increase yield, and the introduction of genes responsible for nitrogen fixation into transgenic plants. If the plants can fix nitrogen, it may alleviate the use of nitrogen fertilizers. These goals are important but difficult to achieve, and it may take 1~20 years of intensive research to achieve them.

Development of Agricultural Biotechnology in China

Over the last ten years or so, China has started to develop agricultural biotechnology. Even though a certain level of success has been achieved, such as developing a number of transgenic plants and planting disease-resistant tobacco and tomato in the field, studies on transgenic cereal plants have just started to show promise. Moreover, China has made very new fundamental contributions to the field of agricultural sciences and agricultural biotechnology. For example, most of the important recent developments cited in this article are the result of research work conducted in developed countries. I believe Chinese biological scientists make fewer fundamental contributions and slower progress in the field of plant biology and agricultural biotechnology because they choose short-term projects instead of carrying out longer-term original research. This is because funding for research and development is insufficient, and a rigorous peer-reviewed system for reviewing progress and renewing support is lacking. A suboptimal working environment, caused by lack of motivation and instability on the part of junior research personnel (partly due to low salaries) and graduate students (most of them try to go abroad), overburdening of senior research scientists with too many non-research related tasks, a lack of willingness for scientists to collaborate with other groups, and time-consuming procedures for ordering supplies and equipment, is also a major contributing factor.

In the next 30 years, scientists in China need to carry out more basic research work in agricultural sciences as well as agricultural biotechnology. China also needs to absorb and make the best use of the new knowledge and technology developed abroad. In order to achieve these goals, leaders in China must substantially increase the budget for agricultural research and development, and introduce a rigorous peer-review system to ensure quality of research output. The top official in agriculture also needs to be familiar with both traditional agricultural sciences and with recently developed biotechnology. It is essential that he understands the value, and also be able to promote international cooperation.

Summary

There is great potential for applying plant biotechnology to improve plants in order to increase yield or nutritional value of grains. Several examples are given for increasing yield of rice or decreasing loss due to diseases, insect pests, drought, salt or low-temperature stresses. The same approach, and in fact, the same beneficial genes are likely to work just as effectively in other cereal grains as in rice. There are other ways to increase yield of grain by applying plant biotechnology, which are not included in this article. Additional means are likely to be discovered in the next ten years through basic research as well as flirther development of biotechnology. By using similar approaches, animal biotechnology can be developed to improve animals, including aquatic animals. Improvements may include faster growth, resistance to diseases or tolerance to low temperature.

References

Camp, W.V., Capiau, K., Montagu, M.V., Inze, D. and Slooten, L. (1996) Enhancement of oxidative stress tolerance in transgenic tobacco plants overproducing F~superoxide dismutase in chloroplasts. Plant Physiol. 112: 1700~1714.

Cheng, W., Su, J., Zhu, B., Jayaprakash, T.L. and Wu, R. (1998) Development of transgenic cereal plants that are tolerant to high salt, drought and low temperature. In *Proceedings of the 26th General Assembly and Symposium of IUBS,* in press.

Duan, X., Li, X., Xue, Q., Ab~El-Saad, M., Xu, D. and Wu, R. (1996) Transgenic riceplants harboring an introduced potato proteinase inhibitor II gene are insect resistant. Nature Biotech. 14: 494~98.

Evenson, R.E. (1996) An application of priority-setting method to the rice biotechnology program. In *Rice Research in Asia: Progress and Priorities* (Evenson, R.E., Herdt, R.W. and Hossain, M., eds.), CAB International, pp. 327-345.

Fang, R.-X., Luo, Z.-L., Chen, X.-Y., Mang, K.-Q., Gao, D.-M., Tian, W.-Z. and Li, L.-C. (1997) Analysis of genetically-engineered resistance to RYSV in progeny rice plants. In *General Meeting of the Int'l Program on Rice Biotechnology,* Malacca,Malaysia, September 15-19, 1997, p.60.

Fauquet, C.M., Huet, H., Ong, C.A., Sivamani, E., Chen, L., Viegas, P., Marmey, P., Wang, J., Mat Daud, H., de Kochko, A. and Beachy, R.N. (1997) Control of the rice tungro disease by genetic engineering is now a reality! In *General Meeting of the Int'l Program on Rice Biotechnology,* Malacca, Malaysia, September 15-19, 1997, p. 59.

Gan, S. and Amasino, R.M. (1995) Inhibition of leaf senescence by auto-regulated production of cytokinin. Science 270: 198~1988.

Gilmour, S.J., Artus, N.N. and Thomashow, M.F. (1992) cDNA sequence analysis and expression of two cold regulated genes of *Arabidopsis thaliana.* Plant Mol. Biol. 18: 13-21.

Kishor, P.B.K., Hong, Z., Miao, G., Hu, C.A. and Verma, D.P.S. (1995) Overexpression of A^1-pyrroline-5-carboxylate synthase increases proline production and confers osmotolerance in transgenic plants. Plant Physiol. 108: 1387-1394.

Lin, W., Anuratha, C.S., Datta, K., Potrykus, S.K. I., Muthukrishnan, S. and Datta, (1995) Genetic engineering of rice for resistance to sheath blight. Bio/Technology 13: 686-691.

Masgrau, C.T., Altabella, T., Furras, R., Flores, D., Thompson, J., Besford, T. and Tiburcio, A.F. (1997) Inducible overexpression of oat arginine decarboxylase in transgenic tobacco plants. Plant J. 11: 465~73.

Roxas, V.P., Smith, K.R., Allen, E.R. and Allen, R.D. (1997) Overexpression of glutathione S-transferase/glutathione peroxidase enhances the growth of transgenic tobacco seedlings during stress. Nature Biotech. 15: 988-991.

Shen, B., Jensen, R.G. and Bohnert, H.I. (1997) Increased resistance to oxidative stress in transgenic plants by targeting mannitol biosynthesis into chloroplasts. Plant Physiol. 113: 117-1183.

Song, W.-Y., Wang, G.-L., Chen, L., Kim, H-S., Pi, L.-Y., Gardner, J., Wang, B., Hoisten, T., Zhai, W.-Y., Zhu, L.-H., Fauquet, C. and Ronald, P. (1995) A receptor kinase-like protein encoded by the rice disease resistance gene *Xa21*. Science 270: 180~1806.

Toenniessen, G.H. (1991) Potentially useful genes for rice genetic engineering. In *Rice Biotechnology* (Khush, G.S. and Toenniessen, G.H., eds.). C.A.B. International, pp.253-280.

Wu, C., Fan, Y., Zhang, C. and Zhu, L. (1997) Performance of insect resistant transgenic rice progenies containing modified *B. t.* delta~ndotoxin gene and *CpTi* gene. In *General Meeting of the Int'l Program on Rice Biotechnology*, Malacca, Malaysia, September 15-19, 1997, p.5.

Wu, R. (1994) Report of the committee on genetic engineering. Rice Genetics Newsletter 11:59-64.

Wünn, J., Klo~ti, A., BurThardt, P.K., Biswas, E.C.G., Launis, K., Iglesias, V.A. and Potrykus, I. (1996) Transgenic indica rice breeding line 1R58 expressing a synthetic *cryIA(b)* gene from *Bacillus thuringiensis* provides effective insect pest control. Bio/Technology 14:171-176.

Xiao, J., Li, J., Grandillo, S., Ahn, S.N., McCouch, S.R., Tanksley, S.D. and Yuan, L. (1996) A wild species contains genes that may significantly increase the yield of rice. Nature 384: 223-224.

Xu, D., Duan, X., Wang, B., Hong, B., Ho, T.-H.D. and Wu, R. (1996) Expression of a late embryogenesis abundant protein gene, *Hval*, from barley confers tolerance to water deficit and salt stress in transgenic rice. Plant Physiol. 110: 249-257.

CHAPTER 4.11

High-technology and Agricultural Development in China

Yuan Chun Shi
President Emeritus

Formerly
Beijing Agricultural University,
Currently
China Agricultural University,
Beijing 100094 China

This article is to explore the role of high-technology in the development of agriculture in China. The problems discussed are: the science, technical and industrial revolution in agriculture; the demands and selection of high-technology for agricultural development in China; and the high-technology strategies for agricultural development in China.

Science and Technical Revolution and Industrial Revolution In Agriculture

A Review of the Development of Science and Technology and Industry
Viewing back to the two to three hundred years' history of development of human society, there were a few important discoveries in science which brought about a breakthrough in technology, and initiated an extensive and profound industrial revolution. The completion of Newton mechanics and classical physical theory initiated the invention of steam engine of the 18th century and the first technical revolution characterized by popular use of machinery. In the 19th century, the unit theory of electro-magnetic field by Maxwell and the discovery of electro-magnetic wave initiated the invention of electric motor, telephone, electric lamp and telecommunication, and created all new power and electric industry, and there came into the era of electricity. The atomic theory, suggested by Dalton, the element theory by Boyle and periodic table of elements provided by Mendelyleev laid the foundation of modern chemistry, which allowed the development of modern chemical engineering.

AGRICULTURAL DEVELOPMENT IN CHINA

In the 20th century, micro-electronics and computers, information and communication, aeronautics and space-technology, new materials and automation, bio-technology, laser and other modern techniques have been developed one after another, bringing the traditional industries from one generation to another, while new industries continuously appeared. This technical progress may be ascribed to the relativity in a narrow sense and to the discovery of quantum theory of mechanics, according to Dr. Li Zheng-Dao. On the occasion of the Symposium of Scientific and Technical Development Strategy In China in the 21st Century, he said "Up to 1925, the two fields (relativity and quantum theory of mechanics ed.) were completely understood, and henceforth, the atomic structure, molecular structure, nuclear energy, laser, semi-conductor, super-conductor, X-ray, super-computer were developed. If there were no relativity in narrow sense and quantum theory of mechanics, there would be nothing for all of these. From 1925 onward, almost all the material civilization of the 20th century were derived from the discovery of these two physical fundamental sciences." Today, science and technology have been well developed. Science, technology, and industry, the three are not only push forward in one direction step by step, but mutually enhanced. The high-powered accelerator, Hubble telescope, electronic microscope, etc. the large-typed sophisticated apparatus, equipment, facilities strongly support scientific research and new discoveries; social demand and economic competition have become very strong driving motives for the development of science and technology. When stepping into the latter half of the 20th century, the time of conversion of science and technology to industrial products is getting shorter and shorter. In the 18th century the development of steam engine took 100 years, in 19th century motor and telephone took 50 years, electronic tubes and automobiles took 30 years; while radar, TV-set, transistor tubes, atomic energy, and laser apparatus since the 20th century only took 15, 12, 5, 3, and 1 year respectively. To view the development history of science and technology, two points should be mentioned, one is that science, technology and production, the three were pushing and enhancing each other. As the 1994 White Book of Science and Technology of the United States had pointed out, "Science is the basic fuel for technology .and technique is the generator of economic growth." The second point is in the course of development that development is always a process of step by step advances. Accumulation of such advances induced technical breakthroughs, industrial revolution, and economic growth. New scientific discoveries initiated continuous "jumping" development one wave after another. Changes from quantity to quality, accumulative progress and "jumping" progress is the natural law of the world. This also holds true in the development of science, technology, and industry.

Since the 18th century, the technical revolution in machinery and electricity, have pushed economy toward of the industrial system, and created an industrial society and modern material civilization. The impact had covered the whole society, but the effect to each technical and industrial field was widely different. In comparison with concentrated production systems in highly controlled industries, agriculture fell behind mainly due to biological production based on land, and the professional weakness slowed the adaptation and utilization of mechanical, and electric power technique. Diesel engine appeared at the end of the previous century, but tractors with diesel engine were widely used only in the middle of this century, almost 100 years later than used by the industry. However, agriculture has its own history of technical revolution.

The first science, technical, and industrial revolution in agriculture
Agriculture is the most ancient industry. It went through a long process of long natural economic development of self-plowing, self-feeding farming period with low level or little input of science and technology. Rate of progress in productivity was very slow. Up to the first half of this century the grain yield per unit area increased from 930 kg per ha to 1,124 kg, average rate of increase per annum was only 4 kg. When it came into the latter half of the century, agriculture developed with high-speed, from 1949 to 1988, the average rate of increase was 9 times of. that of the previous 50 years per unit area per year, for grain from 1,124 kg/ha to 2,499 kg/ha. According to research results, in the high-speed development period of agriculture, the rate of contribution by science and technology has been 73%, mainly the technique for increasing yield and material input are good varieties, fertilizer, pesticide, and irrigation. The world nitrogen application. in 1913, 1949 and 1994 were 510,000 , 3.6 million and 72.76 million t respectively, i.e. the annual application of the latter half of the century is 17 times higher than that of the former half. Pesticide synthesizing chemical industry came into existence after the 1940's; from 1960 to 1990 the world pesticide sales increased from 8.50 million U.S. dollars to 23.4 billion U.S, dollars, i.e. 27 times of increase within 30 years. Modern technique and material input are the material foundation of high-speed development of agriculture.

As the grain production per hectare, in China, at the beginning of the century, was 960 kg per ha, in 1949 it was 1,027 kilograms, and in 1995 was 4,240 kilogram in the latter half of the century, the average increase of yield per year was 70 kilograms, equivalent to 50 times of the first half of the century, far much higher than average increase per annum of the world grain for the same period. During the same period of time, fertilizer application increased from 6,000 t to 214.2 million t, fertilizer application per ha increased from 1 kg to 1,041 kg. The annual production of pesticides Increased from 93,000 t in 1970 to 200,000 t in 1988. Irrigated area increased from 16 million ha to 48 million ha

While we are pleased to see the high-speed development in agriculture with the help of breeding, fertilizer, pesticide, and irrigation technique in the second half of the century, we are obliged to think of the 19th century Darwin's theory of heterosis *(1859)* and the contributions to modern breeding technique and seed industry by Mendel and Morgan genetics; synthesizing urea through inorganic matter by Friedrich Woehler (1824), and plant mineral nutrition theory by Liebig (1840), all these pushed the fertilizer and modern fertilization technique, and in the 1930's, Muller applied synthetic chemistry to pesticides which lead to the new stage of synthesizing organic pesticides.

The prelude of new technical revolution of agriculture has started
Riding the global tide of new technology revolution with great momentum in the latter half of this century, agriculture also developed with high speed achieving new scientific technical breakthroughs. In 1953, the discovery of double helix structure of DNA pushed the biology to molecular epoch. In 1973, the success of DNA recombination opened the new era of biological technology and genetic engineering. The traditional breeding technique using heterosis may now be done in laboratory operations. Introduction of bio-technology makes it possible to transfer genes in animals, plants, and micro-organisms, make use of all organisms efficiently. Problems which were very difficult to solve in the past become now possible. In addition, molecular biology and biotechnology will exercise great impact in broad areas of agriculture, including growth and development of animals and plants, biological pesticides, biological fertilizer, animal vaccine, and preventing diseases, fermenting engineering, enzyme engineering, as well as environmental protection.

Beginning in the 1950s, but accelerated in the 1980s, the development of micro-electronics, computers and information technique greatly improved many traditional industries. Agriculture uses land as its basic means of production, and make use of photosynthetic capability and local resources of light, heat, and water, that all influence biological production. These factors are widely diversified, location and season specific, and information about them was mostly based on experience with little possibility for control or stabilization. For generations it was rather difficult to quantify, standardize, or intensify and concentrate such information. However, the week information gathering now can be changed by the advanced technical means of information collection, processing and transmission from scattered and small-scale agricultural industry. By their competent capacity of computing and intelligent software technique, making the very complex and variant factors of production in agriculture to be quantified, standardized, collected, so as to overcome the. weak points such as time and location variables that much rely on tradition and experience. In addition the development of aviation and space technology (RS, GIS, GPS etc.), greatly strengthened the ability of monitoring, warning, and forecasting the natural environment. It also allows the forecasting of meteorological /biological catastrophes which affecting agricultural production, which in turn raises the degree of controllability and increases stability for scientific management of the process of production. Advanced manufacturing processes, refined chemical engineering, automation, new materials, and other modern technique are bringing about changes day after day to hasten agricultural processes. If machinery, electric power, and chemical engineering played important role in the previous technical revolution, which promoted industry and left agriculture behind the new technical revolution information technique and biological technique will bring agriculture to the mainstream, providing traditional agriculture and agrotechnique new vitality and establish a new agro-technical and production system.

Essential Elements and High-technology Requirements in Agricultural Development

The current agriculture in China is facing two fronts of challenges and opportunities. One is that the State is changing from the controlled, planned economy system to a market oriented economy system, that requires an unified managerial model for Nationwide, integrating agriculture, industry and trade. The other is the tide of new technical revolution in agriculture mentioned above. The former is moving toward advanced types of management, and create favorable environment for improved techniques and the latter is providing a the strong motivation for agricultural development. If this opportunity is properly grasped, it would leave the past difficult situation behind and attain a new breakthroughs for agriculture in China. The essential element and high-tech choices are as fellows:

Seed

Agricultural production begins with seed. Good variety is the most important factor of agricultural increase, and account for 30% of the technical contributions. Within the recent 40 years, 4000 new crop varieties have been brought out and released for extension. The staple crop varieties have been replaced 4-5 times, each replacement may attain about 10% increase of production. Hybrid corn and hybrid rice have made up 75% and 45% of their cropping area respectively. The main difficulties in elite seed production include the shortness of excellent breeding material, insufficient effort to follow up resulting in low rate of of qualified seed production, poor management and not being able to operate farming as an enterprise. In China, traditional breeding manpower is fairly strong, and biotechnology also has laid some foundations. The present urgent task is to establish bio-engineering technology on the bases of traditional breeding in order to achieve breakthroughs for the breeding, of good varieties which are needed for field production. In the area of technology, emphasis should be made on research and application of plant cyto-engineering molecular genetic marker and gene transfer. Objectives in breeding include: "super-rice" that is possibly ahead of the world, super-high-yielding wheat, special corn, insect-resistant cotton and corn, rice-blast resistant and rice fulgorid resistant rice, yellow-dwarf and mildew-resistant wheat, and drought-tolerant and alkali and saline-tolerant wheat and corn are of top priority and hopefully may get great advance.

Animal and Aquatic Culture.
In the recent decade, animal culture have been developing at a high-speed. In 1995, the production of meat was 6.1 times of that of 1978 (i.e. 8.56 million t and 52.60 million t); the total production of aquatic products of the same period from 4.66 million t rose to 25.17 million t, i.e. an increase of 5.4 times, the average rate of increase per annum of the two were 30% and 26% respectively. However, efficiency of production, level of technique and rearing are still low, the number of pigs kept In stock per annum was 6 times of that of the United States, but the meat production was only twice. The number of chickens kept in stock every

year were about 4 billion, 90% of the elite hens rely on foreign countries. China is a pork-liking country, 69% of meat consumption are pork, the annual number of pigs kept in stock (about 460 million heads) make up more than half of the world. Now, the "super pig program" by means of gene engineering technique combined with traditional breeding technique, is proceeding, possibly can catch up the level of the world. As to cattle, 140 million heads of field labor animals are going to be shifted to meat and milk production. Super ovulation, nucleus-transfer, getting oozoid through living animal, etc. and embryo engineering technique will strongly push this program. The research on gene-transfer aims to the introduction of "whole fish" gene, especially the breeding of fish appropriate to the rice field amounting to millions of hectares. In addition, making use of animal bio-reactor technique to produce protein for medical use and veterinary vaccine will make great success.

Pests and diseases
In the period of 1973-1989, pests and diseases caused the loss of grain on the average per annum 9.2 million t, and cotton 220,000 t. In 1993, cotton boll-worm broke out, direct economic loss costs more than 1 billion U.S. dollar. The cause of exterminating disease to shrimp has not yet been determined. The amount of chemical pesticides used annually have been the second place of the world. Over-reliance and inappropriate use of chemical pesticides usually lead to destroy the ecological system of farms, to raise the resistance of harmful organism, to reduce the beneficiary organism and natural enemy, and to increase pollution of environment. The European Community propose that by the end of the century, the quantity of chemical pesticides should reduce by 20-25%, and In 1992, the World Conference on the Environment and Development suggested that up to the year 2000, the biological pesticides should make up 60% of the total pesticide, while in China, it is less than 1%. Each year the direct economic loss is more than 3 billion U.S. dollars by death of animal or aquatic cultures from epidemic disease and the economic loss of about 8 billion U.S. dollars are due to slowing of productive functions. The annual demand for vaccine amounts to 80 billion head dosage, yet the productive capacity are only 13 billion head dosage. Furthermore, there are many problems concerning quality, technique, and market management. The biological pesticides are difficult to develop, because their effects are slow and weak, their effective period is short; and their own resistance to stress are not strong. It is difficult to find plants that contain chemicals involved in resistance. Usually these chemicals are not expressed in the young seedling and

seedling plants of resistant cultivars often need to be protected. Fortunately the protective period seedling require is short, however, the protection cost is high. Thisis one of the week points of biological control. Biological technique, especially, micro-organism recombination technique has resulted in important breakthrough and revolution of the traditional production and construction, and new momentum has emerged. Advances such as construction of engineered bacteria with high toxicity, broad spectrum, and strong stress resistance closely at hand The research for quick effect, biological pesticides without public harm, reduction of toxicity by way or gene-deficiency, live carrier and superior quality, high efficient gene engineering vaccine and matching molecular differential diagnostic technique, and SAR protein Haropin, Zearalanol, rpGH, bST etc. and other growth regulator agent of the new generation for plants and animals have shown their strong vitality. The recent blooming industry of agro-biological agent is surely to have a great development, and make important contributions to enhance the plant and animal production.

Water

Insufficient fresh water supply is an important limiting factor to the development of agriculture in China, and it will be getting more serious In the next century. The per capita fresh water resources is only 1/4 of the world average, and it is extremely unevenly distributed in time and space. The arable land in Yangtze River valley and to the south of it make up 36% of that of the whole country, and possess 83% of the fresh water resources. On the other hand 64% of the arable land in the north only have 17% of fresh water resources. The rainfall of monsoon season varies very much year after year, thus lead to frequent problems of drought and flooding. In the period of 1949-1993, the average area of drought and flooding was 23 million ha, causing loss of grain annually 10 million t and economic loss 30 billion yuan. Serious insufficiency of fresh water resources and low efficiency of water usage have perplexed agriculture in China. In the irrigated area, the water utility coefficient generally are 0.4, in the field the grain productive efficiency of one cubic meter of net water consumption are less than one kg, which is equivalent to 1/3 of general level of the world. China is determined to improve the field irrigation facilities and water management. In those agricultural areas when conditions are available will raise the efficiency of field water as a centeral project making great effort to extend low pressure pipe conveyance technique, dripp irrigation and osmotic irrigation technique, field covering technique, transpiration reduction technique, and other advanced system of irrigation, etc. and modern agricultural water-saving engineering technique.

Fertilizer

Fertilizer, especially steady increase in the use of chemical fertilizer is the important factor of increasing agricultural production in China, In 1994, the fertilizer production and consumption in China have been the second (21.88 million t) and first place (33.18 million t) respectively, making up 16.6% and 27.5% of the world total respectively. The energy consumption for fertilization have been 82% of the total energy consumption in agriculture (1988). In the year of 1992/1993, the average net nitrogen used per hectare have been 184 kg, equivalent to 3 times of that or the world and the United States (54.7 kilo and 55.5 kg respectively). The problem of high application and low efficiency of use is getting even more serious: in the years of 1952-1994, fertilizer grain ratio have been dropping from 16.1 to 2.4. The utility efficiency of fertilizer is generally around 35%, lower than the world general level about 15 percentage points; the rate of loss of nitrogen from rice, corn and wheat fields are 50%, 40%, and 30% respectively, each year the loss of fertilizer by way of leaching and evaporating are approximately 9 million t, cost 5 billion U.S. dollar, and also causing serious pollution of environment. The portion of organic manure in the field nutrient input has been greatly reduced, thus causing people great concern, in the years of 1957, 1980, and 1990, the organic nutrients make up 95%, 49% and 37% of the total nutrient input, respectively. It is important and urgent that the efficiency of fertilizer use to be raised and environmental protection be emphasized in the course of agricultural development in China. In addition to adjust irrational nutrient proportion and improve supply and distribution management by way of government measures, China must apply modern technique, promote production, research (apply fertilizer in balanced manner), and use in combination, develop concentrated, superior quality and for specific use compound fertilizer (in the developed country making up 80% of fertilizer in kind, but in China less than 10%), and release control specific use compound fertilizer CAF. The recently developed 3S, (RS; GIS, GPS) exact fertilization technique will have broad prospect of application In China. The Chinese rural villages are scattered widely, the grass-root managers and farmers' quality of science and literacy are fairly low,. It is important that in addition to combine modern advanced fertilization technique with fertilizer product, to develop and produce new fertilizers which are convenient for the farmers to use.

AGRICULTURAL DEVELOPMENT IN CHINA

Vegetable and Animal Facilities Improvement.
In the period of 1986-1995, the urban population in China increased from 225 million to 295 million, it is forecasted that by the year 2010, that will increase to about 400 million. Under the condition of fast increase of urban population, conservation of energy and resources and factory-like agriculture will need great development. At present, vegetables production through marketing channels reached 120 million tons, 8% of which is produced in greenhouses and plastic shelters. Now, there are 300,000 hectare of greenhouses and plastic shelters, the majority is simple plastic shelter. They have to be replaced by better facilities which are technically sound and will provide higher benefits. In recent years, the outskirts of large cities are very active to buy foreign greenhouses or shelters, only because of the complexity of situation in China, there are many problems. Therefore, it is urgently needed to study the construction material of greenhouse and shelter, and control of environment, machinery operations and energy-saving, specific varieties and cultivating and improving technique, to develop greenhouses and shelters which will can meet the special conditions different agricultural area.

The number of pigs in the stock annually in China is about 460 million, almost making up over 1/2 of the world production. However, the pig-raising farms which sold 3,000 heads or more per year are only 7%. In the development of rural economy agriculture must be industrialized, and need modern facilities, high efficiency cultivation and management.

Information
China has very large territory, natural conditions are very complex, climatic and biological catastrophe were frequent. Farm families are small-scale, and widely scattered, techniques, production and management are much relied on traditional experience. The information technology may evidently improve those professional weak points, and to change the traditional agriculture to scientific management. The first is to use software technique and intelligent technique (system of agricultural expert, system of assisting decision-making, etc.) to quantify, standardize, and assemble agricultural techniques to help production management, decision making and personnel-training in order to raise their level and efficiency. The second is to transmit various information, including market information, and services, by means of data bank technique and network technique. The third Is to combine aeronautic and space technology to monitor agricultural resources and environment, productive conditions meteorological and biological disaster, in a macroscopic way

to provide advance warnings and forecasts, for scientific and efficient response in order to achieve exact agriculture development. It is important to speed up agricultural information system in relation to elevate the level of production and management scale, develop agricultural economy, and realize modernization in agriculture.

Sustainability
On the development of sustainable farming, China is confronted with an acute challenge. The per capita arable land is only one third (1/3) of the world average. Among 130 million ha arable land those with unfavorable factors like soil erosion, salinisation, desertification, gleying, etc. reached 50 million hectare. Destroying of forests, land deterioration, excessive use of slopes, grass land over-grazing etc., all these factors make soil erosion and desertification continued to expand. In addition 59% of the arable land are deficient of phosphorus, 23% deficient of potassium and basic fertility reduced because of organic manure were reduced. In the next 10-20 years, 3-4 million ha of arable land will be lost to other uses, and the reclaimable resources reserves are only about 10 million ha. The leaching of N fertilizer is as high as 45%. The chemical pesticide sprayed area reached 150 million ha. Nine million ha farm land have been polluted by agricultural chemicals. Farm land pollution again bring calamity to the rivers, lake,, coastal regions, enriching the water with nutrients resulted in destruction of the water ecological equilibrium and damaged the environment. Confronting serious challenge of sustainable farming, Chinese agriculture is implementing "Agenda of 21st Century in China" and adopting relevant policy, system and management at the same time. China should be active in adopting modern technique which are helpful to improve the efficiency of land, water, and fertilizer use, to protect resources and ecological environment, especially, through biotechniques be combined with conventional technique to bring about forest, grass, crop varieties which may be used in preventing , with strong stress resistance. Other measures should also be conducted, including use biological pesticides to replace chemical pesticides; produce new types of fertilizers, use water-saving facilities, planting/cultivating facilities for factory-like production, and reduce resources pressure and protect environment. By means of information technique and aviation and space technology mainly 3S be combined with agriculture, it would strengthen monitoring the managing, and protection of environment of agricultural resources. Sustainable development is a national policy, it needs strict guidelines, policy measures, and plans together with comprehensive technology to guaranty its success.

Strategy of High-tech Agricultural Development in China

The solution of Chinese agriculture depends on science and technology
In more than 40 years, China with its 7% of the world land has produced 24% of grain, fostered 22% of world population, this is a well-known achievement. The heavy burden of population on resources, and environment together with its relatively weak foundation, little input, low level of production and technique, make Chinese agriculture a heavy and continuously difficult task. To compensate the scarce resources the solution is to develop science and technology. Sci-tech may improve the level of technology and productivity and make up some deficiencies due to insufficient funding and manpower. In the primary stage of industrial society, the wealth came from raw material, land and capital (Keynes). Marx suggested labor-value theory, but in current stage in which sci-tech and social economy have been highly developed, manpower capital and technical knowledge have become the main body of creating social wealth, Romer suggested that world economic growth should be mainly ascribed to knowledge, i.e, "Theory of New Economic Growth", Tofler suggested that knowledge will be the final "replacement resources" of all economic activities. Confronting the global exiting economic competition, in 1994, the European Community have compiled "The Fourth Research Frame Plan of Sci-tech Development"; the United States have suggested" Sci-Tech White Book", and "National Institute of Standards and Technology" (NIST), and "Advanced Technology Plan" (ATP), all are marked with the important thought "economic development will be made through technology", China has compiled "Science and Education will make the country prosperous" strategy. China has to pay attention to, but not to be surrounded in the worry of resources, foundation, and funds, and should clearly recognize that time has come to give sci-tech an opportunity which is more important and precious to develop in China than other considerations. China should stand on strategic point of view pool technical and manpower capital, and make haste to change tracks from traditional agriculture to high efficiency and low cost sustainable farming. The achievements of development of agriculture in Holland and Israel whose resources conditions are not good have given us very beneficiary enlightenment.

Grasp "Opportunity for change of quality", and start "Development by jumping"
As mentioned above, sci-tech is the means to promote economic growth, a process step by step forward and jumping toward in attaining quality

and quantity It is essential to grasp the important opportunity which in the course of development may cause great breakthrough, and push the economy develop by jumping. Watt did not limit the steam engine invention to one specific use, but it was "an engine of big industries covering wide uses." Siemens invented dynamo and wrote in a letter to his younger brother, "electric power technique has promising prospects, it will create a new era." It is because they have insights this quality change and revolution, also with their far sight which projected indefinitely wide and broad prospect and opportunity. The Nobel prize winner, the inventor of tunnel-scanning microscope, Dr. Rorel suggested, "those countries who paid attention to micrometer sci-tech have all become developed countries, while nanometer sci tech provided the new opportunity of development, today, the country who pay attention to nanometer sci-tech, would singly enjoy the achievements." Reviewing the modern history, from Europe to the United States, all the developed countries have benefited from sci-tech, and push their economy and society progress. Consider, in the development of industrial society, who arc in the leading position in steel making, machinery manufacturing, automobiles and aviation, electricity and petroleum; and in the development of modern society In micro-electronics, computers, information and communication, aviation and space technology, nuclear energy and laser, and bio-technology, etc. who then gains superiority in the development of technique and economy, in competition. Agriculture is the same, in the first sci-tech revolution, who kept the leading position in the field of hybrid corn, dwarf and hybrid rice, urea synthesis and fertilizer production, organic phosphorous and pseudo-pyrethrin pesticide, and fine irrigation, and in the push of modern bio-technique facilities agriculture, exact agriculture, and new technique of macroscopic monitoring, who will be able most efficiently to push agriculture increase rapidly. Every new important breakthrough gives an opportunity, China and agriculture in China should grasp and make use of these opportunities in developing sci-tech to get double the mileage with half the effort.

High-tech is the key of development strategy
In the long run, at any time, there must be a group of high-tech standing at the top of the peak, leading the way of pushing forward. In the economic and market competition which is getting more fierce, each country uses the high-tech as the strategic highland for pushing development of technique and economy. In 1991, approximately 31% of capital in the United States have been put to high-tech industry, in the

first season of 1996 it already reached 42%; Korea, having gained high speed development in economy in 1982, compiled "Development Plan for Pioneer Industries", in 1991 again suggested the "G-7 Plan"; Israel "Establishing the Country on High-tech" policy advanced the level of the whole country. This is because the innovation and advancement are popular they bring up other industries to high-tech. Industrial development has the obvious superiority in costs and benefits, in competition among products and markets and have been the important pushing force for reforming traditional industry. The Chinese agricultural production and sci-tech have attained considerable progress and good foundation for further development, if don't restrict itself to the regular step by step process, but make use of the new technical wave to develop and apply high-tech, then the agricultural technique and economy would be in jumping advance. Be active to develop cell (embryo) engineering, molecular genetic markers, gene-transfer, recombination of microbes, and other bio-technique and its up-reach research, then it will give new variety or breed, new types of bio pesticides, animal vaccine and animal growth-regulator a new breakthrough and important progress. The traditional fertilizer production system could be shifted to the research and production of high efficiency, superior, specific compound fertilizer. The traditional field irrigation could be transferred to the modern water-saving engineering technique to use pipe conveyance and exact quantity irrigation, covering and improving soil, reduce soil evaporation and plant transpiration. Devote great effort to study and to produce the facilities for factory-like plant production and cultivation which will be adaptable to various agricultural regions. Through the better facilities a highly efficient use of resources, and environmental protection can be achieved. In addition, to make haste to establish the information system in agriculture, and 3S agricultural macroscopic warning and forecast system and pushing exact farming will greatly raise the capacity of agriculture and benefit. High-tech are mostly developed in recent years, China has certain foundation, at the start, not far lagging behind the developed country, and some new technique can be introduced. Therefore, it is not only necessary but also possible to make full use of the superiority of high-tech, to push agro-technique and production a jumping advance. This should be a strategic choice of agricultural development.

To develop agro-technique industry is the urgent task of today
As the sci-tech developed, the transfer of technical results in agriculture to be quantified, collected, and materialized, more and more applied products are produced, and the market developed, information

transferring modernized, the leading role of agro-technic enterprise to transfer the results to the producers is more striking.

Fertilizer industry have developed so far for more than hundred years, the German BASF Company producing compound fertilizer containing almost 10 nutrients elements, the Japanese Mitsubishi Company producing more than 80 specific compound fertilizers, the American AG-CHEM Agro-chemical Equipment Company, managing 3S exact fertilization facilities. Other agricultural technical enterprises have strongly pushed the fertilizer production and application forward. Swiss chemist H. Staudinger in 1924 published the chemical structure of effective content for natural pyrethrum, within the next several decades, The Japanese Sumitomo, American FMC and other chemical companies have studied more than 40 commodity varieties. By the end of preceding century, Bill suggested the possibility to use Fl hybrid of the species of corn, Wallace in 1926 created the first hybrid corn seed company which so far have taken up 42% of hybrid corn seed market in the United States, and the American Pioneer Seed Company supply 22% of seed for the world hybrid corn planting area. Actually, it is a real pioneer in pushing hybrid corn business. With hundred years history, the American Monsanto company manufactured pesticides and chemicals In the early 1980's, when most people still doubted the potential of bio-technology', Monsanto's chairman of the Board, J. Hanley invested 100 million U.S. dollars for developing bio-technology products. Later again invested 200 million U.S dollar to establish a research center for life sciences. So far this center released and put to the market worm-resistant cotton, worm-resistant corn, natural hormone BST and PST for cattle and pigs, more than 10 bio-technic products, and it will be the new pioneer of sci-tech revolution in agriculture. Chai Tai Group of Thailand in the 1930's marketed seed and fertilizer and by the 1980's occupied more than 30% of the feed market in China. This Group now became a transnational corporation, possessing assets worth several billion U.S. dollars, mainly by marketing agro-technic products. The success of these enterprises provide good examples for China to learn. Those agro-technic enterprises place sci-tech, production, and marketing into one organization, with sufficient funding support, advanced technology, flexible structure, competent marketing system, and instant market response to consumer demands. To meet market competition, some big companies devote more than 10% of their gross sales to technical and product development in order to keep their techniques always in the forefront. Those enterprises are not only the main forces of technological transfer, but also important forces in promoting sci-tech advancement. In China there was a nation-wide, competent network of agricultural extension. However, at the beginning of changing toward market oriented economy, the system was broken and people scattered. The current organization of social service does not work efficiently, and the agricultural sci-tech industry is still in its cradle. This is the important cause of low conversion rate of sci-tech achievements into products. In moving toward market oriented economy through a continuous economical developing, the transfer of sci-tech achievements in China will have to develop. The main force promoting economical and society progress is the transfer of technology into products and various commodities. Technical level of a country should not be limited to scientific papers and laboratory results, but reflected in the productive technique of enterprise, costs of energy consumed, quality of product, and competition on the market. China has to make great effort to remove the obstacles which separate research from production developed under the old system of planned economy. Instead China should learn from the experience of the developed countries, greatly develop agricultural sci-tech industry, especially the high-tech industry. It is the best way to develop agriculture through develement of sci-tech, to grasp this revolutionary opportunity. It is the most important component of practicing high-tech strategy.

Problem is still on "R & D"

The above discussion described strategic consideration and potential approaches. They require an adequate support system, policy, plan and well-developed environment, or. "software" conditions, but most important is the non-failing funding input. Since the reform and opening-up policy have been adopted, achievements of the Chinese economy are excellent, but the proportion of R&D in proportion of GDP has been fairly low, and the tendency is further decrease. In 1990,the share of R&D in GDP was only 0.7%, in 1996, dropped to 0.48%, of which the share of agriculture is even more limited. At present, R&D in developed countries generally is between 2.5-2.8% of the GDP, while in developing countries R&D is around 1.5% of GDP. In a high-tech country, such as Israel, the share of R&D is as high as 3% of GDP. In Korea in the period of 1991-1996, the share of R&D has increased from 1.9% to 3.5% of the GDP Recently, in the 9th session of Trade and Development, United Nations published material have pointed out, thatn the share of high-tech contribution to the economic growth is 70% in the developed countries, but in East Asia this is only 10-40%. UNESCO's

"World Science Report 1996" mentioned that the R&D input of 27 member countries of the World Economic Cooperative Organization makes up 85% of the world's R&D expenditures . It is clear that competition in sci-tech and economic development are getting more fierce, the strong ones are getting stronger and the weak ones become weaker, the rich ones are getting richer, the poor one becoming poorer, polarization is ever more evident. China has adopted the policy that "sci-tech is the primary productive force" and "sci-tech will make the country prosperous," which pointed out the correct approach of development. At present, the government and various circles all ask the question "sci-tech will make the country prosperous, but who will make sci-tech prosperous?" We believe that in the development of economy and sci-tech greater progress in scientific knowledge and application application of this knowledge can be assured.

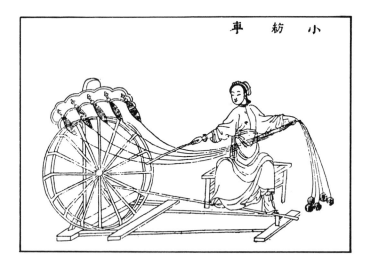

CHAPTER 4.12

Agricultural Education, Research, and Extension in China

Recent History and Future Prospects 1

T. L. Yuan

Professor Emeritus, University of Florida

In its goal of self-sufficiency in feeding its people, China has only recently recognized the pivotal role of agricultural science in any national policy. For much of the last 150 years, China was effectively prevented from adopting improved agricultural practices developed in other countries, due to foreign aggressions, conflicts between warlords, and civil wars. Although the founding of the People's Republic in 1949 led to the initiation of many attempts at agricultural reform, mismanagement and disastrous political movements hampered such efforts. It was not until the late 1970's that the political situation became sufficiently stable for agricultural modernization to proceed.

The last twenty years have witnessed great strides in Chinese agriculture. The State Statistical Bureau of China recently reported (1) that staple food (cereal) production reached record levels in 1996. This level of production,

1 The author wishes to express his appreciation to: Professor Li You-Kai, Retired Professor of the Beijing Agricultural University for valuable constant consultation; Prof. Mao Dai-Ru, President of the China Agricultural University, for his generosity to send the author a copy of his book on Chinese agricultural colleges and universities; Dr. H.C. Chiang, Emeritus Professor of the University of Minnesota for reviewing the first draft of this paper; Dr. T. C. Tso, Board Chairman of the IDEALS for his continuous encouragement and useful information materials on the subjects; Dr. J. M. Davidson, Vice President for Agriculture and Natural Resources of the University of Florida for providing the personnel statistics of the Institute of Food and Agricultural Sciences of the University; and the author's sons, Daniel Shih-En, M.D., and Jason Shih-Ning, M.D., for reading and editing the manuscript.

amounting to 490 million tons, represented a 5.0% increase over the preceding year. The improvement was credited to favorable weather conditions and attention to technology transfer. Production of meats, seafood, fibers, and other agricultural products also generally increased. In the meantime, ongoing birth control measures led to a population increase of 12.68 million over the same period to 1.22 billion, representing only a 1% increase.

Despite continuing improvements in agricultural productivity relative to population growth, however, there remain reasons for concern about China's ultimate ability to meet its goal of agricultural self-sufficiency unless the agricultural system as a whole is reformed. China's arable land is extremely limited and has been further reduced in recent years by the rapid expansion of industrial and housing developments. This shrinkage of arable land for agricultural use will undoubtedly continue. Although the current birth rate is relatively low, the average life span of the people has increased to 70.8 years (1). There will therefore be more people to feed in the future with already limited land and resources than the low birth rate might indicate. In the year 2030, the population is predicted to reach 1.6 billion, about 31% over that of 1995 (2). Accordingly, the demand for food will continue to rise. It seems inevitable that the goal of agricultural self-sufficiency will be attainable in the long term only with continued increases in agricultural productivity. Such increases may be achieved by using advanced science and technology. They will be needed not only in improving crop yields, quality, and nutritional content and in the development of land that is currently considered poor or non-productive, but also in the creation of agriculturally-related enterprises capable of offering new economic opportunities for the hundreds of millions of Chinese people presently engaged in agriculture.

This paper addresses the educational infrastructure which must be in place in order to build a more efficient, highly competent, and innovative agricultural workforce, with individual attention to agricultural education, research, and extension. The most urgent need is for more scientists, technical workers, and administrative managers in all areas of agriculture, especially at the top levels, but attention must be paid to ensuring that these workers are sufficiently trained. Clearly, agricultural scientists capable of solving problems in agriculture, be they fundamental or applied, will require a broad background in the basic sciences. Workers in agricultural extension will need modern communications and managerial skills if they are to function effectively in conveying new knowledge to workers in the field and to the general public. In addition, personal traits of open-mindedness and dedication will be important for each of these professional roles. To achieve this level of training will require high-quality, efficiently managed education at all levels. Interests, ideals, and personal character are instilled and the fundamentals of culture are best taught in primary and secondary schools, while specialized knowledge and skills are most efficiently transmitted in institutions of higher learning. Improvements in China's educational system are therefore mandatory if Chinese agriculture is to remain strong.

Higher Education in China Since 1949

A Brief Retrospective

Shortly after the founding of the People's Republic in 1949, almost all private colleges and universities were taken over by the government (3). In order to build up the scientific and engineering manpower more rapidly for the economic reconstruction, the education system was thoroughly reorganized and the curricula revised. Starting in 1952, all institutions of higher learning were disbanded and similar colleges from different universities were reorganized into single-discipline independent colleges, such as colleges for engineering, agriculture, medicine, arts and sciences, teachers, forestry, etc., as patterned from the then Soviet Union. Academic departments were also involved in this reorganization. Over time, these specialized colleges came to be called 'universities', despite the fact that they lacked the multi-disciplinary breadth that defines universities in the traditional sense. Although various academic departments did exist, each department was divided into several specialties. Each specialty formed a small unit called 'education/research' group. In this way, the usual three levels of administration, i.e., university, colleges, and departments, became the college ('university'), departments, and specialties.

The purpose of this reorganization was to intensify the learning process and thereby accelerate the training and production of skilled manpower. Each university or college was charged by the government a year in advance with the mission to produce a certain number of future workers in certain areas of specialization according to projected needs. To accomplish this goal, courses that were deemed 'unnecessary' were eliminated, the scope of the curriculum was reduced, and course contents were modified and standardized so that all students in the same specialty would learn the same material throughout the country. Contributions from a professor's particular expertise were discouraged. Some new practical courses were added. Because students were trained in specialty units, they developed narrow fields of specialization. Upon their graduation, students were assigned to a certain job at a certain location as the government saw fit. Over the years, a large number of such

417

specialists wereproduced to suit the purpose of the government for what it considered to be its needs. This emphasis on the production of skillful manpower led to a rapid increase in enrollment in institutions of higher learning. Although the number of institutions only increased from 191 to 227 in the years between 1949 and 1958, full-time student enrollment increased from 155,000 to 441,000 over the same period. Graduate student enrollment increased from 424, the highest before 1949, to 4800 in the 1955-56 school year (3).

In 1958, the government launched the Great Leap Forward movement and established the commune system in its attempt to advance the simultaneous development of industry and agriculture. The production target of grain, steel, and other major items was set to increase by 100% in that first year.Teachers and students were sent to factories and farms to engage in manual labor as well as to set up new schools. With an emphasis on quantity as a measure of productivity, 612 new universities and colleges were established in 1958 alone. Most of the students admitted to these new schools were poorly prepared, if not unqualified. In addition, a large number of the faculty members in these institutions did not have adequate training and needed facilities were lacking. The initiative proved detrimental not only to the quality of manpower produced but to the educational system as a whole. The policy also caused false claims of productivity in all areas. Because of bad weather, coupled with mismanagement of the Great Leap Forward movement, severe food shortages occurred that brought the nation to starvation.

The failure of the Great Leap Forward prompted the government to take corrective measures. In education, the system was readjusted and many of those new universities were abolished. These measures slowly led the country to recovery. In 1966, however, an even more devastating movement, the Great Proletarian Cultural Revolution, took place. It lasted until 1976. During this 10-year period, especially from 1966 to 1969, disruptions of schooling occurred everywhere. Educational institutions were under siege, the administrative officers were removed, and faculties and staffs dispersed. All universities were closed in the late 1960's and remained closed until the early 1970's. Some of them did not fully reopen until the late 1970's. The damaging effects of the Great Leap Forward and the Great Proletarian Cultural Revolution on morale and education are being felt even today.

In 1978, in an attempt to revive the country, the government initiated a new movement called the Four Modernizations. That is, the government resumed its focus on science and technology, as well as agriculture, industry, and national defense. A number of leading institutions were designated as 'key universities'. Certain technical institutes, as well as a few specialized departments in provincial universities, also received such designations. These institutions and departments generally had better qualified faculties, facilities, and accomplishments. They received special attention and preferential funding so that they would be able to produce better scientists and technologists. In the 1970's, after reopening its doors to the world, China sent scientists, scholars, and administrators abroad, mainly to the West, for short visits and for technical meetings. These initiatives were made to help remedy the educational lapses that resulted from the complete shutdown of educational institutions during the Great Proletarian Cultural Revolution. Since the middle of the 1980's, China has also been sending better-prepared college graduates abroad, especially to the United States, for advanced study. The number of students abroad has increased tremendously from mid-1980's to the 90's, although many of them, after finishing their training and/or receiving their advanced degrees, have chosen to remain abroad because of poorworking environments at home. In the meantime, graduate programs in Chinese universities have been expanded, and foreign scholars and scientists have frequently been invited to China to lecture, to train personnel, and to help or cooperate in various scientific projects, including agriculture.

Present-Day Organization of the Education System

China's national educational and research policies are set by the State Education Commission and the State Commission for Science and Technology of the State Council. They are executed by the education and science/technology departments of various ministries, provinces, counties, and lower governmental units. The universities in China may be classified into three groups according to the level of the government that they are intended to serve, namely, the State, the provinces, or the local governments (e.g., municipality, communities, or lower government agencies). Most, if not all, of their funding comes from these respective governments.

National educational institutions fall under the jurisdiction of specific ministries of the central government as well as that of the Academia Sinica or the Chinese Academy of Sciences, the highest national research institution in China. Institutions of arts and sciences and three major engineering universities were administered by the Ministry of Education until 1985, when the Ministry of Education was abolished and their control passed to a State Education Commission. Other specialized or single-discipline institutions at the national level have been administered by various individual ministries. Almost every ministry of the national government has one or more universities under its control and provides the funding for these universities or colleges. Agricultural and agriculture-related universities, for an example, are under the jurisdiction of the Ministry of Agriculture. It should be mentioned that various ministries not only have specialized educational institutions under their jurisdiction but also have a number of research institutions in specific areas of endeavor. These research institutions are administered by Science/Technology Departments in their respective ministries. The Academia Sinica has three universities under its control.

The second group of universities or colleges is under the control of provincial governments. As with the institutions administered at the national level, these institutions are administered and funded by specialized divisions or departments responsible for education, science/technology, agriculture, etc. However, special funds may be granted by the corresponding ministries in the national government. The provinces, likewise, have designated a few universities or colleges as key institutions which receive preferential funding, generally on the basis of their reputation for high-quality faculty, relatively better facilities, and past accomplishments. However, funding levels vary from province to province. Those universities in relatively affluent provinces are generally better funded, better equipped, and better staffed.

The third level of institutions of higher learning are locally established, either by municipal governments or communities, to fill their individual needs for manpower. They finance their own institutions with the permission and blessing of higher levels of government. Thus, the production-oriented local enterprises of a given ministry may establish workers' universities or colleges for a particular purpose. Night and TV 'universities' or 'colleges' as well as various vocational 'universities' and 'colleges' have also appeared. These institutions, though still called universities and colleges, generally lack the caliber of faculty, facilities, and course offerings found in more traditional institutions of higher learning.

Education Needs for the Future

Ever since its first years as the People's Republic, China has emphasized the development of specialized expertise as a priority of its educational system. The rationale for this emphasis is understandable in view of China's urgent needs at the time for skilled personnel designated for specific tasks. Training was more practical and efficient, without wasting much time in unrelated or unnecessary schoolwork. However, the kind of expertise that this training engendered was often narrow and superficial. Specialists were able to solve specific problems but were not able to work outside a narrow focus of expertise. These difficulties were not initially obvious because China was not an industrialized nation. The success of the market-based economy in recent years, however, has revealed that top-level scientists and technologists require more than the present educational system can offer. The specialists that have been trained are perhaps best considered high-level technicians, not scientists or technologists in a true sense. A shortage of higher levels of experts will soon be felt and is already becoming manifest. While the present setup may be useful in training middle-level technical personnel, it is only transitional at best. The production of top scientists and high-level technologists require a much improved educational system. In response, China has expanded existing graduate programs in national and provincial key universities and created new ones. Master and doctor degrees have been awarded in recent years. Many students have been sent abroad for advanced study. Universities have also been allowed to diversify and strengthen their academic disciplines. However, further development will be necessary.

Because research funding in the universities has been scarce, advanced training at the graduate level has sometimes been carried out jointly by the universities and related research institutes, both at national and provincial levels. However, the success of this arrangement depends greatly on cooperation and coordination among institutions. Rivalries and conflicts have often arisen, obstructing effective collaboration. For the last twenty years, farsighted educators and scientists in China, and those visitors invited to China to lecture and consult, have suggested that the specialized universities and colleges be reunited to form multi-disciplinary 'comprehensive universities' in the manner that they existed before the 1952 reorganization. Forceful and compelling justifications for reunification have been presented, but serious attempts at reunification have not been made until very recently.

As briefly mentioned above, primary and secondary schools should not be neglected as an opportunity to further strengthen the educational system. They are crucial for laying down fundamental knowledge of science, as well as for nurturing other indispensable subjects, such as language, history, and culture. A review of educational systems for primary and secondary schools in China is beyond the scope of the present discussion but is available elsewhere (4, 5). However, a unique occurrence should be mentioned here. Chinese universities generally have their own primary and secondary schools. Unlike the laboratory schools of the colleges of education (i.e., teacher's colleges) in American universities, these are regular schools provided by the university for the children of faculty and staff as part of a comprehensive university community.

Agricultural Education, Research, and Extension: Past and Present

Education

Like other educational fields, agricultural education in China is administered at three levels of government. Thus, agricultural universities and colleges are specifically administered by the Ministry of Agriculture, the provinces, and in some cases, the cities. A few of them have the 'key universities' designations. Before the founding of the People's Republic, 23 Chinese universities contained colleges of agriculture. There were also 20 other high-level agricultural schools. In the 1952 reorganization of educational institutions, these colleges were separated from the universities and grouped them into 30 independent specialized agricultural and agriculture-related colleges, including those for agriculture, forestry, animal husbandry, fisheries, agricultural mechanization, etc. They were relocated from mostly the coastal and urban areas and distributed to all provinces and special administrative districts except four. The school system, setup, and operations completely followed those in then Soviet schools. Attempts were made to produce rapidly a large number of personnel with specific skills for dissemination in all geographic areas. In 1957, the numbers of students in agricultural colleges reached 36,189 at the undergraduate level and 183 at the graduate level, as compared with 10,726 and 20 in 1949, respectively. The number of faculty members increased about seven times probably from the participation of agricultural workers who had received advanced training abroad (6), mostly from the former Soviet-bloc countries. When the United States lifted the ban in 1956 for overseas Chinese nationals to return to China, many such scientists and scholars did so, further increasing the numbers of faculty, although the number in agriculture is not known. During the Great Leap Forward movement, the number of agricultural institutions increased to 171 and new students in the freshman class each year to more than 26,800. They constituted about 5.3 and 3.8 times as many as in 1957, respectively. However, as discussed in the previous section of this paper, the training of agricultural manpower actually suffered tremendously. In addition, many highly respected college professors and research workers were purged in the anti-rightists movement.They were disgraced and their names were not cleared until the 1980's. In 1961, the government made a corrective measure in educational policy to weed out many low-grade institutions. The number of agricultural universities and colleges was cut down to 45 by 1965 and the annual admission of freshman students to 13,849. The agricultural education gradually returned to normal and the quality of agricultural education improved. The impetus did not last long because the Great Proletarian Cultural Revolution took place. As has already been discussed, the damage to the agricultural education and education in general was beyond calculation. The agricultural and agriculture-related institutions were either disbanded or forced to move to the rural areas to be with the farmers. There were no adequate classrooms and facilities and the life of faculty and staff was extremely hard. Manual labor on the farms was a part of teaching and learning. New schools were established. Like those of other disciplines, the students enrolled were without adequate background. No tests and examinations were required and time for graduation was shortened. In the meantime, the original campuses were occupied, equipment and laboratories damaged, and library materials destroyed. During 1966-1969, all agricultural universities and colleges were completely closed. They started to open again in 1970 with a total student enrollment of only 1090. However, the agricultural institutions struggled and were not able to return to their campuses until much later, because they had no campus to return to. Even at the end of the 1970's, several returning agricultural universities found that their campuses were still being occupied by other agencies. During the ten-year period of the Great Proletarian Cultural Revolution, the agricultural institutions produced a total of 75,261 graduates. As stated previously, despite the fact that these workers were considered to have graduated from college, they lacked the training and skills to be considered qualified technical workers. The worst which came out of the Cultural Revolution was the attitude of the young people that school education would not contribute very much to their future life.

In the years following the end of the Cultural Revolution in 1976, the government reemphasized the importance of education, set up new education policies, issued new directives, changed its attitude toward intellectuals, raised the standard for admitting new students, and resumed graduate programs. Particular attention was also given to agricultural education, enabling the agricultural institutions that were forced to close, divide, and move to the rural areas during the Cultural Revolution to return to their original campuses. A small number of institutions were provided new locations to rebuild. Student enrollment gradually returned to normal. As a result of all these efforts, the number of higher agricultural institutions increased from 31 in 1971 to 38 in 1976 and to 64 in 1993. The total enrollments in these three particular years were 5,112. 52,987, and 143,329, respectively (6). The greatest increase occurred in the early 1980's after the government instituted its 'Four Modernizations' policy in its attempt to modernize the country. Since then, continuing efforts have been made to build up the training programs of agricultural workers by consolidating and expanding the existing and new agricultural schools and enrolling new and better qualified students. By 1994, the number of institutions had increased to 104. The total student enrollment was up to 229,984 (7), Table 4.12.1. The large discrepancy between these figures and those quoted above for the year of 1993 is due to the inclusion of institutions which are not directly under the Ministry of Agriculture but under agriculture-related ministries.

Graduate programs also grew rapidly with the implementation of the Four Modernization policy. In 1981, advanced degree programs were reinstated. The enrollment in graduate studies increased from 314 in 1978 to 1,666 students in 1993 (6). From 1981 to 1993, 12,392 students completed their graduate studies, including 578 students who received the Ph.D. degree. In the meantime, selected graduate students and junior faculty members were sent abroad to pursue advanced studies in agriculture. The restrictions were loosened for students who acquired financial supports themselves to study in foreign countries. However, like in other fields, many students have chosen to remain in the foreign lands for a better life after the completion of their studies. Some of them returned. They were extremely helpful for replacing the aged faculty members and senior scientists who guided the Chinese education and scientific research enterprises through those difficult periods of political movements.

With the exceptions of about 18 agricultural institutions of higher learning, including eight key universities, which are under the direct control of the Ministry of Agriculture, the rest of the institutions is administered by provincial or local governments and at the same time, guided by the Ministry. In rural and remote areas, specialized agricultural and secondary agricultural schools have been set up to train local graduates from junior and senior middle (high) schools as lower- and mid-level technical workers. In 1994, there were 82,391 such graduates from 574 schools and the total enrollment reached 421,100.

In addition to the aforementioned educational institutions for the training of agricultural workers, national and international forums and special training sessions have been conducted for the advancement of middle- and high-level agricultural workers. These forums and training sessions are usually sponsored and financed by the Ministry of Agriculture and conducted by key universities at national and provincial levels. Such efforts have contributed greatly to agriculture for its lost times.

Within the agricultural universities and colleges, improvements and reforms have also been made or attempted. The curricula and course contents have been strengthened. Although an emphasis on specialization still exists, the distinction of closely related specialties is not as sharply defined. After

China was re-exposed to the western system of education in the late 1970's and early 1980's, Chinese universities began to add colleges with the aim of forming multi-college institutions. However, these new colleges were mostly created by promoting some academic departments to the college level with the existing faculty and facilities, as a result of lack of qualified personnel and budget. In 1995, the agricultural branch of the State Science and Technology Commission approved several specialized agricultural universities and colleges under the Ministry of Agriculture to combine and become a multi-college university. This represents a remarkable progress in agricultural education toward modernization, although the multi-disciplinarity achieved to date falls short of that found in full-fledged multi-disciplinary institutions.

Research

Research was generally conducted by university faculty as well as the agricultural research institutes before the founding of the People's Republic. The research in the universities then was mostly in fundamental scientific areas and of academic interest. Since the universities were reorganized in 1952, their mission has been limited primarily to teaching. Funding for university research has been extremely limited and only provided perhaps in conjunction with teaching needs. In recent years, however, faculty members with high levels of training, ability, and interest for research have received funds from related ministries, provincial governments, and communities. In addition, the State Council has recently established a State Foundation for Sciences to which university faculty and other institutional staff may apply for research funds. However, these and other sources of government research funds seem to go more often to research academies. Only a minor amount has gone to the universities. As the market economy in recent years has flourished, funds from production agencies and units have also been sought by individual faculty members to support their research while solving problems of these agencies. However, the amounts they receive from these sources are usually small. Therefore, the universities cannot be really considered the major centers for agricultural research under the present Chinese education system.

Agricultural research under the aegis of government is sponsored by the Science/Technology Department of the Ministry of Agriculture as the Chinese Academy of Agricultural Sciences. This Academy contains a number of institutes, such as those of crop breeding and cultivation, plants protection, soil and fertilizer, utilization of atomic energy, natural resources, crop germplasm resources, biological control, agricultural meteorology, agricultural economic, science and technological information, etc. These institutes are more or less equivalent to the academic departments of a university. It is noted that the Academy has also a graduate program, presumably in cooperation with local agricultural universities. As agriculture has regional and climatic variations, the Chinese Academy of Agricultural Sciences has a considerable number of research institutes around the country which deal with the research of specific soils, crops, animals, insects and diseases, etc. Like education, each individual province and major city has its own divisions of agriculture and science/technology which manage the provincial or city's academy of agricultural sciences. These provincial and cities' research institutions have several institutes under them to study specific problems of their concern. The research in the academies is generally more applied than basic. Below the province level, there are prefectures and the counties, under which there are also research institutes or extension-oriented agro-technical stations and groups to deal with practical problems. This research system is parallel to the education system but there has not been a close relationship between the two.

Aside from the agricultural research conducted by the Chinese Academy of Agricultural Sciences and provincial and sub-provincial administrative regions' agricultural research institutions, the Chinese Academy of Sciences has also several institutes engaged in agricultural research, dealing with the nation's soils, plants, forestry, and perhaps in other areas too. It is assumed that research in the institutes of the Chinese Academy of Sciences is more basic in nature. However, there may not be a close relation and an exchange of information between the two separate systems.

Extension

Agricultural extension work is coordinated by the Ministry of Agriculture. There were formerly four main areas of extension: plant protection, soils and fertilizers, seeds, and agricultural technology applications. They are presently combined into one unit, the 'National Extension and Service Center of Agricultural Technology', under which there are 20 subdivisions. Among them, 12 are technical and the rest are administrative. The technical areas were developed from the original four areas. They are soils, fertilizers (formerly soils and fertilizers), agricultural chemicals, disease and insect forecasts, pest control, plant quarantine (formerly plant protection), seed propagation, seed inspection, seed professional advisory center (formerly seeds), food and oil crops, economic crops, and farm system (formerly agricultural technology extension). However, most agricultural departments in the provincial governments still have the original four extension areas, as do the prefectures. In a few places, these four areas of extension are incorporated into an extension center of agricultural technology, much like that in the Ministry.

In the counties, there are service centers of agricultural technology. Under these centers, there are extension stations of soils and fertilizers, plant protection, seeds, and crop cultivation. Other special extension stations, such as those for fiber crops, fruits, tea, sericulture, etc. have been established to serve the particular needs of the districts. Before the commune system was abandoned in 1984, there was a three-tier network of extension below the county level, the experimental farm in the commune (presently, xiang or town), experimental plots in the production brigade (village), and the service of farmer technicians (team or workers' group). Presently in towns and villages, there are agricultural technology centers stationed by seven to eight people. In the countryside, the heads of the small villages are responsible for the agricultural production as well as the extension work, though not in an official capacity. When needed, they will be able to assign the production people to set up demonstration plots to show the new technology. As it has been stated, the provincial academies of agricultural sciences deal largely with practical problems. The results from their research may be field-tested by cooperating with various levels of governments and extension units and/or setting up experimental and demonstration plots in the production fields. Through local extension people, they transmit the scientific and technological information to the field people and obviously become an important arm of the extension system. Although administratively, the Ministry of Agriculture and the agricultural divisions of the provincial and local governments are responsible for the transfer of the new information or technology to the field people, the active role is played by those of extension stations and academies of agricultural sciences. To some extent, agricultural university personnel are also involved in extension work.

It is understood that the agriculture branch of the State Science Commission, the highest agricultural policy-making body, and the Academia Sinica also have their own extension systems for agriculture-related matters.

Agricultural Education, Research, and Extension System:

Toward the Future

Since the late 1970's, China has recovered from national chaos and starvation to the bumper crop production of today. This is quite an accomplishment in 20 years and may be attributed to the hard work of the agricultural policy makers and the workers at all levels in agricultural education, research, and extension as well as that of Chinese farmers. However, problems exist in the education, research, and extension systems in agriculture as they currently stand. The most important, albeit difficult, tasks to be immediately undertaken include removal of the redundant administrative structure, reorganization of the institutions, and promotion of cooperation among institutes and personnel in parallel systems as well as within the system. The production of high-caliber personnel, both in quality and in quantity, and full development of their work potential depend on these changes. It is also imperative that the government make a commitment to support fully funding needs and create a better working environment so that the morale of the workers could be boosted. In the following paragraphs, attempts are made to discuss ways, as the author sees it, to improve the systems for education and training, for research, and for technology transfer. It is hoped that these will be taken into consideration while implementing present systems. Otherwise, the profession of agriculture, which is so important to a country like China, will fail to attract the brightest young people when compared against other professions of higher visibility and prestige.

Education System

Agriculture may be considered as an applied science whose foundations lie in a wide variety of scientific disciplines. Chemistry, physics, engineering, geology, biology, microbiology, botany, entomology, meteorology, economics, sociology, law, medicine, environmental science, and management are but a few of these disciplines. Because agriculture depends on such a wide variety of scientific disciplines, scientific versatility is vital if agricultural practice is to advance. None of the physical, biological, social, and other sciences is dispensable in the education of the modern agriculture students. In order for future workers to meet new challenges with modern ideas and technology, students must develop a broad scientific background. As discussed previously, this is probably best taught in the setting of true multi-disciplinary, comprehensive universities.

Despite the fact that the present education system has produced a large number of technical workers and helped China to progress in all directions, the training for more advanced personnel should be a priority. China urgently needs a large number of top-level scientists and technologists in all fields, including agriculture. Statistics from 1993 (6) show that among agricultural university faculty, 20% had postgraduate training. There were a total of

7,886 faculty in the upper ranks, including 1,593 full professors and 6,293 associate professors. The number of faculty in these two ranks and the percentage of them with postgraduate training can certainly be considered low. It may not be unexpected that the same is true for the research staff in the agricultural research academies. Most of these workers are said to be relatively young. Among those on the faculty, 17,579 members or 71.2% were below the age of 50 in 1993. In the case of the upper two ranks, full and associate professors, the figure is 1,696 or 21.4% (6). Even if these young faculty members have had postgraduate training and are dedicated and energetic, a large proportion might have been the product of the single-discipline or specialized institutions. They might be hindered by their way of thinking, narrow academic background and lack of a strong basic science foundation to meet future challenges.

Since the inadequacy of the present education system to train high-level personnel can be demonstrated, the return to the multi-disciplinary comprehensive universities as those before 1952 is necessary and should be an immediate goal. In recent years, the government has made a series of policy changes. In their national report to the 39th session of the international conference on education (8), a new set of goals for all levels of education was listed and plans to achieve them were outlined. What followed was the national education reform of 1985. A number of improvements have been made since then, with an attempt to fulfill modern needs, including the establishment of new areas of specialties and enlargement of science departments, etc. in the universities. Courses in the humanities and social sciences were also added to the curricula in many science and technology institutions. Indeed, this is one of the alternatives to transform the present setup to the multi-discipline comprehensive universities, i.e., by gradually adding and expanding new areas of endeavor to existing institutions. Some universities even created new colleges. Due to the shortages of highly qualified people and funding for facilities, these changes may prove not to have lasting benefit.

A better way to build a multi-disciplinary university under the present Chinese situation would be to recombine and consolidate those specialized universities and colleges which were separated in 1952-54 from the same university. Many of these are now coexisting in the same cities or even sharing the same former campuses. Nevertheless, it would seem to be an unthinkable task to recombine them, since they have been separated for so long and each has had a big and independent administrative structure. They certainly would resist such plans. Even when these institutions are combined, there would be many conflicts and frictions. However, anything could be done if the government and the national leaders have the vision and strong determination. The government should make a great effort to overcome all expected and unexpected problems to make it work, because in the foreseeable future, China may not have additionally qualified faculty, facilities, and funding to produce the top-level personnel otherwise. In the meantime, administrative costs could be cut down tremendously to make the universities more cost effective. It is gratifying to know that the government has been trying to work towards this end. In 1995, the State Council and the Ministry of Agriculture approved the merger of Beijing Agricultural University and Beijing Agricultural Engineering University, forming a new China Agricultural University which has now fourteen colleges. Several other agriculture-related universities and colleges of different specialties have also been restructured. The Ministry of Agriculture and other ministries as well as various provinces have joined hands to establish and manage a few schools of higher learning for mutual benefit (6, 7). Although many difficulties and problems have been encountered and are as yet to be resolved, it is the first step towards the ultimate goal of forming comprehensive multi-disciplinary educational institutions.

It is known that the State Council in 1996 also approved an attempt to combine one national and three key provincial specialized schools, agriculture, arts and sciences, engineering, and medicine, in Hangzhou, Zhejiang Province into one. These four schools were separated in 1952 mainly from a single well-known university, strong almost in all colleges and being called 'the Cambridge of the East' during the World War II. Though now separated, they have all developed into major academic institutions in China because of their previous school tradition, highly qualified and dedicated faculty, and hardworking students. However, the planned integration in 1997 of these four schools was 'postponed' before the unification was actually announced. The reasons are not known. The government should recognize the urgent need for the reorganization of educational institutions and make preparations to remove as soon as possible the unwarranted obstacles. The government should set a target date for them to reunite and in the meantime, appoint a committee composed of nationally respected educators and scientists with vision and ideals from outside the four universities to start planning carefully but actively and eventually take over the administration of the integrated university. Hopefully, the reunification would be completed before the end of this century or in the beginning of the 21st century. It should not be delayed and should be definitely made successful because it serves as an example for the other specialized universities to follow. The year 2010 should be set as the date for all specialized universities to convert and adjust to the true multi-disciplinary universities. Otherwise, China would be continuously lagging behind her needs in top-level scientists and technologists not only in agriculture but also in other areas of reconstruction. In addition to the formation of multi-disciplinary universities, the restructuring of the whole system of higher education must also be attempted. At present, ministries of the central government, the Chinese Academy of Sciences, the provincial governments, and cities/communities all have 'their' own universities and colleges to train needed personnel. They each have an administrative structure to direct these institutions. Among the educational institutions and their personnel, there have been little cooperation and communication. This setup is much like the self-governing city-states of ancient times. Time has changed and needs are different. This system has out-lived its usefulness. It could be made more efficient and effective to produce better scientists and technologists for these respective governmental units by putting all the universities under one administration. The way to do it is to recreate the Ministry of Education and give this ministry the sole authority and needed funds to oversee all the universities. The management of universities may be delegated to the provinces where the universities are located. Guidelines could be set up by the State Education Commission and/or the Ministry of Education, and be based on a national educational policy to train personnel needed in all areas. Such change may save considerable amounts of administrative cost as well as facilitate the operation of truly multi-disciplinary universities. The present funding differences among different specialized disciplinary areas and among different levels of government institutions could also be avoided. This change may increase the responsibility of the provincial education department. However, a special committee, similar to the Board of Control or the Board of Trustees in western countries, may be established to coordinate and oversee the operation of various institutions of higher education within each province. The authority of running the school could be completely delegated to the president of the university with an advisory committee, if needed, to help make the decisions. Although the ministries would seem to lose their direct control over the schools, they could still maintain close and frequent contacts with the university, its related college or colleges, or even a particular department. Through such contacts, the ministries could channel their education funds for the training of particularly needed personnel and for the use of university talent to advance their interests.

A number of community-established institutions that call themselves 'universities' or 'colleges' do not have adequate facilities, qualified faculty, and sufficient courses. The names are misnomers. Many of them are actually vocational schools and others offer only amelioration or adult education courses. While the purpose is commendable and such schools are needed, they should be renamed, regulated, and given guidelines to set up requirements and upgrade their standards. These 'universities' or 'colleges' operated by the community, business, factories, etc. could be transformed to a two-year junior college system with additional courses to provide a stronger foundation in humanities, science, technology, business, and management. This would broaden the students' interest and help them to upgrade their working skills for better jobs in the future. This junior college system would be particularly valuable if provincial and local governments could establish a number of such colleges in strategic locations in rural and remote areas, thus providing the young people in those areas with access to a better education.

What has just been discussed may not seem on the surface to be closely related to agriculture. However, it should be kept in mind that in China, agricultural education cannot be separated from education in general. The future of agriculture and education as a whole in China depends greatly on the improvements suggested. In addition to the institutions of higher learning, there exist a large number of agricultural secondary specialized schools. They are equivalent to high schools and/or junior colleges and have been set up for the training of agricultural workers in the rural and remote areas. If and when the agricultural universities are integrated into a comprehensive multi-disciplinary university, the senior and junior agricultural secondary schools could still be jointly managed by the provincial and local agricultural education or education departments. However, a mechanism should be set up to bring the college of agriculture of the integrated multi-disciplinary university into the system. Thus, the technical and faculty support of these schools could be more conveniently rendered by the resources available in the university. In the aforementioned junior college system, agricultural courses may be included or grouped together as an area of interest for an associate degree upon graduation. This degree is somewhat equivalent to the diploma of an agricultural technical school. The inclusion of agriculture in the junior college system has the advantages of getting non-agricultural people's attention and understanding of agriculture which are long overdue.

While the shortage of top-level and highly skillful manpower is real, it is also true that present human resources that already exist have not been used to their potential. As shown in Table 1, the ratio of students to teaching personnel is equivalent to 6.85 students per full-time faculty member. Among the three levels of full-time faculties, the lowest rank, lecturers and instructors, constitutes 63.8% of the total. As previously mentioned, because of budget constraints, only a very small number of faculty members are able to engage in research and find funds somewhere for the purpose. The large majority would be in teaching only. Certainly there are not that many courses to teach and laboratory sessions to direct. Besides, new courses are not likely to be offered under the system. It is not known what the assignments for most of the faculty are, particularly for those at the lecturer and instructor levels. Assuming that most are not idle or doing nothing, a large number of them must not be working at their potential. In the University of Florida where this author has been an active faculty member for 33 years and an emeritus professor for 10 years, the full-time student to the full-time teacher ratios in 1995 (9) were 41.2, 29.4, and 13.9, respectively, in the lower division (first two years), upper division (the last two years), and in graduate school. For the college of agriculture, the ratios were 96.0, 24.3, and 8.7, respectively. The high student to teacher ratio for agricultural students in the first two years of education is due to the fact that their courses are mostly foundation courses offered by the departments in other colleges. Also, the basic courses offered by the college may be taken by students of other disciplines. The overall ratios for the university and for agriculture were 26.2 and 19.6, respectively. These ratios are three or four times higher than that of Chinese agricultural universities and seem to be more reasonable. It is also remarkable that the student enrollment to the total number of the faculty, staff, and workers formed a 3.01 ratio (Table 1), showing that 25% of the total university population was composed of faculty, staff, and other workers. Obviously, the agricultural universities in China are over-staffed, reflecting inadequate uses of human resources and thus, of funding. The excessive number of faculty members may result in unequal work assignments which would also affect the efficiency and morale of the faculty and staff. It is believed that a similar situation occurs in all other specialized universities. The government must be aware that such superfluous existence of lecturers and instructors would not help education but rather would hinder the efficient development of the universities. A critical review of this situation by the government and universities themselves should be promptly attempted and corrective measures be planned and executed, making full use of these people in which the country has invested.

One area might find those excessive university personnel helpful. The lower levels of Chinese schools need more and better-trained teachers. It would be a tremendous asset if these excessive number of lecturers and instructors from various specialized universities can be transferred to the junior colleges and high schools to bring up their academic standards. The agricultural secondary schools certainly would be benefited by those from the agricultural universities. However, looking at the statistics in Table 2, the student to full time teacher ratio in the agricultural secondary specialized schools is even narrower, 3.48 students for each full-time teacher. These statistics are

Table 4.12.1. Number of faculty members and students in Chinese agricultural institutions of high learning in 1994 (7).

Types of Institutions	Number	Graduates	Enrollment	Faculty & Staff Workers	Professors	Full-time - Associate Professors	Faculty Lecturer Instructor	Total
		/------------------in thousands----------------------------/						
Agriculture	53	29.3	132.1	42.8	1.5	5.3	11.9	18.7
Forestry	12	5.81	24.39	8.8	0.39	1.15	2.49	4.04
Fisheries	5	1.80	8.78	2.7	0.04	0.30	0.93	1.28
Agriculture & Reclamation	6	2.24	11.69	4.9	0.12	0.48	1.30	1.91
Water Conservancy	23	8.69	44.27	14.4	0.58	1.79	4.05	6.43
Meteorology	3	0.84	2.48	1.29	0.02	0.13	0.30	0.46
Others	2	1.21	6.21	1.21	0.04	0.19	0.40	0.64
Total	104	49.9	229.9	76.4	2.72	9.43	21.4	33.5

puzzling. China needs a great many skilled people but is not making full use of the existing ones. The excessive personnel from these schools should be reassigned to appropriate positions or further trained in other needed areas.

After consulting with former faculty members and students of the Chinese universities now residing in the United States, it is found that the excessive personnel is, in effect, 'necessary'. Other than teaching and research, the university provides on-campus living quarters, transportation, security, cooperative stores, nursery to high schools, and all other community services related to living needs to all its people and their families. Many of these excessive personnel are assigned full time or part time to such duties, as these services are not necessarily run by the professionals. In China, not only the universities, but also research institutions, government units, factories, and other large production units are said to form such inclusive

Table 4.12.2. Number of faculty members and students of Chinese agricultural Secondary Specialized Schools, 1994 (7).

Types of Institutions	Number	Gradua-tes	Enroll-ment	Teachers Staff, & Workers	Full-time Teachers
Agriculture	298	44,734	216,717	109,465	64,889
Forestry	52	7,570	40,734	20,198	11,918
Fisheries	22	2,354	13,972	6,677	3,682
Land Reclamation	11	1,296	6,647	3,196	1,416
Agricultural Machinery	100	13,589	77,723	36,791	19,931
Water Conservancy & Electric Power	75	11,693	58,483	29,536	17,836
Meteorology	16	1,155	6,824	3,592	1,402
Total	574	82,391	421,100	209,455	121,074

society to furnish the needs of their workers. While the formation of many small 'societies' within the city or county is a widespread practice, it seems to be a huge waste of useful resources and manpower because the living needs could be more cost-effectively managed by the city or county governments where the institutions and government agencies are located.

The amounts of time and energy of the universities and other government agencies spent on fulfilling the daily needs of its personnel are tremendous. As it is, the mission and goals of an educational institution to teach and train the young for the benefit of the society and the nation are distracted and hindered. Instead of devoting their energy to advance education and research and using their vision and creativity to elevated the university to a better institution, the administration has merely become the house-keeper worrying about things which are not the main responsibility of a university. Looking into future, to make the education and research systems more effective, something has to be done to simplify the operation of all institutions and let them concentrate on their designated charges and real responsibilities.

As modern technology requires workers to have a broad foundation of scientific and management training, the existing narrow specialty policy in the university should be abandoned or at least downplayed. A broad, multi-disciplinary university curriculum should be developed so that students may have choices of courses in different colleges to suit their individual interests and develop what they do best. Course materials used should not be limited to those in the pre-prepared text books so that the students may take advantages of the expertise and experience of the faculty in their specific fields of various disciplines. Guidance could be provided by departmental faculty advisors. Other improvements can be made in many areas. In 1982, a group of agricultural scientists of Chinese origin in the United States was invited by the Ministry of Agriculture to participate in a week-long forum on China's higher agricultural education. Prior to the forum, the group spent two weeks to review thoroughly the programs of four most prominent agricultural universities in China. Its report (10) listed a number of recommendations which were equally applicable to education in China as a whole. These recommendations are still useful and should be carefully considered.

Modern computer and network technologies have led the world into the information age. These technologies have a huge potential to improve all levels of education including agriculture and should be embraced. In addition to their obvious role in scientific research, the ability of computers and networks to expedite word processing and communications has a special relevance to agricultural education. Access to such capabilities, whether by collections of printed material or by on-line services, would speed the rate at which new findings and other urgent information are disseminated, not only in university classrooms but also in outlying and rural areas. This invaluable tool could be effectively used to expand the teaching programs of the universities, including agriculture, to educate the young and the old alike who otherwise would not have been able to reach the institutions or to attend the school because of various reasons. In agriculture, the creation of a comprehensive agricultural education network would open the way for all agricultural workers in the country to access information from all education, research and extension units within China and from all over the world. One way to create and sustain such a network, a computer nerve center for agricultural and related information, would be to establish a Chinese national agricultural library, if this does not already exist. Extension of this network to remote units of agricultural research and extension would also help create an 'information superhighway' for Chinese agriculture. Apparently an infrastructure for the manufacture of the necessary technologies already exists. What is really most needed is the government financial support for the introduction of computer hardware into the research institutes and universities, on the one hand, and increased recognition of computer science as an essential component of a general education, on the other.

Despite the emphasis on advanced science and technology in this review, one thing which should not be neglected is the teaching of traditional Chinese values, social norms, humanities, and culture, in addition to political ideology. Cultivation of these virtues seems particularly imperative today, at least to this author, if China wants young people to have a desire to serve the society, the country, and mankind. Such virtues are actually the ultimate objective of 'education'. It is unfortunate that many 'educators' and national leaders in industrialized countries tend to overlook these virtues and mistakenly define education as the learning of 'language, mathematics, and sciences'. Overemphasis on modern science and technology without an appreciation of the qualities of citizenship will lead to self-centeredness and selfishness, traits whose corrosive effects can destroy the colleagiality that is so important to cooperation among the various components of agriculture.

Research System

At the center of agricultural research in China is the research arm of the Ministry of Agriculture, the Chinese Academy of Agricultural Sciences. This institution is comprised of a large number of specific research institutes that conduct research in all areas of agriculture. Their aims should go beyond scientific agricultural research in the traditional sense and must include policy issues such as national and regional agricultural development, crop distribution, protection of natural resources and the environment, identification of marketing and trade opportunities, etc.

The overriding goal of research in the Academy should also be the provision of technical leadership at the national level, such as in the preparation and planning for food and environmental emergencies. The various research institutes in the Academy should take better advantage of their national standing by focusing their efforts on problems that fulfill a criterion of broad relevance to regional and national agricultural needs. In addition, because research in the Academy is as a whole more fully funded than that in provincial or local research institutes, the Academy is best placed to undertake research on problems that are more specialized and difficult, especially those problems that may be termed 'basic' or 'theoretical', as opposed to 'applied'. Such research is limited in agriculture but is nevertheless important to support, since the general scientific principles that arise from this type of research will ultimately become the foundation for new methods for solving problems.

In addition to the Ministry of Agriculture, other Ministries and research academies in the central government also sponsor several of their own agricultural research institutions. Although such apparently redundant agricultural research probably reflects, at least in part, the special perspectives of those agencies towards agriculture, it may be useful to examine the degree to which duplication of effort exists, since such duplications represent wastage of scarce financial resources.

At the provincial level, research is sponsored by provincial academies of agricultural sciences. The aim of the research is to increase agricultural productivity, both in quality and in quantity, relevant to the special characteristics of each province. Therefore, most of the research is necessarily 'applied' in character, in which the aim is to use scientific knowledge to solve practical problems, as opposed to developing systematic knowledge for its own sake. Nevertheless, some investment in basic science is justified, since the technical information that goes out to extension services in the field must be solidly grounded in science. One way to promote research at the provincial level is to involve staff from the universities. Although the mission of these universities has presently been one of education and, to a much lesser extent, research, the missions of research and education are in fact mutually reinforcing, as embodied by the concept in the United States of the 'research university'. With respect to agriculture, it is likely that those at the forefront of creating new knowledge will be the most qualified to teach others state-of-the-art techniques and knowledge, while those in education will be especially well-placed to recognize important new problems to solve, by virtue of their contact with the recipients of new information.

An excellent example of the interdependence of education and research is the development of new agriculture procedures that protecting natural resources and the environment. This research requires close communication between the research community and those who work to put the new information to practical use. In recent years, to meet food production quotas, the government has pursued a 'Three-H' policy ("high yield, high quality, and high efficiency"), but has tended to overlook some necessary precautions. When a country develops and the technology is used indiscreetly, problems likely will follow. In agriculture, such problems have already appeared in the form of soil, water, and air pollution, as well as the eutrophication of rivers, lakes, and waterways. These problems can be attributed to the overuse of pesticides and fertilizers and the improper management of industrial and animal wastes. Their adverse effects threaten the well-being of human and animal life. Similarly, excessive use of energy and water depletes natural resources. Problems have already occurred in China and will become worse as more advanced technology is developed and used. Effective and sound management of such potential problems need to be considered and addressed in advance, because they would eventually affect the quality of crops and cause adverse effects on environment as well as human and animals. Therefore, knowledge of how to protect the environment must be included in the education, research, and extension programs.

It would, therefore, be mutually beneficial for research and education units within each province to be brought together under a single administration as a college of agriculture within a multi-disciplinary university. Such a merger would make it possible for research staff to broaden their knowledge in allied fields, to promote advanced education, and to expedite the communication of research results to workers in the field. These potential benefits have long been recognized. About twenty years ago, the unification of a key provincial agricultural university and the academy of agricultural sciences was attempted in a more agriculturally advanced coastal province. The two institutions united but later again separated. They have remained parted. The failure of this reorganization was probably due in large part to conflicts between the two administrations and among the egos of individuals. In this particular case, many of the research and teaching staff were also against the reorganization, because of resistance to change. In modern times, individualism and self-interest seem to have been obstacles to any reform in China. This likely was one of the major causes for the postponement of the reunification of the four key universities in the province, as mentioned in the education section. It takes the determination and power of the government and the unequivocal support and hard work of farsighted policy makers, administrators, educators, and researchers to make such changes work. If, after serious consideration, any systematic reform or reorganization is to be made, such as the integration of the aforementioned institutions in both cases, it would always be wise to appoint an outside team of highly respected scholars/administrators with vision and ability to execute the integration of the systems and to head and guide the integrated unit.

Extension System

The present network of agricultural extension has been briefly described. The term "extension" can be thought of referring to the "extended" arm of education and research. Its function is to educate those in agricultural production and agriculturally related industries and businesses on one hand and to transfer information on new technology and its effect on environment on the other. In the meantime, the needs of the farming community are identified and communicated through extension personnel to the educators and researchers, who may not be familiar with the local situation and actual problems. Because of these functions, agricultural extension services should also become integrated into a unified research and education entity. Such a unification not only will enable the extension staff to execute their work more effectively but also enhance their science and technology background by closely associating with educators and researchers. The most important of all, however, is to promote teamwork in attaining the common goals of educating rural people, feeding the nation, helping agriculture-related business, and maintaining a good natural environment. To meet the challenges of the future in China, teamwork is a necessity in all areas of endeavor. This is particularly important in agriculture, whether in education, research, or extension itself, or among these three branches.

In universities of western countries, tight integration of education, research, and extension has been accomplished by designating varied amounts of teaching, research, and extension responsibilities to each of the the integrated agricultural faculty in the university. Taking the system of the University of Florida's Institute of Food and Agricultural Sciences for an example, all members of the faculty have joint appointments from teaching, research, and extension branches in proportions based on his or her interest and expertise. There are 930.7 full-time-equivalent faculty positions budgeted for the 1997-98 academic year. Each full-time-equivalent or FTE represents a 100% workload. Except for 8 full-time administrative positions, 135.0 of the remaining 922.7 FTE positions are assigned to teaching, 395.8 to research, and 140.7 to extension in 23 academic departments on the main campus and in 21 specialized research and education centers in the strategic locations throughout the State. The other 251.2 FTE are assigned to those as county extension agents stationed in each of Florida 67 counties and 4 demonstration units to provide their services to local residents in various areas of expertise. The decimals in FTE indicate the percentage workload of a position reflecting that the position is shared by other people. The actual numbers of faculty involved in teaching, research, and extension (not including the county extension agents) are 379, 578, and 254.0, respectively, in the 135.0, 395.8, and 140.7 FTE positions budgeted. In other words, each teaching, research, or extension position is shared by several faculty members and each faculty member has different proportions of teaching, research, and extension responsibilities. This distribution includes the academic department chairs and directors of the research and education centers, who also have administrative responsibilities. Such a system intertwines the three branches of agriculture and form an integrated system which benefits all those involved. In addition, the expertise available in the agricultural college and other colleges of a multi-disciplinary university would also become more accessible for consultation.

From what is known, the present extension system in China is dedicated to improving food production. However, the technology being introduced is mostly relatively old.

Few new technologies have been introduced, probably because of the limitations suffered by the extension personnel. In addition, little attention has been paid to promoting the quality of rural life and to protect the environment. With the integration of three branches of agriculture, more people could be involved in the extension service. Programs could also be broadened to include, for example, services to people in farming communities; education for the young and the elderly; support for rural enterprises; and information about nutrition, food preserve, and clothing. Through the university's as yet to be integrated agricultural education-research-extension system, the governments, both national and provincial, could also offer environmental education programs for pesticide management, water resource protection, food processing and safety, as well as telecommunication, electric programs, and rural construction.

Summary Statement

The discussion presented above on China's agricultural education, research, and extension systems since 1949 may be reiterated. China has an obvious ongoing deficiency of top-caliber personnel in each of these areas and thus, there is a pressing need to train a large number of top-notch scientists and technologists to meet the challenges of the 21st century. These people must have a very solid background in all branches of sciences, natural, biological, agricultural, and social, as well as language, management, communication skill, and others and should be trained together in an integrated agricultural education-research-extension system in universities which are truly multi-disciplinary. Their training may be enriched with different additional course work, based on their personal traits and interests. They should also be trained to appreciate teamwork and to have the enthusiasm to work together, since it is only through cooperative effort that the processes of education, research, and extension can move productively from one to another. Success in future food production, rural social and economic development, and improvements in the quality of life will depend on these educational advances.

Although many efforts have been made in China, the country's educational and agricultural systems remain as they have been. Thorough reforms need to be planned by farsighted policy-makers and proceeded as soon as possible. The target year set for discussion in all the papers of this monograph is 2030. This author finds it difficult to suggest a timetable for what may be accomplished from now to 2030 for China to feed 1.6 billions and raise their living standard while still maintaining the environment and natural resources. The size and composition of the workforce are actually dictated by the policies of the Chinese government. However, the integration or recombination of existing specialized universities should be completed and well adjusted before 2010. The establishment of an agricultural education-research-extension system should be done in the same time frame. During this period, graduate programs need to be strengthened, not merely in number but in substance. Shortages of top-quality scientists and technologists in the universities may be resolved in part by creating well-funded programs to send promising graduates with commitment and dedication to well-known foreign universities for further training. It is essential that a much needed improvement of working environment and living conditions at home should be made so that it will become attractive to the students after completing their training, to return to homeland and make contributions. In this period, they should all be assigned, upon their return, to the universities to teach and to do research in the course of training the young. Outstanding foreign scholars who have a genuine interest to help China may be invited to engage in all three areas of agricultural activities without being treated with superficial and unwarranted discriminative attitude. In the author's opinion, the first decade of the 21st century should be a period of rebuilding for institutions of Chinese agricultural education, a true "Agricultural Modernization". In the next twenty years, with a full complement of high level technical personnel, agriculture in China should enjoy rapid development. There will then be little doubt that China will be able to feed and clothe her 1.6 billion people in 2030 and compete in the world food market.

Some Final Comments

Starting from the end of the 1980's, China has adopted a market economy. It has been very successful. However, national investments in education and agriculture are still extremely low, despite the fact that the government repeatedly indicates their importance to the country. The vice president of the Chinese Academy of Social Sciences, Teng Teng, recently told news reporters that government investment in education last year (1996) was only a little more than 7 billion Ren-Min-Bi (12). Ren-Min-Bi (RMB) is the unit of Chinese currency, approximately equivalent to 12 cents of the U.S. dollar at the present time. The figure quoted by the newspaper, however, is apparently in error and should probably be in U.S. dollars. The total spending on agriculture was 33.6 billion RMB or 4 billion in U.S. currency, which was about 1.9% of the GDP in 1996 (13). These investments were extremely low when compared with those of the United States. The budget for the U. S. Department of Agriculture alone is 83.7 billion dollars for 1997 and that projected for 1998 is 88.4 billions, not including the budgets for agriculture in the 50 States. Mr. Teng strongly pleaded that the government should look into the future and invest more in education. He had been a former university president, a vice president of the Chinese Academy of Sciences, and a vice director of the State Education Commission. His statement should serve as a testimony of the present situation. It is imperative that the government has to shift its funding from other projects to education and agriculture, as they were correctly pointed out as the prime areas for the Four Modernization.

It is true that China is still a developing country and that funds are needed everywhere. As Teng pointed out, the distribution of funding for many programs was not well thought out and pre-planning was lacking (12). From what has been discussed, huge saving could be obtained by reorganizing duplicated administrative units among all levels of government agencies and down-sizing their excessive and misused staff of both skilled and unskilled workers. Such saving could be more efficiently used where they are needed. Furthermore, it would reduce the bureaucratic structure of these agencies and create a better working environment. Admittedly this is a government policy matter. However, the government has to change the status quo and cut out large amounts of such fat in order to finance the education, agriculture, and other worthy areas and meet the challenges of the next century.

The market economy has enriched the businesses and industries owned by various government agencies and present private investors that have taken advantages of the fruits toiled out of the educators and scientists' efforts and have never repaid. It is suggested that the central and provincial governments should start to collect fees, if possible, on their huge profits from these big businesses and industries to help finance the education and agriculture which the country's future rests upon. In the area of agriculture, companies like seeds, feeds, fertilizers, pesticides, etc. should also be encouraged to make contributions and share their profits with those units engaged in agricultural education, research, and extension activities. These fees, if collected, should be used exclusively for the universities to supplement the teaching, research, and extension needs as well as the professors' salaries. It will help the universities to go a long way, whether in teaching, research, or extension programs. However, this money is not to replace the budgets of the central and provincial governments for education and agriculture. The central and provincial governments still have to raise the faculty's salary and sufficient funds for facilities and expenses. It is the responsibility of the government to see to it that the faculty is content and works to their best ability. It is to the advantage of the country to do so, since these are the people who must train new technical leaders, create new ideas, and transfer these benefits to where they are needed. Only in this way can Chinese agriculture fully benefit all Chinese people and successfully compete with the rest of the world in the next century.

It should be mentioned that the government did initiate an ambitious plan for higher education, 'The 211 Engineering Project'. The aim of this project is to build 100 first-class universities in the 21st century. The State Education Commission has been taking applications from the nation's best universities and several have passed initial review, including two or three agricultural universities. Those who win the final approval are said to receive 60 million RMB each in five years for capital improvement. The rest of the needed funds is to be provided by the respective ministries and provinces. The details of this project are not known to the author. While this is undoubtedly a very plausible project, the author has wondered if the funds provided for these universities are just for the buildings and other capital improvement, or there are additional funds for other urgently needed expenses. It is known that the general funding situation for the universities in China has been very inadequate. The professors' salary is so low that they can hardly make the ends meet. A great many of them is said to have to look for supplementary income somewhere else to meet living needs. It cannot be imagined how these professors stay on the job struggling through their hardships, and in the meantime, try to educate the country's young and do research for the betterment of the country. Physical improvement is certainly needed but it is these professors who will be the builders of the first-class universities. Substantial increases in professors' pay are essential if the government really wants to achieve its goals. The government should also be aware that scholars in China have always been respected and their traditional prestige in the society has now slided shapely. This fact has already discouraged the bright students to enter the teaching and research professions. In addition to the funding in the '211 Engineering Project', the author also wonders if the government has included in the project the second phase and third phase to upgrade those institutions other than the selected 100 for the 21st century. Better education is for all the youths of the country, not merely for those in the 100 elite universities. China needs more and better manpower in all fields. Education is the key and requires continuous attention. Long range plans and firm commitment are necessary. After all, the prime beneficiary is none other than the country itself. While attempting to build the 100 first-class universities as an investment on higher education, it is hoped that the government would also take what has been said in this paper into serious consideration.

Literature Cited

1. State Statistical Bureau. 1997. A statistical report on national economy and social development in 1996 (in Chinese).
2. Brown, L. R. 1995. Who will feed China? W. W. Norton & Company, N.Y.
3. Cheng, C. Y. 1965. Scientific and engineering manpower in Communist China. NSF 65-14, National Science Foundation, Superintendent of Documents, U. S. Government Printing Office.
4. Chaffee, F. H., et al. 1967. Area handbook for Communist China. DA Pam 550-60, Superintendent of Document, U. S. Government Printing Office.
5. World Bank. 1985. China: Issues and Prospects in education. A world Bank Country Study.
6. Mao, D. R. 1995. A guide to agricultural colleges and universities in China. China Agricultural University Press (in Chinese).
7. China Agricultural Yearbook. 1995. China Agricultural Press (in English).
8. Ministry of Education. 1984. The latest development in China's education (1981-1983). Ministry of Education, PRC.
9. University of Florida. 1997. Fact Book. U.F. Office of Academic Affairs.
10. Tso, T. C., et al. 1982. A report on the Agricultural Forum to the Ministry of Agriculture, Animal Husbandry, and Fisheries. Delegation of Agricultural Scientists from the United States.
11. State University System of Florida. 1997. Fact Book, 1995-1996. Board of Regents, State of Florida.
12. Teng, T. 1997. A news quotation from the September 11, 1997 issue of the World Journal in the Mainland News Section (in Chinese), N.Y.
13. Tso, T. C. 1998. General Issues. In this Monograph.

CHAPTER 4.13

From Green Revolution to Gene Revolution

Samuel S.M. Sun
The Chinese University of Hong Kong

Shain-Dow Kung
Hong Kong University of Science and Technology

"Food is the first necessity of the people and agriculture is the foundation of the country." This ancient Chinese motto makes plain the importance of food to life. The development of civilization was clearly tied to agriculture as our forefathers moved from hunting and gathering to cultivating and improving food crops for their primary sustenance. The selection and upgrading of plant products constituted the first step toward modern plant breeding since all food is derived directly or indirectly from plants. Plant breeding became an important branch of science immediately after the rediscovery of Mendel's laws of genetics (Borlaug, 1983).

Among the estimated total of 250,000 plant species, over 1,000 are used for food. Of these one thousand plus species, only 29 are major crops. They include eight cereals, seven legumes, seven oilseeds, three root crops, two sugar crops, and two tree crops (Harlan, 1976). These basic food crops, supplemented by about 15 major species of vegetables and 15 major species of fruits (Borlaug, 1983), collectively constitute the human diet. In particular, four major crops, namely wheat, rice, maize, and potato contribute more tonnage to the world total food supply than the rest combined. Given that our food supply depends on the success of only a small number of plant species, the failure of one may mean starvation for millions of people. The 1970 corn leaf blight epidemic, which caused the loss of an estimated 15% of the U.S. corn crop, is a compelling example of this dependency.

By definition, domesticated plants are dependent on man for survival since domestication is the result of selections that make plants better adapted to man-made environments than to natural ones. For example, once the natural mechanism for seed dispersal is lost, plants can only rely on man for pollination and reproduction. Though the dramatic crop improvements brought about by conventional plant

446

breeding techniques (Borlaug, 1983; Goodman *et al.*, 1987) are expected to continue in the future, there are strong pressures for further improvements in crop quality and quantity. These pressure are exerted by population growth, housing establishment, highway construction, social demands, health requirements, environmental stresses, and ecological considerations. World population is projected to reach eight billion by the year 2010, an alarming doubling in only 35 years. While we anticipate a continual reduction in agricultural land and other resources, the feeding of an additional two billion of people in the next twelve years calls for a huge increase in crop production. This will certainly require the creation of technology applicable to agriculture on top of continuous improvements of the conventional techniques.

In this chapter, we would like to describe, from the perspective of plant molecular biologists, recent steps taken from the era of the Green Revolution to that of the Gene Revolution. In our view, the present era of the Gene Revolution had emerged in response to the limitations of conventional plant breeding techniques. The Gene Revolution made use of biotechnology-based agricultural science to overcome the barrier of sexual incompatibility which has in the past limited the genetic pool available for crop improvements. The later development of somatic hybrid plants, though transcending this limit (Carlson, 1973; Shepard *et al.*, 1983; Gilmelius *et al.*, 1991), has not yielded results with immediate economic value. However, the Gene Revolution holds the promise of introducing crop improvements at the molecular level, with a power and control previously inconceivable, thereby producing plants with far greater ranges of genetic variability and selectivity.

PLANT BREEDING

Mendel's laws of genetics has provided the scientific basis for plant breeding since 1900. Plant breeding concerns with identifying and selecting a combination of desirable traits for an individual plant. Since all traits are controlled by genes located on chromosomes, plant breeding can be considered as the manipulation of chromosomes. In general, there are four major ways to manipulate plant chromosomes. First is pure-line selection. Similar chromosomes can be sorted out and retained in one plant species or cultivar to reach a homozygous state. Second, different chromosomes can be combined together to

447

obtain a heterozygous state. This method is termed hybridization. Third, new genetic variability can be introduced through spontaneous or artificially induced mutations. Finally, polyploidy also contributes to crop improvement.

Pure-line Selection

Pure-line selection begins with selection for the desired phenotype from the mixture. In genetic terms, a selected and inbred population is more homozygous than its wild relatives. Pure-line selection generally involves three distinct steps. First, selections are made from the genetically variable original population. The number of initial selections should be as great as time, expense, space, and competitive plant breeding projects will permit. Second, progeny rows are grown from the individual plant selections for observational purposes. After obvious elimination, the selections are grown over a shorter or longer period of years to permit observations of performance under different environmental conditions for making further elimination. Finally, when decision between lines can no longer be made based solely on observation, the breeder must turn to replicated traits. The remaining selections are compared with established commercial varieties in relative yielding ability and other aspects of performance. This stage of evaluation takes at least three years. If a large number instead of a single pure-line are likely to be retained, this procedure is usually referred to as mass selection.

One of the most important features of pure-lines is the great precision with which they re-produce themselves. Along with this advantage are potential problems in unmasking some harmful genes as well as eliminating some desirable genes

Hybridization

Hybridization is the most frequently employed plant breeding technique. It was demonstrated that crops such as maize could be inbred for several generations until there was no further reduction in vigor or size. When these highly inbred plants were hybridized with other inbred varieties, very vigorous, large-sized, large-fruited plants were produced. This led to the origin of hybrid maize in 1919 (Chrispeels and Sadava, 1977), the most significant improvement in American agriculture at that time. The term "heterosis" was used to describe this phenomenon of hybrid vigor.

The first step in hybrid production is to generate homozygous inbred lines. This is normally done using self-pollinating plants where pollen from male flowers fertilizes female flowers on the same plant. Once the pure-lines are generated, they are outcrossed. A problem with such hybridization is that the farmers must buy new hybrid seed each year.

Another useful breeding technique is backcrossing. Through backcrossing, it is possible to transfer specific genes from one plant variety to another. In this way, desirable characteristics of one variety can be combined with those of another variety. This trick circumvents the problem of trying to select for many a trait simultaneously in one variety.

Mutations

An alternative to the introduction of genetic variability from the wild species gene pool is the introduction of mutations. Spontaneous mutation or those induced either by chemicals or radiation is applied to both modes of pollination. The mutant obtained is tested and further selected to meet standards of an established cultivar. It may also be used as a donor in a crossing program. Nonetheless, mutation has not been widely used in breeding programs. It is because the great majority of mutants carry undesirable traits.

Polyploidy

Normal plants are diploid. In contrast to animals, plants with three or more complete sets of chromosomes are common, and are referred to as polyploids. The increase of chromosomes sets can be induced by applying colchicine. This chemical disrupts spindle formation during cell division so that daughter chromosome sets remain in the same cell, doubling the chromosome number. Generally, the main effect of the chromosome doubling is size increase.

Morphologically, polyploids tend to be larger than diploids, with thicker leaves, and they may respond differently to environmental conditions. Genetically, the plants may also differ. For example, an autotetraploid with two allelic forms of a gene at a locus (A and a) can have five genotypes in the population (AAAA, AAAa, AAaa, Aaaa, aaaa). This increases its genetic variability. Autopolyploids contain genes similar to their diploid progenitors, whereas allopolyploids combine the gene content of two different species. The latter possesses higher potential capacity for variation.

HYBRID PLANTS

It is quite possible that the first drastically improved crop species was a hybrid plant. Plant breeding was recently described as the selection of plants with desired traits after sexual exchange of genes by cross-fertilization between two parents (Goodman et al., 1987). This clearly indicates that hybridization is an essential technique in plant breeding which gives the resulting hybrid plants. There are, in general, two types of hybrid plants : interspecific and intergeneric hybrids. Beyond this biological boundary, hybridization cannot be accomplished due to sexual incompatibility. The successful breeding of wheat (McFadden, 1930), tomato (Bohn and Tucker, 1939), and soybean (Newell and Hymowitz, 1982) are noted examples of interspecific hybridizations. The origin of many tobacco species was believed to arise from the interspecific hybridizations followed by the doubling of the chromosomes (Gray et al., 1974). On the other hand, some modern crop.species, such as rapeseed and certain wheat strains originated in nature by hybridization between different species or genera (Simmonds, 1976). The hybrid produced between different genera is referred as intergeneric hybrid. The most notable and successful hybrid plant ever produced is the hybrid maize bred more than half a century ago (Borlaug, 1983). Many improved elite hybrids with continually higher yields, increased disease and insect resistance, and shorter and stronger stalks suitable for mechanical harvesting have since been developed.

The first hybrid maize plant was developed after recognition that inbreeding in maize leads to reduced vigor in the following generation and that vigor can be restored by crossing. In 1908 Shull reported that the hybrid plants produced by a cross between two different pure-lines quadrupled the yield, from 20 to 80 bushels per acre (Chrispeels and Sadava, 1977). By 1919, the first commercial hybrid maize was available in the United States, and two decades later, nearly all the maize was hybrid, as it is to this day.

There are two technical steps in the production of hybrid maize plants: the production of desirable homozygous lines and the crossing of these lines. Hybrid maize plants are the result of double crossing. One parental plant possessing the unique property of cytoplasmic male sterility was used to avoid inbreeding. Currently, hybrid maize plants produced in the United States can have yields as high as 130 bushels per acre.

Although interspecific as well as intergeneric hybrid plants can be produced in certain species, there are still severe limitations in obtaining hybrid plants between many genera and even species because of sexual incompatibility. In most cases, sexual incompatibility is genetically controlled and regulated by the pollen-stigma recognition processes (Heslop-Harrison, 1978). This recognition is vitally important in preserving species identity by preventing cross-hybridization between genetically unrelated species. Otherwise, distinctions among plant families, genera, and species would disappear. However, this incompatibility limits our ability to widen genetic variability within a given plant species because it precludes the natural genetic exchange between distinct or unrelated species. Even within similar species, there may be self- or cross-incompatibility

GREEN REVOLUTION

In recent history, the success in applying conventional plant breeding principles and agricultural practices to crop improvement reached its peak when high-yielding wheat and rice varieties were cultivated in the 1960s, with a profound impact on agricultural production (Borlaug, 1983). Before the introduction of high-yielding semi-dwarf varieties of Mexican wheat into India between 1966 and 1968, the country's annual wheat production was 11.39 million metric tons. By 1981, after widespread adoption of the high-yielding varieties, annual wheat production increased to 36.5 million metric tons. This increase in production in 15 years provided sufficient gain to feed 184 million additional people at 375 g of wheat per person per day (Borlaug, 1983). The wheat production gains in Argentina, Bangladesh, China, Pakistan, and Turkey were equally impressive. In China, agricultural production has risen 8% yearly for the ten years from 1982 to 1991. This gives China the distinction of being the largest food producer in the world. For this remarkable accomplishment, He Kang, Minister of Agriculture of the People's Republic of China from 1983 to 1990, was awarded an honorary doctoral degree by the University of Maryland at College Park in 1986, and the "Food Prize" in 1993.

In the developing countries, wheat and rice production increased by about 75% between 1965 and 1980. Individual performance of course varied from country to country. From 1950 to 1984, wheat production in Mexico increased 400% while Indonesia doubled its rice production. Improved seed and agricultural practices have brought about 300-400% increase in the yield of maize, sorghum, and millet in the Sudan, Ghana, Tanzania, and Zambia. The doubling or tripling of rice and wheat production in Asia was termed the Green Revolution to describe the social, economic, and nutritional impact (Chrispeels and Sadava, 1977). Norman Borlaug was awarded the Nobel Peace Prize in 1970 for his remarkable contributions.

The Green Revolution has been praised and at the same time damned. As the use of new varieties spread rapidly in Asia, some experts claimed that many underdeveloped countries would soon be self-sufficient in cereal grains. However, the new varieties were highly responsive and required far more intensive fertilization and irrigation than traditional crops. Consequently, poor weather, high energy costs, and global economic constraints slowed the progress of the green revolution considerably. Critics argued that the Green Revolution's emphasis on intensive agriculture, large farms, and prime crop land damaged the environment and offered little to the poor farmers who should be the intended beneficiaries. The history of the Green Revolution illustrates both the potential and limitations of conventional plant breeding technology.

PROTOPLAST FUSION

Being totipotent, each living plant cell is capable of regenerating into an entire plant identical to the one from which the cell was obtained. The term "cell culture" is used to describe *in vitro* culture of plant cells, cell suspension, and protoplast. In the 1950s, F.C. Stewart and

colleagues pioneered plant cell culture and regeneration techniques based on their discovery that plants could be developed from single cells (Steward *et al.*, 1983). For example, they produced seed-forming plants using specially treated cells that broke away from pieces of cells derived from carrot root phloem.

Perhaps the greatest contributions of cell culture has been its subsequent role in enabling the genetic engineering of plants. Cell and tissue cultures allow the rapid propagation and regeneration of genetically engineered cells, an essential step in producing transgenic plants.

Successful cell culture techniques were soon extended to the production of protoplasts from plant cells by stripping off their walls (Shepard *et al.*, 1983).

Protoplasts are studied as simple cellular entities, like microorganisms. Isolated protoplasts can be used for fusion, DNA uptake, cell wall studies, and other cellular investigations. In the 1960s, Takebe and his co-workers prepared tobacco protoplasts for efficient viral infection experiments. Protoplasts were quickly adapted for regeneration (Nagata and Takebe, 1971). Success in regenerating complete plants from protoplasts eventually led to attempts to combine cells with different genetic backgrounds. In 1972, Carlson *et al.* were the first group to succeed in fusing tobacco protoplasts from two genetically compatible *Nicotiana* species, *N. glauca* and *N. langsdorffii*. The fused products were initially termed parasexual hybrids (Carlson *et al.*, 1972) and later termed somatic hybrid (Shepard *et al.*, 1983).

SOMATIC HYBRID PLANTS

The development of somatic hybrids was a new approach to plant breeding based on the unique property of plant cells. By definition, somatic hybrid plants are hybrids derived from the fusion of somatic cells. Since all kinds of fusion are possible in principle, regardless of the extent of genetic relatedness, this breeding method theoretically offer unlimited possibilities for genetic exchanges.

Historically speaking, the dramatic success in producing somatic hybrids generated great excitement and great expectations in the

1970s. Hopes were high that this technology could permit unlimited generic exchange. Plant biologists speculated about the possibility of producing a somatic hybrid between soybean and maize that would fix nitrogen, contain high protein context, and produce high yield. Furthermore, a somatic hybrid combining tomato, potato, and tobacco could produce an efficient plant with leaves, fruits, and tubers to be harvested and economically utilized. To move toward this goal, many fusions between protoplasts of phylogenetically unrelated species were attempted. Notable examples included fusion between soybean and tobacco, soybean and barley, soybean and maize, soybean and clover, soybean and alfalfa, soybean and rapeseed, carrot and barley, potato and tomato, carrot and parsley, soghum and maize, and even plant and human cells (Lima-De Faria et al., 1977). In most cases, not surprisingly, the attempt failed. No progress could made beyond the simple fusion stage. Following the initial fusion between two protoplasts of different background, division and growth of the fused product is almost impossible.

The success in fusing two sexually compatible *Nicotiana* species led to the fusion between two sexually incompatible petunia species, *Petunia parodii* and *P.* as a conceptual extension of interspecific sexual crosses of incompatible species, a substantial contribution. Beyond this plant breeding technology, success of as a conceptual extension of interspecific sexual crosses of incompatible species, a substantial contribution. Beyond this plant breeding technology, success of as a conceptual extension of interspecific sexual crosses of incompatible species, a substantial contribution. Beyond this plant breeding technology, success of intergeneric protoplast fusion was very limited. Many fertile intergeneric somatic hybrid plants were generated in the Brassicaceae family (Glimelius et al., 1991). They have been used for backcrossing to the cultivated species. Nevertheless, the well-publicized somatic hybrids between potato and tomato deserve special mention.

Potato and tomato are members of the same family (Solanaceae) but different genera. Potato belongs to the genus *Solanum* and tomato to *Lycopersicon* and their somatic chromosome numbers are 48 and 24 respectively (Shepard et al., 1983). They are, of course, not sexually compatible. In 1978, Melchers et al. were the first research group to produce an intergeneric somatic hybrid between potato and tomato. They fused protoplasts between cultured diploid potato line and tomato leaf cells. The resulting hybrid plants, named "pomatoes" (Melchers *et al.*, 1978), displayed morphological features of both parents like most hybrids. Some of them formed "tuber-like stolons", but none set fertile flowers or fruits or produced true tubers (Shepard *et al.*, 1983). Although somatic hybrid plants produced to date, like pomatoes, are not of immediate economic value, they serve as the starting point of a genetic introgression scheme. Moreover, it should be mentioned that the contributions made from these fusion studies on the interaction between the nucleus and organelles (Kung *et al.*, 1975) and on the regeneration and development of cell wall and others are invaluable.

The evidence from the era of hybrid plants to the era of somatic hybrid plants permitted expansion of the scope of plant breeding to both the organismic and cellular levels. Further advances in plant breeding — to the molecular level — are the subject of the next section. It brings us the new era of biotechnology or genetic engineering.

GENETIC ENGINEERING

The discovery of the DNA structure in 1953 triggered off rapid development of DNA related technologies. The advances were especially breathtaking during the 1970s. The discovery of the first restriction enzyme and DNA transformed into E. coli in 1970 was followed by molecular cloning in plasmids and gene transfer into cells (1973), and Southern analysis and DNA sequencing (1975). Ended with the generation of transgenic mice from microinjected embryos in 1980, these advancements established the technical foundation of engineering living organisms at the gene level, and induced the birth of the modern biotechnology. Indeed, by 1976, biotechnology had become a reality when the first expression of a human protein, somatostatin, from recombinant DNA was achieved.

In the world of green plants, the decade of 1970s was also marked with molecular technology developments. Thus, methods and techniques for mRNA and DNA isolation, cloning, and sequencing; for cell culture and transformation; and for plant regeneration from transformed cells were studied, developed, and improved. Though lagging considerably behind the rapidly advancing microbial and animal molecular biology, by the end of 1970s, plant molecular technology nevertheless was sufficiently developed for plant scientists to attempt the transfer of genes into plant cells and the generation of transgenic plants. (Kung & Wu 1993 a & b) The first transgenic plants were achieved in the early 1980s. In 1983, the (-phaseolin gene from French bean (Sun *et al.*,1981) was transferred into sunflower cells and tobacco plants (Murai *et al.*, 1983). In the following year, several transgenic plants harboring foreign genes were generated through the use of *Agrobacterium tumefaciens* vectors (Horsch *et al.*, 1984; De Block *et al.*, 1984). In 1986, the transfer of the luciferase gene from firefly into tobacco received much press attention (Ow *et al.*, 1986).

Plant protoplasts were often used in the early gene transfer experiments until the leaf disk transformation technique (Horsch *et al.*, 1985) offered a substantial simplification to the procedures and enhanced the efficiency of transformation. For gene transfer, there are currently at least five systems available. They are microinjection, electroporation, direct DNA uptake, particle bombardment, and the

Agrobacterium-mediated vector system. Among them, the Agrobacterium procedure is the choice for high efficiency, reproducibility and stability. However, the use of this biotechnological approach on important cereal and legume crops had encountered much difficulty and became a bottleneck in the engineering of these plants. Such a major obstacle for crop improvement was removed when recent advances in transformation technologies permit the generation of transgenic rice, wheat and maize, and to a lesser extend, soybean. Molecular tools for plant gene cloning and for generation of transgenic plants became available in the early 1980s. They allow the transfer of specific genes from diverse sources, including microbes, animals and plants, into any target plant species. Eliminating the limitation of hybridization barrier and sidesteping the need of time-consuming backcross selection in conventional plant breeding, these two major biotechnological capabilities propelled agriculture into the era of gene revolution.

GENE REVOLUTION

From early 1980s to now (1997), the development in agricultural biotechnology can be characterized by four distinct technologies, namely gene addition, gene subtraction, pathway redirection, and quantitative trait loci (QTLs) analysis and

marker-assisted selection (MAS). Gene revolution unfolds through the development and application of these technologies.

Gene Addition

The establishment of transgenic technologies by the end of 1970s generated much enthusiasm to experiment with the transfer of single genes for crop improvement. The approach is straightforward: To identify, transfer, and express a specific gene that would confer a desirable trait in the target transgenic plant. This not only resulted in fuller testing of the new transgenic technologies but also demonstrated the feasibility to generate transgenic plants that harbor desirable traits. Examples of gene addition are many, including plants resistant to insect, herbicide, virus, or other pathogens, and plants with improved product properties.

As estimated by the FAO, about one sixth of world agricultural products is consumed by insect pests. Two main biotechnological strategies have been used to generate insect-resistant plants. The first makes use of a gene from the bacterium Bacillus thuringiensis (Bt) that codes for an insecticidal protein (Gill *et al.*, 1992). The other employs genes for protease inhibitors to restrict the insects from food digestion (Boulter *et al.*, 1989). These genes have been transferred into various crop plants including Bt cotton and Bt corn, and shown to be effective in conferring insect-resistance to the transgenic plants (Brunke and Meeusen, 1991; Gatehouse *et al.*, 1991). Weed infestation causes about 10% loss in world crop production each year. Generation of herbicide-resistant crops will allow post-emergence application of herbicide in weed control. Various approaches used to generate transgenic herbicide-resistant plants using appropriate transgenes include :

i. overproduction of the herbicide (glyphosate)-sensitive enzyme (5-enolpyruvylshikimate-3-phosphate synthase, EPSPS);

ii. expression of mutant enzyme (acetolactate synthase, ALS) nolonger sensitive to the herbicide (sulfonyluea compounds); and iii introduction of enzyme (nitrilase) which can inactivate the herbicide (bromoxynil) (Oxtoby, and Huges, 1990; Quinn, 1990).

Plant viruses cause considerable crop damage and significantly reduce yields. Unfortunately, there is often no effective chemical treatment for virus infection. In 1986, Beachy's group showed that significant resistance to tobacco mosaic virus (TMV) could be achieved by expressing only the TMV coat protein gene in transgenic plants (Beachy et al., 1990). This "coat protein-mediated protection" has since been shown to produce similar results in a variety of transgenic plants against a broad spectrum of plant viruses.

Enhancement of the quality of plant products has also been demonstrated by the approach of gene addition. As plant proteins are generally deficient in certain essential amino acids, they are nutritionally incomplete. To correct for the deficiency of sulfur amino acids in plant proteins, Sun's group isolated from Brazil nut the gene encoding a 2S protein exceptionally rich in the sulfur amino acids (18 mol% methionine, Met, and 8 mol% cysteine, Cys) and transfered this gene into tobacco test plants. Under the regulation of the seed-specific phaseolin promoter, total transgenic tobacco seed protein, resulting in a 30% increase in the Met it was shown that the Brazil nut sulfur-rich protein was expressed as 8% of the content. This ascertained the feasibility to use trasngenic approach to enhance the nutritional quality of plant products (Sun and Larkins, 1993). Using gene addition technology, Penarrubia et al. (1992) attempted to enhance the taste of tomato and lettuce by adding a gene encoding sweet protein monellin. Only the tomato fruits harboring the fruit-ripening specific promoter (E8) gave a faint sweet taste after subjected to ethylene treatment. Witty (1990) reported earlier that potato hairy roots containing the sweet protein thaumatin transgene acquired a new characteristic taste and after taste of thaumatin.

Gene Subtraction

Molecular and transgenic technologies may also reduce or eliminate the product, i.e. mRNA or protein, of a gene in transgenic plants, bringing about a desirable crop performance or trait. This subtraction could be accomplished by using the antisense technology.

For example, in order to alter the ripening of tomatoes, antisense technology was used to suppress the synthesis of ethylene, the hormone involved in the ripening process. As a result, the ripening of the transformed fruits took a longer time and the shelf life of the fruit was extended (Hamilton *et al.*, 1990; Oeller *et al.*, 1991). To alter the softening of tomatoes, antisense technology could be used to suppress polygalacturonase, a cell-wall-degrading enzyme involved in fruit softening. The suppression resulted in tomatoes that could be left on the vine longer than normal ones (Sheehy *et al.*, 1988). Calgene, an agricultural biotechnology company in Davis, California, markets these tomatoes as FlavrSavr tomatoes, which are generally regarded as the first agricultural biotechnology product entering market. Similar strategy had been used to generate longer-life cut flowers. By suppressing the enzyme, 1-aminocyclopropane-1-carboxylic acid (ACC) oxidase, involved in ethylene synthesis, it was shown that the transgenic carnation flowers produced low levels of ethylene and exhibited a marked delay in petal senescence (Mol *et al.*, 1995).

Through antisense technology, gene subtraction has been used to improve other agronomic and horticultural traits including, for example, viral resistance (Cuozzo *et al.*, 1988; Kim *et al.*, 1993); flower color modification (Mol *et al.*, 1995); and pathway redirection, as will be discussed in the following section.

Pathway Redirection
By manipulating the gene(s) controlling the key enzyme(s) in a specific pathway, through gene addition and/or subtraction, a target product can be produced with desirable quantity, composition, and functional properties. In essence, this is a technology that re-directs metabolic pathways.

Carbohydrates. Starch is the major polysaccharide reserves in green plants and the second most abundant polysaccharide, next to cellulose, in nature. Given that starch is used in food, beverage, paper, packaging, and textile industries, developing technologies to improve starch content and property is of nutritional and economic importance. To do so through pathway redirection, the genes involved in the synthesis and determination of the structure and property of starch are to be cloned and analyzed. Starch consists of two types of polymers, the straight chain amylose and the branched chain amylopectin. The biosynthesis of starch in higher plants involves three enzymes: ADP-glucose pyrophosphorylase (ADP-G-pp), starch synthase (SS), and branching enzyme (BC). The genes encoding these enzymes have been cloned from many plant sources.

Starke *et al.* (1992) transferred into potato a mutant E. coli ADP-G-pp gene, glgC16, encoding an enzyme less sensitive to fructose-1,6-bisphosphate — to minimize the interference from allosteric regulation, under the tuber-specific patatin promoter regulation. The tubers of the tansgenic plants were found to contain 35% more starch than the non-transformed tubers, demonstrating that the content of starch can be increased through pathway redirection.

Starch synthase (SS) activity locates both in the stroma of plastids (soluble SS) and in association with the starch granules (granule-bound starch synthase, GBSS). GBSS involves in the synthesis of amylose. The wild-type potato GBSSI cDNA, in antisense orientation and under the regulation of the CaMV promoter, was introduced into potato plants (Visser *et al.*, 1991). Results showed that 70 to 100% of the GBSS activity was inhibited and the inhibition was accompanied by reduced levels of amylose. Therefore, it is clearly feasible to manipulate the composition of starch through redirecting the starch synthesis pathway.

Lipids. Plant lipids are a major component of human diets. The major source of plant lipids is oilseed crops, which accumulate and store lipids in the form of triacylglycerol (TAGs). During lipid biosynthesis, fatty acids (FAs) are the precursors of TAGs. Their structure and composition thus determines the physical and chemical properties, and nutritional value of plant lipids. Chain length and degree of saturation are two main composition / structural features of FAs. The ability to alter them is of importance in lipid nutrition as well as utilization. The enzymes and their genes that play crucial roles in determining the saturation and chain length of FAs are targets for alterations through pathway redirection.

Knutzon *et al.* (1992) used antisense technology to suppress the expression of the stearoyl-ACP desaturase in Brassica napus and found that the suppression caused the accumulation of stearate (18:0) in the transgenic seeds. Some of the antisense engineered seeds contained stearate up to 40% of the total fatty acids with a corresponding decrease in oleate, while the nontransformed seeds contained 1.2% stearate. This study provided the first example of genetically engineering the composition of seed oil. In this instance, the saturation of the fatty acids. High stearate oil can be used for production of margarine and cocoa butter substitutes.

In plants, while the end products of FAs synthesized are usually 16- or 18-carbon (long-chain FAs), some species of higher plants accumulate large amounts of predominantly 8- to 14-carbon medium-chain FAs. This is due to the presence of a medium-chain specific thioesterase. Voelker *et al.* (1992) isolated and transferred the medium-chain 12:0-ACP thioesterase of California bay into Arabidopsis thaliana. Results showed that in the transgenic plants, the 12:0-ACP thioesterase activity was elevated up to 70-fold over the control plants. When the lipid composition of mature trasngenic seeds was determined, laurate (12:0), which was absent from the untransformed plants, accumulated in all the transformants that showed 12:0-ACP thioesterase activity. While seeds of control plants showed near-exclusive accumulation of long-chain FAs, one transgenic plant contained laurate predominantly. These studies demonstrated a mechanism for medium-chain FA synthesis in plants and the feasibility of altering the chain lengths of FAs.

Proteins.
The transfer of a gene into target plant for production of a specific protein product is an obvious process of pathway redirection. In 1988, human interferon was expressed in turnip plants as an attempt to test if the presence of interferon would inhibit viral infection (Dezoeten *et al.*, 1988). It did not, but the experiment demonstrated, by accident, that plants can be directed to produce foreign proteins with pharmaceutical functions. Other examples of directing plants to produce useful proteins are many, such as leu-enkephalin 9human brain opiates), in rapeseed and Arabidopsis (Vandekerckhove *et al.*, 1989); human serum albumin in potato and tobacco (Sijmons *et al.*, 1990); and α-amylase in tobacco (Pen *et al.*, 1992).

Other compounds. Plants could also be redirected to produce other compounds such as anthocyanin, scopolamine, resveratrol, and terpenoids, and polyhydroxybutyrate (Poirier *et al.*, 1992).

Plants as Bioreactors. As the pathways in plants can be redirected to produce diverse arrays of compounds, plants can be exploited as bioreactors, or chemical factories, for production of "tailor-made" compounds of plant or non-plant origin. In comparison with bacteria, yeast, and animal cell culture, the advantages of using plants as bioreactors are multifold. They include low costs; easy to scale up; devoid of animal pathogen (e.g. HIV) contamination; with post-translational modifications; and having similar codon usage as eukaryotic cells. In recent years, production of a variety of lipids, carbohydrates, pharmaceutical polypeptides, industrial enzymes, and other compounds in many different species of plants had been shown (Goddijn and Pen, 1995), demonstrating the potential and promise of using plants as bioreactors.

Quantitative Trait Loci (QTLs) Analysis and Marker-Assisted Selection (MAS)

With the advances of molecular technologies, selections in plant breeding can now be made based on genotype through the use of DNA-based molecular markers linked to the phenotype of interest. The technologies involve the detection of polymorphism through DNA hybridization or PCR-based methods, followed by identifying markers closely linked to traits of interest, either by molecular linkage maps or tagging experiments. The use of these maps to assist the selection for monogenic traits offers saving in cost, labor and time over conventional phenotypic selection.

However, applying molecular markers to select for polygenic traits (quantitative trait loci or QTLs) is not as simple and lacks practical proof thus far. Recently, Tanksley and McCouch (1997) reported that by using molecular linkage maps and a breeding technique, referred to as the advanced backcross QTL method, QTLs in the wild tomato species could be identified and brought into the existing elite tomato lines, resulting in 48, 22, and 33% increase in yield, soluble solid content, and fruit color, respectively, over the original elite variety. When similar method was applied to elite Chinese rice hybrids, two QTLs in the wild species wereidentified and each could increase the yield about 17% over the original hybrid rice. These improvements are substantial when consider normal yearly improvement of these traits through conventional breeding is about 1%.

CURRENT AND FUTURE DEVELOPMENT OF GENE REVOLUTION IN CHINA

Ten years ago, the second author together with Dean Hamer of US National Institute of Health conducted a survey on Biotecnology in China (Hamer and Kung, 1989). The effort of agricultural biotechnology in China represented only a very small portion of the investment. Certain level of success has been achieved in regard to the control of plant viruses which was one of the major push at that time. However, the overall progress was very slow due to the lack of proper training. This caused some of the major contributors even switched field from agriculture to medicine. Such a situation had generated great concern among us who understand and appreciate the needs of agricultural biotechnology in China It is essential that government at all level: devote strong supports to agricultural research including biotechnology. Agricultural biotechnology is part of the effort in improving agriculture but not a substitute of traditional plant breeding which has made and will continue to make invaluable contributions in food production. It is fair to predict that without the development and application of agricultural biotechnology, China would have difficulty in realizing the goals set for Agricultural Modernization and in feeding her growing population by the year of 2030 Therefore, we would like to see that:

• there is a strong and visionary group of leaders in the government who give high regard to agricultural biotechnology.
• there is strong government support in agricultural biotechnology in terms of personnel and budget considerations.
• there is a national long term plan for the development and implementation of agricultural biotechnology at all levels of government; and
• there is a national review board to monitor the expected progress with practical incentive system.

CONCLUDING REMARKS

For the past 50 some years, agricultural production, through conventional plant breeding and contributions from green revolution, was able to keep pace with the world population increase. However, since 1990, the world yields of the three primary grains, wheat, rice and corn, have slowed or even reached a plateau. The world's population is currently growing by about 2% annually while the average yield of the major grains is at 1.1%, amounting to only 50% of the rate of the population growth. World food security becomes a serious and urgent problem.

The advances in agricultural biotechnology in the past 25 years have brought another revolution to agriculture, the gene revolution (Kung & Wu, 1993a & b). This new technology targets crop improvement at the gene level, through gene addition, subtraction, and pathway redirection. The potential of using this technology to generate crops with improved and new traits/products are tremendous and have been demonstrated. The emerging method of QTLs analysis and MAS combines the capabilities of biotechnology and conventional breeding to tackle multigene traits. This is a much welcome and important technology that could advance the gene revolution to a new height in the future. The promise of using plants as bioreactors, on the other hand, will extend and link the agriculture industry with health and other industry sectors. The developments and contributions of gene revolution to world food security are thus very timely and would be substantial.

ACKNOWLEDGEMENT

We would like thank Betty Law and May Lee for their help in preparing this chapter.

REFERENCES

Beachy, R. N., Loesch-Fries, S. and Tumer, N. E. (1990). *Annu. Rev. Phytopathol.* 28: 451-474.
Bohn, G. W. and Tucker, C. M. (1939). *Science* 89: 603-605.
Borlaug, N. E. (1983). *Science* 219: 689-693.
Boulter, D., Gatehouse, A. M. R. and Hilder, V. (1989). *Biotech. Adv.* 7: 489.

Brunke, K. J. and Meeusen, R. L. (1991). *Trends in Biotech.* **9:** 197-200.

Carlson, P. S. (1973*). Proc. Natl. Acad. Sci., USA.* **70:** 598-602.

Carlson, P. S., Smith, H. H., and Dearing, R. D. (1972). *Proc. Natl. Acad. Sci., USA.* **69:** 2292-2294.

Chrispeels, M. J. and Sadava, D. (1977). In *Plants, Food and People,* W. H. Freeman, San Francisco, p. 192.

Cuozzo, M., O'Connell, K. M., Kaniewski, W., Fang, R. X., Chua, N. H. and Tumer, N. E. (1988). *Bio/Tech.* **6:** 549-557.

De Block, M., Herrera-Estrella, L., Van Montagu, M., Schell, J., and Zambryski, P. (1984). *EMBO J.* **31:** 681-686.

Dezoeten, G. A., Penswick, J. R., Horisberger, M. A., Ahl, P., Schultze, M. and Hohn, T. (1989). *Virology* **172:** 213-222.

Gasser, C. S., and Fraley, R. T. (1989). *Science* **244:** 1293-1299.

Gatehouse, J. A., Hilder, V. A. and Gatehouse, A. M. R. (1991). *Plant Biotech.* **1:** 105-135.

Gill, S., Cowles, E. A. and Pietrantino, P. V. (1992). *Ann. Rev. Entomol.* **37:** 615-636.

Glimelius, K., Fahlesson, J., Landgren, M. Sjödin, C., and Sundberg, E. (1991). *Trends in Biotech.* **9:** 24-30.

Goddijn, O. J. M. and Pen, J. (1995). *TIBTECH* **13:** 379-387.

Goodman, R. M., Hauptli, H., Crossway, A., and Knauf, V. C. (1987). *Science* **236:** 48-54.

Gray, J. C., Kung, S. D., and Wildman, S. G. (1974). *Nature* **252:** 226-227.

Hamer, D.H. and Kung, S.D. (1989) *Biotechnology in China.* National Academy Press.

Hamilton, A. J., Lycett, G. W., Grierson, D. (1990). *Nature* **346:** 284-287.

Harlan, J. R. (1976). *Sci. Amer.* 57-65.

Heslop-Harrison, J. (1978). *Cellular Recognition Systems in Plants.* Studies in Biology. London: Edward Arnold.

Horsch, R. B., Fraley, R. T., Rogers, S. G., Sanders, P. R., Lloyd, A., and Hoffman, N. (1984). *Science* **223:** 496-498.

Horsch, R. B., Fraley, R. T., Rogers, S. G., Sanders, P. R., Lloyd, A. and Hoffman, N. (1984). *Science* **227:** 1229.

Horsch, R. B., Frey, J. E., Hoffman, N. L., Eichholtz, D., Rogers, S. G., and Fraley, R. T. (1985). *Science* **227:** 1229-1231.

Kim, J. W., German, T. L. and Sun, S. S. M. (1993). *Plant Disease* **78:** 615-621.

Knutzon, D. A., Thompson, G. A., Radke, S. E., John, W. B., Knauf,

V. C. and Kridl, J. C. (1992). *Proc. Natl. Acad. Sci., USA.* **89:** 2624-2628.

Kung, S. D., Gray, J. C., Wildman, S. G., and Carlson, P. S. (1975). *Science* **187:** 353-355.

Kung, S.D. and Wu, R. ed. (1993a) *Transgenic Plants.* Academic Press, Volume I

Kung, S.D. and Wu, R. ed. (1993b) *Transgenic Plants.* Academic Press, Volume II

Lima-De-Faria, A., Eriksson, T., and Kjellen, L. (1977). *Hereditas* **87:** 57-61.

McFadden, E. S. (1930). *J. Am. Soc. Agron.* **22:** 1050-1051.

Melchers, G., Sacristan, M. D., and Holder, A. A. (1978). *Carlsberg Res. Commen.* **43:** 203-208.

Mol, J. N. M., Holton, T. A. and Koes, R. E. (1995). *TIBTECH* **13:** 350-355.

Murai, N., Sutton, D. W., Murray, M. E., Slighton, J. L., Merlo, D. J., Reichert, N. A., Sengupta-Gopalan, C., Stock, C. A., Barber, R. F., Kemp, J. D., and Hall, T. C. (1983). *Science* **22:** 476-482.

Nagata, T. and Takebe, I. (1971). *Planta* **99:** 12-16.

Newell, C. A. and Hymowitz, R. (1982). *Crop Sci.* **22:** 1062-1066.

Oeller, P. W., Min-Wong, L., Taylor, L. P., Pike, D. A. and Theologis, A. (1991) *Science* **254:** 437-439.

Ow, D. W., Wood, K. U., DeLuca, M., De Wet, J. R., Helinski, D. R., and Howell, S. H. (1986). *Science* **234:** 856-859.

Oxtoby, E. and Huges, M. A. (1990). *Trends in Biotech.* **8:** 61-65.

Pen, J., Molendijk, L., Quax, W. J., Sijmons, P. C., Ooyen, A. J. J. C., Elzen, P. J. M. V. D., Rietveld, K. and Hoekema, A. (1992). *Bio/Tech.* **10:** 292-297.

Penarrubia, L., Kim, R., Giocannoni, J., Kim, S. H. and Fischer, R. L. (1992). *Bio/Tech.* **10:** 561-564.

Poirier, Y., Dennia, D. E., Klomparens, K. and Sommerville, C. S. (1992). *Science* **256:** 520-523.

Power, J. B., Berry, S. F., Champman, J. V., and Cocking, E. C. (1980). *Theor. Appl. Genet.* **57:** 1-6.

Quinn, J. P. (1990). *Biotech. Adv.* **8:** 321-333.

Ratner, M. (1989). *Biotechnology* **7:** 337-341.

Shepard, J. F., Bidney, D., Barsby, T., and Kemble, R. (1983). *Science* **219:** 683-688.

Sijmons, P.C., Dekker, B. M. M., Schrammeijer, B., Verwoerd, T. C., Elzen, P. J. M. V. D. and Hoekema, A. (1990). *Bio/Tech.* **8:** 217-221.

Simmonds, N. W. (1976). *Evaluation of Crop Plants*, N. W. Simmonds, ed., New York, Longman.

Starke, D. M., Timmerman, K. P., Barry, G. F., Preiss, J. and Kishore, G. M. (1992). *Science* **258:** 287-292.

Stewart, F. C. (1983). In *Handbook of Plant Cell Culture*, P. V. Ammirato, D. A. Evans, W. R. Sharp, and Y. Yamada, eds., Vol. 1, 1-12.

Sun, S. M., Slightom, J. S. and Hall, T. C. (1981). *Nature* **289:** 37.

Sun, S. S. M. and Larkins, B. A. (1993) In *Transgenic Plants*, Kung, S. D. and Wu, R., eds., Academic Press, Vol. 1, pp. 339-372.

Tanksley, S. D. and McCouch, S. R. (1997) *Science* **277:** 1063-1066.

Vandekerckhove, J., Van Damme, J., Van Lijsebettens, M., Botterman, J., De Block, M., Vandewiele, M., De Clercq, A., Leemans, J., Van Montegu, M. and Krebbers, E. (1989). *Biotechnology* **7:** 929-932.

Visser, R. G. F., Somborst, I., Kuipers, G. J., Ruys, N. J., Feenstra, W. J. and Jacobsen, E. (1991). *Mol. Gen. Genet.* **225:** 289-296.

Voelker, T. A., Worrell, A. C., Anderson, L., Bleibaum, J., Fan, C., Hawkins, D. J., Radke, S. E. and Davis, H. M. (1992). *Science* **257:** 72-74.

Witty, M. (1990). New Zealand, *J. Crop Hort. Sci.* **18:** 77-80.

CHAPTER 4.14

Soil and Water Resources for Agricultural Development in China

H. H. Cheng
Dept. of Soil, Water, and Climate,
Univ. Mirmesota, St. Paul, MN.

C. Huang
National Soil Erosion Research Lab,
Purdue Univ., W. Lafayette, IN.

The anticipated increase in population in China calls for a significant increase in food production to meet the growing need and demand. By 2030, the population in China is estimated to plateau around 1.6 billion, which is an addition of 400 million people from the present level of 1.2 billion in a span of mere 33 years. Faced with this daunting task, the Chinese government pledged in its Nineth Five-Year Plan to maintain self-sufficiency in food supply in a sustainable manner. The challenge is to decide how to achieve this goal. To gain a perspective on the magnitude of this task, the anticipated population increase in China in the next 33 years is larger than the total population of the United States, which stands at around 280 million at present.

Furthermore, China is undergoing a tremendous change toward rapid industrialization and urban development. These changes also bring added pressure to food production as the demand is to meet not only the need of a growing population, but also the desire for improvement from a grain-based diet to more meat and poultry consumption. To maintain self-sufficiency, how will the country transform from its traditional farming systems to a sustainable production system to meet the increased food demand will be the crux of the challenge. In reality, this challenge is global in perspective. The crisis in food security is not a problem just in

China, but worldwide. Everyone needs food. As nations of the world become increasingly interdependent, and their economy and resource use become more intertwined, it behooves us to examine all aspects of this looming crisis. In a news article published a year ago in the Chemical & Engineering News, Rouchi (1996) reporting on the future directions of the International Rice Research Institute (IRRI) states: "IRRI estimates that rice production must increase almost 70% over the next 30 years to keep up with demand. Such a huge increase over a short period of time can be achieved, IRRI believes, only with new technology. For IRRI scientists, this means biotechnology. IRRI scientists are using biotechnological tools in conjunction with traditional plant breeding methods to seek solutions to the problems of rice farmers and consumers while ensuring that soil and water resources are conserved for future generations." (Rouchi, 1996)

At about the same time this article appeared in print, there was a major conference taking place in China on the development of technology for Chinese agriculture for the twenty first century, held in Beijing in September 1996. Again, much importance was placed on the need to develop biotechnology to increase agricultural production potential (Sino-American Symposium Editorial Committee, 1996).

The thrust of both the report on IRRI and the conference in China is important in two aspects. One is that it points to the importance of the current activities in biotechnology toward solving the potential world crisis in food security predicted to be forthcoming in the coming century. Major efforts are devoted to the application of molecular technology and genetic engineering to increase the yield potential of crops, with some attention to the quality, especially of the rice crop. There is no question that biotechnology will continue to be an essential component of this effort. While the biotechnology thrust is recognized as a critical component of the strategy for increasing food production, one must also be aware of the limitations and constraints of this approach as well as its potential. Factors other than improved genetics are also involved in crop production.

This brings to the second aspect of the C&EN report. That is what is NOT stated in this article. Little attention has been given to the potential or the constraints of the soil or land and water resources needed to supply the required input to achieve the potential yield. Although the quote from the C&EN article above did point out at the end the need to ensure "that soil and water resources are conserved for future generations", it is a woefully neglected topic in this report as well as a neglected aspect in the current discussion. The question is not merely how to prolong the use of our soil and water resources so that "they are conserved for future generations", as this is still a production or output-oriented approach. What is required is a fundamental shift from output-oriented to input-oriented approach in establishing criteria for decision-making on increasing food production. The basic question to address is whether the production goals should be to maximize production at whatever cost, or they should be balanced by the costs of resource inputs and environmental consequences. This also applies to the current debate in China on the true nature of sustainable agriculture, and how sustainability is measured.

Perhaps, neither "intensive sustainable agriculture" (ISA) nor "sustainable intensive griculture" (SIA) is on target. ISA has been criticized because it is conventional agriculture in disguise. However, the emphasis of SIA is still on production. The basic question that needs addressing is: is "Should we attempt to sustain production, or should we maintain sustainability in resource use?" We submit that the latter should be our goal.

THE CONCERN

Agricultural production has always depended on the use of soil and water resources. China has 21.8% of the world's population, but only 6.80/0 of the world arable land, although recent estimates have increased the amount of arable land in China somewhat. China should be proud of the efficiency of its agriculture, but should also recognize the constraints imposed by the limited amount of arable land available for farming. As the availability of these already limited natural resources is further diminished, due to quantitative loss and/or qualitative degradation, the cost for sustaining production will greatly increase as external resources will be required to make up the loss of natural resources. The impact of rapid industrialization and growth in market economy has drastically transformed China from its traditional farming systems into a more intensive agriculture with higher demand for input. The increase in input is either by adding more fertilizers and agricultural chemicals or by exploiting more of the soil and water resources which are degraded or depleted at an increasingly rapid rate. It is well known that soil and water degradation is an accelerated process; once the process starts, it becomes progressively worse and the cost of reclaiming these degraded resources would become much higher than the marginal benefits which would be derived from their exploitation. To maintain the long-term viability of agriculture will depend upon sustainable use and management of the soil and water resources. Governmental policies cannot solely be based on increasing production to reach certain output goals regardless the cost for input or consequences of the effort on depletion of resources and deterioration of environmental quality.

Most of China's potential agricultural land resources are located in arid or semi-arid regions and the lack of reliable water supply for crop production has been the most significant factor in limiting the agricultural production. Making water available to these drylands, through major water conveyance projects and capital investment in irrigation schemes, is being implemented to improve the agricultural output (Chinese Government, 1996) The concern is not on the immediate benefit of irrigation in increasing the crop yield, but on the long-term effects of soil salinity from irrigation in arid regions. There is sufficient evidence on irrigation-induced soil degradation in northern and western China and throughout the world to warrant the balancing between short-term benefits in increased crop yield and long-term loss of land production potential.

Concern for soil and water resources must be examined from both a quantitative and a qualitative point of view. In quantitative terms, to feed more people will require more land to produce food, or produce more food on the same piece of land. At the same time, more land and water are needed for residential development, industry, and transportation use. Although it may be evident to agronomists that not all lands are equal in quality and productivity, we should be gravely concerned of policy makers for permitting prime agricultural land near the cities to be used for housing, roads, and factories, while advocating the conversion of waste lands into cropland. Villages and cities have traditionally been built in areas of highly productive agricultural lands. The so-called waste lands are wastes because they are ecologically fragile and unproductive. There is a huge price to pay in giving up good land for poor land. Land use policies must encourage houses, roads, and factories be built on agriculturally unproductive land while productive land is preserved for agriculture.

Current policy tends to treat all lands on an equal basis, without differentiating the local and regional constraints which could impose on the productivity of the land. Even on a macro-scale, it is commonly recognized that the North China Plains is a water-deficit region, while the Yangtze River Valley has an over-abundance of water. The infertility and increasing salinity of the soils of the North China Plains, the erodability of the Loess Plateau in the arid and semi-arid northwest, and the fragile and impoverished state of the Red Soils in southern China are obvious examples of the increasingly deteriorated conditions of the agricultural landscape.

EVIDENCE OF DEGRADATION

Evidence is abound in the literature showing the examples of land degradation in China. For instance, a report from the Ministry of Water Resources (1989) states:

"Soil erosion occurs in most of the land of China. In 1950, the area suffering from soil erosion in China was more than 1.5 million square kilometers, mainly covering the Loess Plateau in the Northwest, hilly areas in the South, rock-hilly areas in the North, and the black earth area in the Northeast. Wind erosion affected more than 0.3 million square kilometers, mainly in Northeast, North, and Northwest China. "

In describing how serious the erosion problem is in the Loess Plateau region, the report states: "Some 26% of the total area has an erosion modulus greater than 5,000 t.km^{-2} with some areas as high as 30,000 to 50,000 t.km^{-2} . The sediment concentration of flood water of Yellow River is 37.5 kg.m^{-3}, and annual sediment load of Yellow River is 1.6 billion tons, of which 75% (or 1.2 billion t) was carried into the sea."

Soil erosion does great damages to soil and water resources and causes harm to the environment. Top soil washed away from the land accumulates in rivers, lakes, and reservoirs not only causes loss of resources for agricultural production on land, but also leads to deterioration of aquatic environment, causing problems in flood control, irrigation, water supply, navigation, and electric power supplies.

While soil erosion is the most visible loss of valuable soil resources. Other less visible quality deterioration losses are no less devastating. These include: loss of fertility, loss of ability to hold water, salinization, alkalinization, acidification, waterlogging, desertification, loss of water permeability, and many interrelated properties. These losses in soil quality usually occur gradually and may not be recognized readily. They may be masked by the continuing increase in inputs, such as fertilizers and soil amendrnents, to make up for the loss of soil's own inherent capacity and quality. Irrigation can also be a source of masked soil degradation in arid regions if it is not properly managed for its potential impacts on soil salinity and soil quality.

Soil and water degradation may be related to a number of factors. For instance, an increase in demand for food production would result in increased input of agricultural chemicals and their residues, which can lead to chemical imbalance or degradation of the environment and pollution of the soil and water resources. Decline in soil organic matter contents has resulted in degradation of soil structure, loss of nutrient supply capability, decreased water infiltration and water holding capacity, and decrease of soil biological diversity. Increased industrialization and urbanization have led to increased competition for land and water resources and increased soil pollution caused by random dumping of industrial waste materials such as heavy metals, organic solvent, and untreated raw sewage.

CONSTRAINTS

In addition to the physical causes of soil and water degradation mentioned above, there
are other threats to the proper management of the soil and water resources in China. Because of the current institutional system in China, delays or inaction by resource managers ofien occur when timely measures to prevent resource degradation is needed at critical times. For instance, many agencies are working on similar problems. Yet there is a lack of central coordination to enhance collaboration among the agencies, leading to wasteflil overlapping of efforts among different branches of governments and institutions. Same research projects may be done by several institutions with little communication among them. With funding for research so limited, such efforts should be better coordinated.

There appears in existence any number of database on various aspects of soil and water resources in China. A number of soil and water resource assessments at the national scale have been reported and other similar studies are probably still being conducted. These large-scale resource inventories and assessments have limited practical value at the field scale where conservation practices need to be implemented. Many of these national assessments have no relevant information that can be used at the local scales. Even when various agencies have relevant information which could contribute to an understanding of various processes in the agroecosystems, they are seldom willing to share the data in their possession, thus rendering the usefulness of all of the data limited.

Many policy decisions on land resources did not have a sound scientific basis, and research findings were either not used or ignored. For instance, land reclamation efforts in the Loess Plateau region were often treated as construction projects at huge expenses. Large machineries were used to build terraces for every piece of land in the entire watershed with total disregard of the cost or the benefit in maintaining the natural soil and water resources. Meanwhile, research efforts have already led to the development of comprehensive soil and water management schemes that are sustainable and diverse, and at very little cost. One such scheme for small watersheds of 5-10 km^2 with populations from of 300 to 800 includes a diversity of agricultural enterprises involving pastures, woodlands, and orchards for animal grazing and for production of fuel woods, fruit trees, vegetables, and grains. Under this scheme, steps were made to build productive croplands, restore vegetative cover, and develop higher return fruit orchards and animal husbandry (Zhu and Ren, 1997). However, such an eco-agricultural conservation and production system would take 5 to 10 years to implement.

This gradual, adoptive procedure may be more beneficial to sustain agricultural production but it would not be as dramatic as the appearance of an expensive engineering construction project.

Casual visitors to China have often been impressed by Chinese soil and water conservation demonstration projects. Although the intention of research and demonstration project is to promote their adaptation, often times the amount of capital investment required to implement such measures is beyond the means of the resource-poor farmers in the region. Therefore, it is often not the lack of knowledge and technology for a sustainable soil and water conservation system, instead, it is the lack of resources to implement such a conservation system.

Another factor that may threaten China's soil and water resources is the promotion of rural-based industry or TVE (Town and Village Enterprises) which has been praised for its impact on raising the rural economy and keeping the farm labor moving to urban areas.

Comparing to large industries, these small enterprises tend to have less resources, knowledge and education, and investment capitals and equipment, in dealing with waste materials. Enforcing environmental regulations on these small industries will be a difficult task. But, if not monitored and regulated properly, these rural industries can quickly become a source of soil and water pollution to the most productive farmlands in the country.

APPROACHES

A comprehensive database on China's soil and water resources will be essential for providing valid information basis in the decision making process. This database, in conjunction with a framework of resource management models, needs to be established at a scale in which land users and resource managers can use to evaluate the consequence of different soil and cropping management scenarios and select the best management practice (BMP). These resource management models are not those developed for large scale resource inventories at the regional or national scales by research institutes. Instead, these will be a set of readily useable tools developed for local conditions. These tools allow assessments of soil erosion, soil and water quality, and crop growth under a corninon interface and database to enable users to optimize management over spatially-variable fields. Steps to establish the soil and water resources management tools are listed below.

Identify Regional Priorities:
Identify and prioritize the needs of specific tools for different regions of the country where site specific degradation problems pose most threat to both the environment and agricultural production. For example, soil erosion is the most significant threat to soil and water resources in the Loess Plateau, therefore, an erosion prediction model would be the highest priority tor this region. In irrigated arid and semi-arid area in the northern and western regions, soil water management and salinity models are needed to properly manage the limited water resources. Near urban areas where agricultural, industrial, and domestic demands on soil and water resources are highest, soil and water management models for maintaining the environmental quality are needed.

Assess Suitable Tools and Knowledge Bases:
There are many soil and water resource management tools, each with its

own strengths, limitations, and data requirement. Technology and information tools are emerging at a very rapid pace. A collaborative effort among policy makers, research scientists, and extension educators is needed to assess China's technological needs and develop a delivery system for distributing appropriate information and education to the users.

A usable tool needs to be developed and tested for specific local conditions and with appropriate databases for its parameters in proper scales. Many Chinese scientists have already been using the concepts and frameworks used in these tools, and they recognize the necessity of having a proper database to run these models. Some newer technologies also have more parameters and require larger databases. These databases are not necessarily have to be taken from actual data, judgments from experts can be used as inputs. An example is the US effort in developing erosion prediction models such as the Universal Soil Loss Equation (USLE) and Revised Universal Soil Loss Equation (RUSLE). Many of the USLE and RUSLE parameters were derived by a panel of experts who were able to make judgments on a particular soil and crop from their knowledge and experience. With the input of experts to build the required databases, these two models are now used all over the US. Following the US effort, many countries and regions have adapted their own USLE using local knowledge and experts. It is unlikely for a natural resource model developed elsewhere to be applicable to China directly. It will require Chinese scientists to make these models work for themselves. Some of these technologies may have already been developed by Chinese scientists and published in scientific journals. These research results need to be assimilated into useful tools for general applications.

Recent development in precision agriculture is another prime example of efficient use of resources using available technology. The objective of precision agriculture is to optimize the use of soil and water resources and external inputs such as fertilizers and pesticides on a site-specific basis. Technological advances in variable rate application techniques, digitization of field maps, global positioning systems (GPS), geographical information systems (GIS), and grain yield monitoring have enabled characterization of field variability in crop yields. These new tools have now shown that many crop fields are highly variable, thereby reducing overall field yield and resource use efficiency. However, once one has a measure of the spatial variability, optimizing management requires an understanding of what causes the variability, and a method to determine optimal management over the field. Crop growth models are needed to help determine the pattern of field management that optimizes production or profit. The effective use of these tools requires their evaluation in fields to be optimized, their integration with other information tools such as GIS, geostatistics, remote sensing, and optimization analysis.

Establish a Cooperative Community Beyond Institutional Boundaries:
Under the current government structure in China, there are many overlapping institutions working on the development and delivery of technologies for managing soil and water resources. With the current mechanism that individual researchers have to secure their own research funds, scientists at different institutions become competitors instead of cooperators. Information becomes difficult, and technology delivery to user is hindered. This institutional boundary needs to be dissolved and a collaborative effort to build a cooperative atmosphere among soil and water scientists is urgently needed. In addition to the boundaries within thetechnology development community, i.e., research institutions and universities, the linkage to different levels of government which are responsible for the implementation of soil and water conservation plans needs to be strengthened.

Raise Environmental Awareness Through Education and Extension:
Concepts of soil and water resource conservation and environmental awareness need to be strengthened to the general public through education and extension efforts. A strong extension system will facilitate the technology delivery to the users. The general public needs to be educated that soil and water resource conservation is not just the government's responsibility. Every user of these limited resources is responsible for maintaining the quality of these resources.

Establish Environmental Monitoring Systems:
Certain agricultural practices may have short-term benefits but lead to adverse effects in a long-term basis. Examples of these practices and ill effects are: irrigation and soil salinization and use of agro-chemicals and water pollution. To meet the increasing demand in food production, there may be few choice but to use some of these management practices before better alternatives are developed. In the meantime, the long-term effects of these newer or better management practices might not be known until

sufficient time has been lapsed Under this circumstance, it is essential to have a monitoring scheme to establish the database for assessing the impact of present practice and to detect any signs of resource degradation so that corrective measures can be instituted before it is too late. The rapid expansion of rural enterprises (TVB) makes it more critical to monitor their impacts on soil and water resources.

CONCLUDING REMARKS

Rather than presenting a comprehensive inventory of Chinese soil and water resources for agricultural production in the 21st century, this paper is focused on technologies and issues that may lead to a better management of the soil and water resources for agricultural production. Physical processes causing soil and water degradation have been well documented in the scientific literature. However, institutional issues are less mentioned and need to be corrected. It is probably not the lack of knowledge base that would threaten the Chinese soil and water resources in the next century, it is the lack of technology delivery system that would raise the consciousness an responsibility of resource users. Only coordinated effort with valid scientific basis for policy decisions can China manage its soil and water resources to meet the demand in food production for the 21st century.

LIST OF REFERENCES

Chinese Government. 1996. The Grain Issue in China. Whitepaper issued on Oct 26, 1996.

Ministry of Water Resources. 1989. Terraces in China. Jilin Science and Technology Publishing House, Changchun, China. 117 p.

Rouchi, A. M. 1996. Biotechnology steps up pace of rice research. Chem. Engin. News 74(41): 10-14.7 October issue.

Sino-American Symposium Editorial Committee. 1996. Proceedings of Sino-American Symposium on Agricultural Research and Development in China. China Agriculture Publishers, Beijing. 339 p.

Zhu, X. and M. ken. 1997. Loess Plateau of China: Formation, Soil Erosion and Conservation. (Keynote speech delivered by Guobin Liu during the SEDF97 Symposium).

CHAPTER 4.15

Agriculture in China: Water and Energy Resources

Wen Dazhong

Institute of Applied Ecology Chinese Academy of Sciences
Shenyang 110015, China

David Pimentel

College of Agriculture and Life Sciences
Cornell University Ithaca, New York 14853-0901

A healthy and soundly managed environment is a major benefit to humans and their agriculture in China and in all nations. All life on earth obtains its food and other necessities from this environment.

Humans and indeed all living organisms rely on the basic resources found in the earth's environment for their very survival. Paramount among these resources are fertile land, fresh water, and energy. Already the human population is utilizing more than 50% of the sun's energy captured by all the plant biomass on earth each year (Pimentel et al., 1997a). Humans and other organisms obtain all their food and fiber from this photosynthetic activity.

Currently the world population is about 6 million and is projected to reach 10 billion by 2040 based on the present growth rate of 1.5% per year (PRB, 1996). This excessive number should signal a call for action concerning resource use and management and population limitations. Erosion of farmland, overuse of both surface and ground water, dwindling supplies of finite fossil fuels, and the escalating extinction of plant and animal species threaten the ability of the earth's resources to meet the needs of humans of such enormous numbers. China has 1.2 billion people on a land area equivalent to that of the United States and is facing natural resource shortages especially in land and water (Wen and Pimentel, 1992).

The food situation in China and worldwide is becoming critical. At present, more than two billion humans are malnourished and experience unhealthy living conditions (Nesheim, 1993; McMichael, 1993; Maberly, 1994; Bouis, 1995; WHO, 1995). The number of humans who are diseased is the largest number ever. About 40,000 children die each day from disease and malnutrition (Kutzner, 1991; Tribe, 1994).

Many problems that are now evident emphasize the urgent need to reassess the status of environmental resources in China for agricultural production. Based on the evidence, definite plans must be developed to improve environmental management now and for the future in China.

Human behavior demonstrates a strong will to survive, to reproduce, and to achieve some level of prosperity and quality of life. However, individuals and societies differ in their views of what they consider a satisfactory lifestyle. Contrasting some aspects of life in China, the United States, and the world reveals many disparities in lifestyles which most often are functions of the quantity of natural resources available per person. Furthermore, most of these basic resources (land, water, energy, and biota) are not unlimited in their supplies and many are finite. As human populations continue to expand, prosperity and the quality of life can be expected to decline because resources must be divided among ever more people (UNFPA, 1991; RS and NAS, 1992).

Recent decades have witnessed a great expansion of populations throughout the world. The U.S. population has doubled from 130 million to more than 260 million during the past 50 years (NGS, 1995) and is projected to double again to 520 million during the next 60 years based on the current rate of growth of 1.1% per year (USBC, 1996). In contrast, China currently has a population of 1.2 billion, and despite the governmental policy of permitting only one child per couple, its population is also growing at a rate of 1.1% per year (PRB, 1996). China's desirable population is 650 million, or about one-half of its current level (Qu and Li, 1992). It is projected that China's population will grow to 1.6 billion by the year 2030 (Shi, 1996; Shen and He, 1996). With agricultural resources already stressed, a population of 1.6

billion will further stress the basic resources utilized in food production.

Water and fossil energy resources are the most important natural resources for modern agricultural development. In this paper, the current situations, problems, and trends concerning the use of these resources in China's agriculture are discussed. Some suggestions for improving water and energy use in China are also provided.

Water Resources and Agriculture in China

Fresh water resources

The total amount of fresh water in China is 2800 billion m^3, the sixth largest amount in the world. However, the average quantity of water per capita in China is 2400 m^3 per year, only one-fourth of the world average and the 110th largest amount in the world. China is considered to be one of the world's most water-deficient countries (Jin and Guo, 1996; Jiao et al., 1996).

China has a vast territory and a variety of climatic conditions. The distribution of water resources in China is very uneven because the distribution of precipitation in China decreases progressively from southeast to northwest. About 80% of the country's total water resources are concentrated in southern China even though this part of the country accounts for only 36% of China's total cropland area. Northern China has more cropland area but fewer water resources (Jin and Zhang, 1996). The water shortage is very serious in northern China (Table 4.15.1).

Water use in agriculture

The Chinese government has paid great attention to agricultural irrigation and considers it its most important agricultural capital expenditure. The investment in irrigation accounts for two-thirds of the total investment of agricultural capital in China since 1949. Consequently, irrigated cropland areas have increased rapidly from 15.93 million ha in 1949 to 44.87 million ha in 1980, 47.40 million ha in 1990, and 50.38 million ha in 1993. The percentage of irrigated cropland area has also increased from 18.5% in 1951 to 51.2% in 1993. Currently the irrigated cropland areas produce two-thirds of China's total grain production, 60% of its total cash crop production, and 80% of its total vegetable production each year (Lu and Wang, 1997; Jiang, 1996).

The total water use in China in 1993 was 525.4 billion m^3 and the total amount of water used for irrigation was 385.1 billion m^3, or about 73% of the total water used in China. The total water used for paddy rice irrigation was 215.2 billion m³ accounting for 55.9% of the total irrigation water used in China (Shen and He, 1996). In the most recent decade, the water use for industry and in cities has increased rapidly. Agricultural water use as a percentage of total national water use has decreased. For example, the total irrigation water used accounted for 85.2% of the total amount of water used nationwide in 1980, but only 73.3% in 1993. Total irrigation water use in China has not increased since 1980, but the total irrigated area has increased by about 14% (Shen and He, 1996).

Water use per capita in China averages 0.46 million liters (Table 4.15.2). This water use is less than one-tenth that of the United States which averages 5.1 million liters per person per year (Table 4.15.2). As in China, most of the water in the United States is used for irrigation.

Serious water shortages

Table 4.15.1. The comparison between water resources, cropland, and populations in major river basins in China. The river basin order in the table is from northern China to southern China. (Based on Lou, 1996)

River basin	Annual runoff (%)	Crop land (%)	Average runoff per ha of crop\land (m³)	Popula-tion (%)	Runoff per capita (m³)
National	100	100	28,335	100	2,457
Shunghua-jiang	2.81	11.2	7,140	4.6	476
Liaohe	0.55	4.7	3,315	2.9	221
Haihe and Lanhe	1.06	11.8	2,550	10.1	258
Yellow	2.34	12.8	5,385	8.5	701
Huaihe	2.29	12.8	5,070	13.0	433
Yangzhi	35.08	24.5	40,620	34.9	2,466
Zhujiang	12.31	4.9	71,070	7.5	4,738

China is facing a serious water problem. With 1.2 billion people, China now has about 22% of the world's population but only 8% of the world's water (Postel, 1992). At present, there are 300 cities that have water shortages

(Postel, 1996) and 110 cities that have very serious water shortages. The total water shortage in these cities is 5.8 billion m³ (Jiao et al., 1996), and they are located in northern, northeastern and coastal areas of China. In the Beijing area, the water table is dropping 1 to 2 m per year, and about 33% of the wells in the area are dry (Gleick, 1993). About $15 billion in industrial production output value is lost each year in China because of serious water shortages (Jiao et al. 1996).

About 70 million rural people in China do not have enough drinking water (Jin and Guo, 1996). In the past two decades, the surface runoff coefficients in the northern China plain have decreased 50-70% (Shen and He, 1996). The Yellow River provides water to several cities, especially for extensive agriculture, but now seldom empties into the Bonhai Sea and, in fact, has been dry at the mouth for 70-90 days in the past ten years (Postel, 1996; Tian et al., 1997). It was dry at the mouth for 119 days in 1995 (Tian et al., 1997). The ground water tables have been reduced continuously in some well irrigation areas in the provinces of Hebei, Shanxi, Henan and Inner Mongolia from 1-2 m deep in the early years of irrigation to 20-30 m deep in recent years. The people in these areas have had to introduce more powerful pumping equipment (Shen and He, 1996). Some irrigation areas do not have sufficient water for irrigation (Jin and Zhang, 1996). About 27 million ha of cropland in China suffer from drought each year which accounts for 18% of the total cropland (An, 1995).

Table 4.15.2. Resources used per capita per year in the United States, China, and the world to supply basic needs. (Based on Pimentel et al., 1997c)

Resources	U.S.	China	World
Land			
cropland (ha)	0.71	0.08	0.27
pasture (ha)	0.91	0.33	0.57
forest (ha)	1.00	0.11	0.75
Total (ha)	3.49	0.52	1.59
Water (liters x 10⁶)	5.1	0.46	0.64
Fossil fuel			
oil equivalents (liters)	8,740	700	1 500
Forest products (kg)	1 091	40	70

Water quality problems
Water quality in China has deteriorated because of the large amount of wastes released from cities and industrial regions. Agricultural chemical pollutants and serious soil erosion also contribute to water pollution.

The total waste water released from cities and industrial regions in China in 1993 was 63.6 billion m^3, most of which was untreated and used for agricultural irrigation (Shen and He, 1996). About 50% of the total river length in China is polluted and about 10% is seriously polluted (Jin and Zhang, 1996; Jin and Gou, 1996). More than 70% of the river length in the basins of Taihu Lake, Huaihe River, and Yellow River are polluted, and 565 km of the Yangzi River are polluted (Jin and Gou, 1996). The water quality in about 50% of China's cities does not meet minimum drinking water standards. The underground water in about half of the cities of northwestern China is seriously polluted. About two-thirds of the lakes in China suffer some level of eutrophication while about 10% of these lakes suffer serious eutrophication (Jin and Gou, 1996).

It is estimated that about 10 million ha of cropland are polluted by waste water resulting in the loss of about 12 billion kg of grains each year. About 0.3 million ha of fresh water fishing areas are polluted, causing the loss of 45.5 million kg of fish in 1992 in China (Jin and Gou, 1996). It is estimated that the total economic loss by water pollution in China is $3-3.5 billion each year which accounts for 1-2% of China's total national production (Jin and Gou, 1996).

China's vast mountain plateau areas are suffering serious soil erosion. The total eroded area in China is about 150 million ha including more than 35 million ha of cultivated land which accounts for 27% of China's total cultivated land area. The total soil loss in China is more than 5500 million tons each year, or about 20% of the total soil loss (Wen, 1993). These sediments cause the light tan color of the Yellow River. Sediments are a major problem not only in the Yellow River but also in other rivers (Wen, 1993; Jin and Zhang, 1996). For example, in the Qingtongxia Reservoir on the Yellow River, sediments have filled 84% of the reservoir's capacity just during the past five years (Gleick, 1993). This is the worst example, but other reservoirs are also being filled up relatively rapidly.

Crop production and water consumption

Different crops and regions vary in their water requirements. Rainfall patterns, temperature, soil quality, and vegetative cover all influence soil moisture levels. For ideal growing conditions, soil moisture should not fall below 50% in the root zone (Blackshaw, 1990) but for some crops like rice more than 50% is needed for full yields (Bhuiyan, 1992). Good vegetative cover, high levels of soil organic matter, active soil biota, and slow water runoff increase the percolation of rainfall into the soil for use by growing crops.

The transfer of water to the atmosphere from the terrestrial environment by transpiration through vegetation is estimated to range between 38% and 65% of precipitation depending on the terrestrial ecosystem (Schlesinger, 1991; T. Dawson, personal communication, 1995). The process of carbon dioxide fixation and temperature control require plants to transpire enormous amounts of water. For example, a squash plant transpires 10 times its fresh weight in water per day, and many deciduous trees transpire 2 to 6 times their fresh weight per day (T. Dawson, personal communication, 1995). The water required to grow various food and forage crops ranges from 500 to 2000 liters of water per kilogram of yield produced (Table 4.15.3). For instance, a ha of corn transpires about 4 million liters (4000 m^3/ha) of water during its growing season. At the same time an additional 2 million liters/ha evaporate from the soil (Donahue et al., 1990). Thus, about 6 million liters /ha of rainfall are needed during the growing season for corn production. Even with 800-1000 mm (8 to 10 million liters/ha) of annual rainfall in the U.S. corn belt region, corn usually suffers from a lack of water at some point during the summer growing season (Troeh and Thompson, 1993).

Table 4.15.3. Estimated liters of water required to produce
1 kg of food and forage crops. (Based on Pimentel et al., 1997b)

Crop	Liters/kg
Potatoes	500
Wheat	900
Alfalfa	900
Sorghum	1,110
Corn	1,400
Rice	1,912
Soybeans	2,000
Broiler chicken	3,500
Beef	100,000

High yielding rice requires much more water than does corn. Rice requires from 10 to 18 million liters/ha of water for production (Pimentel et al., 1997b). Up to 50% more rice per ha is produced under flooded conditions than under sprinkler irrigation. Soybeans (2.3 t/ha) and wheat (2.7 t/ha) produce less biomass and total yields than corn (7.6 t/ha) or rice (6.2 t/ha) on average (USDA, 1993). Soybeans are highly consumptive of water, requiring about 4.6 million liters/ha of transpiration. Wheat requires only about 2.4 million liters/ha.

Sorghum and millet productions require only 250 to 300 mm (2.5 to 3 million liters/ha) of annual rainfall (Gleick, 1993), and some cereal production can take place with annual rainfall levels as low as 200 to 250 mm (Rees et al., 1990). Under these relatively arid conditions, crop yields are low (1 to 2.5 t/ha) even with adequate amounts of fertilizers (USDA, 1993).

Chinese agricultural production is projected to expand because of increased food needs and the increase in population numbers. The projected 30% increase in crop and livestock production during the next two decades will significantly stress the water resources system in China. Increasing crop yields imply a parallel increase in fresh water consumption in agriculture.

Improving Agricultural Water Use

The simultaneous existence of water shortages and waste is an important characteristic of the water crisis in China. There is sufficient water to meet increased irrigation needs, but the increased water needs in cities and industrial regions reduce the irrigation water supply. However, agricultural irrigation causes water to be wasted so great potential exists to save water in agriculture. Developing water-saving agricultural techniques and increasing water use efficiency should be the basic approach to improving agricultural water use in China.

Some current irrigation practices waste large amounts of water. Most farmers use flooding or channeling methods to irrigate their crops; thus, irrigation efficiency— the amount of water reaching the crop— is estimated to be less than 40% worldwide. Large amounts of water are lost through pumping and transportation (Postel, 1992). In the U.S. less than 50% of irrigation water actually reaches the crops (van der Leeden et al., 1990). In China about 50% of irrigation water is lost through transport systems (Xu, 1992). Preventing seepage loss in canals is an important approach to reducing water transport loss. Recently, leek-proof irrigation canals have increased water use efficiency from 50% to 70% (Liu and Wang, 1995). Using a low-pressure tube irrigation system could also increase water transport efficiency to 90%. Currently about 2 million ha of cropland areas are using the tube irrigation system in China (Liu and Wang, 1995).

Although improving irrigation efficiency is difficult, conservation technologies can improve irrigation and result in reduced loss of irrigation water for crop production. For example some farmers are turning to "surge flow" irrigation to replace the traditional flooding and channeling irrigation systems (Verplaneke et al., 1992). This involves an automated gated-pipe irrigation system that uses a microprocessor control instead of releasing water in a continuous slow stream in field channels. Using this method, farmers in Texas have been able to reduce water pumping 38-56% compared with continuous flood irrigation of the same area (Sweeten and Jordan, 1987). Some experiments in northern China show that "surge flow" irrigation can save 10-40% of irrigation water (Liu and Wang, 1995).

Another strategy is irrigating at night to reduce evaporation. This technique improves irrigation efficiency by 2 to 3 times (Dubenok and Nesvat, 1992). Evidence suggests that the use of low-pressure sprinklers also may improve water efficiency by 60 to 70% compared to high-pressure sprinklers (Verplaneke et al., 1992). Avoiding overhead watering can reduce evaporation and water needs by 45% (O'Keefe, 1992). Low-Energy Precision Application (LEPA) is another technique for conserving water. LEPA sprinklers deliver water to the crop by drop tubes that extend down from a sprinkler arm (Sweeten and Jordan, 1987). The water application efficiency of the LEPA system ranges from 88-99% (Sweeten and Jordan, 1987). Combined ridge-tillage and LEPA can significantly increase irrigation efficiency (Lal, 1994).

The "drip" or "micro-irrigation" technique developed in the 1960s has spread rapidly worldwide, especially in Israel, Australia, New Zealand, and some regions of the U.S. Drip irrigation delivers water to each individual plant by plastic tubes. This method uses from 30-50% less water than surface irrigation (van Tuijl, 1993). Although drip systems achieve up to 95% efficiency, they are expensive and energy intensive, and relatively clean water is needed to prevent the clogging of the fine delivery tubes (Snyder, 1989). A comparison of drip irrigation with sub-irrigation and seepage for tomato production in Florida indicated that drip irrigation reduced water needs by 50% but added $328/ha to production costs (Pitts and Clark, 1991).

Drip irrigation was introduced in China in 1974. About 13,000 ha of orchards and cash crops currently use drip irrigation. This irrigation technique in northern China has saved 68% of the irrigation water and has increased fruit yields by 27% (Liu and Wang, 1995; Lou, 1996). Currently sprinkler irrigation in China totals approximately 600,000 ha (Lou, 1996).

There is great potential in developing some techniques for dryland crop production in northern China. Sophisticated use of dryland farming techniques such as using more organic matter as fertilizers to improve soil fertility and its water conserving capacity, effectively conserving surface runoff, regulating crop patterns, and using various other conservation methods will increase water production efficiency (Lou, 1996).

Planting trees to serve as shelter belts reduces evaporation and transpiration from the crop ecosystem from 13 to 20% during the growing season (Mari et al., 1985). The resulting increase in crop yields range from 10 to 74% for corn (Gregerson et al., 1989). Furthermore, this practice can reduce wind erosion by as much as 50% (Troeh et al., 1991). Also intercropping crops with trees, if they are "hydraulic lifters" (*Acer* and *Eucalyptus*), may increase water availability for the crop and increase productivity (T. Dawson, personal communication, 1995).

These approaches for conserving water resources, preventing pollution, and using treated waste water for irrigation would relieve the agricultural water crisis in China. Instituting policies to encourage people to conserve agricultural water use and improve water management planning should help to reduce the agricultural water crisis as well.

Energy Use in China's Agriculture

Energy use in agriculture in the early 1950s

The foundation of agriculture rests on the unique capacity of crops to convert solar energy into stored chemical energy. The efficiency of converting solar energy into stored chemical energy in crop systems depends on the human management of the crop system. All forms of management such as planting, weeding, and fertilizing require energy to be productive. In traditional agriculture, labor and animal power are the major energy inputs. Instead of large amounts of labor and animal power inputs, fossil energy inputs in the form of machinery, electricity, synthetic fertilizers, and insecticides are required in intensive agriculture.

Until the early 1950s, hand- and animal-powered organic agriculture was dominant in China. Machinery, synthetic fertilizers, and pesticides were not used in this organic agriculture. Traditional organic agriculture was practiced by peasants using various organic agrotechniques. The major energy inputs in this system included human labor, animal power, simple tools, and crop seeds. For example, an average of 842 hr/ha of labor and an average of 302 hr/ha of horse power were used in crop production in Hailun county in northeastern China in the early 1950s. The only fossil energy input in the crop production was an average of 1.24 kg/ha of hand tools such as spades, picks, sickles, hoes, and some animal draft equipment such as carts and plows (Wen and Pimentel, 1984). All the seeds for the crop were locally grown. Nutrients (P, N, K) for the crop production were obtained from animal and human wastes, crop residues, and nitrogen-fixing legumes like soybeans. Corn yields were about 1300 kg/ha. The kcal crop output/fossil energy input ratio of the crop system was 182 (Table 4.15.4).

Table 4.15.4. Energy input and output for a hand-animal-powered corn production system in Hailun county in northeastern China in the early 1950s. (Based on Wen and Pimentel, 1984)

Item	Quantity/ha		kcal/ha
Input			
labor	842	hr	--
horses or oxen	302	hr	--
seeds	22.5	kg	78,300
tools and implements	1.24	kg	25,683
Output			
corn yield	1,346	kg	4,682,862
kcal output/fossil energy input	182.3		
kcal output/labor hr	5,563		

Energy use in agriculture since the 1980s

Since 1950, China's agricultural improvement included the development of new crop varieties and increased fossil energy based inputs like fertilizers and pesticides. From 1965 to 1983 cereal crop production in China doubled. To achieve this rise in crop yields, synthetic fertilizer inputs increased 7.5-fold, insecticide inputs increased 2-fold, diesel fuel use increased 6-fold, and electricity use increased 11-fold (Wen, 1988). Crop production in China has become energy intensive and relies heavily on fossil energy inputs.

Table 4.15.5 shows the average fossil energy use in crop production in China in the early 1980s. About 300 million farmers and 58 million horses and oxen were involved in agricultural production. It is estimated that an average 2570 hr/ha of labor and 484 hr/ha of animal power were used in China's crop production in the early 1980s. The labor input per ha in China's crop production was 260 times greater than that of the U.S. (table 4.15.5). Although large amounts of labor and animal power were used in China's

Table 4.15.5. A comparison of average energy inputs for cereal crop production in China and the U.S. (Based on Wen, 1988)

Item	China (1982)		U.S. (1975)	
	Quantity/ ha	Kcal/ha	Quantity/ha	Kcal/ha
Input				
Labor	2,570 hr	--	9.8 hr	--
Animal power	480 hr	--	--	--
Seeds	173 kg	594,774	60.5 kg	563,002
Machinery And tools	9.75 kg	196,657	23.8 kg	428,400
Diesel	41.05 L	468,545	57.6 L	657,446
Gasoline	3.56 L	35,988	1.7 L	17,185
Nitrogen	74.47 kg	893,760	61.2 kg	734,400
Phosphorous	25.0 kg	75,000	39.2 kg	117,600
Potassium	5.12 kg	8,191	53.3 kg	85,280
Lime	--	--	166.4 kg	52,416
Irrigation	--	--	271,284 kcal	271,284
Insecticides And Herbicides	5.46 kg	510,019	2.93 kg	288,186
Drying	--	--	148,281 kcal	148,281
Electricity	126.06 kWh	360,910	203,396 kca	203,396
Transporta-Tion	54.21 kg	13,909	136.2 kg	35,003
Total fossil Energy Input	2,562,979		3,038,877	
Output				
Cereal crop yield	3,210 kg	10.726 mil	3,158 kg	11.365 mil
Kcal output/ fossil energy input	4.19		3.74	

crop production, the fossil energy inputs were also extremely large. In 1982, the average diesel and gasoline inputs were 44.6 l/ha which is about 25% lower than those in the U.S. The average synthetic N fertilizer input in

China's crop production was 74.5 kg/ha which is 1.2 times that of the U.S., and the insecticide and herbicide inputs were 5.5 kg/ha, or almost double that of the U.S. The total fossil energy input per sown crop ha in China in the early 1980s was 85% that of the U.S., and the average cereal crop yield per ha in China in 1982 was 3210 kg— almost the same as that of the U.S. in 1975 (table 4.15.5).

Energy use in China's agriculture has increased since the 1980s. First, with rural development, large amounts of rural laborers have moved from agricultural production to industry. There were 443 million rural laborers in China in 1993, 110 million of whom worked in non-agricultural industries (Jiang, 1996). 13.3% of rural laborers worked outside of rural areas (Liu and Wang, 1995). Recently the non-agricultural laborers in China's rural areas accounted for about one-third of all rural laborers. In agricultural production, more and more laborers have moved from crop production to animal husbandry, and to fishery and vegetable production. Thus the labor input in crop production in China has been reduced by about 50% when compared to the early 1980s. At the same time, the draft animal power employed in crop production has declined. Because of reduced labor and draft animal power, fossil energy inputs for agricultural machinery have increased. Recently, about one-half of cultivating operations, one-fifth of planting operations, one-tenth of harvesting operations, and one-tenth of agricultural transportation in China have become mechanized (Jiang, 1996). Agricultural mechanization in China should increase with rural economic development; however, abundant labor resources should limit China's agricultural mechanization. Estimates are that the machinery inputs in crop production, including irrigation, will increase at a rate of 4.2% each year (Wen et al., 1995).

Chemical fertilizers are the major fossil energy inputs in China's agriculture. The total chemical fertilizer use in China's agriculture in 1994 was 33.18 million tons, two times the level in the early 1980s (Jiang, 1996). Recently average chemical fertilizer inputs for cropland in the U.S. have been 99 kg/ha, but in China N use is 246 kg/ha, or almost 2.5 times that in the U.S. The actual cropland area in China receiving N fertilizer is much greater than indicated by the 246 kg/ha. The calculated average chemical fertilizer input in sowing per ha of crop production in China is 99 kg, similar to that in the U.S. (An, 1995).

The chemical fertilizer inputs in some areas of eastern China are extremely high. For example, about 500 kg/ha of fertilizers are applied to crop production in Shanghai. Increased chemical fertilizer inputs in central and western China are needed to increase crop yields. Chemical fertilizers and irrigation are the largest basic fossil energy inputs in China's agriculture. It is estimated that chemical fertilizer use in China will increase at an annual rate of 3.8% (Wen et al., 1995).

The average pesticide input in China is already very high but is not expected to rise. The pesticide input in agricultural areas is expected to decline because of growing integrated pest management techniques. However, herbicide inputs in China will probably increase thereby reducing the labor and machinery necessary for weed control.

In general, China's agriculture will have to produce more food to meet the needs of its increasing population. Increasing fossil energy inputs will be depended upon to increase food production. China's agriculture will therefore become more fossil energy intensive. There is a need, however, to increase fossil energy use efficiency through improved agricultural management.

Improved agriculture to increase fossil energy use efficiency

Poor agricultural management practices waste fossil energy. For example, the utilization of irrigation water in China can be reduced by about 50% if flooding is replaced by water-conserving irrigation technology and canal seepage is stopped (Xu, 1992; Shen and He, 1996). In addition in some regions synthetic fertilizers are applied at excessive rates, although there is a need to use some of the fertilizers in regions that have a shortage (Wen et al., 1995; An, 1995). In some regions, using excessive N and lower rates of P and K has reduced the effectiveness of N. Also, excessive use of insecticides in crop production results in waste and serious pollution (Wen, 1988).

Fortunately, these agricultural problems can be overcome. Extensive training of farmers, using suitable equipment, and improving agricultural management would effectively solve these problems. All of these changes will increase the efficiency of fossil energy use in China's agriculture.

Combining ecological and agricultural technologies

China currently has a fossil energy shortage; therefore, conserving fossil energy use in agriculture is essential. Combining ecological and agricultural technologies should help reduce fossil energy inputs in agricultural production.

China has a long history of using traditional organic agriculture. Some traditional ecological technologies are still used including using crop residues and manure as organic fertilizer; rotation using legume crops as green manure; intercropping; interplanting; multi-cropping; and agroforestry techniques.

Although some labor has left agriculture, China still has rich agricultural labor resources. It appears that the government and people of China are paying more attention to the use of these ecological techniques in agriculture. Employing sound ecological practices should help make more effective use of natural resources, including fossil energy inputs.

Conclusions

I. Irrigation and fossil energy inputs in agriculture are the most important factors for increasing food production in China. Improving the management of water resources and fossil energy inputs in agriculture will be the key to future agricultural development in China.

II. The existence of water shortages and waste is causing a water crisis. First, there is not enough water to meet increased irrigation needs. Second, current agricultural practices are wasting water. The development of water-saving irrigation techniques is essential to improving agricultural production.

III. Since 1950, fossil energy inputs in China's agriculture have increased rapidly. Energy inputs in agriculture will probably continue to increase and agriculture will become more energy intensive.

IV. Sustainable agricultural development in China should focus on both saving water and fossil energy resources by improving the efficiency with which these resources are used. Ecological technologies will help improve the effective use of natural resources and will help agriculture to become sustainable.

References

An, X. 1995. Chinese agriculture: the impact of external materials should be suited to agricultural technical advances. Research in Agricultural Modernization 16: 304-306 (In Chinese).

Bhuiyan, S.I. 1992. Water management in relation to crop production: case study on rice. Outlook on Agriculture 21: 293-299.

Blackshaw, R.E. 1990. Influence of soil temperature, soil mositure, and seed burial depth on the emergence of round-leaved mallow (Malva pusilla). Weed Science 38: 518-521.

Bouis, H.E. 1995. Breeding for nutrition. Journal of the Federation of

American Scientists 48 (4): 1, 8-16.

Donahue, R.H., R.H. Follett, and R.N. Tulloch. 1990. Our Soils and Their Management. Danville, Illinois: Interstate Pub., Inc.

Dubenok, N.N. and A.P. Nesvat. 1992. The effect of irrigation with elements of water saving technology on productivity of alfalfa in South Urals. Izv. Timiryazevskoi. S. Kh. Akad. 1: 21-26.

Gleick, P.H. 1993. Water in Crisis. New York: Oxford University Press.

Gregersen, H.M., S. Draper, and D. Elz. 1989. People and Trees: The Tole of Social Forestry in Sustainable Development. Washington, DC: World Bank.

Jiang, J. 1996. Agriculture, rural development, farmers and modernization. Pages 62-72 in Modern Agricultural Development in China, J. Jiang, ed. Beijing: Agriculture Press (In Chinese).

Jiao, D., Yang, J., Wang, F., and Y. Shi. 1996. Urban water resources measurement, assessment and system analysis for typical cities in China. Hydrology 96: 1-5 (In Chinese).

Jin, C. and Z. Guo. 1996. Brief introduction to water resources quality assessment in China. Hydrology 95: 1-7 (In Chinese).

Jin, L. and C. Zhang. 1996. The problems and protective proposals of water resources in China. Ecological Economy 62: 25-32 (In Chinese).

Kutzner, P.L. 1991. World Hunger: A Reference Handbook. Santa Barbara, CA: ABC-Clio.

Lal, R. 1994. Water management in various crop production systems related to soil tillage. Soil & Tillage Research 30: 169-185.

Liu, C. and H. Wang. 1995. On the connotations of water-saving agriculture. Pages 6-19 in Advances in the Application of Basal Research on Water-Saving Agriculture, C. Liu and Y. Hung, eds. Beijing: Chinese Agricultural Press (In Chinese).

Lou, Q. 1996. Strategy of constructing water-saving agriculture. Research in Agricultural Modernization 17: 136-139 (In Chinese).

Lu, X. and J. Wang. 1997. Supply and demand forecast for grains in China. Research in Agricultural Modernization 18: 12-16 (In Chinese).

Maberly, G.F. 1994. Iodine deficiency disorders: contemporary scientific issues. Journal of Nutrition 124 (8 suppl.): 1473s-1478s.

Mari, H.S., R.N. Rama-Krishna, and S.D. Lall. 1985. Impoving field microclimate and crop yield with temporary low cost shelter belts in the Punjab. Intern. J. Ecol. Environ. Sci. 11: 111-117.

McMichael, A.J. 1993. Planetary Overload: Global Environmental Change and the Health of the Human Species. Cambridge, UK: Cambridge University Press.

Nesheim, M.C. 1993. Human nutrition needs and parasitic infections. s7-s18 in Parasitology: Human Nutrition and Parasitic Infection, D.W.T.

Crompton, ed. Cambridge, UK: Cambridge University Press.

NGS. 1995. Water: A Story of Hope. Washington, DC: National Geographic Society.

O'Keefe, J.M. 1992. Water-Conserving Gardens and Landscapes: Water-Saving Ideas. Pownal, Vermont: Storey Publishing.

Pimentel, D., C. Wilson, C. McCullum, R. Huang, P. Dwen, J. Flack, Q. Tran, T. Saltman, and B. Cliff. 1997a. Economic and environmental benefits of biodiversity. BioScience. In press.

Pimentel, D., J. Houser, E. Preiss, O. White, H. Fang, L. Mesnick, T. Barsky, S. Tariche, J. Schreck, and S. Alpert. 1997b. Water resources: agriculture, the environment, and society. BioScience 47: 97-106.

Pimentel, D., Huang, X., Cordova, A., and M. Pimentel. 1997c. Impact of a growing population on natural resources: the challenge for environmental management. In Environmental Management in Practice, L. Hens and D. Devuyst, eds. Routledge. In Press.

Pitts, D.J. and G.A. Clark. 1991. Comparison of drip irrigation for tomato production in southwest Florida. Applied Engineering in Agriculture 7: 177-184.

Postel, S. 1992. Last Oasis: Facing Water Scarcity. New York: W.W. Norton and Co.

Postel, S. 1996. Dividing the Waters: Food Security, Ecosystem Health, and the New Politics of Scarcity. Worldwatch Pub. 132. Washington, DC: Worldwatch Institute.

PRB. 1996. World Population Data Sheet. Washington, DC: Population Reference Bureau.

Qu, G. and J. Li. 1992. Population and Environment in China. Beijing: Environmental Science Press (In Chinese).

Rees, D.J., A. Samillah, F. Rehman, C.H.R. Kidd, J.D.H. Keating, and S.H. Raza. 1990. Precipitation and temperature regimes in upland Balochistan, Pakistan and their influence on rain-fed crop production. Agricultural Meterology 52: 381-396.

RS and NAS. 1992. The Royal Society and the National Academy of Sciences on Population Growth and Sustainability. Population and Development Review 18 (2): 375-378.

Schlesinger, W.H. 1991. Biogeochemistry: An Analysis of Global Change. San Diego: Academic Press.

Shen, Z. and W. He. 1996 Assessment of China's agricultural water use and approach to the solution of existing problems. Journal of Natural Resources 11 (3): 221-230 (In Chinese).

Shi, S. 1996. To widen our field of vision outside the cultured land: a review of the grain problems in China. Eco-Agriculture Research 4 (2): 1-4 (In Chinese).

Snyder, R.L. 1989. Drought Tips for Vegetable and Field Crop Production . Oakland, CA: University of California.

Sweeten, J.M. and W.R. Jordan. 1987. Irrigation Water Management for the Texas High Plains. College Station, TX: Texas Water Resources Institute, Texas A & M University.

Tian, J., Wang, M., Dou, H., Cai, X., and Y. Jiao. 1997. Studies of the influence of the Yellow River cut-off on the ecological environment and its countermeasures. Chinese Journal of Ecology 16 (3): 39-44 (In Chinese).

Tribe, D. 1994. Feeding and Greening the World. Oxon, UK: CAB International.

Troeh, F.R. and L.M. Thompson. 1993. Soils and Soil Fertility. 5th ed. , New York: Oxford University Press.

Troeh, F.R, J.A Hobbs, and R.L. Donahue. 1991. Soil and Water Conservation. 2nd ed., Englewood Cliffs, NJ: Prentice Hall.

UNFPA. 1991. Population and the Environmet: The Challenges Ahead. New York: United Nations Fund for Population Activities, United Nations Population Fund.

USBC. 1996. Statistical Abstract of the United States 1995. Washington, DC: U.S. Bureau of the Census, U.S. Government Printing Office.

USDA. 1993. Agricultural Statistics. Washington, DC: U.S. Department of Agriculture, Government Printing Office.

van der Leeden, F., F.L. Troise, and D.K. Todd. 1990. The Water Encyclopedia. 2nd ed. Chelsea, MI: Lewis Publishers.

van Tuijl, W. 1993. Improving Water Use in Agriculture: Experience in the Middle East and North Africa. Washington, DC: World Bank.

Verplaneke, H.J.W., E.B.A. DeStooper, and M.F.L. De Boot. 1992. Water Saving Techniques for Plant Growth. Dordrecht, The Netherlands: Kluwer Academic Publishers.

Wen, D. 1988. Energy use in the crop systems of China. Critical Reviews in Plant Sciences 71 (1): 25-53.

Wen, D. 1993. Soil erosion and conservation in China. Pages 63-86 in World Soil Erosion and Conservation, D. Pimentel, ed. Cambridge, UK: Cambridge University Press.

Wen, D and D. Pimentel. 1984. Energy inputs in agricultural systems of China. Agriculture, Ecosystems and Environment 11: 29-35.

Wen, D. and D. Pimentel. 1992. Ecological resource management to achieve a productive, sustainable agricultural system in northeast China. Agriculture, Ecosystems and Environment 41: 215-230.

Wen, S., Jiang, H., and J. Shu. 1995. Macroscopic policy reform and mechanized improvement of physical input in agriculture in China. Research in Agricultural Modernization 16: 228-230 (In Chinese).

WHO. 1995. <u>Bridging the Gaps</u>. Geneva: World Health Organization.

Xu, Y. 1992. The major trends in water-saving agriculture in China. Pages 1-13 in <u>Study of Water-Saving Agriculture</u>, Y. Xu, ed. Beijing: Science Press (In Chinese).

CHAPTER 4.16

CAPABILITY OF WATER RESOURCES IN CHINA

Chen Chuanyou
and
Ma Ming

Commission for Integrated Survey of Natural Resources
Beijing, 100101, China

Water is a vital element for socio-economic development and nothing could exist without water. The imbalance between fresh water supply and demands has been aggravated as the population and economic development in the world are growing rapidly. In recent years, fresh water crisis is commonly seen all over the world. So far, there are three-fourth rural districts and one fifth cities in the world being short of fresh water supply; and about one billion people in the world are using polluted water. UN's Water Resources Convention in 1977 gave a message that 'water resources will become a critical issue as a new social crisis in the near future'. Moreover, it was the admonition made by some international organizations and experts in the 1990s that 'the whole world is entering an era when water resources is in shortage and new conflicts occurring in the world should probably be caused by contending for the control over water resources just as by contending for the control over the oil previously.

China is a country with a mass population. Its grain production and consumption are of decisive to the world. The grain yields in China are inevitably making significant effects not only on the economic development and social stability in China itself but also on the world grain market. Therefore, grain production is a very important issue for China's world status and its sustainable socio-economic development. China is known for its limited arable lands compared to its population with an average arable land per capita being only one third of the world's average and the agricultural potential in terms of cultivated lands is reaching its limitation. In order to meet the growing grain demands due to the increase of population, irrigation agriculture is a indispensable way to increase grain yields.

WATER RESOURCES IN CHINA

China is a country with a great potential of water resources. The total volume of water resources in China is 2800 billion m³ ranking fourth among countries in the world. But China is vast in territory with large population and the average water resources per capita is only one fourth of world's average ranking 110th in the world and is considered as one of 13 countries that suffer water scarcity in the world. Moreover, since China has a monsoon climate that results in a great difference in water resources distribution. There are 18 provinces in China where the average water resources per capita is less than the average of the whole country and among them 9 province are in North China including Beijing and Tianjin with the average water resources per capita below 500 m³. To meet the water demands, the ground water in Beijing has been over-exploited to an accumulated amount of 4 billion m³ since the 1960s which cause a large funnel area of 1000 km². In the plain areas of Hebei province during 1980s, the ground water was over-exploited to an accumulated amount of 3 billion m³ bringing a serious of environmental problems.

Based on the statistics in the early 1990s, the water supply facilities in China offered more than 500 billion m³ water each year, and the annual water consumption in China was also over 500 billion m³. However, imbalance of water supply and demands is still crucial and water shortage is seen not only in dry areas of the northern part of China but also in some humid areas of the southern part of China where water crisis was caused by water pollution and lack of water supply facilities. So far, the water deficit is estimated at 30 billion m³ in agriculture and 6 billion m³ in industry and municipality. It is anticipated that the total water deficit in the 2030s of the next century would be 230 billion m³ with the total population growing to 1.6 billion. Moreover, constrained by the natural limitation of water resources, the water consumption would probably approach to the capacity of water availability and the water allocation between sectors have to be adjusted accordingly as further increase of water supply is impossible. The conflicts between water supply and demands can only be mitigated with advanced water-saving technologies as well as using water wisely. It has become the world wide concerns whether the capability of water resources in China will be able to meet its water demands for grain production and socio-economic development.

Water Resources in China and Its Characteristics

Background and Origination of Water Resources

It is can be seen clearly from the geological history of China that the layout of the China's territory had not been in stability until the Himalayan tectonic movements. Due to the uplift of Qinhai-Tibetan plateau, four large landform units leaning against the Qinhai-Tibetan plateau were formed. They are the eastward Pacific incline, the southward Indian ocean incline, the northward Arctic ocean incline and the plateau itself of inland areas. The largest incline in area is the eastward Pacific incline with a straight length of some 3000 km occupying 60 per cent of China's mainland and dominating the China's general landform that leans to the east where most of the rivers in China are fed. The southward Indian ocean incline has a small area and a steep slope with a mere area in China's territory being 6.5 per cent of Chinese mainland and all rivers on this incline belongs to Indian ocean water system. The northward Arctic ocean incline occupies the smallest area in China's territory located in the north of Xinjiang with a river called Ertix belonging to the Arctic ocean water system.. The Qinhai-Tibetan plateau and inland areas on the north is occupied by many large basins between mountain ranges and every basin has its own inland water system and is a part of the Eurasian continent.

China is dominated by the monsoon climates coming mainly from the Pacific and the Indian ocean. The Arctic ocean and the Atlantic contribute little water vapor to China and affect very limited area, (Table 4.16.1). According to the analysis of "Evaluation of Water Resources in China", water vapor coming into the territory of China's mainland is 18,215 billion m^3 and water vapor going out of the territory is 15,840 billion m^3. The net income of water vapor of China is 2,376 billion m^3 accounting for 13 per cent of the total income. Water vapor from the Pacific on the eastward incline contributes most of the precipitation in China with an annual precipitation volume of 4987 billion m^3 accounting for 80 per cent of the total precipitation in China. The mountain ranges on the eastward incline hinder the moving of water vapor toward the northwest forming the rainfall center on the mountain slope facing wind so as that precipitation become less and less from coast to inland areas. The key element of the process is the Qinling Mountain that lies east-to-west and hampers both humid air moving from south to north and cold wave moving from north to south. This results in an absolutely different landscapes on each side of the Qinling Mountain: its southern side presents humid climate as the tropical and subtropical areas and its northern side features temperate and frigid-temperate climate as the arid and sub-arid areas.

Water vapor from Indian Ocean goes from south to north and northeast via Bangladesh bay. Part of it is stopped by the Himalayan Mountain and produces heavy rainfall on the south slope of the

Table 4.16.1: The water distribution on four land form units in China

Land form units	Territory (%)	Precipitation $(10^9 m^3)$	Runoff $(10^9 m^3)$	Evaporation $(10^9 m^3)$	Runoff Ratio (%)
Pacific incline	56.71	4,986.8	2,132.2	2,854.6	42.7
India Ocean incline	6.52	670	462.9	207.1	69
ArcticOcean incline	0.53	20.8	10	10.8	48.1
Subtotal	63.76	5,677.6	2,605.1	3,072.5	45.8
Inland region	36.24	511.3	106.4	404.9	21
Whole China	100	6,188.9	2,711.5	3,477.4	43.8

Himalayas making it as one of the heaviest rainfall areas in the world. Another part of it crosses the mountain range along river valleys to the north entering into the central plateau and forming much rainfall in the southern area of the plateau, especially in the Yarlong Zangbo River, as hampered by the Gandise-Nianqing Tanggula Mountains. And still some water vapor is able to move across the Hengduan Mountains into the eastward Pacific incline to affect the precipitation in whole Southeast China.

Surface Water Resources

Surface water resources can be calculated with the runoff in rivers. According to the estimates in "Evaluation of Water Resources in China", the total annual volume of runoff in China is 2711 billion m³, converting into a depth of runoff at 284mm, and is displayed in nine water districts in China (see Table 4.16.2). Among these water districts, the Yangtze River basin has the largest annual runoff at 951 billion m³ accounting for 35.1 per cent of the total volume of runoff in China, the Pearl River basin ranks second in annual runoff volume at 468 billion m³ and the Yarlong Zangbo River ranks third at 165 billion m³. The river basins that feed water directly into seas have an area of 4.4 million km², accounting for 46.3 per cent of the China's mainland territory and are all located on the eastward Pacific incline including Liao River, Hai and Nuan Rivers, Yellow River, Huai River, Yangtze River, Pearl River and rivers in Zhejiang and Fujian provinces. The river basins that run into neighboring countries take up the areas of 1.8 million km², 19 per cent of the China's mainland territory, and they are Heilong River in Northeast China, rivers in Southwest China and Etix River in Northwest China and located in the border areas with low level of water utilization. There are many landlocked river basins feeding terminal lakes in the Great Northwest China with a total area of 3.32 million km², constituting 34.7 per cent of the China's mainland territory, such as Caidam inland rivers, Jungger inland rivers, Tarim inland rivers and Qiangtang inland rivers, and they are mostly small and seasonal rivers and vital for sustaining terminal lakes.

Groundwater Resources

Groundwater resources in China can be defined as mountain area groundwater and plain area groundwater. The groundwater in mountain area is usually recharged from precipitation and discharged into river channels forming a base flow of runoff by ways of faults, fissures, and caves. Due to its complex formation dominated by mountain ranges, the groundwater in mountain area is not suitable for mining. The groundwater in plain area is situated in coastal plains or in basins among mountain ranges and is stored in sand and clay soil aquifers with varying thickness. It is recharged not only by the vertical infiltration from precipitation and surface water but also by the lateral movement of groundwater from mountain area, which enriches groundwater in plain areas. For instance, several large plains in China such as Yellow-Huai-Hai Plains, Hetao Plain in mid-Yellow River and Chengdu Plain are major groundwater mining areas in China and the aquifers are thick and rich in groundwater with unit well outflow rate over 2000 m³/d.

Groundwater resources usually refers to dynamic water volume in aquifers and could be represented by recharged water volume to aquifers. For mountain area groundwater is estimated with the infiltrated volume of precipitation while for plain area it equals to the aggregation of the infiltration from precipitation, surface water, lateral recharge from mountain area as well as cross-river-basin recharge. According to the estimates in "Evaluation of Water Resources in China", the annual volume of groundwater resources in mountain area is 676 billion m^3, and the volume in plain area is 187 billion m^3. Making the deduction of overlapped estimation of 34.8 billion m^3 between mountain area and plain area, the total volume of groundwater resources in China is 829 billion m^3. As shown in Table 4.16.3, the five water districts of the northern part of China have 146.8 billion m^3 of groundwater, accounting to 78 per cent of that in plain area of China, among which 100 billion m^3 is exploitable and plays an important role in water supply of the northern part of China.

Total Volume of Water Resources

The total volume of water resources in China is dominated by the amount of precipitation - the original water source which could be repeatedly used by mankind. The average annual volume of precipitation on the lands of China is estimated at 6,189 billion m^3 equivalent to 648mm in depth, in which 56 per cent (about 3500 billion m^3) goes back to atmosphere and 44 per cent (about 2700 billion m^3) runs off into seas through river channels. The total volume of water resources in China is defined as the combination of surface water and groundwater with the deduction of overlapped calculation between them and estimated at 2,812 billion m^3 with an average water module of 294,600 m^3/km^2 (Table 4.16.4).

It can be seen that the water resources in the northern part of China referring to five water districts is limited, with a water volume of 535.8 billion m^3 being 19 per cent of the nation's total water volume and a water module of 88,300 m^3/km^2, while the water resources in the southern part of China referring to four water districts is ample, with a water volume of 2276.6 billion m^3 being 81 per cent of the nation's total and 7.4 times that in the north and a water module of 654,100 m^3/km^2.

Table 4.16.2 - Distribution of surface water resources in China (in yearly amount)

Water districts	Precip. in depth	Precip. in vol.	Runoff in depth	Runoff In vol.	Prop.	Runoff in probabilities $(10^9 m^3)$			
	(mm)	$(10^9 m^3)$	(mm)	$(10^9 m^3)$	(%)	20%	50%	75%	95%
Northeast	510.8	637.7	132.4	165.3	6.1	205	161.7	130.3	90.6
Hai & Nuan rivers	559.8	178.1	90.5	28.8	1.1	38	26.8	19.9	13
Yellow River	464.4	369.1	83.2	66.1	2.4	76	64.9	56.9	47
Huai R. & Shandong	859.6	283	225.1	74.1	2.7	100	68.9	49.6	29.6
Northwest Inland	157.7	532.1	34.5	116.4	4.3	125	115.9	109.1	100
Yangtze River	1,070.5	1936	526	951.3	35.1	1055	941.7	865.6	761
Southeast coast	1,758.1	421.6	1,066.3	255.7	9.4	306	250.7	209.7	161.1
Pearl rivers	1,544.3	896.7	806.9	468.5	17.3	539	464	413	338
Southwest	1,097.7	934.6	687.5	585.3	21.6	643	585.3	538	474.1
Whole China	648.4	6,188.9	284.1	2,711.5	100	2901	2,711	2,549	2,359

Characteristics of Water Resources in China

The water resources in China is plentiful in quantity, but the per capita water resources is considerably low due to the large amount of population. Table 4.16.5
shows the comparison of water resources among the major countries. It can be seen that the per capita water resources in China is only 27 per cent of the world average and the least among the major countries. Based on 1990 population statistics, China ranked 110th in per capita water resources among 149 countries. This has been reflected by the serious conflicts between water supply and demands in China. In order to work out relevant countermeasures, it is necessary to understand the characteristics of water resources in China.

Table 4.16.3 - Distribution of groundwater resources in China (in yearly amount)

Water districts	Mountain		Plain		Overlapped	Whole China	
	Calc. Area (1000 km^2)	Volume (10^9 m^3)	Calc. Area (1000 km^2)	Volume (10^9 m^3)	Volume (10^9 m^3)	Calc. Area (1000 km^2)	Volume (10^{9}m^3)
Northeast	823	31.9	407	33	2.5	1,231	62.5
Hai & Nuan rivers	171	12.5	106	17.8	3.8	277	26.5
Yellow River	608	29.2	167	15.7	4.4	775	40.5
Huai R. & Shandong	127	10.7	169	29.7	1.1	297	39.3
Northwest inland	1,814	56.7	948	50.6	21.1	2,762	86.2
Yangtze River	1,625	221.8	132	26.1	1.4	1,758	246.4
Southeast coast	218	56.2	20	5.2	0.06	239	61.3
Pearl rivers	550	102.8	30	9.3	0.5	5,805	111.6
Southwest	851	154.4				851	154.4
Whole China	6,790	676.2	1,983	187.3	34.8	8,774	828.8

Uneven spatial distribution of water resources and its incongruity over other resources

It is no doubt that disparity in climate between the south and the north is largely responsible for the uneven distribution of water resources in China. Since the precipitation over the mainland decreases from the coastal area of Southeast China to the inland of Northwest China, five climatic zones can be identified as very humid, humid, semi-humid, semi-arid and arid zones accordingly. There is 47 per cent territory of the nation under the control of semi-arid and arid zones in Northwest China with the average annual precipitation below 400 mm and the average annual runoff depth below 50 mm. The regions cover Inner Mongolia, Ningxia, Gansu, Qinghai, Xinjiang and the northern part of Tibetan Plateau, have the economy dominated by animal husbandry or the combination of animal husbandry and agriculture and are sparsely populated.

Table 4.16. 4 - Average annual water resources in districts

Water districts	Surface water resources (10^9m^3)	Ground water resources (10^9m^3)	Over-lapped volume (10^9m^3)	Total Water Resources (10^9m^3)	Water module (1000 $m^3/km^2)$
Northeast	165.3	62.5	34.9	192.9	154.5
Hai & Nuan rivers	28.8	26.5	13.2	42.1	132.4
Yellow River	66.2	40.6	32.4	74.4	93.6
Huai R. & Shandong	74.1	39.3	17.3	96.1	291.9
Northwest inland	116.4	86.2	72.2	130.4	38.6
Yangtze River	951.3	246.4	236.4	961.3	531.6
Southeast coast	255.7	61.3	57.8	259.2	1,080.8
Pearl rivers	468.5	111.5	109.2	470.8	810.0
Southwest	585.3	154.4	154.3	585.3	687.5
5 districts of north	450.7	255.1	170	535.8	88.3
4 districts of south	2260.8	573.7	557.8	2276.6	654.1
Whole China	2711.5	828.8	727.8	2812.4	294.6

4.16. 5 - Water resources in major countries

Country	Annual vol. of runoff (10^9m^3)	Popula-tion (million)	Water volume per capita (m^3)	Cultivated lands $(10^6 ha)$	Water Volume Per unit Area (m^3/ha)
Brazil	6,950	149	46,808	32.3	215,170
former USSR	5,466	280	19,521	226.7	24,111
Canada	2,901	28	103,607	43.6	66,536
China	2,712	1,154	2,350	97.3	27,867
Indonesia	2,530	183	13,850	14.2	178,169
USA	2,478	250	9,912	189.3	13,090
India	2,085	850	2,464	164.7	12,662
Japan	547	124	4,411	4.33	126,328
Whole world	46,800	5,294	8,840	1,326	35,294

However, the water resources there is quite scarce and taking only 7 per cent of the nation's total volume while the abundant land resources suitable for farming, grazing and afforesting presents huge potentials for development as long as water is available.

The very-humid and humid zones in Southeast China constitute 34 per cent nation's territory with the average annual precipitation above 800 mm and the average annual runoff depth above 200 mm, including most of Huai river basin and Yangtze river basin, Pearl river basin and the southeastern coast. The regions exhibit abundant water resources accounting for 81 per cent of the nation's total volume with a rapid economic growth and a well developed agriculture suitable for growing wheat, rice, and other economic plants.

Between above mentioned zones is a transitional zone defined as the semi-humid zone, making up 19 per cent of nation's territory, where the average annual precipitation ranges 400 - 800 mm and the average annual runoff depth 50 - 200 mm. This region appears in long belt from Northeast China through North China plain to the southeast of Qinghai-Tibetan plateau. The western part of the belt is in highland areas with less population and underdeveloped economy. On the contrary, its eastern part is the major agricultural areas of North China and is well developed in economy and communications with high density of population. However, the water resources in this transitional zone is only 12 per cent of nations' total volume, which makes the region as the most crucial place in China regarding the imbalance of water supply and demands, especially in the southern Liaoning province, the plain areas of Beijing, Tianjin, Hebei province and the Shandong Peninsula where water supply is in serious situation. Since the beginning of 1980s, some cities have been suffering deficit of water supply and threatening the local economic development and livelihoods.

As indicated from analysis, the locations of large river basins in China under different climatic zones result in the serious incongruity between water resources and land resources. Table 4.16.6 shows the composition of water resources with population and cultivated lands in China. The five water districts in the north of China have the cultivated lands accounting for 65 per cent of the nation's total cultivated lands with the water resource given only at 19 per cent of the nation's total volume and the water volume per hectare only at

8,689 m³. On the contrary, the four water districts in the south of China get 35 per cent of the nation's total cultivated land and the water resources there is as much as 81 per cent of nation's total volume per hectare of 68,860 m³

Table 4.16.6 - Composition of water resources with population and cultivated land in China

Water districts	Calcu-lated area (10⁶ km²)	Total W. R. (10⁹ m³)	Culti-vated lands (10⁶ ha)	Popu-lation (10⁶)	W. Res. per capita (m³/cap.)	W. Res. Per ha (m³/ha)	Cult. lands per capita (ha/cap.)	Rate of Cult. Lands (ha/km²)
Northeast	1.266	192.8	19.07	115	1677	10110	0.166	15.06
Hai & Nuan rivers	0.320	42.1	10.6	121	348	3972	0.088	33.13
Yellow River	0.798	74.4	12.14	102	729	6129	0.119	15.21
Huai R. & Shandong	0.330	96.1	14.47	194	495	6641	0.075	43.85
Northwest inland	3.400	130.0	5.34	26	5000	24345	0.205	1.57
Subtotal	**6.114**	**535.4**	**61.62**	**558**	**959**	**8689**	**0.110**	**10.08**
North Proportion	**63.7**	**19.0**	**65.1**	**46.5**				
Yangtze River	1.810	961.3	22.34	414	2322	43030	0.054	12.34
Southeast coast	0.240	259.2	2.33	66	3927	111245	0.035	9.71
Pearl rivers	0.583	471.0	6.80	142	3317	69265	0.048	11.66
Southwest	0.853	585.0	1.59	21	27857	367925	0.076	1.86
Subtotal	**3.486**	**2276.5**	**33.06**	**643**	**3540**	**68860**	**0.051**	**9.48**
South Proportion	**36.3**	**81.0**	**34.9**	**53.5**				
Whole China	9.600	2811.9	94.68	1201	2341	29699	0.079	9.86

WATER RESOURCES IN CHINA

Uneven seasonal distribution of water resources and mismatching to water demands

Water resources in China is mostly supplied with precipitation which is largely under the control of monsoon climate. This inevitably leads to an uneven seasonal distribution of water resources within a year and gives the main cause for floods and droughts commonly occurring in China. Such phenomenon can be interpreted by taking Beijing as an example. Beijing has roughly the same amount of annual precipitation as Paris, London and Berlin in Europe do (Table 4.16.7), but the seasonal distribution of rainfall in Beijing is completely different from that in Europe.

The monthly precipitation ratio of maximum to minimum at Beijing comes to 67.4 in contrast to the value in Europe at 1.7 for Paris, 1.8 for London and 2.4 for Berlin. The precipitation during Beijing's monsoon season from July to September accounts for 64 per cent of its annual amount, which suggests that more than half of the annual precipitation in Beijing concentrates in two months. Conversely the amount from December to April next year in Beijing gets only 3.2 per cent of its annual precipitation and almost no rainfall occurs during nearly half of a year. Therefore, aridity during winter and spring is a distinctive highlight of Beijing and also the fundamental reason of why Beijing is much drier than cities in Europe.

Table 4.16.7 - Seasonal distribution of rainfall in Beijing and cities of Europe

Cities	Seasonal distribution of rainfall (per cent)					
	Jan.	Feb.	Mar.	Apr.	May	Jun.
Beijing	0.5	0.9	1.2	2.8	5.4	12.3
Paris	6.7	5.8	6.7	7.6	9.0	9.4
Berlin	8.2	5.6	6.5	7.4	8.2	10.0
London	8.7	6.6	6.4	7.9	7.9	7.0

	Jul.	Aug	Sep	Oct	Nov	Dec.	Total P(mm)
Beijing	33.7	30.3	7.8	3.2	1.5	0.6	630.8
Paris	9.4	9.0	9.0	9.9	9.0	8.5	566.4
Berlin	13.4	9.5	8.2	7.4	7.4	8.2	586.7
. 79	8.7	9.6	7.9	10.0	10.9	8.7	581.7

This is also true in other places of China where runoff in rivers varies considerably due to the great precipitation inconsistency. The more variable the runoff over seasons appears, the more serious the crisis for water supply is. The ratio of the maximum runoff in successive four-months to its annual amount can be taken as an indicator of Ki to measure the degree of runoff concentration within a year. Statistics shows that the whole China can be divided into three classes of runoff concentration: the most concentrated runoff area with Ki greater than 0.7, the highly concentrated runoff area with Ki ranging 0.65 - 0.7, and fairly concentrated runoff area with Ki less than 0.65. The most concentrated runoff is commonly seen in rivers of plain areas, where population and economic activities are also concentrated, and brings about serious floods during summer and heavy droughts in winter and spring as well as water shortages because the increasing water demands mismatch the natural water distribution among seasons. Therefore, the runoff in these areas must be regulated through engineering approaches as a solution to this problem as far as sustainable development in China is concerned. The highly concentrated runoff usually appears in highland and hilly areas including the basins between mountains where inhabitants are not popular and the amount of water use is not much. There is no urgent need to regulate runoff except in some basins and valleys. The same situation is also applicable to fairly concentrated runoff area, mostly in mountainous area, as long as necessary attention is paid to water and soil conservation and rational farming incorporated with the development of small hydroworks.

Great yearly variation of water resources with wet or dry years appearing successively

Another important cause responsible for water crisis in China is the great yearly variation of water resources. It can be evaluated with the ratio of the maximum runoff to the minimum runoff in a yearly runoff sequence represented by Km. Available hydrological records show that the Km in China ranges between 2 - 8 and has a overall trend of increase from southeastern coastal area to northwestern inland area. The vast areas on the south of Qinling Mountain has the Km ranging 2 - 4; Northeast China at 3 - 5; North China at 4 - 6; and the most part of Northwest China reaching 5 - 8. Moreover, in the same region, small rivers usually get big Km. For instance, in the southwestern humid area Km is around 2 on mainstreams and greater than 3 in tributaries, and a Km of 7.64 is found at the Dahuizhuang gauging station on a tributary of Jinsha River. The maximum Km has occurred at the Bangbu gauging station on Huai River at 19.5. Regulation of runoff gets more difficult and applicable water resources gets less as the Km gets bigger.

According to the analysis in "Evaluation of Water Resources in China", runoff in most part of China has the recurrence between dry year and wet year and it means that runoff changes from low flow to high flow then from high flow to low flow. Such recurrence usually has a span of 50 - 70 years in the north and 30 - 60 years in the south. Besides, wet year or dry year in China lasts successively. The rivers in the north has longer successive period of either dry year or wet year than those in the south. Based on the statistics, over five successive years of wet period mostly occur in the north and over five years of dry period are also seen in the north where the dry period lasts for 6 years in Songhua River, 7 years in Yongding River and even 11 years in Yellow River. This is apparently unfavorable for the requirement of increasing and stable water supply in China so that strong capacity against drought is needed. It is obvious that the natural temporal distribution of water resources in China can not conform to the water utilization patterns. Therefore, a mass regulation of river flow aiming to reduce yearly runoff difference must be implemented.

Good natural water quality and much sediment content in rivers

The natural water quality of river flow in China is generally good and acceptable with low mineralization in the mainstreams of major rivers. Among the water districts divided on large drainage basins in China, the lowest mineralization is found in the rivers in southeastern coastal area with a content below 200 mg/L. Highly mineralized water can only be encountered in some inland rivers in the north with a content over 1000 mg/L for both surface water and groundwater, such as in Caidam and Jungger Basins, north of Inner Mongolian plateau and some part of Loess Plateau.

Most of rivers in China are filled with much sediment loads and about 3.5 billion tons of suspended sediment is taken into rivers each year. During flood seasons, a great amount of sediment is deposited in flooding areas or diverted into irrigation areas. The Yellow River, the most sediment carrying river in the world, receives 700 million tons of sediment every year which is deposited on its levee-lined channel bed in lower reach and on irrigation areas and make its lower reach channel to become a suspended river above the surrounding ground. The Yangtze River has 260 million tons of sediment deposited every year in its mainstream below Yichang City and the sediment deposition has raised the river bed in middle and lower reaches and reduced the volume of reservoirs and lakes weakening their flood prevention capability.

Present Capability of Water Supply in China

Great achievement of water projects has been made since the found of People's Republic of China. Before 1980s, China has furnished over 80,000 reservoirs and

600 dykes with a total storage volume of 440 billion m^3; 57 million kW pumping and draining capacity including 20 million pumping wells for irrigation; 25,000 diversion headworks; 170,000 km long levees along rivers; over 90,000 hydropower stations in various size; accumulated soil conservation areas of 440,000 km^2; 100,000 km long river navigation; and over 4.6 million ha water areas for fishery. After 1980s, some large size inter-basin water transfer projects were achieved including the Nuan River to Tianjin water transfer project, Yellow River to Qingdao water transfer project, and Bi river to Dalian water transfer project, and gave a large increment of water supply.

In 1993, the amount of water use in whole China reached 525.5 billion m^3, accounting for18.7 per cent of the China' total water resources, among which 83 per cent is surface water and 17 per cent ground water. The proportion of the water use among sectors is 74.6 per cent for irrigation, 12.6 per cent for industry, 3.8 per cent for urban domesticity and 4.3 percent for rural domesticity (Table 4.16.8). The annual per capita water use in China was 438 m^3, equivalent to one fifth that in United States and three fifth of world average. In addition, the amount of water use in agriculture has tended to be stable with the benefits gained from water saving programs, and this implies that the efficiency of water use in China is increasing gradually.

Water Use in Agriculture

The water use in agriculture consists of two parts: irrigation and rural domestic use. The agriculture in China is subject to irrigation due to

Table 4.16.8 - Water use by various sectors in China in 1993

Water districts	Total W. use (10^9m^3)	Rate of W. use (%)	Popula-tion (10^6)	W. use per cap. (m^3/ca)	Irrigation Volume (10^9m^3)
Northeast	49.82	25.8	115	433	36.77
Hai & Nuan rivers	40.96	97.3	121	339	31.05
Yellow River	38.77	52.1	102	380	31.20
Huai R. & Shandong	56.9	59.2	194	293	45.31
Northwest inland	58.2	44.8	26	2,238	55.94
Subtotal of north	**244.7**	**45.7**	**558**	**438**	**200.26**
Yangtze River	165.9	17.3	414	401	109.30
Southeast coast	28.61	11	66	433	20.72
Pearl Rivers	79.47	16.9	142	560	54.20
Southwest	6.8	1.2	21	324	6.15
Subtotal of south	**280.8**	**12.3**	**643**	**437**	**190.38**
Whole China	525.5	18.7	1,201	438	390.64

Water districts	Rural domestic Volume (10^9m^3)	Prop. %	Industry Volume (10^9m^3)	Prop %	Urban domestic Volume (10^9m^3)	Prop. (%)
Northeast	1.34	2.68	9.67	19.4	2.05	4.12
Hai & Nuan rivers	1.44	3.51	6.44	15.73	2.03	4.96
Yellow River	1.43	3.7	4.89	12.61	1.25	3.22
Huai R. & Shandong	3.41	5.99	6.37	11.19	1.82	3.19
Northwest inland	0.43	0.74	1.46	.5	0.38	0.65
Subtotal of north	**8.05**	**3.29**	**28.82**	**11.78**	**7.53**	**3.08**
Yangtze River	7.63	4.6	41.88	25.24	7.12	4.29
Southeast coast	1.93	6.75	4.99	17.44	0.97	3.38
Pearl Rivers	4.81	6.05	16.53	20.8	3.93	4.95
Southwest	0.23	3.33	0.33	4.8	0.09	1.36
Subtotal of south	**14.60**	**5.20**	**63.73**	**22.69**	**12.11**	**4.31**
Whole China	22.65	4.31	92.55	17.61	19.64	3.74

its natural conditions. The Northwest China belongs to the arid and semi-arid climate and there would be no better agriculture without irrigation. In the southern part of China where dominant crops are mainly paddy that consumes much water, high and stable yields cannot be achieved without irrigation because of the uneven seasonal distribution of rainfall though there is always abundant rainfall. The Northeast China and North China are mostly dominated with semi-humid climate with a fair water condition, but there is still a need to adopt irrigation in dry years or seasons and otherwise the low yields are unavoidable. Therefore, irrigation has been playing a very decisive role in agriculture of China since ancient time. Statistics from nearly 50 year data show that the trend of yearly grain yields has corresponded to that of water consumption of irrigation.

The water for irrigation use increased 2.5 times during the years from 1950s to the beginning of 1980s while the total grain yields doubled in China. From 1980 to 1993, the irrigation water increased by 10 per cent in China with the total grain yields rising by 8 per cent. By the end of 1993, the irrigation areas in China has reached 49.94 ha with a irrigation rate of cultivated lands at 52 per cent, and the total volume of irrigation water use is 391 billion m^3 with the average irrigation quota at 7,822 m^3/ha. The irrigation quota in the south is generally higher than that in the north and calculated at 8,599 m^3/ha on average. The Northeast China and the northwestern inland regions have the highest irrigation quota in China at no less than 10,500 m^3/ha, more than 30 per cent higher than the nation's average. The lowest irrigation quota is found in Hai and Nuan river basins at 4,851 m^3/ha. Irrigated lands contributed a gain yield of 335 million tons accounting for 73 per cent of the nation's total yield. The grain yield per unit water use (irrigation efficiency) was 0.85 kg/m^3., 42 per cent higher than 0.6 kg/ m^3 in 1980, but quite far from the world advanced level at 2.3 - 2.5 kg/m^3. it presents a promising potential for improvement even in comparison with the 1.5 kg/m^3 achieved in Hai river basin of China (Table 4.16.9).

The water volume for rural domestic use at 22.6 billion m^3, though only accounting for 4.3 per cent of the total water use, managed to accommodate the needs of 860 million people and 140 million livestocks in rural areas. But the water use standard was quite low at 70 liters per capita per day and even lower in some places. With the rising of living standards and the extension of tap water systems in rural areas, the water for rural domestic use will increase inevitably but still remains low proportion in agricultural water use.

Water Use in Industry and Urban Domesticity

The urbanization in China has hardly started until 1980s. There are only 191 cities in 1980, but rapid urbanization was seen in resent years with 479 cities in 1990 and 570 in 1993 and is expected to continue. The urban population in China increased from 173 million in 1978 to 343 million in 1993 with an average annual growth rate of 4.4 per cent while the proportion of the urban population to the national population increased from 17.9 per cent to 28.6 per cent. The production value of urban industry also increased from 423.7 billion Yuan in 1978 to 5,269 billion Yuan in 1993 with an average

Table 4.16.9 - Irrigation water use and grain yield of China in 1993

Water districts	Water for irriga-tion (10⁹m³)	Paddy field (10⁶ha)	Wate-ring field (10⁶ ha)	Irriga-tion area (10⁶ ha)	Rain-fed area (10⁶ha)	Cultiva-ted lands (10⁶ha)	Rate of irriga-tion (%)	Water use per ha. (m³/ha)
Northeast	36.77	2.05	1.13	3.18	15.89	19.07	16.7	11,562
Hai & Nuan rivers	31.05	0.19	6.21	6.4	4.2	10.6	60.4	4,851
Yellow River	31.20	0.5	4.05	4.55	7.59	12.14	37.5	6,857
Huai R. & Shandong	45.31	3.73	5.07	8.8	5.67	14.47	60.8	5,149
Northwest inland	55.94	0.8	4.07	4.87	0.47	5.34	91.2	11,486
Subtotal of north	**200.26**	**7.27**	**20.53**	**27.8**	**33.82**	**61.62**	**45.1**	**7,204**
Yangtze River	109.30	12.67	1.8	14.47	7.87	22.34	64.8	7,554
Southeast coast	20.72	1.73	0.13	1.86	0.47	2.33	79.8	11,141
Pearl Rivers	54.20	4.27	0.48	4.75	2.05	6.8	69.9	11,410
Southwest	6.15	0.73	0.33	.06	0.53	1.59	66.7	5,806
Subtotal of south	**190.38**	**19.4**	**2.74**	**22.14**	**10.92**	**33.06**	**67.0**	**8,599**
Whole China	390.64	26.67	23.27	49.94	44.74	94.68	52.7	7,822

Water districts	Total Yield (10⁹kg)	Irrigated yield (10⁹kg)	Irrigated yield per ha. (kg/ha.)	Irrigation Effici ency (kg/m³)	Rain fed yield (10⁹ kg)	Rain fed yield per ha. (kg/ ha.)	Mul-tiple Cropp-ing (%)	Overall Yield Per ha. (kg/ha.)
Northeast	71	27.6	8,679	0.75	43.4	2,731	100	3,723
Hai & Nuan rivers	37	30.4	4,750	0.98	6.6	1,571	130	3,491
Yellow River	34	23.1	5,077	0.74	10.9	1,436	120	2,801
Huai R. & Shandong	84	68	7,727	1.5	16	2,821	175	5,805
Northwest inland	12	11.7	2,402	0.21	0.3	638	95	2,247
Subtotal of north	**238**	**160.8**	**5,784**	**0.8**	**77.2**	**2,283**		**3,862**
Yangtze River	142	109.3	7,554	1	32.7	4,155	200	6,356
Southeast coast	19	17.4	9,355	0.84	1.6	3,404	232	8,155
Pearl Rivers	50.5	42.3	8,905	0.78	8.2	4,000	218	7,426
Southwest	7	5.5	5,189	0.9	1.5	2,830	190	4,403
Subtotal of south	**218.5**	**174.5**	**7,882**	**0.92**	**44**	**4,029**		**6,609**
Whole China	456.5	335.3	6,714	0.85	121.2	2,709	150	4,822

annual growth rate of 14.9 per cent (based on comparable price). As a result, the quantity of urban water supply was growing to a great extent. From 1980 to 1993, the urban domestic use of water grew by 190 per cent with an average annual growth rate of 8.5 per cent and the industrial use by 102 per cent with an average annual growth rate of 5.6 per cent. The total water volume of urban use including industry and domesticity reached 112.1 billion m^3 in 1993, 2.14 times of that in 1980, among which the industrial water use amounted to 92.5 billion m^3, 82.5 per cent of the total urban use. The cubic meters of used water per 10,000 Yuan of industrial output value was 176 $m^3/10, 000$ Yuan at that time, 20 per cent of that in 1980.

The recycling rate of industrial water had been promoted to 50 per cent on average, 1.5 times of that in 1983, and reached over 60 per cent in some cities of the northern part of China suffering water shortage. For instance, Beijing had a recycling rate reaching at 82.4 per cent in 1992, Tianjin at 66.8 per cent in 1989, Dalian at 81 per cent in 1993, and Xian at 67.5 per cent in 1993. In contrast, the recycling rate of industrial water in the cities of the southern part of China was below 40 per cent in general. The urban domestic water use includes commercial use and family use and can be reflected by the quota of water use per capita per day in cities. According to the investigation to 48 large and medium cities in national wide conducted by concerned authority, the urban domestic water use had a great different amount from city to city ranging approximately between 45 - 220 L/cap/day, and the highest was Guangzhou City at 247 L/cap/day, the lowest Yinkou City at 37.5 L/cap/day and most of cities around 100 L/cap/day.

Ecological Consideration for the Development of Water Resources

Water needed for environmental consideration has received less attention in the past, but the environmental situations appear critical as water is a key and indispensable factor functioning with other environmental factors in an ecosystem. Hence, water utilization and development should be guided on the basis of giving no adverse effects on the balance of ecosystem. In other word, a certain amount of water should be kept in its natural conditions to sustain ecosystem. There has been a sign in China that environment tends to be deteriorated because of the over-withdrawal of water resources for socio-economic activities. For example, in some places of China, over-exploitation of groundwater has resulted in a vast decline of groundwater table and even surface depression.

Key Issues in Development of Water Resources

Severe Water Shortage and Worsened Water Environment in the North Part of China

Severe water shortage has been encountered in the northern part of China as water distribution is unfit to land resources and population, particularly more serious in North China and Northwest China . The North China covering an area over 400,000 km^2 is the main production base of grain, cotton and energy as well as the center of politics, economy and culture. The population and cultivated lands in this area account for 11 percent and 15 per cent respectively of the nation's total but only with some 50 billion m^3 water accounting for 1.8 per cent of nation's total water resources, and the water volumes per capita and per unit cultivated area are no more than one-sixth and one-tenth respectively of nation's average, even lower than those in Israel known as one of the driest places in the world. The deficit between water supply and demands has become the main constraint to further social-economic development. Since 1980s, many cities in north China have encountered water shortage. In order to ensure the urban water supply, irrigation water had to be reallocated to urban use in some regions, this has intensified the water competition between sectors and regions. About one-quarter of cultivated lands were affected to some extent by water shortage every year in the plain areas of North China. In 1982, Beijing, Tianjin and Hebei province together had near half cultivated lands struck by droughts reducing the grain production by hundreds thousands tons.

To accommodate the continued increase of water demands, the over exploitation of groundwater had to be undertaken in some regions of North China and this has seriously undermined the sustainable use of its renewable resources and resulted in a series of water environment issues such as dried wells and sea water intrusions. During 1980s in Hebei province, an accumulation of 3 billion m^3 groundwater was over-exploited and caused the decline of groundwater table covering areas of more than 20,000 km^2. The groundwater table in Cangzhou Prefecture located in the center of the decline went down to 90 meters below the surface and over 80 meters lower than the sea level.

The great Northwest China occupies almost half of China's territory including Xinjing, Qinghai, Gansu, Ningxia and part of Tibet and Inner Mongolia with a sparse population no more than 100 million. It is situated in the dry and semi-dry inland regions with a very vulnerable ecosystem, and it is evident that the major environmental issues are desertification, water and soil losses, shrinkage of lakes and glaciers. The water volume entering Qinghai Lake decreased by 400 million m^3 each year on average and the eastern part of Qinghai Province will be under the jeopardy of desertification without the protection offered by the Qinghuai Lake. Unless viable countermeasures for Northwest China is carefully addressed, the deteriorated ecosystem in this region will not only threaten the sustainable development in local societies but also make adverse effects on the ecosystem in the eastern coastal area of China. One of the typical cases is the interruption of flow in Yellow River. In the past 24 years, there were 18 years during which the Yellow River's lower reach ran dry with 27 days of interruption on average due to the irrational and increasing water use in its middle and upper reach areas. In 1995, the interruption to the river flow extended up to the territory of Henan Province and lasted for 122 days and this gives a sign that the interrupted time is getting longer and earlier and the interrupted reach is going more upper. It is undoubted that an effective remedy to these environmental problems in Northwest China is water.

Widespread Pollution of Water Resources

With the progress of urbanization, the effluents of cities in whole China have increased greatly and come to 35.5 billion tons in 1993 about 100 million tons a day, in which 80 per cent was released to water bodies without treatment causing water pollution to different degree in rivers, lakes, reservoirs and groundwater. Based on the monitoring records, the water authority conducted a water quality assessment in 1980 on national wide involving 874 rivers with the river channel lengths of 92,115 km, 34 lakes and 111 reservoirs. The results showed that high organic pollution was found in the north of China with 73 per cent of assessed river channels polluted to different degree mostly in Heilong River, Liao River and Hai and Huai Rivers.

The river channels of water quality conform to drinking standard made up 33.6 per cent of the total assessed river channels and that of water quality conform to irrigation was 89.1 per cent. About 5,281 km river channels were polluted with toxic materials and exceeded over the national standards. The statistics from 47 cities showed that there were 10 and 20 cities with groundwater pollution in heavy level and middle level respectively. Moreover, the general trends of water pollution have not been stopped effectively as more point sources of pollution are added to water systems due to the rapid growth of township enterprises. In the summer of 1990, the Tai lake in Jiangsu Province was seriously polluted with 0.5 meter thick accumulation of algae along its northern bank, resulting in the close of the tap water plants of Wuxi City and hundreds million Yuan economic losses. The water pollution has added more pressure on the crucial water supply of China. Therefore, prevention and control of water pollution will be one of the major countermeasures in the water resources development in China.

Wasteful use of water resourcws

Wasteful water use is prevailing in China while water supply is in shortage. In agricultural sector, irrigation canal systems in China have lower efficiency and most of canals are unlined with serious water leakage. Investigation shows that the canal efficiency of nation's major irrigation areas is averaged at 0.4 - 0.6, in which the canal efficiency is 0.4 - 0.5 in North China and 0.4 - 0.45 in the northeast and no more than 0.6 - 0.7 in pumping irrigation areas. It is estimated on the basis of field findings in Henan Province that the water losses from canal leakage constitute 80 per cent of total water losses. Water losses also come from outmoded irrigation methods. Flooding irrigation is still adopted on over 80 per cent irrigation areas in China and requires higher irrigation quota due to unleveled lands, uneven watering and high rate of infiltration. The irrigation quota in the northwest ranks high in China at 11486 m^3/ ha on average, 1.5 times of nation's average, and in some places in Ningxia Province the quota even higher at 30,000 m^3/ha. In the south of China, paddy fields are often over-irrigated causing a large amount of water and fertilizer to be wasted and rise of groundwater table as well as soil salinization.

In industrial sector, there has been a high water consumption per unit products due to aged equipment and outdated techniques. Water consumption for producing one ton of steel is below 10 m^3 in developed countries while it is above 20 m^3 in China in general, and even more than 50 m^3 with some manufactures; water consumption for producing one ton of pulp is only 30 m^3 in developed countries while it is as high as 100 m^3 in China. In addition, the water recycling rate in industrial use is 50 per cent on average, and there are about 40 per cent of cities with the recycling rate below 30 per cent and 18 per cent of cities no more than 10 per cent. The water consumption per 10,000 Yuan of output value was 176 m^3/10, 000 Yuan on average, and more than 500 m^3/10,000 Yuan in some cities. It is evident that the potential for water-saving is promising. Moreover, the urban domestic water use is also wasteful. Among the over 35 million sets of water utensil used in whole nation, 25 per cent has the problem of leakage causing 400 million m^3 water losses. Water conservation is also one of the important aspects in water resources development.

Rampant Disasters of Floods and Droughts

The inhabitants in China are mainly concentrated in lowland areas where floods and droughts occur frequently and threaten the property and security of people. According to the statistics, from 1951 to 1990, China encountered 7.5 droughts on average each year with maximum at 11 a year and minimum at 3 a year, and 5.9 floods with maximum at 10 and minimum at 3. They usually cover a wide range of areas with high intensity. In the past 40 years, China had 17 per cent of cultivated lands affected by droughts each year and about 7 per cent in drought disasters, and also near 7 per cent of cultivated lands under the suffering of floods as well. It can be seen from analysis that there are five drought centers where droughts happen frequently and continuously and they are Yellow, Huai and Hai river basins, southeastern coastal region, western part of Southwest China, western part of Northeast China, and Northwest China. The flooding areas mainly appear in southeastern coastal region, southern areas to Yangtze River and Huai river basin and flood disasters generally reduce from. the southeast to northwest. In resent years, human activities have enhanced the formation of flooding disasters due to the extension of land reclamation by deforestation and the irrational land uses in upper reach areas.

The only way to eliminate drought disasters is the rational development of water resources. On the basis of water conservation,

WATER RESOURCES IN CHINA

intensified regulation and reallocation of water resources should be carried out to change the spatial and temporal distribution of natural runoff so as to make full use of water resources in China. Flooding disasters can only be under control by firstly adopting biological measures to prevent soil erosion in upper reach areas coupled with the modern flood prevention systems..

Trend Analysis of Growing Water Demand in Next 30 - 50 Years

China is the largest country of the world in population and also a developing country with the great demands for various resources. Sufficient water is indispensable as a security to its social-economic development. But how much water will China need to satisfy such development 30 - 50 years late, particularly to ensure the self-supplied grain products? Is there enough quantity of fresh water in China available for exploitation and use? These issues must be clearly addressed in the future.

Since the 1980s, many experts and scientists have done many studies on such issues and come up with a series of predictions. "China 21st century agenda" predicted that, based on the requirement of the triple growth of national economy by 2000, the prerequisite of water saving and rational water use and the condition of fair dry year, the total water demands of China would be 600 billion m^3 in 2000, and 720 billion m^3 in 2010 under the water demand growing rate of 2 - 3 per cent from 2000 to 2010; "Utilization of Water Resources in China" suggested that the total water demands of China in 2000 would reach 709.6 billion m^3, in which 130.2 billion m^3 is allocated for industry with an integrated water intake per 10,000 Yuan output value at 650 m^3, 484.7 billion is used for the irrigation of 54.4 million ha farming areas, 21.7 billion m^3 for the irrigation of 2.5 million ha forestry and grass areas, 51 billion m^3 is provided for the rural domestic use with a 877 million population and 21.9 billion for the urban domestic use with a 320 million population; "The General View of Water Issues of 21st Century in China" estimated that in 2030 the domestic water use for both rural and urban use is approximated at 90 billion m^3, the industrial water use at 200 billion m^3 and the irrigation at 500 billion m^3, resulting in a total water demands of 790 billion m^3; "Population, Grain and Water" analyzed that the water demands of China in the middle of next century would be 960 - 1040 billion m^3 as the population reaches its peak at 1.6 billion, with a water use increment of 420 - 550 billion m^3 above the 1995's level; "Well Managing,

Rational Using and Sound Protecting the Limited Water Resources" indicated that the integrated norm of per capita water demand in developed countries is generally above 1000 m^3 though a norm of 500 m^3 be likely to be accepted in water shortage areas with the water conservation and management properly reinforced, therefore the water supply of China in 2040 should be increased to at least 800 billion m^3, 275 billion m^3 higher than that in 1993, if the norm of per capita water use is raised from 450 m^3 in 1993 to 500 m^3.

Although the above predictions were made with different points of view and methods, a general conclusions can be summarized as following:

- Around the middle of next century, China's total water demands will inevitably increase by the extent of 250 - 600 billion m^3.
- Industry will contribute most of the increment in water consumption and agriculture and domestic use take low percentage in the increment.
- All the predictions were made on the basis of rational water use and conservation but environment water use is rarely taken into account.
- Emphasis of the predictions was place on the subsistence for 1.6 billion population of China and the estimates of water use were based on the growth rate of population fixed by family planning norm, while water use for industry was identified with certain growth rates.

Following are the further studies on the forecasts made on the basis of above predictions.

Forecast of Population and Grain Demands

The population in China has increased by 15 million a year on average. Though China has tried to control the growth rate under 1.25 per cent, this limitation could not be followed in some places. Experts estimate that the population will be likely to reach 1.6 billion during the 2030s or 2040s with the 1.3 million population in 2000. Still someone believe that the peak population of 1.6 billion would probably be delayed to the 2050s with the promotion of the socio-economic development and the knowledge of civilization and sanitation in rural areas. It is evident that China is now relatively young in population composition and the growth rate will remain high in near future, and as the population turn to be in old age , the grow rate will decrease accordingly. It is expected that by 2030 China will be likely to achieve the zero growth on population at the peak of 1.6 billion.

In resent fifteen years the living and consumption level of Chinese people is close to the world's average, and all people are basically subsisted with increased quality of food. In 1993 the amount of major foods consumed per capita in China is 235 kg of grain, 37.5 kg of meat, 12.3 kg of egg, 5.4 kg of milk and 18.6 of aquatic. Such consumption can provide Chinese people with necessary calorie, but direct use of grain is relatively high while protein and fat are in low consumption. The annual grain consumption per capita in China is 380 kg higher than India, lower than Japan and equal to South Korea, and it is not low for a developing country under the constraints of high population and limited cropping lands. Therefore, it is practicable to maintain the per capita grain consumption at the level of 400 kg by the 2000 and 450 kg by the 2030, and it will result in a grain demand of 500 - 550 million tons during 2000 - 2010 and about 720 million tons by the year 2030.

Forecast of Cultivated Lands

By the year 1993, China had the cultivated lands of 94.68 million ha with a per capita area of 0.08 ha. There are some 33.3 million ha reclaimable wastelands in China, mostly distributed in the northeast and the northwest inland region, in which about 16.7 million ha wastelands are suitable for agricultural development. It is anticipated that the decrease of cultivated lands imposed by the occupancy of growing population and urban construction could be presumably compensated by the increase of reclaimed wasteland in future, so that the cultivated lands in China around 2030 are expected to remain constant at some 94 million ha. In 1993 the irrigation areas in China were 49.94 million ha with an irrigation rate of 52.7 per cent, and achieved 75 per cent of the national grain production presenting a decisive function in ensuring the grain production. There are 26.7 million ha cultivated lands situated in hilly areas and unsuitable to be irrigated. According to the investigation conducted by Ministry of Water Resources, in the existing cultivated lands the maximum irrigation areas are most likely to be extended up to about 60 million ha. Therefore, the land potential for irrigation would be some 10 million ha except the existing irrigation areas. If reclaimed lands that can be developed only with irrigation are included, the maximum land potential for irrigation would be some 26.7 million ha.

Despite the abundant water favorable for irrigation in the south of China, the cultivated lands have been irrigated in high level and the rain-fed lands are hilly giving little potential for further irrigation development. There are vast areas of arable wastelands in the north of China with a favorable conditions of landform and radiation for agricultural development, and yet the extension of irrigation is largely constrained by locally limited water resources. Nevertheless, in order to meet the grain demands in the next century, there is a need to increase the irrigation rate in the north of China to a great extent. The Northeast China now has a low irrigation rate only at 16 per cent, which is expected to be raised to 43 per cent. The Hai and Nuan river basins as well as Huai and Shandong Peninsula encounter a growing conflicts between water supply and demands due to the scarcity in water resources, and since these regions are also the important grain production bases of China, it is expected to increase the irrigation rate from about 60 per cent to 70 per cent under the condition of implementing the inter-basin water transfer. The northwest inland regions are characterized by the oasis agriculture which can not exist without irrigation and the vulnerable ecosystems, and there are abundant arable wastelands over 30 per cent of the nation's arable wastelands, which are likely to become the additional bases of gain production in next century. At present, as the irrigation quota in the regions is much higher than its actual needs and if water conservation and inter-basin water transfer are fulfilled, it is possible to augment additional irrigation areas of 3.13 million ha. As above discussed, under the pressure of growing population and grain demand, the application of varied water saving irrigation technologies and the implementation of inter-basin water transfer projects are to be speed up, and water supply to the north of China can be greatly increased. Therefore, it is possible to achieve an extension of 14 million ha irrigation areas in China by 2030s, which are mainly concentrated in the north of China (Table 4.16.10).

3.   Forecast of Water Demands in Agriculture

The water demands from agriculture are dependent on the total irrigation areas and the irrigation quota. Wasteful irrigation is now popular in China's agriculture with much higher water demands than the actual needs of crops. The water saving irrigation must be adopted to take the advantage of great water saving potential in

Table 4.16.10 - Forecast of cultivated lands of China around 2030

Water districts	Irriga tion area (10^6 ha)	Rain-fed area (10^6 ha)	Arab le waste land (10^6 ha)	Rate of irriga tion (%)	Expan Ded irrig. Area (10^6 ha)	After exp. irrig. area (10^6 ha)	After exp. Rain-fed (10^6 ha)	Rate of irriga tion (%)
Northeast	3.18	15.89	8.53	16.7	5.00	8.18	10.89	42.9
Hai & Nuan rivers	6.4	4.2	0.8	60.4	1.00	7.40	3.20	69.8
Yellow River	4.55	7.59	2	37.5	0.67	5.22	6.92	43.0
Huai R. & Shandong	8.8	5.67	0	60.8	1.67	10.47	4.00	72.4
Northwest inland	4.87	0.47	4.33	91.2	3.13	8.00	0.00	100.0
Subtotal of north	**27.8**	**33.82**	**15.66**	**45.1**	**11.47**	**39.27**	**25.01**	**61.1**
Yangtze River	14.4	7.87	0.67	64.8	2.33	16.80	5.54	75.2
Southeast coast	1.86	0.47	0	79.8	0.00	1.86	0.47	79.8
Pearl Rivers	4.75	2.05	0.07	69.9	0.00	4.75	2.05	69.9
Southwest	1.06	0.53	0.27	66.7	0.20	1.26	0.33	79.2
Subtotal of south	**22.1**	**10.92**	**1.01**	**67.0**	**2.53**	**24.67**	**8.39**	**74.6**
Whole China	49.9	44.74	16.67	52.7	14.00	63.94	33.4	65.7

China's agriculture. The water saving irrigation means that the amount of water used for irrigation should not exceed the supplementary amount actually needed for crops. In order to evaluate the water saving potential in agriculture, the water saving irrigation quota of crops can be roughly estimated according to the maximum evapotranspiration of crops in different places of China that meteorologists have worked out through field studies. The maximum evapotranspiration refers to the total evaporation needed from both plants and soils under the conditions of sufficient water supply to the soils. The difference of the maximum evapotranspiration to the effective rainfall in a certain area should be presumably considered to be the actual water amount (the water saving irrigation quota) needed for supplement to crops, and the effective rainfall is equal to the total precipitation subtracting the average depth of runoff and will be eventually consumed to actual land evapotranspiration. Table 11 shows the evaluation process of the supplementary water need for crops in water districts, and the total precipitation and runoff depth used in the process are taken from moderate dry year (with the guaranteed reliability of 70 per cent). The comparison of the estimated supplementary water need for irrigation with the current water use quota exhibits that water saving in irrigation can be achieved by 10 - 60 per cent with different water districts, and the average water saving potential would be 28 per cent. It is can be seen that the northeast and the northwest inland areas give a much higher water saving potential due to the wasteful water use in irrigation. It should be noted that the supplementary water need for irrigation is estimated to keep the soils with sufficient water supply for full evaporation. If the soil evaporation can be lowered by using water saving irrigation technologies such as drip irrigation, buried irrigation and pipe irrigation, the supplementary water need for crops can be further reduced.

At present, low irrigation efficiency has been the major problems in China's irrigation areas, and was caused mainly by incompatible irrigation facilities, outdated irrigation methods and the serious water losses of conveyance systems. The emphasis in implementing water conservation should be placed on the rehabilitation and improvement of existing irrigation facilities and conveyance systems. It is tested that lined canals can reduce water losses from seepage by 50 - 90 per cent. A recently developed field conveyance system by using low pressure pipe can increase water utilization efficiency by 20 - 30 per cent with higher water saving rate and lower costs. The irrigation areas equipped with pipe conveyance systems are more than half inthe United States and 90 per cent in Israel. China has only about 2.7 million ha using the pipe conveyance system and about 5.3 per cent of its total irrigation areas. With all the efforts made to increase the water utilization efficiency, it is anticipated that about 50 per cent of irrigation areas will be likely to achieve the above mentioned water saving irrigation quota by the year around 2030, and about 65 billion m^3 water can be saved by that time accordingly. The saved water can be either used for expansion of effective irrigation areas or transfer to urban and industrial use.

Table 4.16.11 - Estimation of rational irrigation quota for crops

Water districts	W. Requ. of crops (mm)	Pre cip. P= 75% (mm)	Run off P= 75% (mm)	Effec tive pre cip. (mm)	Wa ter Shor tage (mm)	Wa ter quota (m³/ ha)	Ratio nal quota (m³/ ha)	Cur rent quota (m³/ ha)	Rate of wa. Save (%)
North East	810	419	104	315	495	4,950	4,950	11,562	57.2
Hai & Nuan rivers	855	426	63	364	492	4,915	4,800	4,851	1.1
Yellow River	810	329	72	258	552	5,522	5,550	6,857	19.1
Huai R & Shan Dong	900	671	151	520	380	3,802	3,900	5,149	24.3
North west inland	810	101	32	69	741	7,412	7,500	11,486	34.7
Yangtze River	990	877	479	398	592	5,916	6,000	7,554	20.6
South East coast	1,200	1,443	874	569	631	6,308	6,750	11,141	39.4
Pearl Rivers	1,320	1,328	711	617	703	7,032	7,500	11,410	34.3
South West	720	944	632	312	408	4,077	4,500	5,806	22.5

Based on the estimated water saving irrigation quota and the forecast irrigation areas, the irrigation water use in the next century can be approximately estimated. Suppose that 50 per cent of the existing 49.94 million ha irrigation areas can be improved to meet the above water saving irrigation quota after next 30 year efforts, the newly developed 14 million ha irrigation area will also meet the same quota and the remaining 50 per cent of existing irrigation areas use the current irrigation quota, the total irrigation water use around 2030 will probably reach 414.6 billion m³ (Table 4.16.12). The result indicates a little increase over the current level of irrigation water use. The northern water districts get more increase in irrigation water use due to the large newly developed irrigation areas, while some southern districts show no increase and some others even decrease in irrigation water use.

Table 4.16.12 - Forecast of irrigation water requirement of China around 2030

Water districts	W. save area (10⁶ ha)	W. save quota (m³/ha)	W. save w. req. (10⁹ m³)	Ordi nary area (10⁶ ha)	Ordi nary quota (m³/ha)	Ordi nary w. req. (10⁹ m³)	Total w req. (10⁹ m³)	Cur rent w. req. (10⁹ m³)	Ad ded w. req. (10⁹ m³)
Northeast	6.59	4,950	32.62	1.59	11562	18.38	51.00	36.77	14.24
Hai & Nuan Rivers	4.2	4,800	20.16	3.2	4851	15.52	35.68	31.05	4.64
Yellow River	2.94	5,550	16.34	2.27	6,857	15.60	31.94	31.20	0.75
Huai R. & Shan dong	6.07	3,900	23.67	4.4	5,149	22.65	46.33	45.31	1.02
North-west inland	5.565	7,500	41.74	2.43	11486	27.97	69.71	55.94	13.77
Subtotal of north	**25.37**	**5,303**	**134.5**	**13.9**	**7,204**	**100.2**	**234.7**	**200.3**	**34.41**
Yangtze River	9.565	6,000	57.39	7.23	7,554	54.65	112.0	109.3	2.74
Southeast coast	0.93	6,750	6.28	0.93	11141	10.36	16.64	20.72	-4.08
Pearl Rivers	2.375	7,500	17.81	2.37	11410	27.10	44.91	54.20	-9.29
South west	0.73	4,500	3.29	0.53	5,806	3.08	6.36	6.15	0.21
Subtotal of south	**13.6**	**6,233**	**84.77**	**11.0**	**8,599**	**95.19**	**180.0**	**190.4**	**-10.4**
Whole China	38.97	5,627	219.3	24.9	7,822	195.3	414.6	390.6	23.98

Forecast of Grain Production

With the national maps of radiation and temperature crop productive potential, the grain production under above irrigation conditions can be estimated. The radiation and temperature crop productive potential refers to the crop yields controlled by local radiation and temperature under the most suitability of water, soil fertility and proper farming and it can be possibly achieved through adequate efforts. Crops can grow all year around in the south of China with high multiple cropping index above 200 per cent and the radiation and temperature crop productive potential can be over 24,000 kg/ha. The North China has a multiple cropping index around 150 per cent with two harvests a year or three harvests in two years and a radiation and temperature crop productive potential around 18,000 kg/ha. The Northeast China has cold climate with only one harvest a year, and the radiation and temperature crop productive potential is about 13,500 kg/ha. According to the statistic data provided in "Agricultural Climate Resources in China", the actual grain production in high yield farmlands can be 70 - 90 per cent of local radiation and temperature crop productive potential. This forecast assumes, by the year around 2030, that the grain yields on the water saving irrigation farmlands is required to reach 70 per cent of local radiation and temperature potential and the grain yield on non-water-saving (ordinary) irrigation lands still use the current average grain yields, that the unit grain yield on rain-fed farmlands is supposed to be boosted by 50 per cent through rain irrigation and advanced farming technologies, and in addition, that the water saving irrigation farmlands will have 80 per cent area for growing grain crops and non-water saving and rain-fed farmlands will have 90 per cent area for growing crops. Based on above presumptions, it is estimated that the irrigation areas will contribute the grain production of 605 million tons and the rain-fed farmland will produce 119 million tons, and the total grain production will be 724 million tons which can meet the grain demands in next century. From above results, the average irrigation efficiency of whole nation can be calculated at 1.46 kg/m^3 (Table 4.16.13).

Forecast of Water Demands in Industry and Urban Domesticity

The water demands of industrial use are mainly dependent on the growth rate of national economy and the implementation of industrial

Table 4.16.13 - Forecast of grain production of China around 2030

Water districts	Water-saving irrigation areas				Non-water-saving irrigation areras			
	High yield poten tial (kg/ha.)	Area (10^6 ha)	80% for grain crop (10^6 ha)	Grain yield (10^9 kg)	Ordi nary yield (kg/ ha.)	Area (10^6 ha)	90% For Grain Crop (10^6 ha)	Grain yield (10^9 kg)
Northeast	9,450	6.59	5.27	49.8	8,679	1.59	1.431	12.4
Hai & Nuan Rivers	15,750	4.20	3.36	52.9	4,750	3.20	2.880	13.7
Yellow River	12,600	2.95	2.36	29.7	5,077	2.28	2.048	10.4
Huai R. & Shan dong	17,850	6.07 .	4.86	86.7	7,727	4.40	3.960	30.6
North-west inland	10,500	5.57	4.45	46.7	2,402	2.44	2.192	5.3
Subtotal of north		25.37	20.30	265.9		13.90	12.51	72.4
Yangtze River	16,800	9.57	7.65	128.6	7,554	7.24	6.512	49.2
Southeast coast	19,950	0.93	0.74	14.8	9,355	0.93	0.837	7.8
Pearl Rivers	21,000	2.38	1.90	39.9	8,905	2.38	2.138	19.0
South west	9,450	0.73	0.58	5.5	5,189	0.53	0.477	2.5
Subtotal of south		13.60	10.88	188.8		11.07	9.96	78.5
Whole China		38.97	31.18	454.7		24.97	22.47	150.9

Continuation Table 4.16.13

Water	Improved rain fed farming areas						
Districts	Impro Ved Yield (kg/ ha.)	Area (10^6 ha)	90% for grain crop (10^6 ha)	Grain yield (10^9 kg)	Irriga ted yield (10^9 kg)	Total yield (10^9 kg)	Irriga Tion Effici Ency (kg/ m^3)
Northeast	4,097	10.89	9.80	40.1	62.2	102.4	1.22
Hai & Nuan rivers	2,357	3.20	2.88	6.8	66.6	73.4	1.87
Yellow River	2,154	6.92	6.23	13.4	40.1	53.5	1.25
Huai R. & Shandong	4,232	4.00	3.60	15.2	117.3	132.5	2.53
Northwest Inland	957	0.00	0.00	0.0	52.0	52.0	0.75
Subtotal Of north		**25.01**	**22.51**	**75.6**	**338.2**	**413.8**	**1.44**
Yangtze River	6,233	5.54	4.99	31.1	177.7	208.8	1.59
Southeast Coast	5,106	0.47	0.42	2.2	22.7	24.8	1.36
Pearl Rivers	6,000	2.05	1.85	11.1	58.9	70.0	1.31
South West	4,245	0.33	0.30	1.3	8.0	9.3	1.26
Subtotal of south		**8.39**	**7.55**	**45.6**	**267.3**	**312.9**	**1.49**
Whole China		33.40	30.06	121.2	605.6	726.7	1.46

water conservation technologies. According to the macro blueprint of national economic development formulated by Chinese government, the total value of GNP in 2000 will doubled over the 1990's level. With the progress of far-reaching economic reform in 1990s, Chinese government has set another goal for 2010 with another doubled GNP over the 2000's level. The growth of GNP is mainly contributed by the growth of industrial output value. Therefore, there must be a great water demand for industrial use as industry will be kept in a high growth rate which, in turn, cause a great increase of urbanization adding demands on the domestic water use.

Since the 1990s, the national economy exhibited a rapid growth and the industrial output value accordingly increased with a great extent, representing 2392.4 billion Yuan in 1990, 5269.2 billion Yuan in 1993 and 9189.3 billion Yuan in 1995 (in current price) with an average annual growth rate of 15 per cent on the basis of the constant price of 1990. Chinese government has made every effort to regulate the macro economy and to adjust the growth rate of national economy under a reasonable level. It is expected that by 2000 the growth rate of industrial output value can be controlled to 10.6 per cent and by 2030 it will be finally adjusted to 6.8 per cent, when such value will be 34 times of that in 1990. Even if advanced water saving technologies are adopted in industry, the growing water demands in industry are unavoidable.

It is known from the history of industrial development in developed countries that the industrial water use shows a great increase during initial stages of industrialization due to the establishment of fundamental industry. After industrialization, as the industry goes to deal with high tech and high value enterprises, the demands of industrial water use will tend to slow and eventually reach a stable level. Furthermore, it is likely to be true that the water supply will be unable to expand as water demands is approaching the limitation of exploitable water resources of China. In order to maintain a smooth growth of national economy with the little growth of water supply, great efforts have to be made to promote advanced water saving measures in industry in the next decades and to achieve a considerable reduction in water intake quota in industry. As analyzed before, there is a great water saving potential existing in industry, but such potential is not boundless and will be limited by economic and financial constraints. It is possible to set the industrial water recycling rate at 75 per cent (not including electric power industry) as a long term goal by the year 2010. There are a large number of outmoded and high water consuming equipment still in use in many enterprises and it is a fundamental measure to fulfill the goal by replacing the equipment with new and water saving equipment. Other measures include the relocation of high water consuming industry to coastal areas where making a full use of sea water is possible, and the adjustment of low water price as China is entering market economy.

In the forecast of industrial water use, the water intake quota is represented by the water intake per 10,000 Yuan of industrial output value which is forecast on the basis of the constant price of 1990 (Table 4.16.14). According to the statistics, the industrial output value of the whole nation was 2392.4 billion Yuan in 1990 with an average water intake of 334 m^3/10,000Yuan. The industrial output value of the whole nation around 2030 is estimated at 76,323 billion Yuan. Based on the analysis of the actual industrial water intake before 1995, it is anticipated that the average water intake will be able to be reduced to 30 m^3/10,000 Yuan around 2030 through varied water saving measures. This means that an annual reduction rate of 6 per cent on industrial water intake will be required during the three decades and is considered practicable. Therefore, it can be concluded that the total industrial water intake around 2030 can be possibly limited to 230 billion m^3.

Table 4.16.14 - Forecast of industrial output value and growth of water intake around 2030

Years	Growth rate of ind. Output (%)	Output value (Yuan in constant price in 1990)	Indust rial water intake (10^9 m^3)	Growth rate of water intake (%)	Water intake per 10^4 Yuan ($m^3/10^4$ Yuan)	Decline rate Of Water intake Per 10^4 Yuan
1980		9,895	45.7		462	
1985	9.7	15,747	74.3	10.2	472	0.4
1990	8.7	23,924	80.0	1.5	334	-6.7
1993	17.2	38,506	92.5	5.0	240	-10.5
1995	11.7	48,137	103.0	5.5	214	-8.5
2000	10.6	79,662	119.5	3.0	150	-6.9
2010	9.5	197,421	157.9	2.8	80	-6.1
2020	8.0	426,217	191.8	2.0	45	-5.6
2030	6.0	763,290	229.0	1.8	30	-4.0

With the rapid urbanization in the next century, the urban population using tap water will greatly increase. The domestic water use standard will arise gradually despite the application of water saving measures. According to the study from the ministry of civil affair, the proportion of urban population in China will increase from 28 per cent in 1993 to 33 per cent in 2000 and 42 per cent in 2010. With the extrapolation of the trend, it is estimated that the proportion of urban population around 2030 will reach 50 per cent. If the urban population is 0.8 billion with an integrated water use standard of 250 liters/capita/day, the total domestic water use will be about 73 billion m³.

Forecast of Water Requirement for Ecosystem

Much more attention has been paid to the resolution of water demand and supply problems during the prior water resources development, while water for ecosystem maintenance was somehow ignored. River flow is one of very important elements in ecosystems, and if it is over-used, ecological balance could not be maintained. The water resources in the north of China has been exploited to an excessive extent resulting in serious environmental problems, such as decline of ground water table, river flow interruption, land desertification and lake shrinkage. Therefore, limitation should be set for the utilization of both surface water and groundwater. Some scientists believe that an utilization rate of over 20 per cent in river flow is likely to make effects on water environment and an over 50 per cent utilization rate will cause adverse effects. The authors suggest that the limitation of water utilization is set based on the following consideration. Due to the effects of monsoon climate, the river flow in China varies both annually and seasonally, and in particular, the great seasonal variation of river flow makes the runoff in low flow season to be only 30 per cent of the annual runoff on average. Such 30 per cent of annual runoff in low flow season is the minimum requirement for ecosystem maintenance and the adverse environmental impacts will occur if the river flow in low flow season fail to be kept to the proportion. Therefore, it is can be concluded that the utilization rate of river flow should not exceed 40 per cent of the annual runoff for a river basin.

However, it is can be seen from Table 4.16.8 that all water districts in the north of China except the northeast have their water utilization rates exceeding the limitation. In favor of maintaining and restoring the ecosystems of the river basins in the north of China as well as satisfying sustainable utilization of water resources, it is necessary to replenish water to the river basins or to reduce the water utilization rate to a level below the limitation. From this sense, it is reasonable to regard the portion of utilized water exceeding the limitation as the amount of water to restore or replenish in the river basins in the north of China, which can also be considered as environmental water requirement. Thus the environmental water requirement in the north of China can be approximately estimated according to the water utilization rate in each water district. Among the five water districts in the north of China, there are four water districts excluding the northeast with their water utilization rate exceeding 40 per cent of their annual water resources, and the excessive amount of water totals about 57.8 billion m^3, and will be taken as the environmental water requirement in the north of China. This considerable amount of water can be though partly compensated by the reduction of water use through the water conservation and the rearrangement of industrial structures. But for a fundamental resolution, inter-basin water transfer, e.g. the south to north water transfer or the Tibet to northwest water transfer, is likely to be the only option to replenish the environmental water requirement in the north of China as the growing water demands of economic development are forecast.

Analysis of Water Resources Capability and Development Strategies

Capability of Water Resources

The capability of water resources refers to the supply capacity of exploitable water resources from varied water bodies which meets the demands in socio-economic activities under a certain stage and level of economic development and technologies. It is not only dependent on the exploitable water resources but also related to the level of socio-economic development and the water use efficiency of various sectors. There has been an evidence of being short of water supply in China, and the effective ways to tackle the problems can only be the water exploitation and water conservation which are closely related. Water conservation is the prerequisite of water exploitation, and new water sources can only be exploited feasibly and necessarily when available water supply is used to its fullest extent in terms of both efficiency and rationality. Hence, the water demands of socio-economic activities should be carefully evaluated with the full consideration of water conservation and sustainable water use as far as the capability of water resources in China is concerned. For this reason, the water demands of various sectors forecast in this article are the actual needs imposed by socio-economic activities within a certain time scale and based on the possible implementation of water conservation.

The water demands of irrigation are dependent on the growth of irrigation areas of China as well as the irrigation efficiency. As analyzed in the forecast of water demands, under the constraints of limited expansion of irrigation areas, by the year 2030 the total irrigation areas of China will be stabilized about to 63.9 million ha. with the total irrigation water demands up to 414.6 billion m^3 with a certain degree of water saving irrigation. But with the continued increase of water use efficiency in irrigation, the irrigation water use will tend to fall rather than rise.

For industry water use in China, the amount of water intake will tend to be stable on account of a continuous rise of water use efficiency. The industrial production may reach its maximum when the total population is in its zero growth around the

year 2030. The total water intake in industry will be most likely stabilized at some 230 billion m^3 as the industrial structures turn into high-tech industry and the water recycling rate reaches a certain level. It is evident from the industrial development of Japan that the industrial water demands will no longer increase with a certain level of water recycling rate.

The water demands of urban and rural domestic use will increase as the total population grows. By the year around 2030 the total population will reach its peak at 1.6 billion while the urban domestic use is forecast at 73 billion m^3 and the rural domestic use will be able to reach 37 billion m^3 if the water use norm in rural areas is taken as half of the norm of urban areas. Later on the water demands of both urban and rural domestic use will appear in slow growth with the continuous rise of water use norm. But during the same time, the irrigation and industrial water use will tend to be in slow decline with the increase of water use efficiency. Therefore, it is believed that by the middle of next century the zero growth of total water demands in China can be realized under the prerequisite that China also realizes the zero growth of population as well as the highly efficient and sustainable water utilization supported by huge finance and technologies, and the total water demands in China should be stabilized at some 755 billion m^3 (Table 4.16.15).

Table 4.16.15 - Comparison of water usage of current situation and the year around 2030

Years	Irrigation use		Rural domestic use		Industrial use		Urban domestic use		Total water
	Vol. $(10^9$ $m^3)$	Prop. (%)	Vol. $(10^9$ $m^3)$	Prop. (%)	Vol. $(10^9$ $m^3)$	Prop. (%)	Vol. $(10^9$ $m^3)$	Prop. (%)	use $(10^9$ $m^3)$
1980	369.9	83.4	21.3	4.8	45.7	10.3	6.8	1.5	443.7
1993	390.6	74.3	22.6	4.3	92.6	17.7	19.6	3.7	525.5
2030	415.0	55.0	37.0	4.9	230	30.5	73.0	9.6	755.0

Based on the above analysis, as the zero growth of water demands is realized, the capability of water resources will be mainly dominated with the amount of exploitable water resources. At present, the water supply capacity in China is 525.5 billion m^3 with 74 per cent provided by diversion from river channels and groundwater and only 26 per cent provided from reservoirs. If the final water demands are taken as 755 billion m^3, an increase of 230 billion m^3 will have to be implemented in water supply capacity. Hence, Is there enough exploitable water resources in China to meet the water demands in socio-economic development? If the answer is yes, how to meet such water demands?

Potential of Exploitable Water Resources

In order to increase the water supply capacity, the construction of new water facilities is indispensable besides the water supply potential of existing water facilities. But additional water supply should be based on sufficient water resources that is exploitable.

The fresh water resources in China is 2,812.4 billion m^3, in which 2,711.5 billion m^3 is surface water and 828.8 billion m^3 is groundwater (including overlapped calculation of 727.8 billion m^3). At present, the exploited water resources in China is 525.5 billion m^3, accounting for 18.7 per cent of the total water resources. It indicate that the water resources utilization rate is not high compared with that in some major countries. United States had 27 per cent of its water resources in utilization in 1975, Italy had 23.2 per cent in 1970, India had 23.5 per cent and Japan had the same degree of water utilization as China. Moreover, there is a big difference in the degree of water utilization between the north and the south, and the north has higher degree of water utilization than the south in general. For instance, the water utilization rate in Hai and Nuan river basins has greatly exceeded the limitation allowed for ecosystem consideration.

As discussed above, the water utilization rate of river flow should not exceed 40 per cent of its runoff on account of ecosystem maintenance. Thus, the possibly exploitable runoff in China would be 1084.6 billion m^3. Plus some 100 billion m^3 exploitable groundwater estimated in "Evaluation of Water Resources in China", the possibly exploitable water resources in China is around 1,184.6 billion m^3, 2.25 times the exploited water resources. However, the actual exploitable water resources should be less than the above value if water usage in river channels is taken into account, and it requires certain water depth or amount kept in river channel for the purpose of hydropower generation, navigation and aquatic breeds although they do not consume water. followings are the approximate estimates of exploitable water resources in each water districts.

The Yangtze river basin has a total water resources near 1,000 billion m^3, and its 40 per cent is 400 billion m^3 possibly allowed for exploitation. The exploited water resources for water supply has been 170 billion m^3, leaving 230 billion m^3. After the implementation of the Three Gorge Project, it will give rise to flow discharge in low flow season so as to increase the capacity of navigation and hydropower generation. Aside from the increased flow in low flow season, an approximation of 100 billion m^3 is left for further exploitation.

The Pearl river basin has a total water resources at 470 billion m^3 with 188 billion m^3 possibly allowed for exploitation. About 80 billion m^3 has been utilized for water supply leaving 108 billion m^3. Since its upper and middle reaches are rich in hydroenergy with a series of cascade hydropower stations planed or constructed which need additional water to promote their guaranteed output, it is estimated that further exploitation of water resources should not more than 100 billion m^3.

The southwest water district involves a few large rivers with the total water resources at 585 billion m^3 among which the Yarlong Zangbo River is the largest with a runoff of 165.4 billion m^3.(within the boarder), Lancang River ranks second with a runoff of 74 billion m^3, and Nu River is the third with a runoff of 70.3 billion m³. The possibly exploitable water resources is about to 240 billion m³, and the current water use is no more than 10 billion m³ remaining 230 billion m³ for further exploitation. Since this water district covering highland area will exhibit little water demands in view of its sparse population, it is the water district that offers the largest potential for water resources exploitation allowed for water transfer to other places of China.

The southeast coast rivers have a total water resources around 260 billion m³ with the possibly exploitable water resources about to 100 billion m³. There remains 70 billion m³ apart from the 30 billion m³ exploited. The rivers in this coastal area appear in middle and small size with the runoff varying dramatically between flooding and low flow seasons and being in need of regulation to ensure hydropower generation and navigation. Therefore, the actual exploitable water left should be no more than 40 billion m³.

The northeast water district has a total water resources near 200 billion m³, in which 80 billion m³ is possibly allowed for exploitation. The exploited water resources for water supply has been 50 billion m³, leaving 30 billion m³. The rivers in the northeast are highly furnish with reservoirs, and by 1985 there were 1,483 reservoirs built in the Songhua river basin with a total storage volume of 20.6 billion m³ accounting for 35 per cent of its annual runoff. The water resources for further exploitation is estimated between 15 - 20 billion m³.

The Yellow river, Hai and Nuan rivers, Huai and Shandong Peninsula and northwest inland are featured with over 40 per cent water utilization rate, especially the Hai and Nuan rivers with the water utilization rate up to 96 per cent. For the consideration of ecosystem maintenance, theses rivers should be replenished to increase river flow rather than exploited for water supply.

As discussed from the general points of view, it can be concluded that the overall potential for further water resources exploitation is approximated at 510 billion m³, more than enough in comparison with the China' water demand increment of 230 billion m³ in the next century. However, it should be recognized that most of the potential is located in the south of China and can not be supplied to the north without large scale engineering facilities. In particular, there is about 140 billion m³ potential in the Pearl river and the southeast coastal area, and the geographic condition would not allow for water transfer to the north or northwest. Fortunately, water transfer to the north and northwest can be made possible only in the Yangtze river basin and the southwest water district in view of both geographic conditions and exploitable water potential. Therefore, the capability of water resources in China can roughly match its water demands in the next century.

Key Measures for Achieving the Capability of Water Resources

According to the estimated 755 billion m³ water demands in the next century, China will require an additional 230 billion m³ water supply in the future, in which about 100 billion m³ water is able to be provided within individual river basin; and about 130 billion m³ water must be provided through the water transfer of inter-basins or regions. In the next 30 - 50 years, the key water transfer projects that need to study and implement in China are south-to-north water transfer project (from Yangtze River), Tibet-to-northwest water transfer project, water transfer project from Songhua River to Liao River and water transfer project from Ertix to north Tianshan area in Xinjiang.

<u>South to north water transfer project (from Yangtze River)</u>

The south to north water transfer project is a huge project that transfers water from Yangtze River to the north of China. After a few decade study conducted by Huai River Water Commission, Yangtze River Water Commission and Yellow River Water Commission, a three route water transfer scheme has been conceived for the project including the east route, middle route and west route.

The east route: The water supply of the east route water transfer covers the areas of 183,000 km², involving Tianjin, Hebei, Jiangsu and part of Anhui with more than 100 million population and 8.8 million ha cultivated lands. The east rout water transfer starts from the lower reach of Yangtze River near Yangzhou, conveys water to north by make use of the Great Beijing-to-Hangzhou Canal and its paralleled rivers, then crosses the Yellow River near Weishan to the Weiling Canal and South Canal and ends in Tianjin. The total length of the east route is 1150 km with 660 km located on the south of Yellow River and 490 km on the north of Yellow River. There will be 13 pumping stations installed along the east route with a total lifting height of 65 m. Several lakes and reservoirs along the route connected either in series or in parallel by the conveying canals will form a water storage system for regulation. The east route water transfer project will be constructed on the basis of the existing Yangtze River diversion project in Jiangsu Province and expanded to its final size in three stages according to the plan. After its final stage is implemented, the east route will divert 17.6 billion m^3 water a year from the Yangtze River, in which about two-fifth will be transferred across the Yellow River to Hebei Province and Tianjin.

The middle route: The middle route water transfer aims to supply water to Hubei, Henan, Hebei, Beijing and Tianjin covering a total area of 155,000 km^2 with 8.4 million ha cultivated lands and 160 million population. At present, these areas are in serious water shortage. The middle route water transfer will take water from the Danjiangkou Reservoir on Han River, a tributary of Yangtze River. The reservoir was built in 1973 in its initial size with a 163 m high dam and a 17.4 billion m^3 storage, but the foundation of the dam and power generation was constructed according to its final size designed to meet the requirement of water transfer. The Han River has a total runoff of 59.1 billion m^3 in which 40 billion m^3 is produced in the catchment above the Danjiangkou Reservoir. It is estimated that the average transferable water amount will be 14.5 billion m^3 a year with the maximum at 20.7 billion m^3 and the minimum at 6.1 billion m^3. In this amount of water to be transferred, about 1.1 billion m^3 is supplied to Qingquangou irrigation system in Hubei Province, 5.38 billion m^3 will go to Henan Province, 4.84 billion m^3 to Hebei Province and 1.6 billion m^3 to Beijing and Tianjin respectively. To fulfill the water transfer, the Danjiangkou Reservoir will have to be enlarged to its full size with a dam height of 176.6 m and a storage volume of 29.05 billion m^3. The main conveying canal will be constructed starting from the Taocha gate which was already built near the reservoir, going northwards along the east side of Taihang Mountain through cities of Anyang, Handan, Xingtai, Shijiazhuang, Baoding and finally to Beijing, and will be 1245 km in total length. After heightened to the new level, the Danjiangkou Dam will also solve the problem of 100 year floods in the lower reach of Han River and to have hundreds thousands hectare cultivated lands protected from flooding.

The west route: The west route water transfer is mainly dedicated to replenish water to Yellow River for its growing water deficits and to solve the water shortage problems in Northwest China. The overall benefited areas are 700,000 km^2 with over 6 million ha cultivated land and 110 million population, and currently short of water at 1.54 billion m^3 which is expected to grow to 26.4 billion m^3 in the next 30 years according to the estimates made by the Yellow River Water Commission. The west route water transfer is designed to implement with two sub-routes. One is the sub-route of Tongtian River to Yalong River then to Yellow River which take water from Tongtian River and Yalong River at 10 billion m^3 and 5 billion m^3 respectively. A 302 m high dam is planed at Tongjia on Tongtian River and raises water to a 158 km tunnel entering to the reservoir at Changxu on the Yalong River with a dam of 175 m, and then the water is diverted from Changxu Reservoir through 131 km tunnel to the Yellow River, and the total length of the sub-route is 289 km. Another one is the Dadu River to Jiaqu sub-route which has a water transfer capacity of 5 billion m^3. This sub-route will require a 296 m dam built at Kerga on Dadu River forming a reservoir of 9 billion m^3 in storage, and a pumping station with a pumping height of 458 m, and will involve constructing a 30 km long tunnel. It is believed that, after the completion of the west route, the water shortage in the Yellow River basin can be fundamentally changed and about 1.7 million ha additional irrigation area can be created. Other benefits include promoting local socio-economic development, improving environment condition and stopping the desertification in the great Northwest China.

Tibet to northwest water transfer project

The Tibet to northwest water transfer project is a newly proposed program now in a stage of tentative study and it plans to bring water from the rivers on Qin-Tibetan Plateau such as Jiansha River, Lancang River, Nu River, and Yarlong Zangbo River to the Yellow River, Qinghai Province, and Xinjiang Uygur Autonomous Region. The great northwest covers the areas of Xinjiang, Qinghai, Gansu, Ningxia, and part of Inner Mongolia and constitutes almost the half of China's territory with population only one-twelfth of China's total. The GDP of the northwest is no more than 7 per cent of the national GDP presenting a low level of socio-economic development. However, there are vast wastelands in Tarim basin, Caidam Basin and upper and middle Yellow River catchment reclaimable for farming, grazing and afforesting with huge agricultural potential as long as water is accessed. The great northwest is also known as the rich reserves of coal, petrol, and other mineral resources and will comes to an important support for the rise of economic development of China in the next century. In addition, the ecosystem in the northwest becomes fragile due to water shortage as desertification is aggravated tending to threaten the eastern part of China, and the Yellow River's lower reaches run dry due to the growing water demands.

However, there are limited sources of water around the great northwest except its southeastern neighboring area - the Qin-Tibetan Plateau, where four large rivers (Yarlong Zangbo River, Lancang River, Nu River and Jinsha River) are located with a total annual runoff over 400 billion m^3. Since there are only little water demands in these mountainous areas, the abundant water resources makes the water transfer to be possible. The Yarlong Zangbo River is the third largest river in China in terms of water quantity with an annual runoff of 165.4 billion m^3. It originates from Tschemajungdung Glacier in the middle of Himalaya and runs from west to east through south Tibet. The river course turns to northeast near Pai in Milin County and then take a sudden turn to the south forming the well-known great U-turn in the lower reach of the river. After the great U-turn, the Yarlong Zangbo River enters India and is known as Brahmaputra River. The great U-turn starts from Pai with el. 2880 m In Milin County to Beiben with el. 630 m in Motuo County. The river course between them is 240 km and the direct distance between them is about 40 km with a drop of 2250 m. The average annual flow discharge is 1900 m^3/s with an annual runoff of 60 billion m^3 allowing for erecting a huge hydropower station with 40,000 MW installed capacity. It is planed to build a diversion tunnel that goes from Pai through Doxiongla Pass (el. 4500 m) and arrives at the upstream (el. 2880 m) of the Doxiongla River, a tributary of Yarlong Zangbo River. The total length of the tunnel is estimated to be 16 km long, of which 11 km is under a top cover less than 1000 m deep and 5km under a top cover greater than 1000 m deep. Since the tunnel will not go through under the Namjagbarwa Summit, the maximum depth of top cover is no more than 1500 m and not deeper than that expected before. The Doxiongla River runs 40 km from northwest to southeast and joins the Yarlong Zangbo River at Beiben where the hydropower plant called the Great U-turn Hydropower Station is located. It is suggested that the diversion tunnel should be followed by a 20 km long steel penstock connected with the power house on account of the steep slope of the Doxiongla River.

The size of the Great U-turn Hydropower Station will be dominated by the degree of the flow regulation in the Yarlong Zangbo River. According to the hydrological records from 1956 to 1989 at Nuxia Gauging Station, the lowest flow recorded over the years is 282 m^3/s corresponding to a 5,400 MW installed capacity while the highest flow is 8030 m^3/s corresponding to a 153,600 MW installed capacity being 28 times the lowest flow. However, full regulation of river flow requires building larger and more reservoirs that will cost much and are constrained by on-going technologies. Therefore, it would be feasible to implement the huge hydropower station in several stages according to different development conditions and in this way many difficulties can be avoided such as huge generating units, transportation, engineering geology and finance. The first stage is suggested to deal with the major works of a diversion tunnel and a hydropower station with an installed capacity of 10,000 MW which requires low degree of flow regulation. With the progress of national economy and technologies, the hydropower will be exploited to its fullest extent.

Based on the primary estimation, the Great U-turn Hydropower Station is expected to use an amount of 40 - 45 billion m^3 water to generate electrical power used mainly for pumping water from the four rivers. It is planed that one third of the power is to be used for water pumping from the Yarlong Zangbo River for lifting 20 billion m^3 water; one ninth of the power is for water pumping from the Jinsha River with a volume of some 8 billion m^3; another one third of power is for water pumping from the Lancang and Nu rivers for pumping 16 billion m^3 water; and two ninth of the power is for electrical power supply to both local and neighboring countries.

According to the latest maps and field investigations, the conveying route of water transfer is possible. Covered canals are constructed on the highland with the elevation ranging 4,050–4,400 m, and the route starts from the middle reach of Yarlong Zangbo River near Yongda, goes northeast across the Nu River, Lancang River and Jinsha River and arrives in Zaling and Eling lakes situated in the headwaters of Yellow River. The total length of the route is 1,113 km, in which tunnels make up 265 km, 23.8 per cent of the total route with the longest tunnel in 35 km, and inverse siphons make up 99 km, 8 per cent of the total route. There are six pumping stations arranged along the route on the rivers and they are the Yongda Pumping Station on the main stream of the Yarlong Zangbo River with a lifting height of 400 m, the Pangduo Pumping Station on the Lasa River with a lifting height of 200 m, the Songqu Pumping Station on the Palongzangbu River with a lifting height of 400 m, the Reyu Pumping Station on the Nu River with a lifting height of 700 m, the Nanqian Pumping Station on the Lancang River with a lifting height of 600 m and the Yeduo Pumping Station on the Jinsha River with a lifting height of 400 m. The total water catchment areas above the pumping stations are 224,000 km^2 with the annual runoff of 63.43 billion m^3, and the total amount of water to be transferred is 43.5 billion m^3 about 10 per cent of the total runoff in the four rivers and 68 per cent of the runoff above the pumping stations (Table 4.`6.16).

Table 4.16.16 - Location of water to be transfered in Tibet to north water transfer project

Location	Rivers	Eleva tion (m)	Catch ment area (km^2)	Runoff (10^{9}m^3)	Trans Fered Water (10^{9}m^3)	Remainig water (10^{9}m^3)
Yeduo	Jinsha River	3,990	125,000	9.93	8.00	1.93
Nangian	Lancang River	3,700	13,800	4.50	3.00	1.50
Reyu	Nu River	3,500	65,000	21.00	13.00	8.00
Songqu	Palong zangbu	3,950	3,862	4.00	3.50	0.50
Pangduo	Lasa River	4,000	16,700	6.00	3.00	3.00
Yongda	Yalong Zangbu	3,680	111,400	18.00	13.00	5.00
Total			335,762	63.43	43.50	19.93

After the whole project is implemented, hundreds thousands hectares of irrigation areas can be added to the great northwest, a large area of afforestation on wastelands can be made possible and sustained flow in Yellow River can be ensured in addition to the full water supply to the industry in the upper and middle reach of Yellow River. Further water transfer to the Caidam and Tarim basins can also made possible from the Zaling and Eling lakes. Moreover, by making use of the dropes over 2000 m along the water transfer route, additional four or five huge hydropower stations can be built which can replace thermal power plants to save coal and improve the quality of atmosphere.

Water transfer from Songhua River to Liao River

The Northeast China is quite rich in water resources and there two major river basins, the Heilong River on the north and the Liao River on the south, with a total annual runoff of 193 billion m^3. But the water use in the two river basins is different. The Songhua River, a major tributary of Heilong River, has low water utilization rate with about 16 billion m^3 water exploited out of its 76.2 billion m^3 water potential. On the contrary, The Liao River has its water resources exploited to more than 80 per cent, and further exploitation is almost impossible. The lower reaches of the Liao River are the major economic development areas in the northeast, where water shortage is in crucial situation with a current water deficit up to 1.14 billion m^3. In addition, the Songhua River basin is also suffering water shortage due to the low degree of water exploitation especially in cities. More important is that the flood prevention capability is lower in the Songhua River basin with most levees capable of preventing only 5 - 10 year floods.

To mitigate the water shortage in the northeast of China, a general plan of water transfer from Songhua River to Liao River has been conceived and involves the integrated water resources development for both Songhua River and Liao River. This plan requires constructing the Buxu and Dalou Reservoirs on the Nen River, the upper reach of Songhua River and the Hadashan Reservoir on the Second Songhua River. On account of the complex problems of the inundation losses in Dalou Reservoir area, it is suggest to build the Buxi and Hadashan reservoirs in the first stage. The Buxi Reservoir controls a catchment area of 65,095 km^2 with an average annual runoff of 10.4 billion m^3, has a total storage volume of 9.88 billion m^3 with a normal storage level of 219 m and involves inundating 27,300 ha farming land and resettling 58,500 residents; The Hadashan Reservoir controls a catchment area of 72,800 km^2 with an average annual runoff of 15.4 billion m^3, has a total storage volume

of 4.21 billion m^3 with a normal storage level of 147 m and involves inundating 11,700 ha farming land and resettling 41,800 residents. The diversion canal can easily excavated and makes its way through the lowland areas in the western part of Jilin Province and the gentle divide between Songhua River and Liao River with a total length of 400 km, and conveys an amount of 7.82 billion m³ water in 75 per cent dry year providing reliable water source for the socio-economic development in the Liao River basin.

Water transfer from Ertix River to north Tianshan area in Xinjiang

The Ertix water diversion project is a plan which brings the water from Ertix River in the north of Xinjiang to the Urumqi economic zone and Klamay Region in north Tainshan Mountain area. The diversion project begin from the middle reach of the Ertix River at the mountain outlet where a reservoir, called "635" water project, will be constructed serving as the water source for the diversion. Then the water transfer route goes to south across the Ulungur River and the desert area of the Junggar Basin and ends at the Mengjin Reservoir near Urumqi. The total length of the route is 530 km. The diversion project aims at meeting the increased competing demands of the industrial water use in the Urumqi economic zone and the north Xinjiang oil fields, and water supply will also be extended to the agricultural use and pasture irrigation in the farms adjacent to the diversion route.

The north Tianshan Mountain area is located along the southern periphery of the Junggar Basin characterized by inland arid climate and has limited water resources only capable of maintaining oasis economy and the low level water use in industry. The north Xinjing oil fields, with its base in Karamay - the key industry of the north Tianshan Mountain area, have suffered from water shortage which constraints the oil exploitation and economic development. In recent year the Urumqi economic zone has seen a rapid growth in economic development and urbanization with a high utilization degree of local water resources. With the total water resources within the Urumqi economic zone only at 2.1 billion m³, the water utilization rate is over 70 per cent for surface water and up to 140 per cent for groundwater implying difficulties for further water supply. It is anticipated that the Urumqi economic zone will have a water deficit of 0.25 billion m³ in 2000, 0.474 billion m³ in 2010 and 0.748 billion m³ in 2020. Therefore, in view of either near future or long term future, the water transfer to the Urumqi economic zone and Karamay Region is necessary and urgent.

The Ertix River basin is situated in the northern periphery of the Junggar Basin in Altai Prefecture and one of the three international rivers in Xinjiang. It drains a catchment area of 52,730 km^2 on the territory of China with a total annual runoff of 11.2 billion m^3. Since there are no sizable water facilities on the Ertix River, the river flow can hardly be harnessed with up to 9.49 billion m^3 water running out of the country each year. The "635" water project planned on the Ertix River controls a catchment area of 16,000 km^2 with an average annual runoff of 3.27 billion m^3, and provides favorable conditions in terms of geographic location and water quantity for the water transfer.

For the huge costs of the whole water transfer scheme, it is planned to implement it in three stage. The first stage will involve constructing the "635" Reservoir and a 136 km main conveying canal to Dingshan distribution gate followed by two branch canals to Urumqi and Karamay individually. The "635" Reservoir will have a storage volume of 0.282 billion m^3 with a normal water level at 645 m and will supply water to Urumqi at an amount of 0.357 billion m^3 and to Karamay at an amount of 0.13 billion m^3. The second stage is arranged to erect the Kalasuke Reservoir 10 km upstream of "635" Reservoir. The Kalasuke Reservoir will have a storage volume of 2.5 billion m^3 with a normal water level of 740 m which can better harness the runoff above the "635" Reservoir, and will increase the water supply to 0.748b billion m^3 for Urumqi and to 0.188 billion m^3 for Karamay. The third stage is to build a reservoir (the Shankou water project) on the Buerjin River, a tributary of the Ertix River, which will harness the 4.28 billion m^3 water of Buerjin River. It aims to transfer water from Buerjin River to the "635" Reservoir having the whole water transfer scheme expanded to its final size with the targets of water supply at 1.5 billion m^3 for Urumqi and 0.5 billion m^3 for Karamay.

Conclusion

At present, the exploited water resources in China is 525.5 billion m^3. As discussed in this article, the total water demands in 2030s will be up to 755 billion m^3 exhibiting a water deficit of 230 billion m^3, while the water resources potential left for further exploitation is approximated at 510 billion m^3. It will be certain from overall points of view that the exploitable water resources potential is likely to meet the water demands of China's socio-economic development in next entury. However, it must be pointed out that the spatial and temporal distribution of water resources in China exhibits a great unevenness and does not match that of population and land resources. Unless the necessary and effective measures are adopted, the water supply and demands in China will still be in crucial situation. Above analysis reveals that the exploitable water resources left for further exploitation in China is mostly concentrated in the south of China, and particularly about half of the exploitable potential is located in the southwest water district where the water exploitation seems with difficulties. On the contrary, about 11.5 million ha irrigation areas out of the new increase of 14 million ha irrigation areas in the next century is located in the north of China. The irrigation water use will be growing inevitably even under the conditions of the widespread extension of water saving irrigation technologies in China. Plus the increase of industrial and urbane water use, at least 130 billion m^3 of additional water supply will have to be added to the north of China. Therefore, the fundamental measure to achieve the capability of water resources in China is to transfer the abundant water from the southwest water district to the north of China. Otherwise, the water shortage in the north will be unavoidable.

The development of irrigation agriculture is the fundamental way to meet the grain demands of 1.6 billion population in the next century. At present, the irrigation water use constitutes over 70 per cent among the entire water use sectors. With the rapid progress of industrialization in China, the water competition between agriculture and industry will be continuously intensified which will cause the decrease of the proportion of irrigation water use. Subject to the constraints of new exploitable water resources, the possible extension of irrigation areas should be mainly dependent on the water saving irrigation technologies. As estimated in this article, with the promotion of water saving irrigation technologies and the optimal regional reallocation of water resources, great efforts must be made to increase the irrigation areas from current 49.9 million ha to 63.9 million ha in 2030s. In addition, with the help of advanced farming technologies, the irrigation efficiency (grain yield per unit water usage) is expected to boost from current 0.85 kg/m^3 to 1.46 kg/m^3 in 2030s. if these two targets are met, the expected grain yields of 726 billion kg should be fulfilled in 2030s without much increase of irrigation water use.

Acknowledgments

The facts and data about the south to north water transfer project come from the relevant study reports conducted by Huai River Water Commission, Yangtze River Water Commission and Yellow River Water Commission; The data about the water transfer from Songhua River to Liao River come from "Water Transfer from North to South is a Strategic Project in the Northeast of China" written by Wang Danyu; The data concerning Ertix water diversion project come from "Feasibility Study of the First Stage of Ertix Water Diversion Project", Xinjing Water Resources Department. Thanks.

References

[1] Hydrology Bureau of Ministry of Water Resources, 1987, Evaluation of Water Resources in China, National Water Resources and Hydropower Press.
[2] Planning and Design Institute of Water Resources and Hydropower, Ministry of Water Resources, 1989, Utilization of Water Resources in China, National Water Resources and Hydropower Press.
[3] Chen, Jiaqi & Wang, Hao, 1996, Outline of Water Resources Study, National Water Resources and Hydropower Press.
[4] Zhang, Hongren, 1990, Optimal Reallocation of Surface Water and Groundwater in North China.
[5] Chen, Zhikai, 1996, Well Managing, Rational Using and Sound Protecting the Limited Water Resources, Forum of Water Problems, No. 2, 1996.
[6] Liu, Shanjian, 1997, Population, Grain and Water, Forum of Water Problems, No. 2, 1997.
[7] Liu, Changming & He, Xiwu, 1996, The General View of Water Issues of 21st Century in China, Science Press.
[8] Zhao, Jubao & Li, Kehuang, 1995, Aridity and Agriculture, National Agriculture Press.
[9] Hou, Guangliang, Li, Jiyou & Zhang, Yiguang, 1993, Agricultural Climate Resources in China, Press of China Peoples University.
[10] He, Xiwu, 1993, On the Strategy of Water Resources in China in 21st Century, In: Resources in China: Potentials, Trend and Strategies, Proceedings of Geo-science Forum of Chinese Academy of Science, Beijing Press.

[11] Shi, Yulin, 1985, Arable Wasteland Resources in China, Beijing Science and Technology Press.

[12] Liu, Changming & Ren, Hongzun, 1995, Sustainable Development and Utilization of Water Resources in China, CCAST-WL Workshop Series Vol. 49.

[13] Zhang, Yue, 1995, Development Strategy of Water Resources in China, National Water Resources and Hydropower Press.

[14] China Year Book of Statistics in 1996, 1997, National Statistics Press.

CHAPTER 4.17

New City Designation in China

Mei-Ling Hsu
University of Minnesota

When the People's Republic of China was founded in 1949, there were only 132 cities in this very large country (Hsu, 1994). Since then the number of Chinese cities has risen almost yearly to reach a total of 663 in 1996 (Zhang, 1997). The era of reform of the Chinese economy began in 1978. In recent years, the economy has grown steadily at a commendable annual rate of between six and ten percent and with a tolerable factor of inflation (Liu, 1996, pp. 1-26). What we witness in China today, then, is a period of considerable economic growth and even more impressive urban expansion.

One must be careful, however, in interpreting the relationship between these two phenomena. To what extent are they correlated; to what extent has the former stimulated the latter? What we do know is that, for example, in some parts of Africa and Latin America, rapid urban expansion has occurred without the support of adequate economic development. Has this also happened in China? The Chinese have a set of official procedures for establishing new cities which, as may be expected, has been revised from time to time. Two recently published documents which outline these procedures will be discussed below.The rapid increase in the number of urban places in China is a recent phenomenon, especially concentrated in the reform era; it remains to be seen, however, whether this increase is a positive factor in the country's drive for modernization. In this paper, I shall examine the following topics: (1) the criteria used in the official city designation process, (2) the increasing numbers of Chinese cities in different periods of time between 1953 and 1997, (3) other important Chinese policies and view points concerning the establistunent of new cities, and (4) Chinese urban development in relation to other aspects of the socioeconomic progress in that country.

The Designation of New Cities

All new cities in China are established through an official procedure of designation. In 1986, during the first decade of the economic reform the government issued a State Council Circular entitled "On Adjustment of Standards for City Designation and Conditions for Cities to Administer Counties" (Hsu, 1994; Zhu, 1987). It specified the requirements for designating a new city in terms of the city's GNP and the size of its nonagricultural population.

In the Chinese literature, the term "nonagricultural" usually is not used to denote a person's occupation. Rather, an individual is identified as "nonagricultural," if his or her household registration is under the jurisdiction of an urban place. This is an important distinction in China, because until recently, two very different kinds of employment have existed: that offered by the state run (national) agencies mostly in urban areas, and that offered mostly by local agencies in small towns and in rural areas. The former has provided much better benefits and has been more stable, making it highly desirable. Recently, with the implementation of reform policies, some of these benefits have been eliminated, however.

Furthermore, some exceptions to the 1986 official standards have been made in special areas as follows: in places close to the international border, at special tourist sites, and in areas where a large proportion of the population consist of "national minorities" (Zhu, 1987). The 1986 Circular, however, did not stipulate that the city have particular urban amenities and facilities. For example, the question of how well the city could or should function as a socioeconomic center to serve the community was not raised.

The city designation standards have been revised since the 1986 circular, and, more recently, a major statement was issued entitled "The 1993 Standards For City Establishment" (SSB, 1994, 216-17). It provided one short section on the standards for establishing cities at the prefecture level (diji), and a rather lengthy section on the requirements for a county (xian) to be up - graded to a county - level city (xianji shi). The complete set of specifications is too lengthy to be quoted in this paper; a brief discussion is given below.

The prefecture - level city must have a population of 250,000 or more, of which at least 200,000 are in nonagricultural households having non - famiing occupations. In addition, the city should meet a set of measures of development with regard to secondary and tertiary industries.

Three types of county level cities can be established each with a different population density: over 400 persons, 100 to 400, and below 100 persons per square kilometer. Furthermore, in order for a county (xian) to be reclassified a city, it must meet other criteria. For example: (1) a required percentage of people who are registered as "nonagricultural" must actually be engaged in such economic activities, (2) between 60 to 80 percent of a place's "total industrial and agricultural output value' must be generated from the industrial sector, (3) cities must meet a number of basic budget requirements, and lastly (4) new cities must provide for their residents the basic urban infrastructure, roads, water supply, sewerage system, etc.

This set of specifications reflects the govenunent's attempt to sharpen the definition of an urban place. The density specification should lead to a decrease in the degree of overboundedness, which currently exists in many Chinese cities. Likewise, these specifications aim at upgrading the economic activities in cities by developing more of the secondary and tertiary industries. Lastly, we note that this recent document does include a number of requirements concerning the question of how well a new and existing city can function to properly service its population.

The Expansion of Urban China

As stated earlier, in 1949 there was a total of 132 cities in China. This number increased to 163 in 1953, at the beginning of the First Five-year Plan. Thus there was a net gain of 31 cities within a four - year period (Hsu, 1994). From 1953 to 1990, a net of 304 new cities were established (Table 1); in fact, 274 of them were founded within the last 12 years of this period, between 1978 - 1990. There were in all 467 cities in 1990. The expansion trend has continued, as there were 663 cities in 1996, a net increase of 196 new cities within a short period of six years from 1990. In sum, the total 531 new cities were added within one half of a century, a six - fold increase!

A supplementary note should be provided as follows: the figure of the total number of Chinese cities in a given year should be arrived by subtracting from the total number of From 1953 to date, most years have shown an increase in the number of cities (Table 4.17.1); however, there have been some important exceptions. For example, in 1961 and 1964, following the Great Leap Forward movement (GLF, 1958-61), the total numbers of cities was 208 and 167 respectively, i.e., an increase before 1958 followed by a reduction. This was a period of time during which the Chinese had serious economic and political difficulties, and it has been referred to be "three disastrous years' in the literature. The govenunent statistics from this period also are known to be less reliable.

During the early 1960s, a large number of farmers migrated into the cities, a flow termed "mang liu" (blind flows); meanwhile the government also drafted some 10 million farm laborers for city construction work (Zhou and Guo, 1996, pp. 1-46). Later many of these people were forced to return home, because the city government could not accommodate them. As a result, the urban population increased rapidly for a short period of time and then decreased somewhat.

In sum, during the Great Leap Forward period, the goveniment established some 24 new cities. The urban population was 19.7 percent of the total population, a "degree of urbanization that was viewed too high for China's level of development at that time (Ye, 1994, pp. 28-30). It also was quite expensive for the government to provide for the daily needs of the burgeoning city populations such things as vegetables, meat, etc. The food supply had to be done, because it was important for the socialist government to maintain proper order in the cities, a lesson it learned from the failure of the Nationalist government some decades before.

Shortly afterwards, the government changed its policy and began to control the number of urban places and the size of their populations. Between 1961 and 1965, the net outmigration rate from the cities was 17.6 per thousand (Ye, 1994, p. 29). We note, in fact, in the early 1960s, that there was a decline in both the number of cities and the total urban population (Table 4.17.1).

The cultural revolution which took place between 1966 and 1976 is well known for its anti urban stand. As expected, during these ten years many cities suffered losses from outmigration; moreover the creation of new cities was limited. These difficulties are reflected in the fact that, during a period of eleven years, the total number of Chinese cities increased only by seventeen.

Toward the end of the 1970s, the country tried to recover from the cultural revolution and other related difficulfies. In 1978, the nation entered the era of reform and since then the process of urban expansion has accelerated. In that year, there was a total of 193 cities in China. In comparison with the previous decades, the 1980s represented a period of more orderly and rapid urban expansion, and this trend has continued into the current decade. The total number of cities in China in the years 1990, 1993 and 1996 was 467, 570 and 663 cities, respectively (Hsu, 1994; SSB, 1994; Thang, 1997,) In sum, between 1978 and 19%, there was an increase of 470 cities, an average of 26 cities per year! To date, such a sustained record of urban expansion has no precedent in world history. Furthermore, based on recent experience, one can only predict that the number of Chinese cities will continue to increase in the near future!

The Distribution of New Cities

Figure 4.17.1 shows a total of 146 new cities which were established between 1993 - 1996 (Note: the map does not indicate the cities which were abolished or combined during this period). The four provinces which gained the largest numbers of new cities were Guangdong (29), Jiangsu (13), Sichuan (II), and Henan (10) (SSB, 1996). Since 1978, the communist government has begun to take the "capitalist road"[9] to modernization, thus a rapid development has been encouraged in the coastal provinces. For example, many special privileges have been granted to Guangdong Province, and particularly to its capital, Guangzhou, because of its proximity to Hong Kong, and the fact that it is the "native home" of a large number of overseas Chinese. These have been important investors in the Chinese economy. The southern part of Jiangsu Province, between Nanjing and Shanghai, has been the area of fastest urban-economic growth in the nation (Fig. 4.17.1). Also a number of secondary centers were selected earlier as foci for development, e.g., central Henan, during the period of the First Five-year Plan, and central Szechuan, during the "third - front" development (1962 - 1965), thus emphasizing the development of cities in central and interior China (Hsu, 1996).

Cities as socioeconomic Centers

We have just discussed the impressive expansion of urban China during the past half century. We know much less about how well these recently designated cities Itinction as socioeconomic centers to serve the society at large. In an earlier study, the author made an effort to inquire into this question. Social data for cities designated in two different periods were compared, i.e., those established in 1986 /1987 and those established in 1994. It was assumed that cities which had a longer time in which to develop would become better socioeconomic providers than those designated only recently.

Table 4.17.1 -- URBAN PLACES AND POPULATIONS 1949—1990

Year	New ly desig nated	Re - desig nated	Ca n cel led	Com bi- ned	Num ber of cities	(1)	(2)	(3)	Num ber of towns
1949					132	39,491	27,406	7.3	2,000
1950	28		14	2	144				
1951	15	2	4	1	156				
1952	5	1	9		153	47,880	34,910	8.3	
1953	7	6	2	1	163				5,402
1954	4		1	1	165				
1955	2		2	1	164				4,487
1956	11			1	174				3,742
1957	4		1	1	176	70,773	54.127	10.9	3,596
1958	16	1	9		184	92,184	60,667	14	
1959	2		7		179				
1960	20	3	1	2	199				
1961	10	3	4		208	101,325	69,063	15.4	4,429
1962	1	14	1		194	94,611	64,152	14.1	
1963			16	1	177	90,494	64,921	13.1	4,219
1964		2	11	1	167	89,782	66,031	12.7	3,148
1965	3				168	88,576	66,906	12.2	
1966	1	2			171				
1967		1			172				
1968					172				
1969	1	3			176	93,241	66,449	11.2	
1970	1				177				
1971	1	3			181				
1972	1				182				
1973				1	181				
1974					181				
1975	3	2		1	185	105,357	74,019	11.4	
1976	3				188				3,260
1977		2			190				
1978	3				193	116,571	79,867	12.1	
1979	12	11			216				3,228
1080	1	6			223	134,184	90,830	13.6	2,879
1981	10				233				2,843
1982	8	4			243				2,819
1983	38	10		4	289				2,781
1984	10	2		1	300				6,211
1985	19	5			324	212,312	118,217	20.2	
1986	28	1			353	233,177	122,002	21.9	
1987	31	2			381	262,450	129,747	24.3	
1988	53				434	298,218	140,341	27.2	
1989	16				450	317,622	146,256	28.6	
1900	17				467	335,428	150,378	29.3	9,583

(1) City populations (In thousands); (2) Nonagricultural population in cities (In thousands); (3) As percent of China's total population. Sources: Zhang, 1990; Ren, 1990 and 1991

557

The types of data needed for this Idnd of inquiry are difficult to obtain; those which were available to the author for analysis: were per thousand of the city population having: (1) numbers of middle school teachers (2) numbers of students in the same schools, (3) the amount of technical personnel (ke ji ren yuan), and (4) the number of health workers. Unfortunately the results of this study were not conclusive; perhaps the data quality was not good enough. They did show that the spatial factor was important, i.e., the coastal cities fared better than irjand places. Otherwise, the data generally revealed a low level of this kind of socioeconomic development within Chinese cities.

New City Establishment: Beyond the Numbers

The recent increase in the number of urban places in China should mean that the degree of urbanization in the country has also risen rapidly. Usually one could confirm this by noting the percentage of urban population of the total population reported in the national censuses of different years. The first census of the People's Republic was taken in 1953, but unfortunately, since that year the definition of the Chinese urban population has altered from one census to another.

Fortunately, however, in one recent publication, China *Population Statistics Yearbook 1996*, the editors made an effort to redefine the term "urban population' slightly and provide a common base so that it is possible to compare the data from 1951 to 1995 (Table 4.17.2) (SSB, 1996, p.364). First of all, this study did not use the earlier census definitions nor the urban data in the censuses. In this publication, it indicates: (1) figures prior to 1981 are taken from the annual report of the Ministry of Public Security. The term "urban population" refers to the total population living within the areas under the jurisdiction of the city and town. Some towns are under the jurisdiction of the county.

(2) "Figures of 1982 -1995 are adjusted based on the 1990 National Population Census. "Urban population' refers to people living in city districts (in cities are so divided), or street conunittees (in cities which are not divided into districts), as well as those living in neighborhood committees in towns under cities and counties.

We note that since 1951 in general the percent of urban population in China has increased (Table 4.17.2).

TABLE 4.17.2 URBAN AND RURAL POPULATION, 1951-1995

Year	Total population	Urban population	Population proportion	Rural population	Population proportion
1951	56300	6632	11.78	49668	88.22
1952	57482	7163	12.46	50319	87.54
1955	61465	8285	13.48	53180	86.52
1960	66207	13073	19.75	53134	80.25
1961	65859	12707	19.29	53152	80.71
1962	67295	11659	17.33	55636	82.67
1963	69172	11646	16.84	57526	83.16
1964	70499	12950	18.37	57549	81.63
1965	72538	13045	17.98	59493	82.02
1966	74542	13313	17.86	61229	82.14
1967	76368	13548	17.74	62820	82.26
1968	78534	13838	17.62	64696	82.38
1969	80671	14117	17.50	66554	82.5
1970	82992	14424	17.38	68568	82.62
1971	85229	14711	17.26	70518	82.74
1972	87177	14935	17.13	72242	82.87
1973	89211	15345	17.20	75866	82.80
1974	90859	15595	17.16	75264	82. 84
1975	92420	16030	17.34	76390	82.66
1976	93717	16341	17.44	77376	82.56
1977	94974	16669	17.55	78305	82.45
1978	96259	17245	17.92	7901	82.08
1979	97542	18495	18.96	79047	81.04
1980	98705	19140	19.39	79565	80.61
1981	100072	20171	20.16	79901	79.84
1982	101654	21480	21.13	80174	78.87
1983	103008	22274	21.62	80734	78.38
1984	104357	24017	23.01	80340	76.99
1985	105851	25094	23.71	80757	75.48.
1986	107507	26366	24.52	81141	76.29
1987	109300	27674	25.32	81626	74.68
1988	111026	28661	25.81	82365	74.19
1989	112704	29540	26.21	83146	73.79
1990	114333	30191	26.41	84142	73.59
1991	115823	30543	26.37	85280	73.63
1992	117171	32372	27.63	84799	72.37
1993	118617	33351	28.14	85166	71.86
1994	119850	34301	28.62	85549	71.38
1995	121121	35174	29.04	85947	70.96

China Population Statistics Yearbook 1996, p. 364.

One obvious exception was the 1962 - 63 period, during and after the Great Leap Forward Movement. As mentioned earlier, this period has often been referred to in the Chinese literature as the "three disastrous years." Lastly, we note that the yearly total and the percentage of the Chinese rural population simply who are not included in the urban population (Table 4.17.2).

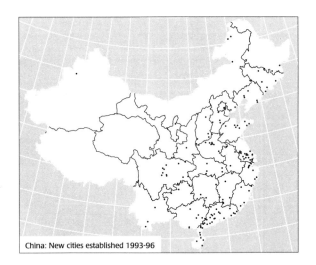

China: New cities established 1993-96

Fig. 4.17.1

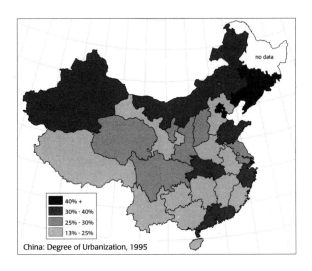

40% +
30% - 40%
25% - 30%
13% - 25%

China: Degree of Urbanization, 1995

Figure 4.17.2

560

City: A 'Valuable Entity"

In contrast to the complexity involved in defining Chinese urban population, the types of cities are surprisingly few. At the top of the Chinese urban system are the three provincial - rank cities: Beijing, Shanghai and Tianjin. These cities are the most difficult places for an outsider to become a permanent resident, i.e., to be registered in the city household system. Below this top rank are two other levels of cities, the prefecture (diqu) and the county (xian) level cities. Lastly, there are two types of towns (zhen), one under city jurisdiction and the other under the county jurisdiction. This study does not consider town development.

Why has the number of Chinese cities increased almost yearly? One simple answer is that nearly every official and average Chinese would like to live and work in a city. Cities offer many opportunities for growth and development both for the citizenry and for the local government. in a developing country, such as China, cities are more convenient places to live than the countryside. A large city which has a large budget can provide more benefits and services to its residents. It is not very surprising that for many Chinese living in a large city is a "valuable" asset and a status symbol as well. The status of a city is particularly important in this era of reform. Cities which are located along the coast, large rivers and trunk transportation lines are often granted special status and / or economic and political privileges by the Beijing Government. These include pernfission to receive foreign investments, to participate in joint - venture businesses with foreign partners and / or with overseas Chinese. In short, the "city" is so important a status symbol and vehicle for development that some administrators and scholars in Guangzhou City, for example, have suggested abolishing the xian and combining the city and countryside into one entity (Situ, 1996, 1-12). Guangzhou, which is located along the Chinese southern coast and near Hong Kong, today has one of the fastest growing economies in China. Briefly, the conversion process would include two actions:

(1) "che xian she shi," (abolishing the xian and replacing them with newly established cities, and
(2) "cheng xiang yi ti hua," (merging the city and the 3dang governments into one entity).

If such a plan were to be drawn up for the Guangzhou area, it could be successful, because in this area the degree of urbanization is higher and econon-fic reform has been more thorough than elsewhere in China. Nevertheless, this is just one more example of the common tendency of the Chinese government to avoid a direct approach to problem solving rather than to change the administrative system! Abolishing The Yian and Expanding the City Function?

In a short but informative article entitled "Will there be a system of xian's in the Twentyfirst Century China?" (Yi, Ling, 1996, 28-30). The author discussed several points which already covered in this paper, but some additional points are revealing. He cites the hundreds of new cities that were established in recent decades and predicts that, by the year 2000, there will be over 1,000 cities in China. Some of the existing cities also will be up-graded from the xian to that of the prefecture level. Why there has been such an urban expansion? He notes the importance of the expansion of administration in terms of both the stuff and office space. In addition, there were the increase of several types of expenses which were related to the city employees' pay and benefits. In short, we are witnessing an expensive exercise which benefits the bureaucracy but not the city itself and its citizenry.

New Cities Functioning as Urban Places: Example I

We have teamed about the rapid process of Chinese city designation in the 1990's; the next question to be asked should be "how have these new cities been functioning as urban places?" To date there is surprisingly little information concerning about this. The author has been able to find only a few recent studies which examine the effects of urban expansion and city establishments in the period of late 1980s and the 1990s. One such study analyzes the new urban places in the area of Zhu Jiang delta, near the city of Guangzhou (Wei, 1996, pp. 1-8).

On the eve of the reform in 1978, there were only five cities and 32 towns in this delta area, with a density of nine cities and towns per 10,000 kM2. In 1994, there were 25 cities and 392 towns in this area, with a density of 100 cities - towns per 10,000 km2. The average distance between two cities / towns was approximately just IO km. This delta is quite close to Hong Kong; it is not an average region in China. Other reasons for its recent rapid urban expansion in this area have been the rapid local economic growth and modernization, and administrative changes that have related to urban expansion. The latter include the abolishment of xian and xiang (township), and the establishment of many new cities and towns (Wei, 1996, pp. 1-8). More discussions on this topic as follows.

During the study period, there were twenty new cities being established which included such well known ones as Shenchen (now over one million population) and Zhuhai (400,000 people), the former once a small town and the other, a small fishing village. In fact, within many new cities, there are extensive area of agricultural fields and a number of practicing farmers. Meanwhile, many of the officials of these new cities have been quite causal in making decisions on converting farmland to other uses. The city of Zhongshan provides another kind of example for urban expansion. Within this city there are only ten streets and 237,000 inhabitants in the central district. Yet in the official statistical records this city has a total area of 12,000 km2 and includes 27 towns (chen's) and a total population of 1,216,000. This is indeed a description of a large city. The author of this study calls this phenomenon a "false degree of urbanizatioif" (Wei, 1996, 1-8).

New Cities Functioning as Urban Places: Example H

Limited to the availability of the reference materials concerning this topic, the location of the second example which may be cited also a study of the cities in Guangdong Province (Li, 1996, 1-1 1). Nevertheless, many of the observations made in the study do have national applications.

In the 1980s, during the first decade of the reform era, the establislunent of urban places accelerated in China, particularly in some more progressive provinces in the southern part of the country, such as Guangdong. In these areas, both agricultural and as well as rural commercial activities expanded rapidly. Meanwhile the general problem of over - population in the countryside could no longer be ignored. In 1984, the government issued two documents to relaxed the restrictions of farmers' movement. They were allowed to reside, work and set up businesses in towns (not in cities). However, those who did so were required to supply their own food, "zi li kou liang," This meant that these immigrants were not to be dependent on the government food supplies. We note once more the association between food supply and the Chinese urban development.

The number of cities in the province also was increased rapidly. In 1953, there were eight cities and 13.7 percent of the total population was urban. In 1982, 13 cities and an urban population of 18.3 percent. By 1990 the urban population was 29.6 percent of the total and in 1995 39.3 percent, according to the 1990 National Census and the 1995 sampling survey. The total numbers of cities in 1996 was 54 (Li, p. 4). Between 1978 - 1995, more than 300 xian were converted to cities, and in Guangdong Province, some 40 xian were so converted. (Li, p. 4).

The increase in the number of cities is impressive enough, however, some of the city areas are too large, including inside the city sizable agricultural areas and populations. In 1990, within seven of the 19 district - level cities, over 50 percent of the population was agricultural, and in another three cities of this level, it was over 80 percent (Li, p.8). Moreover, within these 19 cities together, the percent of agricultural population was 54.4 percent!

The Degree of Urbanization

Figure 4.17.2 shows the level of urbanization of China in 1995, based on figures reported by each of the provincial statistical bureaus (SSB, 1996, in various pages). The areas with the higher degrees of urbanization are of three types. One consists of the three provincial rank cities and two provinces in northeastern China, all well industrialized. The second includes two types of places: (a) locales which during the 1950s were selected to be developed into new industrial bases, for example, Mnjiang and inner Mongolia, and (b) locales which have developed since 1978 due to their favorable locations for industrial development and/or due to the special supports given by the Beijing government or by private capitals. The third consists of provinces in which there are mineral resources or other advantages.

Concluding Remarks

During the recent decades of rapid expansion of the Chinese city system, there was no a rational central direction or effective coordination between the urban development and other types of socioeconomic development within the country. This was and to a certain extent, still is a country governed by planning, centrally as well as regionally! Instead, national and regional politics and that of the *relationship* between the Beijing and the provincial governments have played an important role in all affairs, including city building.

Although the Communist Party had its early political base in the countryside, her social policies to date show clear biases toward cities and urbanites. More attention should have been paid to the demographic realities of the country, i.e., the uneven distribution of the population and the large surplus agricultural population in the countryside. This concern should have been an important consideration in the process of formulating the course of urban development.

Throughout Chinese history, urban development has always been a close concern of the government and has shared the fate of the various regimes (Chang, S.D., 1963; Gu, 1992; Kirky, 1985). As the end of the century approaches, one should require that the Chinese cities serve the people, not the other way around. It is important that we carefully study the positive and negative results of the official city designation processes of the five decades and examine how well the rapidly established new cities have functioned as centers for properly conducting a broad based socioecononuc activities and servicing well the citizenry. In this era of reform, we hope the Chinese would try to depoliticize the process of urban development. To put it simply, one should let the socioeconomic and geographical realities lead the growth or decline of an urban place. In this way those urban places which have the greatest number of advantages WHl rise and expand, those less favored will likely decline or grow only through the natural increase of population. These urban advantage include a host of economic, geographical, and political variables.

REFERENCES

Chang, Sen-dou. 1963. Historical Trend of Chinese Urbanization. Annals of the Association of American Geographers 53: 109-143.
Gu, Chaoling. 1992. The Chinese Urban System. Shangwu Yinshuguan, Beijing.
Hsu, Mei-Ling. 1994. The Expansion of The Chinese Urban SysteM 1953 - 1990. Urban -Geography 5: 514-536.
Hsu, Mei-Ling. 1996. China's Urban Development: A Case Study of Luoyang and Guiyang. Urban Studies 33: 895-910.
Kirkby, R.J.R. 1985. Urbanization in China, Columbia University Press.
Li, Ling. 1996. Administration System Changes and Population Urbanization in Guangdong In: The Rural - Urban Transition and Development in China. p. 1-5. Zhongshan University, Guangzhou, China.
Liu, Guokuang, et al. (eds.). 1996. Zhongguo Jingji Xingshi Fensi Yu Yuce. Shehui Kexue Wenxian Chubanse, Beijing.
Situ, Shangji. 1996. The County Transferring City and Urban-rural Integration –
Citing an Instance with the Reform of Guangdong Admistrative Area. p. I- 1 2 In: The Rural Urban-Transition and Development in China. Zhongshan University, Guangzhou, China.
SSB (State Statistical Bureau). 1994. China Development Report 1994. China Statistical Press, Beijing.
SSB (State Statistical Bureau). 1996. China Population Statistics

Yearbook 1996. China Statistical Press, Beijing.

Y.i, Ling. 1996. Ershi - yi Shiji Zhongguo Huanyu Yian Zhi Ma? Urban and Rural Development 261:28-30.

Wei, Qingquan. 1996. Administrative Division and Rural - Urban Transfomiation. In: The Rural -Urban Transition and Development in China, Zhongshan University, Guangzhou, China.

Ye, Shunzan (ed.). 1994. Chengshi Hua ji Chengshi Tixi. Kexue Chubanshe, Beijing.

Zhang, Mngliang (ed.). 1997. Zhonghua Renming Gangheguo Xingcheng Xunhua Jiance. p. 1.

Zhou, Shulian and Kuo, Kesha (eds.). 1996. Zhongguo Chenoang Jingji Ji Shehui Yietiao Fazhan Yanju. pp. 1-46. Jingji Guanli Chubanse, Beijing.

Zhu, Tiezhen (ed.). 1987 A Handbook on Chinese Cities. Jingii Kexue Chubanshe, Beijing.

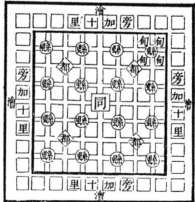

CHAPTER 4.18

Sustainable Agriculture in China: Review Prospects and Future

Cheng Xu
China Agricultural University
and
Science & Technology Committee,
Ministry of Agriculture, China

In recent years, articles about China's sustainable agriculture appear more and more frequently, suggesting the issue has been arousing unprecedented attention. However, along Lester Brown's argument about China's inability to feed itself, the focus has been concentrated further on the issue of "sustainable food security".

Today in China, almost all scholars have accepted the concept of 'sustainable agriculture', including those who do not agree the notion and campaign of 'Chinese ecological agriculture'. They had been and. to a certain extent, are still partly the so-called 'mainstream' in China's agricultural community, with strong agronomy background. Their extreme anxiety is generated by the concern that any new concept, including sustainable agriculture, could possibly causes 'misunderstanding' for physical input and resulted in reducing the inputs, particular the input of agrochemicals. In fact, they argue they raise the issue that in developed countries, some conventional agricultural modernization practices are being questioned for certain negative effects. Facing world-wide[1] prevailing situation of sustainable development, they can not oppose the concept of 'sustainable agriculture', but tend to explain it according to their own ideas, rather than based on internationally accepted ones. A typical example is that they recently raised a formulation of so called 'Intensified Sustainable Agriculture' (ISA), so as to demonstrate they are in fashion, and at the same time to emphasize 'intensified' input to agriculture. just like the western model of conventional modernization of agriculture does. Some of them even insist on advocating the assertion that there is no problem now in China in term of agricultural sustainability.

At the other extreme, there are a few foreign scholars who asserted China can never realize the sustainable food security. Represented by Lester Brown and William Rivera, they tend to make simple conclusion toward such a very complicated issue, although they know little about China's agriculture. For example, Mr. Brown's basic ground in his article ' Who will feed China ' came from a great rise 41% of China's grain price happened in first quarter of 1994.

To those who are familiar with the distorted grain price during the central planning system period, it is not strange the very low base price had to be raised, so that peasants can get necessary incentives to produce more grain. Thus, this large increase in grain price, even though the adjusted price still lower than international market price, did by no means imply any false relationship between grain supply and demand. Whether blindly optimistic view or groundless pessimistic attitude, neither are helpful for China's sustainable agriculture. This article will give a concrete analysis, trying to tell some truth.

The model of conventional modernization of agriculture can never fit China's condition. This conclusion has been drawn by many agro-scientists, also reflected by my personal experience of 10 years onsite research & experiment in the countryside. During 1977-1986, I led a research group working at a village named Doudian of *Fangshan County*. Our mission was to find a way of modernizing China's agriculture. We really got the great success at the first stage: productivity of crop cultivation and livestock raised several folds, we even realized some degree of agricultural mechanization. Table 4.18.1 shows all these achievements.

While this experimental demonstration site village was appraised as a 'prototype of Chinese agricultural modernization' by the State top leaders in 1985, we were sensitive to know that some agricultural scholars in developed countries advocated a rethinking to conventional modernization. Based on an extensive review, the judgment was made that there are a lot of limitations for extending Doudian model:

- Duodian village is fortunate enough to be located at a district of ample aquifer, thus canyse ground water without any limit

- Doudian Village's annual consumption of about 60 ton diesel oil was guaranteed at a subsidized price. Besides the subsidized diesel oil, the village was located close to a coal mine, where cheap coal was always available, and the annual coal consumption reached as high as 1.5 tons per capita, being quadruple more that of the national average.

• Higher dose of chemical fertilizer, and the ideal proportion (40%) of high quality chemical fertilizer (complex fertilizer) can not be available for other villages, too. Also, higher application rate of nitrogen fertilizer (300 kg/ha) , together with confinement feeding of swine and chicken, began to threat the water quality with nitrate leaching problem.

Doudian Village has to purchase a great deal of protein feed, including 60 tons of fishmeal and 320 tons of bean meal annually. It is unrealistic for other villages to follow Doudian's example since China has to depend mainly on importing protein feed.

Table 4.18.1. Comparison of Agricultural Productivity of Doudian Village

	1977	1984	Growth rate %
Grain unit yield (kg/ha)	5,745	11,475	100
Procurment meat(t)	225	620	175.6
Productivity I.(hectares borne by per labor)	0.26	1.4	438.5
Productivity ll.(grain prod. t,per labor per year	9	225	2,400
Production value (U.S. dollar, million)	0.54	38.5	7,030

In order to overcome the above mentioned shortcomings, we decided to adopt principles for the so called 'Chinese Ecological Agriculture', to search for the way of saving water, recycling nutrients, and making full use of photosynthetic energy, etc.. One of our accomplishment was setting up a physical recycling system in this village (Figure 4.18. 1). It means that we finally abandoned the model of conventional modernization of agriculture, featured by intensified inputs as well as neglect for negative effects, and turned to 'Sustainable Intensified Agriculture' (SIA). 'Sustainable Intensified Agriculture' concept implies that we understand that intensification is a common rule for any advance in agriculture, but the essential divergence between two definitions of sustainable agriculture, i.e. SIA and ISA, lies where the emphasis should be placed. For China, the real challenge is by no means just emphasize increase of physical inputs but the key issue: efficiencies for these inputs. Each country should decide its own emphasis of intensification based on nation's condition, for example, intensified labor input, or intensified physical input, or intensified knowledge input, etc.. However, everyone has to obey the common principle: *sustainable* when making a decision.

Sustainable agriculture is by no means a fancy for China, although we are facing so many problems in this field. Chinese top leaders has announced repeatedly that sustainable development should be a fundamental strategy for China, and decided to avert the traditional manner of economic growth, i.e., from the course of developing economy by depending upon 'extensive strategy'--just to increase input, investment and production scale and search for higher growth rate, turn to the 'intensive strategy'--depending upon technical advances and higher quality labor and searching for the quality of growth. Such a statement verified the correctness of our judgment that although in the past 4 decades, China gains great achievements of feeding 21 per cent world population by means of only 8 per cent of world arable land, also this achievement is indeed worthy of pride. However, the success is gained, to a certain extent, at a heavy expense of environment and resource deterioration. The necessity of altering (economic) growth manner seems not unfamiliar to westerners, since this topic had long been debated in the 1970's, headed by the Club of Rome and its rival. But very few Chinese people have known such a controversy and understood its significance until now. If China could adopt open policy earlier, we would have avoided detours and errors on the growth issue.

Nevertheless, there is an urgent need to clarify some facts related to China's sustainable agriculture and sustainable food security. Concerning the following three issues, our viewpoints are quite differ from some foreign scholars:

1. Does China's grain productivity already reached the top and will be inevitably decline? (as Mr. Lester Brown asserted)

2. Is China's agro-natural resources in crisis and China's perspective of stabling its arable land being hopeless?

3. Are China's poverty-stricken areas, such as the arid Northwest, no value for agriculture, and should be an area for outmigration ? (as Mr. Rivera asserted)

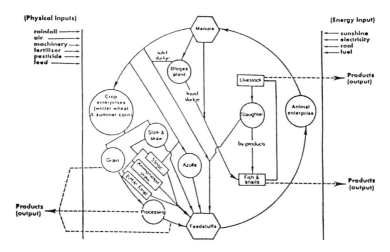

(Physical Inputs)

rainfall
air
machinery
fertilizer
pesticide
feed

(Energy Input)

sunshine
electricity
coal
fuel

Products (output)

Products (output)

Products (output)

Figure 4.18.1. Principal food-chain (physical recycling) system in Doudian Village, China

It is clear that Mr. Brown greatly underestimated the wisdom and diligence of Chinese agrotechnicians/scientists and peasants, also his information was quite wrong. The national average paddy rice unit yield was 6.2 ton per ha (convert to milled rice: x 0.75= 4.7 ton/ha) in 1995.

And 'Chinese super rice', a revolutionary new variety, which is capable of yielding 13-15 ton per ha, is projected to put into production in 3-5 years to come. Do not forget, that Chinese rice breeders were who created the first batch of semi-dwarf higher yield variety in the end of 50's, 8 years earlier than their IRRI colleagues. And one of the two parent lines of IRRI latest 'super rice' comes from China. Also, China's grain yield has been continuously increased since 1978 (Table 4.18.2).

Undoubtedly, China is facing a very serious problem of diverting arable land into non-agriculture utilization. However, scientists and government people are all aware of the necessity of stopping this land diversion trend. China has promulgated the land Management Law in 1991, and is preparing to implement it in an utmost strict manner.

In this year, the government take an unprecedented stem step of quitting any appropriation of arable land. Besides, Mr. Brown perhaps neglected a fact that a great parcel of grain land have converted to cash crops, such as vegetables and fruit tree. According to the statistics, from 1979 through 1994, the grain acreage decreased by 10 million hectares, but merely the increasing of vegetable and apple tree land amounted 8.5 million hectares. Mr. Brown might be unaware that it is such a diversion of grain land that bring a great deal cash income for farmers, and then they can increase their technical & physical input to grain production, so as to raise grain yield.

Table 4.18.2. China's Total *Grain Production and Unit Yield (1978-1996)

Y e a r	Total Production (million t)	Unit Yield (kg/ha)
1978	304.8	2,527
1980	320.6	2,735
1984	407.3	3,608
1990	446.2	3,932
1991	435.3	3,876
1992	442.7	4,004
1993	456.5	4,131
1994	445.1	4,063
1995	466.6	4,239
1996	490.0	4,353

Note:Grain includes sweet potato and potato, which usually accounts for 6 per cent of total grain production

As to Northwest area is concerned, the situation is entirely not as being worsen as Mr. Rivera's description. According to a recent survey for northern part of Shaanxi Province, which was done by authentic MOA experts team, the ecological condition has been greatly improved due to the persistent artificial (by airplane) grass/brush tree planting for 30 years, thus reducing the soil erosion of Loess Plateau. One of a encouraging innovation for land/water conservation: broad ranged (810 meter) terrace field, is displaying powerful potentials. Such kind of terrace field can be constructed through bigger tractors, instead of just labor, it can collect and reserve more rain than the traditional one (narrow range), and at same time reduce erosion. It is reported that the unit yield of corn (plastic film mulched) or soybean, planted on this new type of terrace field, can be doubled or tripled. Furthermore, since the majority part of Northwest area is within the 'semi-drought zone' with an annual precipitation of 250-550 mm, rather than 200 mm, as Mr. Rivera stated in his article, thus the most recent innovation, 'runoff (or water-catchment) irrigation' should be of a great breakthrough significance.

Residents in Northwest countryside have a long history of using 'water cellar (pit)' to collect very limited but also very intensive (in summer) rain for their only living water resource. The recent advances shows the vital value of '(modem) water cellar --- water saving irrigation --- residue/film mulch system' for greatly improving grain production. Under such a system, the collected water is used to irrigate- field crops, and the unit yield of corn and wheat could be reached as high as 12 ton per hectare and 7 ton per hectare respectively, compared with the ordinary corn yield of 2-3 ton per hectare in this area! Considering the fact that Northwestern farmers own biggest 'size ratio' of land to arable land (8-10:1), there is a huge potential to increase irrigated (using water-saving technology such as drip irrigation) land through vast but separated mini rain-catchment areas. Till now, about 1 million modern water cellars (with cemented raincollecting and inside cellar wall surface , with a capacity of reserving 80 m^3 water) are planning to be constructed by 3 provinces within the next 3 years. For these reasons, agriculture will be the reliable back-up, instead of a barrier, for the newly erected energy/heavy industry site in Northwest Area, since huge mines of coal, oil and natural gas which are the biggest ones in China, have been found in this area. And in this process of 'with the double approach', the rural poverty will eventually eradicated at the end.

In order to transfer the yield-increasing potential into practice, while to avoid negative impact upon environment and resource basis along with these efforts, China has to search for a variety of solutions focusing on sustainable food security. Following issues might be considered with priorities:

1. To Popularize the concept of sustainable agriculture among hundred millions peasants. The only way peasants get awareness for sustainability might be the indigenous examples that could show vividly the relation between sustainable farming and peasants' as well as their descendants' vital benefit. Nevertheless, the problem remains that no popular publications are available for masses peasant, even no appropriate edition for teachers and students above primary middle schools. Under China's countryside conditions, the most important and effective way of sustainable agriculture education is to educate millions of local cadres at the grassroots level.

2. Institutional and policy reform are imperative for China's sustainable agriculture. Cadres keep playing vital role in various level of rural community in China. Based on current evaluation system (decide one's promotion) for them, the construction and economic success become

only 'good' criteria. Such a system lacks the least incentives encouraging cares to think of the long term social benefit and environment/resource conservation. Secondly, reprising the agricultural resources is bound to conduct, for current prices encourage waste of resource. Today irrational prices have become major barrier to sustainable usage of resources. Take water as a typical example: because the water price is too low, also there is a lack of progressive accounting regulation, it is no surprise that peasants arbitrarily waste water in their irrigation practice, and without any incentive to adopt water-saving technologies. Another more important mission should be aiming at change the dispersion of administrative forces. Sustainable development needs a concerted and always cross- department/region efforts. But current situation is each does things in his own way. Since there are so many administrative departments, bureaus and ministries, there is an urgent need to follow a successful example like 'Mountain, River and Lake Program' of Jiangxi Province. In this comprehensive harnessing program, every related department and region related have been thoroughly organized, and acts under unified orders from a powerful interdepartment office which had never been existed in the past.

3. Technical advancement will play specially important role for China's capacity building of sustainable agriculture. China's agriculture bears more population & higher demand than any other country, it inherited a poor resource /environment endowment in 1949, and suffered post-natal misconduct in resource/environment management till 1978. China is still a developing country, farmers' purchasing power remains very limited. All these factors make the sustainable practice extreme difficult, gaining the major reliance on special technologies. For this reason, China needs very much international cooperation in agricultural science & technology field. Following are some technologies requiring prior consideration:

Water-saving technics & equipment Some water-saving technologies such as central pivot sprinkler irrigation and drip irrigation have been shown no wide perspective in China. The reason stems on the different objectives and conditions: China is now laying priority of water-saving technology utilization on grain /cotton production, instead of high value horticulture production like developed countries have been done.

Degradable plastic film for agriculture Plastic film has been made incalculable contribution to China's agriculture, causing 'white revolution'. Mulching greatly changes the poor match situation of temperature with other conditions such as enough sunlight, ample monsoon rain and favorable temperature difference in day and night time, those are all beneficial to higher crop yield .

Plastic film tunnels can been seen everywhere right now, and contribute a great deal of vegetable production and now almost all kind of vegetables can be grown in anywhere and anytime. Plastic film mulching make the northern edge of corn planting being streched northward over 10 degrees of latitude, reaching $53^{o}N$, enlarging corn area by at least 15,000 km^2. It also make the height of corn planting climbing to a mountainous area of 3200 meter above sea level. Both record can not be paralleled all over the world. However, due to the fact that the Chinise plastic film is too thin to recover, and being too frequently used for raising corn and even wheat yield, 'white pollution' becomes more and more serious.

Raising efficiency of chemical fertilizer Current efficiencies remain very low, for example, nitrogen that absorbed by crops in the same year can account for only 30 per cent of the total amount. It is impossible for China to adopt the system of company consultant. Facing hundred millions of production unit-peasant families, the realistic way might be a very simple but applicable equipment for on-site testing, and assitanting decision-making. Besides, "fertigation" might be another valuable way, together with the popularization of simplified drip irrigation technology.

Artificial rearing of corn seedlinigs and transplant machine. Multiple cropping and prolonged growing season in high-latitude areas can contribute greatly to corn production in China. However, foreign technologies and equipment in this category fit only for vegetables.

Agricultural ecological engineering Biogas plant has been shown its central role in recycling agriculture and Chinese Ecological Agriculture. But since the majority of China's territory located in temperate instead of tropical or subtropical area, thus high thermal efficiency design and reliable & cost-efficient construction/maintenance technologies are waiting for new breakthrough.

Pollution-free technologies China's township and village enterprises (TVEs) have provided hundred million employment opportunities for countryside labors. It is characteristic of miscellaneous kind of industry, and most of them still adopt backward technologies. Taking recent events happens in Haihe River as an example. In Haihe River watershed the densities of population , tributary and small medium plant (most of TVES) are all highest in the world. Even if the government has taken a very severe steps of closing a huge number of small TVEs in order to improve water pollution, there still exist more than 1500 factories, each one may discharge more than 100 tons of polluted water per day.

Among them, the number of chemical factories and paper making factories cover up 19% separately, brewing factories: 10%, besides, there are also leather tanning, fiber dyeing, starch making and meat processing plants, etc.. A huge gap between demand and supply for water pollution preventing technology should be bridged as soon as possible.

In conclusion, today in China, the opportunity and challenge to sustainable agriculture are obviously co-existed. Since China has been established the fundamental strategy of sustainable development, it can be expected that just as its agriculture has been already sustained for five thousand years under the correct ancient philosophy, and since China is preparing to conduct a battle of 'new agricultural science & technology revolution' for which President Jiang Zeming called in 1996, the future Chinese agriculture will be further sustained depending upon support of correct modern philosophy and unprecedented new technologies.

REFERENCES

Aurelio Peccei , *One Hundred Pages for the Future* Pergamon Press Inc., 1981) pp. 127-135

Cheng Xu, and James R. Simpson, " Biological Recycle Farming in the People's Republic of China: The Doudian Village Experiment", *American Journal of alternative Agriculture,* Vol. 4, No. 1(1989) pp.3-7

Cheng Xu, Han Chunru and Donald Taylor, "Sustainable Agricultural Development in China", *World Development,* Vol. 20, No. 8 (1992), pp. 1127-1144

Cheng Xu, "'Hard' & 'Soft' Restraints for China to Sustain Agricultural Development and to Follow the Conventional Modernization of Agriculture and Deserved Alternative Way", Paper presented at International Conference on Integrated Resource Management for Sustainable Agriculture (Beijing, September 5-13, 1994) .

Cheng Xu and Zeng Xiaoguang, *Introduction of Sustainable Agriculture,* (in Chinese) (Beijing : Agriculture Publishing House, 1997)

Lester R. Brown, " Who Will Feed China?", *World Watch Journal,* September/October (I 994), pp. 10- 1 9

L.S. Pereira et. al, *Sustainability of Irrigated Agriculture,* Dordrecht: KluwerAcademic Publishers, 1996, pp. 309-321

UNDP, Agriculture, Forestry and Fisheries, Capacity Building for Sustainable Development, Canada: Dickson Communications Inc., 1997.

Vernon Ruttan, " Sustainable Growth in Agricultural Production. Poetry, Policy and Science", Paper presented at International Conference on Agricultural Sustainability, Growth, and Poverty Alleviation: Issues and Policies, (Feldafing, September 23-27, 1991), pp. 13-29

Worldwatch Institute, *State of the World, 1996* (New York: W.W. Norton

& Company, Inc., 1996), pp. 3-20

William M. Rivera, "China on the Road to Unsustainability: Agriculture and Natural Resources in the Northwest" , *Final Report of UNDP Project CPRI911110* Beijing: LJNDP Bureau for Asia and the Pacific, 1996.

Zhao Songling, *Introduction to Catchment Agriculture,* (in Chinese) Xi'an: Shaanxi Science & Technology Press, 1996, New York:

CHAPTER 4.19

Chinese Agriculture Systems:
Sustainability versus Intensification

Shu Geng, Jill S. Auburn Charles E. Hess and Yixing Zhou

University of California, Davis, CA

In the Priority Program for China's Agenda 21, which was developed by the State Planning Commission and the State Science and Technology Commission of China, there is a list of 69 priority programs in 9 groups. Future developments and plans on agriculture are presented in Group-2, programs that are closely related to agriculture in Conservation and Sustainable Utilization of Natural Resources are listed in Group-3, and programs in Global Change and Biodiversity Protection are listed in Group-9. These programs will shape and direct future social and economic developments in China.

The first program in Agriculture calls for a strategy for further development of the so called Intensive-sustainable Agricultural Systems in China. Ten counties will be selected to demonstrate the concept, with emphases on high yielding and high quality varieties development, multiple cropping, high management efficiency and good environmental practices in different ecological regions (China Agenda 21). Practices to be examined include intensive cultivation, inter-cropping, relay cropping and multiple cropping, crop rotation, rotational use of organic and inorganic fertilizers, techniques suitable for dry-land farming etc. As stated in the Activities Section, new technologies will be incorporated into traditional and conventional farming practices to enhance the productivity. The document contains no formal definition of Intensive-sustainable Agricultural System, but it indicates that any practices that will further improve the productivity and sustainability of the traditional intensive system will be adapted.

In this paper, we will analyze the implications of further intensification of

China's agricultural systems on its sustainability and conduct a review of the progress made in the U.S. on sustainable agriculture research and practices in the last 10 years. These examples, we hope, provide ideas for learning and areas for potential collaboration between the two countries, as we all strive for furthering the sustainability of our food production systems.

Intensity and Sustainability of China's Grain Production System

China's agriculture has made enormous progress in the last 20 years. Cereal yields has increased with an annual rate 3.3% and the total production increased with an annual rate 2.7% since 1978. But these growth rates were mostly achieved between 1978 to 1984. Since 1984, the yield and production growth rates have decreased to 0.9% and 0.8% respectively. In fact, grain yield and production were stalled during 1980-1984 and again between 1990-1994. Grain sown area accounts about 74% of the total sown areas of all crops; agriculture is almost synonymous to grain production in China. Given that the population is still increasing, the sustainability of the grain production system is, therefore, the most important concern of all issues in China. Between 1978 and 1994, the total chemical fertilizer use increased from 8.84 million metric tons (mt) to 33.18 mt with an annual increase rate of 8.6%. The increased fertilizer input greatly contributed to the increased grain yield: 2.5 tons per hectare in 1978 to 4.1 tons per hectare in 1994. Green manure, which used to be a very important organic component in Chinese agricultural system, has been almost completely replaced by chemical fertilizers since 1985. Today, China's fertilizer input is at least 5 times higher than the US per unit arable land, and the Chinese agricultural system is now basically a chemical-based production system. At the same time, the arable land and grain sown area are decreasing and the multiple cropping index is increasing (Geng et al, 1998). Intercropping and relay cropping are commonly practiced. More than 51% of the total arable land were irrigated in 1994, with an increasing rate of 0.5% per year for the last 20 years. All these practices make China's cropping system one of the most intensive and complicated cropping systems in the world. The efficiency of the system in terms of chemical fertilizer inputs, however, is rapidly decreasing: the amount of chemical fertilizer input for per ton grain production was 35 kg in 1978 but was 82 kg in 1994.

The intensification comes from three types of inputs: labor, material and technology, and natural resources. With a multiple cropping index above 1.57 and an irrigated area over 51% of arable land in 1995, the pressure on land and water use were extremely intense. The technology inputs of breeding (hybred seed production and/or early mature cultivar development) and cultivation method (plastic cover, transplanting etc.) for inter-cropping, relay-cropping and rotation is very sophisticated and labor intensive. With an available rural labor force of 450 million (in 1995), China is indeed the largest agriculture country on earth and has perhaps the most intensive agricultural production system in all aspects of the inputs. However, the characteristics of the Chinese agricultural system is rapidly changing. The changes are particulalrly noticable since 1988 (Yao, 1994) The main changes as observed by Yoa in a village about 450 kilometers to the east of Guangzhou, are the drastic reductions of male labors in farming and the conversion of cultivated land to housing and industrial uses. He estimated that at least a 50% reduction of male labor force or a 20% total labor force reduction has occurred during the last 20 years in the village. This is a direct result of economic growth and increased job opportunities in non-farming areas. Yao's observation is by no means isolated and limited to Guangzhou area. In a regional analysis, Geng et al (1998) divided the provinces into three groups according the rural incomes in 1995. The high income group included Jiangsu, Zhejiang, Fujian, and Guangdong. The middle income group included the other 10 eastern provinces. The low income group included the provinces that extend to the central and western two thirds of China. Yao's observation is typical for the high and middle income regions in China. One implication for farming operation is that more chemical, mechanical and technological inputs will be required to offset the reduced labor force and cultivated land area in order to maintain or increase the production. The question is, how sustainable are these intensive systems? It may be helpful for the purpose of discussion to review some of the early development of the concept and recent progress made in the US in sustainable agriculture research and education programs.

Sustainable Agriculture in the United States

While the roots of the concept of sustainable agriculture have been traced back to the mid-1500s (Gates, 1988), Sustainable agriculture can mean different things to different people (Douglass, 1986; Francis et al., 1987). Geng et al (1990) provided a conceptual definition of sustainable agriculture that conveys certain goals and expectations regarding the agricultural production systems. Sustainable agriculture is currently defined in national legislation (U.S. Government Printing Office, 1990) as: "an integrated system of plant and animal production practices having a site specific application that will, over the long term— (a) satisfy human food and fiber needs; (b) enhance environmental quality and the natural resource base upon which the agricultural economy depends; (c) make the most efficient use of nonrenewable resources and on-farm resources and integrate, where appropriate, natural biological cycles and controls; (d) sustain the economic viability of farm operations; and (e) enhance the quality of life for farmers and society as a whole"

This broad definition, and the emphasis on "quality of life" in particular, is an evolution from the 1985 language that first authorized a sustainable agriculture program at the U.S. Department of Agriculture (Food Security Act of 1985), which spoke primarily in terms of production systems that would be both productive and beneficial to the environment.

Emphasis on Crop Production. Since much of the early interest in sustainable agriculture was due to concerns about environmental consequences of modern agriculture (National Research Council, 1989), most of the early emphasis in sustainable agriculture research programs was on environmentally sound crop production methods. The study of soil management methods based on internal, biologically-based inputs rather than on external, chemically-based inputs, was an early priority. Cover crops and green manures received particular attention due to their multiple benefits as on-farm sources of nutrients, contributors to water infiltration and other aspects of soil physical condition, and habitat for beneficial insects (Anon., 1992; Hargrove, 1991; Sarrantonio, 1991). The measurement and enhancement of soil quality is another major recent high-priority topic. The Soil Quality Network (formerly the Soil Quality Working Group) includes scientists and other interested parties who have met annually since 1991 to share, among other things, definitions and indicators of soil quality. The U.S. Department of Agriculture's Natural Resources Conservation Service has formed a Soil Quality Institute at Iowa State University that has published a number of practical and scientific papers and is expanding a Soil Health Kit initially developed by the Agricultural Research Service (Mausbach and Tugel, 1997). Ecologically-based pest management has also received considerable attention—usually in the context of the entire soil-crop-pest management system. Examples include a study of vineyard management of cover crops in relation to insect and weed pests as well as plant nutrition (Hanna, Zalom & Elmore, 1995), and a study of the interaction between tree nutrition and susceptibility to insects and diseases in stone fruits (Daane et al., 1995).

Animal Agriculture Topics. Sustainability of animal agriculture has received increasing attention in recent years. In particular, rotational grazing or management-intensive grazing, based on the concept pioneered by Voisin (1959), has received considerable attention. In this system, pastures are subdivided into paddocks, and animals are moved from one paddock to the next in accordance with the condition of the forage. These systems often result in greater productivity, fewer weeds, and lower costs than either traditional grazing or confinement in dairy cows (Murphy and Kunkel, 1993), beef cattle (George et al., 1989), and even poultry (Salatin, 1993). Recently, public concern has grown about the environmental impacts of concentrated livestock production and the effects of a more concentrated and vertically integrated production and marketing system for livestock, hogs in particular. (Benjamin, 1997; Schildgen, 1996; Stith & Warrick, 1996) A number of promising smaller-scale, low-cost hog production methods are being studied by researchers and farmers, including deep-litter systems, low-cost structures such as hoop houses, and farrow-to-finish on pasture. (Honeyman, 1996; Practical Farmers of Iowa)

Organic Farming and Sustainable Agriculture. Organic farming had received positive attention in the early 1980s from a few leading scientists in government and academia (Youngberg, 1980; Kral, 1984), but "most considered it, incorrectly, as a [political] 'cause' rather than as an alternative that had the potential to equal the productivity and profitability of conventional agriculture." (Schaller, 1991, p. 23). As the US Sustainable Agriucltural Research and Education Program approaches its tenth anniversary, and the concept of sustainable agriculture is more widely understood to include a number of environmental and social/community/quality of life issues, there is a growing understanding that organic farming is a significant component of, but not equivalent to, sustainable agriculture. And as the organic farming movement grows in both market share (estimates of 20-25% growth per year are common) and in official recognition (e.g. the Organic Foods Production Act of 1990, which authorized the National Organic Program within the U.S. Department of Agriculture), there is a renewed discussion about more explicit recognition and support of research and education on organic farming. (Youngberg, Schaller and Merrigan, 1993; Lipson, 1997)

Over-Arching Themes and Approaches. A "whole-systems" approach has been a consistent theme of sustainable agriculture research. A participatory approach, involving farmers and ranchers in all phases of the research—from problem identification through design, execution, analysis and interpretation of results gained popularity in recent sustainable agriculture research. Emphasis on a participatory approach involving a wide range of partners and practical, on-farm demonstrations has emerged in extension as well. One particularly successful example is the Biologically Integrated Orchard System (BIOS) program in California, in which groups of 20-30 growers work together with scientists and private consultants to make voluntary reductions in pesticides and synthetic fertilizers on their farms. (Dlott et al., 1996)

The United States Experience. The United States experience in sustainable agriculture research has demonstrated that the biologically-based crop and livestock systems can be developed which are both environmentally sound and economically profitable for the producer. Traditional farming systems that depend on chemical inputs are considered less sustainable because their documented negative impacts on environment, water resources, and food safety. The evolutionary development of China's agricultural systems is, however, going opposite the direction to that of the US. While US systems seek alternative approaches to reduce chemical inputs; the same inputs are aggressively applied in China's systems under the name of intensified sustainable agriculture system. Admittedly, China's major concern at this time is the quantity of the grain production and is not focused on the quality aspects of the production or environment. But that emphasis will change as it's per capita income advances. It should not be surprising if the environmental and food safety concerns that have so powerfully brought about the movement of sustainable practices in US agriculture once again shape and determine the direction of Chinese agriculture in the next century. By then, however, the magnitude of the problem of environmental pollution and chemical residues in the food chain that may be generated by the intensive agricultural practices in China could far exceed those any country has had to deal with so far. At the same time, there is a growing understanding that "agricultural sustainability" ia a much broader issue than "sustainable agriculture", which may be influenced by the impact of global warming, international trade and policy. We will discuss a few below.

External Impacts on Sustainability of China's Agricultural Systems

Global Climate Change impact. Agricultural practices affect environment. However, environment determines the existence of agricultural systems. There is evidence that global environment is becoming less stable, which will have tremendous impact on agricultural productivity and sustainability. Results of simulation studies showed that wheat and yield variations in Pacific Rim countries would be greatly intensified and exacerbated by weather variations (Geng and Cady, 1991). In assessing the impact of doubling the atmosphere CO_2 concentration on US wheat yield, Barry and Geng (1992, 1995a, 1995b) concluded that the probability of crop failure would increase by 20%. They also found that the yield variance would increase at locations where annual rainfall was low. Seven Global Circulation Models were used to simulate climatic changes for China in 2050 (Wang et al., 1992). In general, winter temperature changed more (1.5°C) than summer (1.2°C) and night temperature increased more than day temperature. The impact of the climate change on yields in central and southern regions of China will probably be negative. This is because: (a) Higher rates of evaporation offset possible improved water use efficiency. (b) Incidence of extreme droughts and heat waves would probably rise. (c) Increase in respiration will out-balance increased photosynthesis. (d) Higher temperature will accelerate phase development, which may reduce yield if it occurs during critical growth stages (Wang, 1992). In addition to theoretical studies, there is empirical evidence of the negative weather impact on stability of grain production. In the last 32 years, world cereal production increased from 885 million t in 1961 to · 1917 million t in 1992, a 2.1 fold increase.(Figure 4.19.1) According to the yearly variation, the trend of production can roughly be divided into three periods: 1961-69; 1970-79; 1980-1992; The annual rate of production increased about 39 million t per year for the first two periods, but was only 26 million t a year since 1980s. More important is the variability of change. Considerably more variability was observed in recent periods, which was partially reflected from the decreasing R-squares of the lines fitted for the three periods. The three regression lines for the three periods respectively are: Period-1 = 849 + 39 Year ; R^2 = 97%; Period-2 = 1183 +39 Year ; R^2 = 89%; Period-3 = 1590 +26 Year; R^2 =74%. Furthermore, if the absolute deviations from the fitted lines are calculated, the averages of these deviations for the three sections are respectively 13.8, 36.0 and 51.6 mmt. Among all causes, the increasingly unstable weather conditions in recent years is highly plausible.

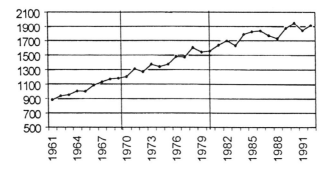

Figure 4.19.1 World cereal production in 100 million t.

China's disaster data revealed a similar trend of increased weather impact. An area is claimed to be disaster if the yields of the area are reduced by at least 30%. Twelve of the twenty-three years between 1952 - 1974 had disaster areas less than 10 million hectares. The disaster areas exceeded 10 million ha for all years since 1975. In fact in 11 of the 17 years between 1978 to 1994 disaster areas exceeded 20 million hectares. The disaster areas increased despite the fact that the irrigation areas were expanding during the same period: pointing to the possibility that China's agricultural systems is increasingly more vulnerable to weather fluctuations. (The irrigation area increased from 20 million ha in 1952 to 49 million ha in 1994.) Ironically, the middle and lower Yangze River, which covers the most fertile land in China, is an area where floods occurred most often in China during recent years. This is also the area where supposedly the most advanced irrigation systems were constructed. It has been postulated that the reclamation movements in the 1960s had indeed destroyed much of the ability of the hills and lakes and therefore, the ecosystems along the middle and lower Yangze River to store and adjust water flow, resulting in more frequent floods. Whether the gigantic project of the Three Gorges Dam will alleviate the flood problem along the middle and lower Yangze river in addition to generate 18,200 megawatts of electricity remains to be seen. Water problems would only compound problems associated with global climate change. Thus, climate change, including increased weather fluctuations, would be a significant threat to agricultural production in general and to grain yields and production in particular.

Economic Impact. Economically, China has been one of the fastest growing countries in the world for the last 20 years. It's per capita annual consumption increased 311% from 1978 to 1994 (China Statistical Yearbook, 1995). The per capita GDP increased from 376 yuan in 1978 to 5,568 yuan in 1996 (Asian Development Bank, Internet site: http://www.internotes.asaindevbank.org). Other estimates, however, showed a much higher per capita GDP. For example, the Office of the Director of the US Central Intelligence (http://www.odci.gov) estimated the per capita GDP for China was $2,900 (or 23,200 yuan) in 1995, when corrected for purchasing power. This rapid economic growth has a negative impact on China's grain production. Geng et al (1998) predicted that the economic growth would reduce farm labor force, arable land, grain sown area and possibly, water allocation to agriculture. On the demand side, with an increased income and disposable money, people will have more purchasing choices and more diversified diets. Direct per capita grain consumption will decrease significantly but the demand for feed grain will increase. By 2030, Chinese will be more concerned with environmental issues and quality of life and will be willing to grow less but import more those crops that are less profitable, such as grain. Under these conditions, policy and economics would have stronger influence on sustainability of agriculture than technology.

Conclusion

China's agricultural systems are primarily production-oriented currently, and pay little attention to long-term effects on environment and natural resources. The concept of sustainable agriculture that was defined by researchers in the late 1980s, and later by the US national legislation, reflect a strong commitment and high value on environmental well being. The US research programs in sustainable agriculture point to two directions. First, a strong shift or substitution of environmental friendly technologies (such as no-till, cover cropping, Integrated Pest Management, disease resistant cultivars, precision irrigation and operation, etc.) with chemical technology which can cause water and soil contamination or toxic residues on food. The second trend in sustainable agriculture development is that the US system is becoming knowledge intensive instead material intensive. Researchers, extension specialists, governmental officials, consumers and farmers often work as a team to address specific issues and problems that are significant for local concerns. The collective effort can maximize the chance of success in accomplishing the specific and general goals of a production system. The most appropriate available technologies will be applied not only to benefit the goals of production, but also to advance long-term environmental goals and ultimately the quality of life for the community and for future generations. The sustainable system is a knowledge-transformed system: the type, the amount and the timing of the inputs are all very critical parts of the process to create a system that would satisfy both short- and long-term goals. Though currently China's system is mostly focused on production goals, we predict China will soon re-prioritize its goals, with more emphasis on environmental and quality of life concerns. The shift of priority is necessary to balance the new demands that will be brought about by the increased per capita income and the desire for better quality of life in China.

References

Anonymous. 1992. *Managing Cover Crops Profitably.* Washington, DC: Sustainable Agriculture Publications, U.S. Department of Agriculture. 114 pages.

Barry, T.A. and S. Geng. 1992 The Impact of weather and climate change on wheat yield in the United States. World Resource Review 4:419-450.

Barry, T. and S. Geng. 1995 The effect of climate change on United States rice yield and California wheat yield. American Society of Agronomy Special Publication no. 59. Page 183-205.

Barry, T. and S. Geng. 1995 Risk assessment under current and double CO2 conditions for US wheat Yield. International Journal of World Resource Review 7:25-46.

Benjamin, Gary L. 1997. Industrialization in hog production: implications for Midwest agriculture. *Economic Perspectives* 21(1):2-13.

Dlott, J., T. Nelson, R. Bugg, M. Spezia, R. Eck, J. Redmond, J. Klein, and L. Lewis. 1996. California, USA: Merced County BIOS Project. In: L.A. Thrupp (Ed.)

New Partnerships For Sustainable Agriculture. Washington, DC: World Resources Institute

Geng, S., C. E. Hess and J. S. Auburn. 1990. Sustainable agricultural systems: Concepts and definitions. Journal of Agronomy and Crop Science 165:73-85.

Geng, S. and Y. Zhou. 1990. China's agricultural resources: From the present to the year 2000. p. 199-231. In: T. C. Tso (ed.) Agricultural Reform and Development in China, Sixth Colloquium Proceedings. IDEALS, Inc., Beltsville, MD.

Geng, S., and C. W. Cady (eds.) 1991 Climatic variation and change: implications for agriculture in the Pacific Rim. Public Service Research and Dissemination Program, the University of California, Davis.

Geng, Shu, H. Kawashima, C. Carter, Y. Guo. 1998 Food Demand And Supply In China For The 21st Century in this monograph.

George, M.R., R.S. Knight, P.B. Sands, and M.W. Demment. 1989. Intensive grazing increases beef production. *California Agriculture* 43(5):16-18.

Hanna, Rachid, Frank G. Zalom, and Clyde L. Elmore. 1995. Integrating cover crops into grapevine pest and nutrition management: The transition phase.

Sustainable Agriculture: News and Technical Reviews from the U.C. Sustainable Agriculture Research and Education Program 7(3):11-14.

Hargrove, W.L. (Ed.) 1991. *Cover Crops for Clean Water.* Ankeny, IA: Soil and Water Conservation Society. 198 pp.

Kral, D.M. (Managing Editor) 1984. *Organic Farming: Current Technology and Its Role in a Sustainable Agriculture.* ASA Special Publication No. 46. Madison, WI: American Society of Agronomy. 192 pp.

Honeyman, Mark S. 1996. Sustainability issues of U.S. swine production. *J. Animal Science* 74(6):1410-1417.

Lipson, Mark. 1997. *Searching for the "O-Word:" Analyzing the USDA Current Research Information System for Pertinence to Organic Farming.* Santa Cruz, CA: Organic Farming Research Foundation. 83 pp.

Mausbach, Maurice J. and Arlene Tugel. 1997. Soil quality a multitude of approaches. Keynote address at Kearney Foundation Symposium, Berkeley, California, March 25, 1997.

Murphy, William M. and John R. Kunkel. 1993. Sustainable agriculture: controlled grazing vs. confinement feeding of dairy cows. Pp. 113-130 in: William C.

Liebhardt (Ed.), *The Dairy Debate: Consequences of Bovine Growth Hormone and Rotational Grazing Technologies.* Davis, CA: University of California Sustainable Agriculture Research and Education Program.

National Research Council, Board on Agriculture. 1989. *Alternative Agriculture.* Washington, DC: National Academy Press. 448 pp.

Salatin, Joel. 1993. *Pastured Poultry Profits.* Swoope, VA: Polyface . 330 pp.

Sarrantonio, Marianne. 1991. *Methodologies for Screening Soil-Improving Legumes.* Kutztown, PA: Rodale Institute. 310 pp.

Schaller, Neill 1991. Background and status of the low-input sustainable agriculture program. pp. 22-31 In: *Sustainable Agriculture Research and Education in the Field,* Board on Agriculture, National Research Council. Washington, DC: National Academy Press.

Schildgen, Bob. 1996. Murphy's laws: 1. hogs rule, 2. You pay (environmental hazards of hog farms). *Sierra* 81(3):29-30.

Stith, Pat and Joby Warrick. 1996. Boss hog: North Carolina's pork revolution. *Amicus Journal* 18(1):36-40.

U.S. Government Printing Office. 1990. *The Food, Agriculture, and Conservation and Trade Act of 1990* (PL101-624), Title XVI, Subtitle A, Section 1603. Washington, DC: U.S. Government Printing Office.

U.S. Government Printing Office. 1996a. *The Food and Agriculture Improvement Act of 1996.* ****check title, get full ref. Washington, DC: U.S. Government Printing Office.

U.S. Government Printing Office. 1996b. *Sustainable America: A New Consensus for Prosperity, Opportunity, and a Healthy Environment for the Future.* Washington, DC: U.S. Government Printing Office. Publication 06100008578.

U.S. Government Printing Office. 1996c. *Sustainable Agriculture Task Force Report.* Washington, DC: U.S. Government Printing Office. 1996-404-680:20025.

van Bruggen, Ariena H.C. 1995. Plant disease severity in high-input compared to reduced-input and organic farming systems. *Plant Disease* 79(10):976-980.

Yao, Shujie. 1994. Agricultural Rdform and Grain Production in China. St Martin's Press, Inc., New York, NY 10010.

Youngberg, Garth (Chair of USDA Study Team on Organic Farming). 1980. *Report and Recommendations on Organic Farming.*

Washington, DC: United States partment of Agriculture. 94 pp.
Youngberg, Garth, Neill Schaller, and Kathleen Merrigan. 1993. The
sustainable agriculture policy agenda in the United States:
politics and prospects. In P. Allen (Ed.): *Food for the Future:
Conditions and Contradictions of Sustainability.*. New York:
John Wiley.

CHAPTER 4.20

Prospect of Future Grain Production
Based on Current Status

Zhou Yi Xing ,
Agricultural Regional Planning Center Wuhan,
Hubei, China

Ever since Lester Brown raised the question 'Who is Going to Feed China', in 1995, it led to much discussion. This report projects the future food grain production situation based on currently available data. This may clarify some of the issues.

Per capita grain yield

The utmost indicator in evaluating grain problem is the domestic production of food grain per capita. This is the base of measuring food grain supply in any nation. Food grain production per capita is the function of per capita cropland (land area that produces cereal grain) and unit yield. There is a dynamic relationship among the two variables:

Variable index of food grain production per capita = (variable index of cropland per capita) x (variable index of unit yield) [1].

First of all, let us examine their relationship from macroview point. The peak of birth rate (4.33%)[2] in China occurred in 1963 and the rate of natural increase was 3.33%.

Since then the rate decreased sharply at 3.53% annually. In 1995, the rate of natural increase has reduced to 1.05% but the total population still rapidly increased at the rate of 1.79%, primarily due to the after-effect of the population peak value. From 1963 to 1995, the net increase of the population is 522.5 million, which is twice as many as the population in the United States. During the same period, the cropland decreases at a rate of 0.29% annually, and China has lost 10.68 million hectares of arable land, which is equal to the total land area for cereal grains in Thailand. Because of the population increase and the loss of cropland, cropland per capita has reduced from 0.177 h to 0.0913 h[3] , or a loss of 48%.

Variable index of cropland per capita, 0.9795, reducing at the rate of 2.05%, is a disastrous change. In other words, if the unit yield remains unchanged, the food grain per capita will decrease at the same rate. The food grain per capita in 1995 would be only 129 kg, 40% less than the amount of 215 kg during the great famine years like early 1960s. If the annual increase of unit yield may reach 2.1%, it can only balance the loss of cropland per capita. Thus, food grain per capita in 1995 could only maintain at the level of 1963, or 249 kg per person, which is 36% less than the actual food grain per capita of 387 kg today.

Imagine that if food grain per capita in China today were only 129-249 kg, then China must import 160 to 300 million tons annually to satisfy the current demand, which would be a nightmare. If the increase of unit yield is less than 2% during the period of 1963 to 1995, China would have been in great famine. Annual unit-yield increase worldwide during 1963 to 1995 was only 2.3% and China's food grain production could have fallen below that figure.

It is very interesting to discover that Brown's prediction and the assumed potential nightmare in China are so close. Brown predicted that by 2030, cropland per capita in China would be reduced to half, the annual rate of decrease would reach 2.4%; and additional increase of unit yield is unlikely, which would lead to food grain shortage in the neighborhood of hundreds million tons. Statistics had shown that in the past 32 years, there is a 48.4% loss of cropland per capita at the rate of decrease 2.05%, the numbers are very close to Brown's prediction and could have served as the basis of his argument.

The deteriorating relationship between people and land resources is a matter of great concern. We are worrying about what might happen in the next 30 years. The experience of what could had happened during the last 30 years would be a nightmare. However, the success of Chinese agricultural production has compensated for the loss of land resources. In reality, unit yield of food grain in China has increased by 3.5%, a figure higher than the worldwide average. The yield increase means an addition of 259 kg food grain per capita. Forty seven percent of the increase, or 121 kg, compensates for the loss of cropland and 53%, or 138 kg, is for the consumption by each individual, including the addition of 500 million people in China since then. Food grain per capita was increased from 249 kg in 1963 to 387 kg in 1995; thus, quality of life for Chinese has significantly improved.

However, we should not become complacent. From what we have learned in the past 32 years, there are serious problems concerning two basic variables and annual change rates. There still is a momentum to continuously losing per capita cropland at the annual rate of 2.05%, even though the cropland per capita is only 0.0913 h; also the annual rate of increase of yield per unit of cropland, 3.5% currently, is slowing down. The opposite tendency of the two factors and their perspective are indeed worry some and that has become the key problem for us to discuss.

Available arable land area

Food supply in China is critical and its basic problem is resource control and its focus is centered on the fluctuation of cropland per capita. The fluctuation is the result of that of population and that of total cropland area. Fluctuation of cropland is under the control of three factors: basic cultivated area, crop index, and proportion of cropland for food grain in proportion to total planting area. Let's first examine their factual relationship.

Statistics from 1978 to 1995 indicate that cultivated land has lost 4.42 million hectares during the 17 years, which is magnified to 6.67 million hectares taking multiple crop index into consideration. During the same period, crop index was increased from 151% in 1978 to 157.8% in 1995, or a 6.67% increase; the increase of crop index also translates into an increase of the total planting area by 6.44 million hectares, compensating for the loss of 6.67 million hectares, making the final net loss of total planting area of 225,000 hectares. In other words, 96% of the lost land is substituted by the increased multiple cropping and the final crop planting area did not change much.

Among the total planting area, proportion of cropland for food grains is 80.3%. The loss of total planting area of 225,000 hectares means a loss of 180,000 hectares of cropland for food grain, a minor decrease. However, in the same period, the proportion of land for food grains decreased from 80.3% in 1978 to 73.4% in 1995, a loss of 6.9%, or a decrease of 10.34 million hectares, or a total loss of 10.52 million hectares including the 180000 hectares mentioned above, which is 8.7% of original cropland for food grains. Therefore, the fluctuation of cropland for food grains is caused primarily (98.3%) by the change of its proportion to total planting area and only 1.7% is due to the fluctuation of planting area and multiple crop index; such a small percentage is somewhat surprising.

The multiple cropping system and planting structure on a cultivated land constitute the utilization space of that land. The significance of the space utilization to food grain production is often ignored. In China, a country of rapid economic expansion, people start to realize how large this space really is only after all those problems concerning food grain crisis and cropland have been pointed out. China has lost a total over 4 million hectares cultivated land and still maintains the same basic planting area; large amount of cropland for food grains has been lost and yet it has shown very little relationship with the total cultivated land and sowing area; all these are peculiar phenomena. Therefore, to examine the future food grain problems, not only we need to carefully study the fluctuation of cropland but also need an in-depth examination of the cropland space utilization and its internal mechanisms and the critical value. The success of China's agricultural program in the past apparently is because the wisdom of how to make use of the land, or spacing of the land.

Fluctuation of the arable land area

Regardless how large the space may be, it is established on the basis of cultivated land area. We need to first analyze the fluctuation of the arable land area. The first feature related to the fluctuation of cultivated land area is that the speed of converting of the area to nonagricultural sectors is closely related to that of investment from national fixed asset. The power function calculated from the 15-year (1981 to 1995) data indicates that the correlation between the two is as high as 0.923. The correlation indicates that loss of cultivated land area is an event of capital shifting and also an unavoidable event during economic progress of the nation; it also explains that the basic problem to control the loss of cultivated land area is to prevent the overheated investment and to select other ways for investment that may save the land resources.

The second characteristic feature is that, since the economic reform, the loss of cultivated land forms a perfect S curve; it has four stages: 'flat', 'peak', 'valley', and 'flat' again. Table 4.20.1 shows the related data collected from 1978 to 1995. During the 'peak' stage, the whole nation started to grab land for industry and economic development in such a zealous fashion; village-and-town industry reached the highest point; real estate started to move forward; plus ignorance of land-management by the government led the land loss at 1.277 million hectares. To analyze the loss by using the power function, it showed that 16% of the loss had exceeded the normal investment. The 'valley' stage was the period of economic reorganization and reduction. The 8-year land and economic undulating did not really represent anything special and we need to only pay attention to the normal period, especially the period from 1991 to 1995 where the rate of loss of

Table 4.20.1.   Fluctuations in cultivated land and its investment from 1978 to 1995.

Dura tion	Patte rn	Annual fluctuations rate (%)			Annual mean fluctuation on cultivated land (1000 hectares)		
		Total cultiva ted land	Origi nal cultiva ted land	Invest ment	Decre Ase	Incre ase	Net loss
1978 -83	Flat	-0.208	-0.928	21.99	-905.76	699.78	-205.98
1983 -87	Peak	-0.634	-1.324	26.32	-1276.6	658.93	-617.72
1987 -91	Val- ley	-0.061	-0.557	10.91	-529.4	470.64	-58.78
1991 -95	Flat	-0.179	-0.740	38.07	-700.22	529.55	-170.68
1978 -95		-0.267	-0.927	24.22	- 856.11	596.20	-259.92

Note: investment refers to the current value of investment on the national fixed assets since 1981; Original cultivated land refers to the total farmland at the end of last period minus reduced areas in the current period.

cultivated land was 0.179% and the original rate of loss was 0.74%, both numbers are significant for the prediction of the future, because (1) high investment rate (38%) appeared in this period, which is the largest driving force for losing cultivated farmland.   It is estimated that the annual economic expansion will slow down gradually and the future rate of investment set by the government will be within 10%; (2) after learning the lesson of losing farmland, strengthening of land management has begun in 90's leading to significant changes; the future management system will be more strict and better; (3) according to the realistic cumulative calculation, rate of decreasing farmland is only 0.154%.   It is predicted that the decreasing rates of 0.179% and 0.74%, should be the lower limit of the decreasing rate for future cultivated land and the original cultivated land, respectively.   It is possible that certain years may have an absolute decreasing rate greater than these indexes but the overall average rate should not be greater than those numbers.

Comparing to what happened to Japan   and the Asian four little dragons (Hong Kong, Singapore, South Korean, and Taiwan), the absolute value of those indexes is so small and difficult to believe.   Would it be smaller in the future? Is it possible? The third characteristic may explain this phenomenon. The major shift of cultivated land from east to west has occurred in China. Both the amounts of decreasing and increasing of cultivated land in each year are huge but the net decrease or increase is very small. Since 1978, cultivated land has increased 10.135 million acres or 10.2% of the original area; it also decreased 14.55 million hectares of cultivated land or 14.6% of the original; the average rate of decreasing is 0.927%, very similar to Brown's number, 1%. The net decrease of cultivated land is 4.42 million hectares, or 30% of the original reduction in cultivated land. The total decrease of the original cultivated land is 14.55 million hectares, how big is it? The total cultivated land in the six provinces along the east coast is only 12.6 million hectares, the loss of 14.55 million hectares is very scary!

Furthermore, we should also pay attention to the change of location of those cultivated land. If we rank the province-urban districts according to their change in total cultivated land recorded from 1988 to 1995, there are 21 province-urban districts showed a decrease of total cultivated land. The first nine among them are Guangdong, Zhejiang, Shannxi, Hubei, Beijing, Fujian, Shandong, and Jiangsu. On the contrary, there are nine provinces where the total cultivated land has increased; the most important ones are Inner Mongolia, Qinhai, Heilungjiang, Yunnan, Xinjiang, Ningxia, Guangxi, and Jilin. However, the reduction of cultivated land occurred among places where the land is fertile, precious, and have been the sources of Chinese civilization for thousands of years, whereas the places the cultivated land has increased are marginal land in remote border regions, where irrigation water is short of supply, lack of sun energy, and soil is infertile and sandy. The major shift of cultivated land between those regions is rarely seen in history and its far reaching impact is difficult to predict. From the standpoint of irrigation water and heat available for those newly developed cultivated land, it is apparent that the soil quality and tolerance to stresses is decreasing. Combining the unit yield and multiple cropping together and then calculate the total multiple crop yield, the unit yield for single crop in those provinces where cultivated land decreases is 1.97 times higher than those provinces where cultivated land increases. This number (1.97) did not include the differential in yield between the newly developed land and the local average yield; it also did not include the differential in value of those land between provinces showing various degree in economical development. Even the decrease of amount of cultivated land in east is completely compensated for by increase of cultivated land in the west, the productive capability will decrease by at least of a 2:1 ratio. If we divide the newly increased cultivated land, 10.135 million hectares, by 1.97, it will be converted to 5.145 million hectares. The total harvested land area in 1995 should be 90 million hectares, or a 94.8% of the total statistical area. The decreasing rate for the combined rate counting for both quantity and quality should be 0.58%, larger than the progressive decreasing rate of 0.267% calculated from statistical area.

The great shift of cultivated land from east to west is the huge price that Chinese people have been paying for modernization of economy. In Eastern Asia, China is the only one who could afford to pay such an extensive price. There is no comparison between the rate of decreasing cultivated land in China and that of Japan, or with those which had occurred in the four little dragons.

The fourth characteristic is that there is an extremely important responsibility for land reclamation and restoration. China has increased 10 million hectares of arable land in the past 17 years. Although the rate of land decrease has been slowed down, considering the accumulated amount during the past 35 years, it is tremendously frightening. Considering the rate of decrease for the total cultivated land (0.179%) and that of the original cultivated land (0.74%) together, there should be a total of cultivated land of 89.2 million hectares by 2030, among which the original cultivated land was 73.23 million hectares and the actual decrease was 21.74 million hectares and the net decrease of total cultivated land was 5.77 million hectares. The difference between the two, or 15.97 million hectares, should be the amount of reclaimed land in China. The rate of land decrease, 0.179%, could be the lower limit and a land increase of 16 million hectares land could be the upper limit. When the rate of land decrease reduces, then the burden of increasing new land would be lessened. This is a tough job and from now on we should have an increase of 456,000 hectares each year, which is only 76.5% of the average increase of 596,000 hectares between 1978 to 1995. China should be able to accomplish the job. From the survey of potential land resources, it is shown that[4] 6.3 million hectares from a total of 13 million hectares of waste land could be restored, and the number could reach 7 million hectares if we include more of the waste land in the future. Most of those land is located in the economic development zone and should have the priority to be considered and restored. The other 9 million hectares falls in the category of land reclamation. In accordance with the survey of the potential land resources, China has 13.53 million hectares of land that is suitable for agricultural use. Among those, it is possible to choose 9 million hectares of the most suitable land for cultivation and then reclaim 257,000 hectares each year for the next 35 years. During 1985 to 1995, there were 264,000 hectares of land reclaimed each year. Thus, the new responsibility to reclaim 257,000 hectares yearly is less than that of the past. However, the amount of investment is huge for reclaiming marginal land, profit from the reclamation is small, the consequences are not predictable, and such task can only be accomplished by investment and organization from government. The need of 16 million hectares newly reclaimed cultivated land is a surprisingly large number, and make people doubtful whether it can be achieved as it is the total cultivated land area along the pacific coast, except Hebei and Shangdong provinces. It takes 50 years (from 1978 to 2030), during which period China is supposed to complete her modernization, and at the same time , China has to shift one-third of the cultivated land from one region to another. This would be a marvelous task in Chinese history if it can be done.

Multiple crop production

The multiple crop index has been up and down several times in China's history. Beginning from 1978, the multiple crop index decreased from 151% to 146.3% in 1983, reaching the bottom, and then gradually rises to 157.8% in 1995. During these 12 years, cultivated land had a net decrease of 3.4 million hectares and at the same time the multiple crop index increased 11.4%. Thus, the planting area has a net increase of 5.9 million hectares. Obviously, this is the basis to rejuvenate agriculture since 1984.

There are three problems need to be examined. First, what is the future potential? If we consider only the distribution of water, heat, and resource, and their utilization, then multiple crop index is possible to reach even 200%. However, to implement multiple cropping, it involves many other factors, and we should examine the past experience. Historically, those provinces along southeast coast are the regions where three crops can be harvested each year because of favorable climatic conditions. The multiple cropping index started to decrease since 1970s and Table 4.20.2 shows the fluctuations on crop index in those regions during 1980s. A unique feature can be noted in Table 4.20.2. Crop index increased in all regions except those province in the east coast. Such observation countered the Brown's conclusion that multiple crop index will decrease as economy starts to develop, especially in those provinces along the southern coast where economy is booming.

However, crop index in the eastern region is decreasing, particularly in the Yangtze River Delta. Thus, we cannot conclude that the decrease of multiple crop index has nothing to do with economic development. Furthermore, water and heat resources in the Yangtze Valley are not as good as three other regions. Historically, rate of multiple cropping in Yangtze Valley has been lower than that of Yangtze Delta; and those in Anhui, Jiangxi, and Hunan

Table 4.20.2. Fluctuations in utilization space for cultivated land in the southern districts.

District	Early 1980s		Late	1980s		1995
	Prpro por tion of crop land	Multiple cropping index	Pro por tion of crop land	Multiple cropping index	Propor tion of cropland	Multiple cropping index
Eastern Provinces	73.15	208.54	75.53	201.16	72.03	194.68
Southern Provinces	80.37	204.19	72.44	214.16	67.31	228.06
Sum from South-east	75.97	206.84	74.29	206.20	70.04	207.47
Yangtze Valley	75.10	201.61	72.95	207.22	68.84	218.58
Southwest Provinces	83.76	163.68	74.93	175.82	68.22	203.52

Note: (1) Early 1980s refers to the four-year mean from 1978 to 1982; late 1980s refers to the three-year mean from 1988-1990. (2) The division into regions is based on characteristic of water and heat. Eastern provinces: Shanghai, Jiangsu, Zhejiang; Southern provinces: Fujian, Guangdong, Hainan,; Yangtze Valley: Sichuan, Hunan, Hubei, Jiangxi, Anhui; Southwest provinces: Guangxi, Yunan, and Guizhou.

have been traditionally lower than southeast provinces. However, the multiple crop index is the highest in the Yangtze Valley and among which Jiangxi and Hunan are the highest (248%). Therefore, there is a great potential that the other three regions could raise the multiple crop index to that of Yangtze Valley (218%); if so, it would mean a multiple crop index of 163% nation wide. If southeast and southwest regions all could raise the multiple cropping index to those of Jiangxi and Hunan, it would mean a multiple crop index of 166%. Also, if the region along Yangtze River and Huai River also may raise the multiple crop index by a small amount, it would mean a nation wide multiple crop index of 170%. That is, there is plenty room to increase the crop index. Because multiple cropping variable index is inversely related to the cultivated land variable index, then multiple cropping area may remain unchanged. The lower limit of the progressive decrease in cultivated land could possibly be -0.179% and then the upper limit of multiple crop index would be 168%, which is 10% higher than the figure of 157.8% occurred in 1995. This also is the suggestion made by many experts, such as Professors Xiao-he Ma and Ying Du of the Institute of Agricultural Economics, CAAS. Secondly, history has taught us that it is much easier to have a decrease in multiple cropping index than an increase, and the continuous decrease of the index in Yangtze Delta must have basic reasons. First, it is because of the labor shortage, especially during the planting and harvesting which is of short duration. It is labor intensive and hard work. Today, the industry is in great stride providing better pay scales and better chance to be employed in the factories, leading to the loss of farm laborers; and labor shortage has become worse in the farming communities. In Southeast Asian countries it is true that multiple cropping index decreased when the economic development advanced . Since Yangtze Delta has more advanced industries than that of Guangdong and Fujian, it is not surprising that it took the initial dive in the rate of multiple cropping; but it should not necessarily be taken as a model for the future. Situation could be different if Yangtze Delta region could attract a large amount of labor force like that in Zhujiang Delta. We should never compare what happens in small countries to a huge country like China. China has huge amount of surplus labor, which can be a burden as well as an asset. The key is to develop an appropriate policy and strategy to deal with the problems: completely abolish the restrictions on floating labor force, attract the surplus labor force to medium and small cities or towns, encourage the labor force to move to villages in the south and foster the agricultural business in the south to utilize the labor force by employment, leasing and subletting, and other flexible means to absorb the floating labor force from other regions. It is one of the important strategies to fully utilize the water and heat resources in the south for agricultural development. Such strategy is much more economical and logical than artificial heating in the north (plastic covering, glass greenhouse, plastic greenhouse) and land reclamation in the northwest. Secondly, during the crop season, to increase the crop index and investment have little reward. It has occurred repeatedly that an increase of crop yield does not mean an increase of income for farmers, especially when the government lowers the grain prices. Ways to compensate for such occurrences are: increase the efficiency of agricultural production, reduce the cost of input materials for agricultural production, and raise the level of intensive management. The above analysis illustrates that an increase of 10% multiple cropping index is not a matter of available resources, but a matter of economics. Thirdly, the tillering layer of soil as well as the moisture and temperature absorbed by the soil layers constitute a three-dimensional resource of the farmland. Moisture. and temperature are renewable, but soil fertility is very difficult to rejuvenate. There is a severe lacking of energy resources for most of the Chinese farming villages and difficult to return the organic matter (stalks, leaves, etc) to the land leading to shortage of organic matter. In our history, rotation with green manure crops is the major reason for multiple cropping; it is a fertility sustaining system. Green cover crop is now replaced by food crops, over-use of multiple cropping with food grain crops will cause the deterioration of the soil. Raising multiple cropping index in a large scale should only serve as a temporary strategy. We should not raise the multiple cropping index too high at the expense of productivity.

We can make a final analysis of the situation. If we raise the multiple cropping index to the median number of 160%, with 89.2 million hectares of cultivated land in 2030, it would translate into a total 142.72 million hectares of planting area, which is 95% of the planting area of 1995; if we raise the multiple crop index to 168%, then there would be 150 million hectares of planting area, which is slightly higher than that of 1995 or reaching the level in 1978. It is rarely seen that a reduction of cultivated land by 10 million hectares and yet the planting area may remain unchanged for 50 years!

Role of market economy

Under the conditions of market economy, variation of cropland in proportion of total planting area is a market function. Basically, it is a response to the rule of supply-and-demand in the market. Crop structure is adjusted by that of supply and demand through price and rate of profit; when supply of food grain exceeds its demand, market condition forces the proportion of cropland to decline; as demand of food grain exceeds the supply, then the market forces promote the proportion of cropland. Some people consider the continuous decline in the proportion of cropland in China is a consequence of economic development. In fact, such a conclusion is not totally sure, theoretically as well as practically. There are several questions worth to examine. First, China has traditionally a nation with small farms; self-sufficient of food grains has a deep and solid economic foundation and is a historical tradition. Ever since the 1950s, the national policy has led to the emphasis on industrialization, state monopoly of purchasing food grains, lowering the food grain prices, and local responsibility on food grain production. Under the dual impact of historical and governmental action, the proportion of cropland and food grain production formed a peculiar connection. Farmers have to produce food grains at all cost for self-sufficiency and once the grain output reached the target, farmers then changed to non-food grain crops for higher profit. Such a phenomenon was expanded to regions, districts, and then the whole nation. An increase of unit yield resulted in the proportion of cropland to decline, and at the same time the increase of unit yield compensated for the loss of cropland; they form a cause-and-effect interrelationship. From 1952 to 1995, there was a negative correlation between the increase of unit yield and the proportion of cropland to total cultivated land, being -0.952. If we examine region by region since 1980s, the economically well developed southeast region (especially the eastern coast), the proportion of cropland was higher than that of Yangtze Valley. Regional comparison made in 1995 showed that food grain production per capita in Yangtze valley was 408 kg comparing to 329 kg per capita for those along the southeast coast, whereas the proportion of cropland

(0.70) in the Yangtze Valley was lower than that (0.77) of southeast coast. In the north, the food grain per capita was 413 kg comparing to 326 kg of the northwest, whereas the proportion of cropland in the north was 0.75, lower than that of (0.77) northwest. In the northeast, however, both the food grain per capita (575 kg) and the proportion of cropland (0.864) were the highest, while southwest had the lowest in both, 303 kg and 0.682, respectively. This phenomena was properly was due to the difference in natural resources rather than that in economics. Therefore, the continuous decline of proportion of cropland was due to the closed-market system, relative low profit margin for food grain, and policy of self-sufficiency and self-responsibility on food grains.

Secondly, as the market economy develops more and more, food grain production will follow the principle of comparative superiority and deploy accordingly, just like the long term showing of high proportion of cropland in the northeast and the showing of low proportion in the southwest. Reasonable deployment of resources is the consequence of removing limitations, such as government monopoly purchase, local self-responsibility on food grain production, and many other tangible and intangible restrictions, and forming a huge but united market. With the rapid development of the international trade, China finally will join the system of international trade organization. Thus, relative proportion of cropland for food grain is not only the consequence of domestic market, but also that of international grain market. To implement the production policy based on the principle of food grain self-sufficiency is to lower the profit for producers and to balance import and export. The result is undesirable for both, producers and consumers. Thirdly, rapid development of economy and decline of resources per capita will lead to a long term tension on food grain supply. China is a large country, producing food grain locally will have the advantage of lowering the transportation cost. Therefore, following the development and expansion of the market, the mechanism of supply and demand will increase the overall proportion of cropland, especially for those regions where potential of food grain production are superior than others. Although the increase of proportion of cropland varies from Yangtze Valley to Southeast coast and from Northern region to Northwest region, they will all increase the proportion to some degree. If Yangtze Valley and Northern region may restore the proportion of cropland to the level of 1981, the nation-wide proportion will increase by 3.4%, or reaching 77%. In fact, all regions may increase the proportion of cropland except the Northeastern region and it is feasible to restore the level in 1980, or 80%. The proportion of cropland reached 88% in 1953, which was the level without the government pressure guidance It also illustrates that proportion of cropland is not dependent on resources and its suitability is a market problem.

In a final evaluation of the fluctuation of cropland, the following observations are of importance. If we use the medium figure, it shows that with planting area at 142.72 million ha with a multiple cropping index of 160%, and a proportion of cropland of 77%, the total cropland area will be the same as that of 1995. This suggests that it is not a difficult task to maintain the constant cropland area. If we use the high figure, at the multiple cropping index of 168%, then there would be a planting area of 150 million hectares, among which 80% or 120 million hectares would be cropland, resulting in 109% of that in 1995. We need to acknowledge the fact that there is a decline in cropland area per capita if it is compared to 0.0913 hectares per capita in 1995; 0.0688 hectare per capita if the medium figure is used or 0.075 hectares per capita if the upper figure is used. The variable index for the medium figure is 0.992 and the progressive decreasing rate is -0.8%; the variable index for the upper figure is 0.9944 and the progressive decreasing rate is -0.56%.

Comparison of trend between this and the next century

From the analyses mentioned above, we may compare the trend during the past one-third of century to that of future one-third century. The key criterion for the comparison is the variable index on cropland per capita (the progressive decreasing rate). The variable index of cropland per capita is the product of the progressive increase rate per unit yield times the progressive index of food grain production per capita. This value is an impact converting coefficient. When the variable index of cropland per capita is in an inverse relationship to that of unit yield, the variable rate of food grain production remains unchanged. Therefore, we may examine the magnitude of deficiency due to the decrease of average cropland per capita and how much increase per unit land is needed to compensate for the decrease.

The variable index of cropland per capita from 1963 to 1995 was 0.9795 (-2.05%) and its inverse is 1.0209, meaning that an increase of unit yield should reach 2.09% in order to maintain the same amount of food grain per capita. When the number becomes greater than 2.10%, then food grain per capita may increase; if it is lower than 2.1%, the gap will be made up by the base number of food grain per capita. From 1978 to 1995, the variable index of food grain per capita is 0.9812 (-1.88%) indicating that an increase of unit yield by 1.9% is necessary to make up the gap. From 1995 to 2030, the medium index of food grain is 0.992 (-0.8%) and the upper index is 0.9944 (-0.56%); thus, it requires progressive increase rates of 0.8% and 0.56% for the unit yield to make up the gap. It illustrates that the development of agriculture has greatly improved the rate of repayment for food grain consumption. It also indicates that 99.2 to 99.44% of the increase in unit yield has been transferred to the food grain per capita, leading to a great decrease of the resource gap. Under such important changes, if the unit yield may increase by 1%, food grain production per capita will reach 414 to 450 kg and the total food grain can reach 663 to 721 million tons. If the unit yield may increase by 1.5%, food grain production per capita will be 492 to 535 kg and the total yield will be 787 to 857 million tons. If the unit yield may increase by 2%, food grain per capita will be 584 to 636 kg and the total yield will be 935 to 1017 million tons. With the same increase of 2%, China would have fallen into a country with huge import of food grain from 1963 to 1995, while China would have become a country with huge export of food grain from 1995 to 2030. Such a comparison illustrates that the trend in agriculture development for the next 30 years should become better rather than worse. Development of agriculture should promote the efficiency of national economy.

How does such a change occur? We may interpret the case by using the medium figure from 1978 to 1995 to compare with that for 1995 to 2030. First of all, the progressive increase rate of population will decrease from 1.37% to 0.8% and its inverse, or the share index per capita, will increase from 0.9865 to 0.992, meaning an increase of 0.55% per capita. This means a total reduction of pressure or limitations for the nation and is a consequence of strict birth control policy since 1970s. From the absolute population number, the population pressure is still increasing, whereas the population pressure is relatively declining if viewed from the dynamic stand point. Secondly, the progressive decreasing rate of cultivated land is changing from -0.267% to -0.179%; and thirdly, the farmland utilization space, which is constituted by multiple cropping index and proportion of cropland to the total cultivated area, expanded from 99.73% to 100.42%. All those three components are improving and the variable index for food grain per capita that is based on all the three components, has recovered from 0.9812 to 0.992 (the progressive decreasing rate has changed from -1.88% to -0.8%), which is 101.1% of the past level. Assuming the variable food grain crop land per capita as a base of 100, then those of the three components are 54.9%, 8.5%, and 45.1%, respectively. It is clear that the relative variation of population is of most importance, followed by the efficient utilization space of the cultivated land, and the variation of cultivated land becomes the least important factor.

Briefly, if China can increase 16 million hectares of newly cultivated land by 2030, then it will control the progressive decrease rate of cultivated land within -0.179%. If China can maintain 90 million hectares of cultivated land and have a multiple cropping index of 168%, it will keep the total planting area unchanged; with multiple cropping index of 160% and maintain the proportion of cropland for food grain and oil seed at 77%, then the total area of cropland will remain unchanged, but the unit yield must increase by 1 to 2%. At this level, the food grain per capita in China could increase by a large margin and make a comfortable living for people or even become an food grain exporting country.

Potential for continuous yield increase

Finally we need to discuss that China must maintain a 1 to 2% yield increase per unit by 2030. First, what is the current level of unit yield in China? The annual average yield for 1993 and 1994 was 4,528 kg/ h, or about 1.095 times of that in North America, 1.088 times of that in Europe (not including the former Soviet Union), and 1.625 times of the world wide average. Therefore, some consider that the average food grain yield in China has entered the high yielding group and has very little room for further increase. Such calculation is based on the reported lower data of cultivated land in China. According to the investigation of several organizations, the actual area of cultivated land in China is much higher between 120 to 139 million hectares. The land survey in recent years estimated that the actual area is 133 million hectares or 1.38 times of the reported figure of 96 million hectares [5]. This false information resulted in over estimated the unit yield by 27.8%. The actual unit yield should be 3,269 kg/h, which is only 0.79 times of that in North America, 0.78 times of that in Europe, and 1.173 times of that for world wide average. At the current level, it is necessary to increase unit yield by 38.5% in order to reach the current statistical level; that is equivalent of a progressive increase rate of 1%. If the rate of progressive increase is 2%, then the unit yield will be doubled, reaching the current level of food grain production in France, Britain, and Japan. In other words, even if there are no breakthroughs in technology, such an increase still is possible to be realized.

Secondly, Chinese agriculture is in the transition stage changing from traditional agriculture to modern agriculture. The overall scientific technology is not so advanced and severely lacks of coordination. The impact of agricultural technology on increasing of food grain is still is in the 30 to 40% low-level stage. The agricultural technology is 14 to 20 years behind those of developed countries. Mr. Yifu Lin, an economist and professor of Beijing University introduced a project, "Studies on the Priority in Chinese Agricultural Research"[6], which is a large scale investigation involving more than 400 provincial- and ministry-governed agricultural research institutions in 383 districts, and discovered that there are many technical factors limiting the potential for unit-yield increase. Those factors were classified into 42 items in 5 major categories and were evaluated by inviting 2000 experienced experts in agronomy. In addition, the project was consulted with 455 well known breeders nominated by the Ministry of Agriculture for ways to solve those problems. The experts believe that agriculture improvement can be made only by strengthening the investment and the unit yield could be 2 to 3 times of the current level for rice, wheat, and maize as the cultivars are improved through breeding approach. That is, an annual progressive increase rate of 3.5% to 5.5% is necessary. Granted those experts were in error and over estimated by 50%, the rate of progressive increase could still reach 1 to 2%.

Thirdly, Lack of irrigation water resources is a severe obstacle to raise the yield level. We adopted the weather data cumulated from the past 20 years by 590 meteorological stations nationwide and used the "Regional Ecological Approach" recommended by FAO to calculate the requirement of water for crop production. To produce 650 million tons food grain, the evapotranspiration by the crop plants in the field is 658 billion cubic meters; to produce 818 million tons of food grain, it takes 772 billion cubic meters. During the crop season, the cropland receives approximate of 500 billion cubic meter of water from rainfall and thus it needs to have a water supply of 158 to 272 billion cubic meters, which is not an extraordinary amount. However, the ow efficiency of water transportation via aqueduct and low efficiency of water usage in the field, both are estimated to be lower than 30%, the amount of water entering the aqueduct is expanded by 3 to 4 times, leading to a problem that is very difficult to solve. If China may implement the project on modernization of irrigation and water usage at an early date and raise the efficiency on water transportation and utilization by 100%, it is possible to use 500 billion cubic meters of water to produce 800 million tons of food grain. However, it involves the reconstruction of water canal or aqueducts, adopt pipeline for water transportation, using soft-pipe to transport water in arid and semiarid agricultural regions, develop sprinkler and dripping irrigation systems, converting the water shortage problem to a problem of capital. Thus, it could progressively increase the unit yield by 1 to 2%. Fourthly, environment for agricultural ecology is deteriorating. China has been a country suffering lots of agricultural calamity and the victimized crop area is increasing, though the rate of occurrence is decreasing. The progressive rate of damaged area from 1952 to 1994 is 4.64%, among which it was 7.1% during 1965 to 1978 and 2.6% from 1979 to 1994. The unit yield in China is calculated from total planting area instead of harvested area, thus includes the factors of environment deterioration. In the future, the estimation of cultivated area suffering losses from natural calamity should also include those progressive factors concerning natural disasters. As long as there is no serious and sudden disasters, it should not be a problem to include the disaster progressive factors in the estimation. Among the victimized area, flood occurred more often than drought in those years and flood is often concentrated in Yangtze River regions. The construction of three-gorge dam in the Yangtze River should play an important role in flood control and should improve and prevent flood damage in that region and reduce the rate of natural calamity.

Finally or fifth, humans are in the eve of biological revolution and regardless of the disputes or suspicion, agricultural and life sciences and genetic engineering and biotechnology will produce great breakthroughs in this period. It will produce a far reaching and tremendous impact on agricultural production and social economics, raising the biological capability to transform and adaptation, and hopefully changing these traditional limitations on crops due to water, fertility, and soil.

Notes:

[1] Variable index is the geometrical mean of the ratio between current period in proportion to the basic period.

[2] All statistics adopted in the manuscript and without quotation are taken from those concerned years in "Chinese Statistics Yearbook", "Chinese Agricultural Statistics Yearbook," "Chinese Agriculture Yearbook".

[3] The mean values per capita is calculated from annual average population; to be consistent, the population fluctuation also is calculated from annual average population.

[4] Adopted from "Discussions on Potential Increase of Food Grain in China and Strategies", Institute of Agricultural Economics, Chinese Academy of Agricultural Economics, published in "World Management", volume 4, 1995

[5] People's Daily, June 24, 1996.

[6] People's Daily, March 10, 1995.

CHAPTER 4.21

The Potential For Yield Increase In Food Grain

George H. Liang, S. H. Chen, and Y. C. Luo,
Department of Agronomy Kansas State University

An increase in food grain production can be accomplished in three ways: (1) increasing yield per unit area using breeding and (2) management approaches including double-, relay-, and (3) intercropping systems; increasing cultivated land; or both. China has 7% (95.33 million hectares) of the world's cultivated land that has been planted and replanted for nearly 2,000 years and 22% (1.3 billion) of the world population (predicted to increase to 1.6 billion by 2030). Gou (1994) estimated that a total of 500 million tons food grain will be needed by the year 2000. He also estimated that the natural land resources in China are able to sustain a population of 950 million with a reasonable living standard and the upper limit would be around 1.5 to 1.6 billion. Assuming that 400 kg total grain consumption (including grain that is converted into dairy products, meat, and liquor) is necessary for each individual, imagine the quantity required for 1.6 billion people each year -- 6.4×10^8 tons!

Several factors affect maintenance of an adequate grain supply: the lack of a strict birth control in recent years (birth rate increased by 1.5% in 1996); the shortage of irrigation water that could be worsened in the future; the loss of 400,000 to 670,000 hectares (6 to 10 million mu) of cropland each year; the maximum use of fertilizers already implemented in high-yielding areas; the changing diet with more demand on liquor, dairy products, eggs, and meat products; the migration of farming populations to urban areas for better paying jobs; and much of the marginal land already being cultivated. Thus, additional food grain production is becoming the most complex and difficult problem to solve for the future, and a formidable task is facing all the Chinese, including researchers and policy makers, as well as farmers, laborers, and other civilians.

Because of the importance of China having economic and military links to other countries of the world, China's food problems are interlocked intimately with the welfare of other countries , including international food prices, population movement, trade, disease epidemics, or even war. In fact, world population also is rising, although at a slower rate, and is predicted to reach 8 billion by 2025.

The major staple food supply in China are wheat and rice. People in some areas also consume maize, grain sorghum, potato, and foxtail millet. Wheat and rice production in China from 1949 to 1996 is reviewed here and the potential for future yield increases is discussed from the production standpoint with special emphasis on the use of breeding and genetic technologies and cultural approaches and on governmental policies that have a far-reaching impact on farmers' incentives, efficient land use, agriculture output, and development of research and technology.

Current status.

Wheat. In 1995, the planted wheat area was 28.86 million hectares (13% of the world's wheat acreage), a 26.2% of the total food-grain growing area in China), and the grain yield was 102.2 million tons (21.9% of total food-grain production). In China, wheat can be planted in the fall (84.3% of the total wheat-growing area) or spring (15.7% of the total wheat-growing area), depending on the region; some of the wheat planted in winter is spring wheat, the so-called spring -wheat-planted-in-fall. To increase the crop index and efficiency of land use, wheat also is utilized in intercropping, double-cropping, and relay-cropping systems. Thus, early maturity is one of the most important attributes in addition to yield. Quality of wheat grain is not emphasized as it is in the United States or other bread-consuming countries. The number one goal in wheat production is quantity.

Since the 1840s, China has been torn by disastrous internal turmoil, civil wars, and foreign invasions that have severely encumbered agricultural production and improvement. Also, systematic and sustained scientific research on agricultural production was never advocated or implemented by the feudalistic dynasties. In 1949, the year when the People's Republic of China was founded, the average wheat yield in the nation was less than 650 kg/h or 9.5 bu/a (1 hectare = 2.5 acre = 15 mu) (Chinese Agriculture Forges Ahead, 1995)! The highest yield was recorded in Sichuan Province with an average of 990 kg/h, whereas Shandong Province produced an average of 618 kg/h, and the Beijing area, 465 kg/h. With the extended use of chemical fertilizers, availability of irrigation facilities and improved cultivars, and better management, wheat yields reached 1,500 kg/h in the early 1970s. The period of time from 1949 to 1970 was considered as the low wheat yielding era. From 1970 to the early 1980s, wheat yields increased to an average of 1,890 kg/h. From 1980 to 1986, wheat yields increased tremendously, from 1,890 kg/h to 3,045 kg/h. In 1995, wheat yield had reached the highest peak ever with an average of 3,529 kg/h.

The increase of wheat production seems to be related to agriculture-policy, and the percent of increase varied drastically from period to period. For instance, the increase per year from 1978 to 1980 was only 1.17%, whereas it was 9.23% from 1981 to 1985, primarily because of the abolition of the commune system and farmers having greater incentive and more freedom to produce for themselves. From 1986 to 1995, the average annual increase was stabilized to 1.80%.

The wheat planting area from 1978 to 1985 was stable at 29.2 million hectares each year. The planting area increased slightly in 1986 and then increased steadily until 1991, where the peak of 30.95 million hectares was reached. From 1991 to 1995, the wheat planting area gradually decreased by an average loss of 522,000 hectares per year. The wheat growing area was 28.98 million hectares in 1994 (China's Agriculture in Development, 1995) and decreased to the wheat-growing area was 28.86 million hectares in 1995, the smallest area in recent history.

Total wheat production increased each year. The 1995 wheat production was 102 million tons, an increase of 48.367 million tons since 1978, or an average increase of 2.845 million tons (3.84%) per year. The increase of total production corresponded with the yield increase per unit area, with a tremendous jump from 55.21 million tons in 1980 to 90.04 million tons in 1986 (an increase of 8.49% per year). From 1987 to 1995, wheat yield appeared to follow a trend: an increase for two years and a decrease for one year, then an increase for another two years and another decrease for one year.

The unit yield was highest in 1995, whereas the highest total production occurred in 1993 because of the larger planting area. The average yield of 3,529 kg/h in 1995 was far behind the wheat yields of 6000 to 7500 kg/h in countries like UK, France, Germany, and Netherlands.

The wheat growing area is distributed in 29 provinces in China and falls into five major ecological-seasonal regions. A total of 28.86 million hectares were harvested in 1995; winter wheat accounted for 85% of the area and spring wheat for 15%. The general production conditions in those regions are as follows (Yu et al, 1995):

A. Northern temperate ecological region, including regions north of the Great Wall, the northwest, and part of the north central region. Spring or strong spring wheat is grown in this region, which accounts for 15.7% of the total area. The average yield is 2,470 kg/h, with a range from 1,875 to 3,697 kg/h.

B. Temperate ecological region around the Yellow River, Hui River, and Hei River region, including areas south of the Great Wall but north of Qin Mountain (Shanxi Province) and the Huai River. This region is suitable for winter wheat or strong winter wheat cultivars. It is a major wheat growing region and accounts for 53% of the total wheat growing area. The average yield is 4,203 kg/h and the range is 2,566 to 5,830 kg/h. Limiting factors on yield in this area are very low temperatures in winter, frost damage and drought in spring, and high temperature during the grain filling stage.

C. Subtropical region where spring wheat is planted in fall. This includes the area south of the Qin Mountain and Huai River and the depression of Sichuan Province and high plateau of Yunan and Guinzhou Provinces. It has a long summer and no winter, and is suitable for growing short-day and spring type cultivars. The area accounts for 26% of the total area, and the average yield is 2,787 kg/h, with a range of 2,010 to 3,889 kg/h.

D. The inland and Xinjiang region (northwest) where both spring and winter wheat are grown. This area falls in the central and south of the temperate zone with a dry climate. It accounts for 4.2% of the total area, and the average yield is 3,666 kg/h.

E. Qinghai and Tibet high plateau region. This area is known as the high yielding spring wheat region with a long growing season. It accounts for 1% of the total area and has an average yield of 4,137 kg/h.

Because the complex climatic patterns, soil types, and conditions for production vary greatly from region to region, yield levels also are different. However, the yield pattern indicates that yield is high for areas where the economic conditions are better developed, because farmers can afford and are willing to invest more in the land; these includes areas near Beijing, Tianjin, Shanghai, Liaoning Province, and Shandong Province. Likewise, areas where the ecological conditions are suitable for wheat production often yield more than others, such as the temperate region around Yellow River, and Hui River, the Qinghai Province, and the Tibet region because of factors such as a large difference between day and night temperatures; a long growing season; and light, temperature, and rainfall that are adequate for wheat development, even though Qinghai and Tibet are not considered as economically well developed districts.

Rice. Rice is the most important food grain in China and has been cultivated for nearly 7,000 years. Rice cultivation is very labor-intensive and involves planting a nursery and transplanting by hand in most places. It is sown in 30.4 million hectares or 21% of the world's total rice growing area (Xie, 1996) and 29% of the total food-grain production area in China, and produces a total of 178.3 million tons of grain, or 44% of the total grain production in China (Huang et al., 1995) and 33.3% of the world's total rice grain production. Of the total rice production the standard cultivars (both *japonica* and *indica* types) accounted for 46%, hybrid rice for 51% , and sweet rice for 3% (Xiong et al., 1993). The average yield per hectare is 5,854 kg. Most of the rice (90%) is grown south of Yangtze River, and 10% is distributed among three provinces in the northern region. The rice cultivars that are grown in the north belong to *Oryza sativa* subspecies *japonica* and those in the south belong to subspecies *indica*. Based on the growing season, production can be divided into early, medium, and late rice. In 1991, hybrid rice was grown on 17.4 million hectares, or 53.4% of the total planted rice area, twice as much as that planted in 1984. However, rice yield fluctuated somewhat between 1984 to 1991 because of many technical, environment, and political factors, such as the development and planting of hybrid rice, price and fertilizer distribution, labor invested on paddy fields, market price of rice grain, and policies concerning rice farmers in general.

Among factors analyzed, fertilizers and labor investment appeared to be the most important ones influencing rice yield, from either hybrid or standard cultivars. However, effects of fertilizer application vary from south to north. Under the present conditions in the south, addition of 1 kg fertilizers could increase yield of rice grain by 0.67 kg, whereas in the north, each additional kilogram of chemical fertilizer could increase yield by 1.33 kg (Huang et al., 1995), suggesting that applying additional chemical fertilizers in the south does not really increase farmers' income. Actual labor investment in rice production in most provinces is much higher than optimum. With the development of a marketing economy, a low market price for rice will lead to less labor being invested in rice production and could result in serious land abandonment and poor management in economically well developed areas. This already has occurred for provinces along the coast.

Rice production has been increased by using improved conventional culti-vars, developing hybrid rice, and refining management practices, such as the "dry-nursery and sparse-transplanting" technique (Lu et al., 1996; Lu et al., 1997) by which the land area necessary for planting the nursery can be re-duced and saved for production purposes and the number of seedlings to be transplanted also can be minimized without undesirable impact on yield. In the northern region where the *japonica* type is grown, standard cultivars typically are used, whereas in the south, 18 million hectares of hybrid rice was grown *(indica* type) in 1991, and the amount has increased steadily reaching 20.8 million hectares in 1996 (Xie, 1996). For the standard cultivars, vigor, pest resistance, and ideal plant morphology (ideotype), which are interrelated, are the goals in rice breeding programs (Yang et al., 1996). For the hybrid rice, which accounts for about 50% of the total rice area and yields 6,615 kg/h relative to 5,340 kg/h for all rice (Yuan, 1992a), vigor and yield are the main objectives without taking quality (taste) into consideration.

To produce hybrid rice, the "three-line" system involving the cytoplasmic male sterile (A) line, the maintainer (B) line, and the restorer (R) line is still used effectively and will continue to be used in the near future. The "two-line" system, which is based on the use of photoperiod-sensitive genic male sterility (PGMS) or thermo-sensitive genic male sterility (TGMS), has the advantages of eliminating use of the maintainer line and possibly broadening the selection of parents in hybrid combinations. Progress has been made, and the "two-line" system is expected to eventually replace the "three-line" system. The "one-line" system, which is based on using apomixis (agamopspermy or asexual production by seed) that has the potential to fix the heterosis of F1 (producing hybrids from hybrids), requires more research, and apomictic rice materials need to be found for analysis (Yuan, 1992b). Hybrids can be inter-varietal like the existing hybrids; they can be intersubspecies (between subspecies *indica* and *japonica*, *indica* and *java*, *japonica* and *java*), which tend to be semisterile but could exhibit a higher degree of heterosis than intervarietal hybrids; or they can be developed from distantly related (intergeneric) taxa that may have a tremendous amount of heterosis, but long-term research is required to demonstrate such potential in order to implement the practice.

The hybrid rice, already developed, which has the potential for further yield increase, has been grown from 44° north latitude in Jilin Province to 18° in Hainan Island (tropical) and from 125° east longitude (Shanghai) to 95° (Yunan Province). Various maturity types of hybrid rice have been developed to adapt to the local growing conditions. Among the provinces where hybrid rice is grown, Sichuan, Guangdong, and Hunan have the largest growing areas followed byJiangsu, Jiangxi, and Fujian.

Most of the hybrid rice produces 15-20% more than the standard checks. Hybrid vigor primarily stems from the well-developed root systems, large and heavy kernels, strong stalks, more efficient photosynthetic capacity, more tillers, multi-pest resistance, broader genetic background, and better adaptation. However, quality of hybrid rice, such as amylose content (current hybrid rice has too much amylose), low vitreousness of the grain, and taste, needs more attention and effort to improve. Because the endosperm in kernels of hybrid rice is already in the F2 generation and segregating, both parents involved in the hybrid combination must have top quality.

Upland rice that requires only 1/3 to 1/5 of the amount of water to grow accounts for only 2% of the total rice planting area, and its yield is low (1.5 to 3.0 ton/h). At present, upland rice is concentrated in Yunan Province, which is also the major source of upland rice germplasm. The area adaptable to upland rice is around 50 to 60 million hectares, but development of improved cultivars is needed to grow upland rice in about half of this area. Upland rice can be managed like wheat or can be transplanted like paddy rice; it fits intercropping system with other cereal crops, forage crops, vegetables and fruit tress, and not as labor intensive as the paddy rice.

Potential for Additional Production.

Potential for increasing food grain production in China exists on several fronts: increasing yields per hectare by developing high-yielding cultivars and hybrids, particularly those adapted to special areas; properly applying fertilizers with special emphasis on balancing N, P, and K, especially on low-yielding land; searching for additional cropland that needs a minimum amount of irrigation; conditioning seed and establishing a seed certification system for precise control of the seeding rate to avoid excessive use of seed grain; improving the existing irrigation systems; and extending the present distribution and storage facilities. Even more importantly the government must invest much more in agricultural research, production facilities, extension systems, higher the prices for commodities, and education in general to assure farmers' well-being and boost their incentives to work on land.

Wheat

Wheat might produce 12-14 tons with a maximum of 17 tons per hectare under ideal conditions (Zhang and Huang, 1996). The highest wheat yield in the United States, 14 tons per hectare, occurred in the 1960s with the semi dwarf wheat cultivars 'Gaines' and 'Nugaines' of Washington state. In the 1980s, wheat yields of 10 to 13 tons per hectare occurred in the UK and Germany. The highest yield in China was 9 to 12 tons per hectare recorded in Qinghai Province in the 1970s. Optimistically, if we use the average rate of increase during the past 20 years (3.84%) to predict the yield level for the next 10 to 15 years, then the northern temperate spring wheat region might reach 3.7 to 4.5 tons per hectare; and the winter-wheat-producing region around the Yellow River, Huai River, and Hai River might produce 6.0 to 6.7 tons per hectare; the subtropical spring- wheat- plant-in-fall region may produce 4.5 to 5.2 ton per hectare. If these goals are realized, the total wheat yield would be 45 to 76% higher than the current yield level. Realistically, if we use the average rate of increase or 1.8% (from 1986 to 1995) to predict the future wheat yield, then the realized gain by conventional wheat breeding approaches would be around 20% during the next 10 years.

To further increase yield, novel approaches must be explored and investigated thoroughly. Thus, conventional wheat breeding might increase the yield level by 1% per year on the high-yielding land and 2% per year for low-yielding field. These steady increases in yield are realistic and can be accomplished. However, the conventional breeding approach has not brought a breakthrough in yield per unit area in recent years and probably will not in the near future. Production of hybrid wheat using male-sterility-inducing systems, such as cytoplasmic male sterility (T, K, and V types); chemical hybridizing agents (CHAs) (EK, SC2053, BAU 2, and Mon21200), TGMS lines; PGMS lines, and apomixis whereby a hybrid plant is able to produce hybrid seed again all should be investigated. Continued exploration of other new and efficient ways to produce hybrid wheat is essential.

Although the hybrid-wheat growing area is relatively small (less than 1% in the United States), hybrids provide a potential to increase yield greatly. More than 50 research institutions in China are engaged in hybrid wheat studies. Field data in various regions have shown that an average yield increase of 10 to 42% could be achieved compared to the control or standard check cultivars (Zhang and Huang, 1996). Pessimistically, some may point out that why hybrid wheat has been studied for more than 30 years and is not yet widely available. Unfortunately, it may never be, because government investment in agricultural research at present is 0.5-0.7% of the national gross domestic product (GDP), too low an amount to sustain vigorous agricultural research, especially the novel approaches that require substantial amounts of time, effort, and equipment.

However, some progress has been made. Using T-cytoplasm, hybrid wheat 'Ke73-402A/Kehan 10' developed by the Helongjiang Academy of Agricultural Sciences produced yields more than 14.4% higher than those of standard check and has been grown on 200,000 hectares; Ning-Ai 2/R14 developed by the Jiangsu Academy of Agricultural Sciences   yielded 17.8% and 15.4% more than the check in 1990 and 1991, respectively; 83-1A/YH 3 developed by the Shanxi Academy of Agricultural Sciences increased yield by 18.8 to 21.9% over the check in a 3-year test; msTQ431/T-6-3 developed by Northwest Agricultural University increased yield by 16.9% over the check; hybrid C13 developed by the Sichuan Academy of Agricultural Sciences produced 7.3 tons/h or 11-20% higher than the check in 1993 and 1994. The key to successfully using T-cytoplasm for hybrid combinations appears to be selection of high-yielding restorer and maintainer lines in the crosses.

Although T-cytoplasm is a good male sterile source for hybrid production, its sources of R lines are narrow, large number of restorer genes are necessary for complete restoration, and the expression of restorer genes is different under homozygous (in the restorer parent) and heterozygous (in the hybrid) conditions. More than 70 additional male-sterile-inducing cytoplasms have been examined, among which the K cytoplasm from *Aegilops kotschyi* and V cytoplasm from *Ae. ventricosa* have been investigated more thoroughly. Sterility of K-cytoplasm can be maintained by some cultivars containing 1B/1R (chromosome translocation involving wheat chromosome 1B and rye chromosome 1R) and Xiaoyan 6 is one of the better restorer lines. However, when compared to the B lines, K-cytoplasm male sterile lines are weaker during the seedling stage; head early; and have lower kernel weight, shorter height, and fewer spikelets, but the protein content and germination percent are higher. Advantages of V-cytoplasm are that it has many sources of fertility restoration, has no undesirable cytoplasmic effect, and does not produce haploids. However, R lines that can completely restore the fertility are few, such as T-6-3, 87F6820, C1467, CII224, CII115, and 77-65-1R. Nevertheless, the amount of heterosis is high when K or V type cytoplasms are used. For example, K-8222A/T-6-3 was 37.6% better than the parent, and VH8801 yielded 21.6% more than the check; K901, a hybrid using the K-cytoplasm, produced 11.7 ton/h or 15.6% more than the check in multiple-location trials in 1992, 1993, 1994, and 1995.   Hybrid wheat, though is limited in acreage partly by low seed production, presents a realistic potential for yield increase.

Chemical hybridizing agents can substitute for the male sterile and restorer lines and by using CHA, any two selected parents can be placed in combination.   However, some of the CHA agents are toxic and extra care must be taken in large-scale productions. The ideal CHA should have the following characteristics:
- Safe to use, no residual effect, and not toxic to human and livestock

• Low cost.

• Induces complete or nearly complete male sterility for different cultivars under different environmental conditions, but produces no undesirable effect   on female fertility.

• Flexible in application, so the amount of the agent can be adjusted and timing of application can vary.

Application of CHAs simplifies the procedure of hybrid wheat production, places less restrictions on parent selection, and requires less time in producing hybrids. Several institutions have released hybrid wheat and made field demonstrations. Huayou #1 produced more than 9 ton/h in 1992, and established a yield record in Hebei Province; Jinhua #1 was demonstrated in 700 hectares during a 5-year period and produced an average of 7.5 ton/h or a 12.7% more than the check; Nongda 851 produced by the Chinese Agricultural University yielded 15% more than the check; the Hebei Academy of Agricultural Sciences also released hybrids Jiza #89-1 and #90-1, which yielded 9.5 ton/h or 15.2% more than the high-yielding check.

Male sterile lines that are photoperiod and temperature sensitive also have been investigated. Those lines become male sterile under short photoperiod (12 h) and low temperatures (10°C), but are fertile under long days (14 h) and high temperature (18°C). Thus, each of those lines can be used in two ways: to produce hybrids as females under short days and low temperatures and to produce the same genotype by selfing under long days and warm temperatures. Several of those lines have been used to produce hybrid wheat, and yield increases over the check varied from 13.1 to 57.3% depending on the hybrid and the testing site. Such a system apparently can be utilized in areas where the photoperiod and temperatures vary to the degree required for fertility by changing the planting dates.

Although the heterosis in hybrid wheat has been well known since the early 1960s, utilization of hybrid wheat for commercial production has been far from fully realized. Several problems exist. The amount of seed per hectare necessary for wheat planting is large, about 60-70 kg/h; therefore, hybrid wheat seed production must be increased at a cost affordable to farmers. Details such as the female to male ratio, isolation distance, anthesis time and duration of anthesis for female and male parents, amount of pollen produced by the male parents, prevention of contamination, purity of hybrid seed, additional or secondary pollination, and seed production of the female parents should be investigated for each ecological region where hybrid wheat is a feasibility.

Ways to predict and search for hybrid combinations with more hybrid vigor are other problems remaining to be solved. Does wheat have heterotic groups like those in maize, so that selection of parents can be made with reasonable confidence? Because hybrid wheat produce much more grain than standard cultivars, management and cultivation methods could be different: the optimal amount of seed required per hectare, row spacing, fertility including organic and chemical fertilizers, rotation, irrigation, and weed and pest control all should be investigated.

Because wheat is one of the two major food sources, the quality or taste of the hybrid wheat must be considered. Hybrid rice and hybrid sorghum have poor taste and are not desirable as commercial products for human consumption. Would hybrid wheat suffer from the same quality problem? Production of hybrid wheat in China could reach 2-5 million hectares by 2005-2010, so the above mentioned problems should be solved before large amount of areas are planted for hybrid wheat.

Until hybrid wheat can become a reality, improving conventional cultivars must continue, with special emphasis on crosses involving multiple parents with multiple pest resistance; adapting cultivars to special ecological regions, large or small; developing of drought- and heat-tolerant types using a conventional approach and biotechnology; balancing N, P, and K application; and supplying adequate phosphorus and potash, which have become limiting factors for production in certain areas. The soil in northern regions generally lacks phosphorus, and that in the south lacks potash Nitrogen fertilizer has been overused in many fields and accounts for 68.5% of all fertilizers applied. In 1993, the ratio of N:P:K sales in China was 1.0 : 0.37 : 0.12 compared to the worldwide ratio of 1.0 : 0.49 : 0.37. Lack of phosphorus and potash has become a constraint for land where large amounts of chemical fertilizers have been used. Lack of organic matter (0.5-1.0% for most of the cropland) in fields is a serious and common phenomenon in China Overuse of the land for unlimited production to feed the ever increasing population and government policies concerning the right to use the land and land ownership are closely related factors. Appropriate agricultural policies certainly will correct some of the food grain production problems.

Rice

Hybrid rice developed from intervarietal crosses has increased yield by 15-20% over the conventional cultivars, but additional progress has not been made in recent years because of the closeness of genetic backgrounds among the parental lines. Some of the hybrid rice, such as ShanYou 63 released in

1986 by the Fujian Academy of Agricultural Sciences, has been grown for more than 10 years without being replaced by new hybrids. Shan-you 63 has wide adaptability and yielded more than 11 ton/h in multiple testing sites distributed over eight provinces and accounted for 34% of the total hybrid rice growing area in 1996 (Xie, 1996). It is one of the three best hybrid rices grown in China since 1986. However, growing genetically uniform crops over such a large area for such a long time can lead to genetic vulnerability and cause a disastrous loss of yield. The occurrence of southern corn leaf blight on T-cytoplasm maize during the early 1970s in the United States is an example. The current hybrid rice must be replaced on a regular basis with new hybrids possessing genetic diversity and adapted to various regions.

Production of hybrid rice with the three-line system is limited by the availability of superior A and R lines that combine well, so it cannot fully utilize the available germplasm. Thus, the chance of selecting the best combinations is low, and hybrids have the drawback of genetic uniformity. In addition, the cost of producing hybrid rice seed is high because of the complexity of the procedures. Also, the yield level has been nearly the same since 1986 (Zhang et al., 1996).

The two-line system uses photoperiod- and temperature-sensitive lines (A and B lines combined into one line) that are hybridized with the R lines. The combined lines are male sterile under long days and high temperatures, but male fertile under short days and cool temperatures. Thus, the same line can be planted in spring and fall and be male sterile or planted in summer and be fertile to produce hybrid seed. This simplifies the production system, reduces production costs, saves about one-third of the land area, and increases the chances of finding the best combinations because of more freedom in selecting the sterile lines. At present, hybrid rice derived from the two-line system produces 10% more than those from the three-line system.

Research on hybrid rice using crosses between subspecies is also in progress. Although hybrid vigor could be increased by genetic diversity, cross incompatibility presents a problem. Agamospermy, which is an asexual production via seed, is a way to fix hybrid vigor or produces hybrid seed from a hybrid. The possibility of using such an approach also is being explored, and it could become a valuable tool for hybrid rice production where wide crosses are used.

Among the *indica* type, hybrid rice lacks early maturity and high yield combinations, and among the *japonica* type, it lacks high yield and good quality combinations. More than 6 million hectares *japonica* rice paddies and more than 5.3 million hectares of double-crop early- rice paddies around Yangtze River still plant the conventional rice cultivars. These areas have great potential to plant hybrid rice with characters of early maturity, high yield, and good quality.

At present, food grain production in China is between 450 to 490 million tons, and estimated needs are 500 million tons by the year 2000, 550 million tons by 2010, and over 600 million tons by 2030. The planting area of rice is around 30 million hectares annually, or 29% of the total food grain crops, but rice grain production has reached 44% of the total food grain output, indicating the importance of rice as a staple food. Hybrid rice planted on *50%* of the area, produces an average yield of 6.6 tons per hectare, or more than 60% of total rice production. This strongly suggests that hybrids have been making a significant contribution to meet the demand for rice as a staple food. With the declining cropland, use of hybrid rice to increase the unit yield is inevitable.

Conclusions

Although the outlook for agricultural production in China by 2030 does not seem bright, the Chinese people are khown as industrious, resilient, ingenious, tough, and adaptable to adverse environments in which they live. Research has been done in various areas to promote agricultural production. However, like those problems pointed out by Brown (1995), Stover (1996) and Prosterman et al. (1996), the essentials of agricultural production in China are limited: water shortage for irrigation, maximum use of chemical fertilizers; deterioration of soil fertility and physical structure of cropland, a change of diet for urban people; continuing loss of farm land by building roads, airports, factories, houses, and urban expansion; pollution of water and land; double cropping already used on some farmland and scarcity of additional arable land; movement of young males to urban areas causing labor shortage on farms; and lack of reasonable and steady agricultural policy concerning land-users' rights, commodity prices, and farmers' education and well-being, especially after retirement or in sickhess. Whether we agree with these authors or not, many of the statements are facts and should alarm not only the Chinese policy- makers but also those in other countries because of the interrelationship among them. The basic cause of all the hardships, of course, is the unlimited and unreasonable human population growth leading to the environment deterioration, not just in China but throughout the Asia, Africa, and many other places, especially in the developing countries.

Many other ways are available to increase food grain production in addition to breeding and managing approaches. For example, if the amount of seeds to be sown can be reduced by improving the seed quality (e.g., germination rate), then more grain can be saved for consumption instead.

This is not a small quantity, because Chinese farmers tend to over-sow their farm land. Seed quality can be improved by using seed treatment; coating; and better packaging, cleaning, and storage conditions. Although the percentage of loss during storage is not known, it can be a substantial amount. Establishing and enforcing seed laws, establishing transportation systems, and improving storage facilities are essential for modernization of agriculture in the 21st century.

Under the current system established by the Ministry of Agriculture, new varieties or cultivars to be released must out-yield the check or standard ones by 15%, and monetary awards are given to those whose new varieties are grown in large areas. Thus, varieties or hybrids suitable for specific but small ecological areas are ignored. Technically, genotypes showing a genotype-by-environmental interaction should be capitalized on and released to farmers in those specific regions even if the areas are small. China has so many wheat and rice breeders located in various districts, and they are likely to develop cultivars or hybrids specifically adapted to small areas. Why cannot those materials be released?

The soil fertility of most of the cropland, especially the organic matter, is very low. Soil is a dynamic system, and fertility needs to be replenished periodically. Emphasis of double cropping and high yield without fully replacing the fertility, organic and inorganic, is like killing the goose for the golden eggs. Why not establish a green manure system for the high-yielding land similar to the soil conservation system in the United States, which subsidizes the farmers and replaces the crop loss by importing food grains from foreign countries? China has a large amount of national reserves and trade surplus, at least with the United States. Unless the farm land is sustained, keeping farmers on the land and, thus, feeding the people will be difficult. More importantly, completely depending on foreign countries for food grain supplies can lead to disasters.

Government policy and investment dictate the agricultural production, farmers' living standard, institutional research, the morale and stability of the society, and, ultimately, the fate of the country. Agriculture is the cornerstone on which to build China or any country. The investment in recent years in agriculture and education generally has been a meager amount. Some farmers cannot read and still grow wheat cultivars developed 40 years ago; cooperation is lacking between the extension personnel and research units;

salaries are so low that some of university professors and researchers take on second jobs and not able to concentrate on their assignments; the law is not enforced to protect the farmers' land-to-the tiller policy or the land user's right is so changeable that farmers don't have confidence in the policy and refuse to make long-term investment in land; and young high school students are reluctant to select agricultural science as a career. The population is concentrated in the eastern half of China, and, thus, the average land area per capita is becoming smaller and smaller in those productive provinces. The gap between rich and poor is widening without a proper tax system. These and many other problems related to agricultural production can be solved or partially corrected by implementing appropriate government policies.

Increasing China's agriculture production to meet the demand by a population of 1.6 billion by 2030 obviously is a tremendous challenge to the nation and cannot be taken lightly by any standard. However, promising signs indicate that the food supply might be met by then. The statistics on cropland areas reported to the government in most of the places are purposely under estimated by village heads to reduce taxes and to show a higher yield. At least 10% of the cultivated land probably is "hidden". The dedication and sense of responsibility of agricultural researchers in China are unparalleled: the development of hybrid rice; the improvement of cultivation techniques to fit double-cropping systems, such as transplanting maize seedlings, the use of seed coating that contains nutrients, growth regulators, and pesticide to protect seeds and promote the seedling growth; the ingenuity and modesty of research workers who constantly search for new and better ways to produce more food grain using the two-line and one-line production method of hybrid wheat and rice; the establishment of "village industries "to increase farmers' income and increase exports; and support for the one child per couple policy among populations in cities and among the educated classes because they clearly realize the impact of uniimited population growth. More importantly, the old die-hard but not well educated politicians who had been dictating the country's policies, especially rural policies, are dying out, leaving the policy making process in the hands of young, modest, flexible, educated, and pragmatic bureaucrats. Although it is a very serious problem for the future, food grain production in China appears to be as many Chinese scientists have indicated, "a serious and worrisome but not a desperate problem!" We sincerely hope that this is indeed the case.

References

1. Brown, Lester R. 1995. Who Will Feed China - Wake Up Call for a Small Planet. The Worldwatch Environmental Alert Series. W. W. Norton & Company, New York, NY2. China's Agriculture in Development - Report to the United Nations and the Food and Agriculture Organization of the United Nations for the Fiftieth Anniversary of Their Establishment. *1995.* Ministry of Agriculture. Beijing, China

3. Chinese Agriculture Forges Ahead - To the United Nations and the Food and Agriculture Organization of the United Nations for the Fiftieth Anniversary of Their Establishment. 1995. Ministry of Agriculture, The People's Republic of China.

4. Gou, H. C. 1994. Potential and approach to increase an additional 50,000 million kg of food grain by the year of 2,000. In Discussion of Chinese Food Grains for Year 2,000. pp.45-52. Sinica Agricultural Science Technical Press, Beijing, China

5. Huang, J., Q. Wang, and Q. Chen. 1995. Agricultural production resources allocation: nee input and output analysis. Chinese J. Rice Sci. 9: 39-44.

6. Jia, P. H. 1996. Control pest damages in crops - a contribution to increase an additional 50,000 million kg food grains in the year of 2,000. In: Discussion of Increasing 50,000 Million kg Food Grain. Pp.398-402. Sinica Agricultural Technology Press, Beijing, China

7. Lu, X. Y., L., Peng, Y. Tan, X. M. Liu, and Z. M. Luo. 1996. Studies on the biological characteristics of the plants developed from the early rice (oryza sativa L.) Seedlings raised in dry nursery. Jour. Hunan Agr. Univ. 22: 321-325.

8. Lu, X. Y., X. R. Tang, L. S. Peng, X. M. Liu, X. L. Zheng, Z. M. Puo. 1997. The characteristics of morphology, physiology and biochemistry of late rice plants cultivated by raising seedlings with dry nursery management. J. Hunan Agr. Univ. 23:307-315.

9. Prosterman, R. L., T. Haristad, and P. Li. 1996. Can China feed itself? Sci. Amer. 275: 70-76.

10. Stover, Dawn. 1996. The Coming Food Crisis. Popular Sci. 249: 49-54.

11. Xie, H. A. 1996. Breeding theory and practice of "Shanyou 63", the variety with the largest cultivated area in China. pp.59-66. In Proc. Sino-American Symp. on Agr. Res. Develop. in China. China Agricultural Press, Beijing

12. Xiong, C. M., X. D. Chu, Y. K. Luo, and F. S. Huang. 1993. New progress of rice quality research. Rice Review Abstr. 12:1-6.

13. Yang, S. R., L. B. Zhang, W. F. Chen, Z. 3. Xu, and 3. M. Wang. 1996.

Theories and methods of rice breeding for maximum yield. Acta Agronomica Sinica 22:295-304.

14. Yu, H. S., C. H. Nan, and L. C. Tian. 1995. Ecological categories and ecological classification of common wheat in China. Acta Agr. Boreali-Sinica 10:6-13.

15. Yuan, L. P. 1992a. The strategy of the development of hybrid rice breeding. pp.1-S. In Current Status of Two line Hybrid Rice Research (ed. L. P. Yuan), Agricultural Press, Beijing, China

16. Yuan, L. P. 1992b. Progress of two-line system hybrid rice breeding. pp.6-12. In: Current Status of Two Line Hybrid Rice Research (ed. L. P. Yuan), Agricultural Press, Beijing, China

17. Zhang, A. M. and T. C. Huang. 1996. Studies and perspectives of hybrid wheat. Shanxi Wheat Newsletter 16:1-6.

18. Zhang, Q., G. H. Wu, and R. L. Luo. 1996. Perspectives and utilization of hybrid vigor in rice. China Rice Grain 4:2-4.

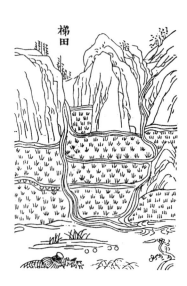

CHAPTER 4.22

FOOD GRAINS--THE POTENTIAL FOR YIELD IMPROVEMENTS

Alexander von der Osten

Consultative Group on International Agricultural Research (CGIAR) The World Bank

The analysis of the potential and prospects for sustainable yield improvements in major food grains should deal with two different aspects:

- the potential for yield improvements per se
- the potential for yield improvements in China.

THE POTENTIAL FOR YIELD IMPROVEMENTS *PER SE*

Any activity that raises yields and enhances their stability can be considered a yieldimprovement, thus raising crop productivity. In principle, future yield increases at the farmer level will depend on three trends:

- the maximum achievable yield under optimal conditions (yield ceiling)
- the yield potential under field conditions
- the farmer's ability to realize the yield potential (reducing the yield gap)

Enhancing the stability of yields (tolerance of biotic and abiotic stress) helps raising the yield potential and the *average* crop productivity, thereby reducing the yield gap and the farmer's risk of harvest failure.

Raising the yield ceiling

In order to raise the yield ceiling, two approaches are possible (separately or in combination):
- to increase the harvest index, i.e. the grain yield at the expense of residual biomass (mainly straw), as exemplified by IRRI's new 'super rice'
- to increase the total biomass, i.e. the plant density per area unit and the plant's efficiency in converting inputs (sunlight, water, nutrients).

IRRI's new rice plant shows a 20-30 percent yield "reserve" hidden in the current plant architecture waiting to be tapped by raising the current harvest index of 0.55 to 0.6 or 0.65. CIMMYT's new wheat plant type could increase yields by 10-25 percent. Changing the plant architecture means another significant step in improving the plant's efficiency in converting inputs into grain.

Past experience shows that strong progress in efficiency has been made. An analysis of the performance of CIMMYT wheat spanning several decades indicates that relatively steady efficiency gains have been achieved. [1]

Efficiency gains improve the sustainability of farming by raising output without increasing inputs. Alternatively, they might be used to maintain output while decreasing inputs. Presumably there are biological limits, for instance a maximum harvest index, to improving input conversion efficiency. The question is: will these natural frontiers affect plant improvement work in the next 40 to 50 years? Will biotechnology help accelerate and facilitate improvement work, thus counteracting the effect of possibly approaching natural frontiers?

Raising the yield potential

Raising a cultivar's actual yield potential under field conditions is of more practical significance than the plant's yield ceiling under optimal conditions. By breeding crop varieties which better resist biotic stress -- usually the main locally important pests and pathogens --and tolerate locally significant abiotic stress -- drought, heat, cold, flooding high and stable yield levels can be achieved in the field with reduced need for chemical plant protection and other protective practices. A hardy cultivar, by offering the farmer a high and stable yield potential, can substantially raise crop productivity and farm income, typically returning many times the expense for improved seed material.

Reducing the yield gap

The size of the yield gap existing between research station test fields and farmers' fields is influenced by a number of factors. On the positive side, in the sense of narrowing the gap, we have
- progress in farm level availability of improved seed and other inputs (e.g. irrigation, fertilizers)
- progress in agronomic practices (crop and farm management).

On the negative side there are a number of economic and environmental factors;
- insufficient profitability of crop production[2]
- plant and soil pests, and spreading of diseases
- spreading of aggressive weeds
- losses in soil fertility, organic matter content and structure, resulting from intensive farming over extended periods; soil nutrient mining, erosion, compaction, improper fertilization, salinization, and acidification due to inappropriate soil, nutrient, and moisture management.

Progress in agronomic practices is, next to profitability, perhaps the single most important factor affecting the yield gap. It comprises a wide range of improvements ranging from more labor input per unit of land -- as a result of rural population growth -- to mechanization, better farmer education and training through schooling and extension.

On the economic side, low grain prices and/or policies discriminating (small) farmers (urban bias) can render grain production so unprofitable that market sales result in little gain or even a net loss for the farmer. A no-gain situation encourages subsistence strategies, discourages investment and the adoption of improved agronomic practices.

On the environmental side, the greatly increased insect and pathogen pressure has put plant breeders on a 'treadmill' in their efforts to stay ahead of the pests by releasing an interminable sequence of more resistant crop varieties. The average useful field life of a variety is about 6 years; then it must be replaced because pests have overcome its resistance threshold.

With the onset of rapid population growth in developing countries in the late 1950s, intensification of food production started, in addition to intensified cash crop production. This intensification has taken its toll on the soil, the water resources, and the environment.[3] Decades of intensified monocropping have exhausted irrigated soils[4]; ill-managed irrigation, loss of forests and neglect of watershed management caused additional problems. Modern crop production, contrary to the conditions prevailing at the onset of the historic Green Revolution, cannot expand into pristine lands; in many places, past degradation needs to be corrected to prevent the yield gap from widening before any work to narrow the gap can be undertaken. Pingali has estimated that over many years virtually the entire incremental biological yield capacity of IRRI's new rice varieties has been "absorbed" by negative environmental factors, resulting in yield stagnation or worse, decline.[5]

Realizing that, under these circumstances, a quantum jump in rice yields could not be achieved, the IRRI researchers opted for developing a new rice plant architecture which promises a one-time yield gain of at least 20 percent.

Plant breeders have also developed varieties that better tolerate abiotic stress. By breeding drought, cold and heat tolerant varieties, the vegetation zone of basic food crops could be expanded into climatically marginal areas. In semi-arid regions, yields could be stabilized by progress in drought proofing crops. Varieties that tolerate toxic or polluted soils (e.g. aluminum toxicity; salinity) effectively enhance the productivity of these lands. Progress in *apomixis* allows farmers to plant higher yielding hybrids without the recurrent need to purchase seed.

In conclusion, it appears near impossible -- due to the large variety of powerful factors at work -- to predict how the yield gap will develop in the future. Average grain yields continue to rise steadily in most countries[6] although crop scientists often fear that the yield ceiling -- especially in rice -- has been stagnating in recent years. That would speak for a gradual narrowing of the yield gap, accentuated by occasional 'breakthroughs', for instance recently in wheat in Egypt. However, in the absence of a clear definition of the yield ceiling and hard data it is difficult to prove this point.

Finally, a methodological point: yield increases are traditionally measured in percent of production per area unit, hence on a logarithmic scale. That gives a skewed perspective since it does not reflect the absolute production gains per land unit. Twenty years ago, average developing country cereal yields rose 2.2 percent per annum (1970-76). By now, the annual rate of increase has dropped to 1.5 percent (1990-96). However, the average annual incremental output in the early 1970s was 33.1 kg/ha per year whereas 20 years later it had risen by 15 percent to 38.0 kg/ha. The decreasing yield growth rate hides the fact that cereal output per hectare in developing countries is now rising faster than during the heyday of the historic Green Revolution[7]. Applying a linear scale to yield increases would therefore help to develop a more balanced view of current achievements as compared to those of the past and, consequently, allow to better gauge long term prospects.

The rapid expansion of agricultural science in and for developing countries since the early 1960s is reflected in the historic productivity gains which have led many to believe in future long term logarithmic yield growth paths[8]. However, there is little logic except limited historical evidence to nourish the expectation of stable yield growth rates over 30-35 years at, say, 1.5 percent per year. It would be more realistic to expect the incremental output to stabilize at, say, 40 kg per hectare and year; this would mean constant productivity gains but a steady decline in the yield growth rate.

IMPROVING GRAIN YIELDS IN CHINA

"In both ancient and modern times, China's leaders tend to define food security as grain security."[9] Grains in China are understood to include cereals, legumes, roots and tubers. China is the leading world grain producer with a per capita consumption of approximately 380 kg, (1949: 210 kg) roughly corresponding to the world average. [10]

China has, early on, developed and introduced semi-dwarf high yielding varieties of rice and wheat, as well as improved varieties of maize and legumes[11]. As a result of its successful Green Revolution, it not only maintained its traditional position as the world's top producer of rice, but also became the largest producer of wheat, and the second largest producer of maize.

Grain policy

China has traditionally and successfully pursued a policy of full or nearly full self sufficiency in grain production (The White Paper indicates a 95

% minimum goal[12]), not surprising in a country that consumes one fifth of the world's grain but in which 65 million people are underfed and subject to a government "poverty alleviating program" [13] Whenever signs of chronic grain underproduction appear, the government launches grain self sufficiency drives such as the current "'governors" grain bag responsibility system", a policy designed to ensure adequate supplies of grain at provincial levels. Grain output and demand balance sheets are drawn up at county and provincial levels and delivered to the Ministry of Internal Trade. Governors are requested to reduce local grain deficits or maximize surpluses for transfer to deficit provinces. [14]

Interventionist policies of this kind tend to distort internal trade and economic development, lead to shortfalls in non-grain agricultural production,[15] raise fertilizer imports, and entail environmental problems from cultivation of marginal land, unsuitable irrigation management, fertilizer and pesticide pollution.

However, in the longer run it is expected that considerations of comparative advantage in production decisions and economic efficiency will prevail over old paradigms of serving urban constituents and maintaining self sufficiency for strategic reasons. [16] "Given China's large population and limited arable land it is likely that in the next decade or two China will begin to pursue a strategy of producing agricultural products which use less land and more labor. China could produce high value fruits, nuts, vegetables, specially processed foods, condiments, and specialty meat products for both the domestic and international market, and could import land-extensive crops such as grains and oilseeds."[17]

Land

"China is undergoing a rapid transformation of its national economy. The comparative advantage of agriculture is declining, and resources like water, labor, and land are rapidly moving out of the agricultural sector.,,[18] This kind of statement sounds plausible but is difficult to prove.

The availability of land, for instance, is a vexing issue in China, and assessments of its availability vary between projections of extreme shortage and moderate expansion. The arable land is believed to have declined from 99.5 million ha in 1979 to 94.3 million ha in 1996.

Thanks to an increase in the cropping index to a record level of 1.61 in

1996 it was possible to stabilize the area harvested at between 148.4 million ha in 1979 and 152.2 million ha in 1996.[19]

While some observers expect a continued decrease in arable land and, at best, no loss in harvested area (cultivated land), government sources paint a rather optimistic picture, according to which China can reclaim 14.7 million ha out of 35 million ha of current wasteland considered suitable for farming, at an annual rate of 300,000 ha, to compensate for future losses of cultivated land to non-farming uses.[20] The area under grains "will be stabilized" at about 110 million ha "through the increase of the multiple cropping index."[21] In addition, about 320 million ha grassland can be used for livestock farming "which places China third in the world in the area of usable grassland."[22] Furthermore, some 65-70 percent of China is mountainous and considered suitable for agroforestry prnduction.[23]

This set of official data needs to be matched with the wild card of allegedly widespread land use underreporting by farmers. According to T.C. Tso, "recent official and nonofficial reports, however, state that total arable land in China is 25-44 percent more than the official report. The State Statistical Bureau (SSB) itself also has acknowledged that this is the case."[24] According to satellite data, China's arable land is 132 million hectares[25].

In fact, an estimate of 120-140 million hectares for China's total arable land would be much more plausible than the official figure of 96 million hectares. Yields and fertilizer inputs are calculated on the basis of the official area data. A revision of the area estimates would reduce yields and fertilizer inputs by a corresponding 25-44 percent, again considerably enhancing credibility.

Irrigation

The enormous past growth of food production came at a price. One was the expansion of irrigated land. China's irrigated area grew 62.5 percent between 1961 and 1994 to 49.4 mill..ha. However, over the same period, the irrigated area in India grew 103 percent to 50.1 mill ha. In 1993, India surpassed China as the world leader in irrigated agriculture. Slow growth rates in China[26] suggest that the country's total irrigated area is approaching a relative maximum size, thus underscoring recent concerns about existing and looming water shortages. After 1990, agricultural water consumption declined from 400 billion cubic meters to 385 billion cu. m[27].

Water is generally considered the main constraint to China's future agricultural development. Again, however, it is difficult to find proof for water shortage at the national level, excluding localized shortages for instance in the northern region. Government authorities plan to expand irrigated land to 56.7 million ha by 2010 and 66.7million ha by 2030 to finally attain a share of 70 percent of total arable land. The area irrigated by water-saving methods is to increase from currently 13 million ha to 40 million ha, and the effective water utilization rate in irrigation is planned to rise from presently 40 percent to 60percent by 2030.[28]

Despite the expected rapid increases in urban and industrial water consumption at the expense of irrigation, the government plans do not appear unrealistic if China succeeds in more correctly pricing irrigation water[29], improving irrigation efficiency[30], reusing treated urban and industrial wastewater, capturing part of the annual monsoon flood runoff, introducing water harvesting techniques in rain-fed areas, and redirecting water from the Yangtze river basin to water-poor northern areas[31]

Cropping intensity

Irrigation, partial mechanization, faster maturing varieties and tighter farming schedules allowed introduction of double cropping in the river valleys on the North China Plain, and triple cropping of paddy in the middle and lower Yangtze valley. However, the third crop requires heavy fertilizer inputs and a fast farming schedule, and its usefulness is therefore subject to discussion.[32]

Fertilizer

The other price China had to pay was the unprecedented surge in chemical fertilizer consumption which continues to rise rapidly (8.1% in 1995 and 6.5% in 1996). With 38.2 million tons (1996, nutrient-weight basis)[33], China is now by far the world's largest fertilizer consumer before Europe (21.0 mill. t, 1994/95) and the USA (19.3 mill. t, 1994/95)[34]. China is also the world's largest fertilizer importer. In 1996, imports of chemical fertilizer accounted for about one-fourth of domestic consumption, the equivalent of half the U.S. consumption.[35] Pesticide consumption is again rising since 1985 although it remains, with 900,000 tons (1995), still well below the peak of the late 1970s and early 1980s (1,582,000 tons in 1982).

China's potential for yield growth

During recent decades, China has concentrated on improving rice and wheat yields, with remarkable success. However, the official yield averages are exaggerated because of underreporting of area harvested. The yield gap is therefore likely to be much larger than official data suggest. Some 50 percent of rice area is currently planted to hybrids which render about 20 percent more than traditional semi-dwarf varieties. Expanding the hybrid acreage would be one way of raising output and the yield average.Compared to other Asian countries, China has a relatively low share of rice in total grain consumption[36] Some 58 percent of the Chinese population live in provinces where rice accounts for less than 50 percent of total grain output. In over half of China's provinces with 42 percent of total population, rice constitutes less than 25 percent of grain production. The share of rice, roots and tubers in total grain production is shrinking while that of wheat (for food) and coarse grains (for feed) is expanding. Maize yields are still low: "Interviews by the authors with breeders from several international companies reveal that even outdated (or third generation) maize technology from North America could provide a substantial improvement over the best, currently available Chinese varieties."[37] While rice will continue to be essential, wheat and maize are not only increasingly important but also appear to offer the larger yield potentials.[38]

The role of research

"Since the founding of the People's Republic, we have gradually established a scientific research system that encompasses almost all fields of study. However, this system was based on the old Soviet model Research was mainly carried out by institutes of the Chinese Academy of Sciences (CAS) and of the various ministries. In developed countries, research is mainly done at universities and by companies. We have encouraged research institutions to establish relations with business enterprises and encouraged large companies to do their own scientific research. CAS has also pioneered in letting research institutes set up enterprises to enter the market Although China has made great strides in development there are also many difficulties, and it will take tens of

years of arduous effort to solve them. We can 't do it alone and would like to establish better cooperation with science and technology circles in the Asia-Pacific region and elsewhere in the world"
Li Peng, Prime Minister of China (1996) [39]

It is obvious from the preceding discussion that future crop improvement and management research in China will have to focus on three main constraints: the need to:

- save land
- save water
- enhance chemical fertilizer use efficiency.

Saving land

Currently, China feeds 22 percent of the world population on less than 7 percent (9% [40]) of the world's cultivated land with 30 percent of the world's agricultural labor force.[41] Some 80 percent of the land harvested is planted to food crops. The per capita availability of cultivated land declined from 0.18 ha in 1949 to 0.11 ha in 1996. In many areas, plots are too small for mechanization. The land shortage finds expression in the small average size of family farms, typically under 1 ha. Under these circumstances, agriculture approaches horticulture, especially in irrigated areas, and the producers of basic food crops are economically disadvantaged if compared to the producers of vegetables, fruit, poultry and livestock who usually manage to achieve higher incomes on their limited land. Measures to save land can aim at:

- increasing productivity per area unit (maximize yield per unit of land)
- avoiding loss of soil fertility (e.g. nutrient mining, erosion, desertification)
- improving soil status and structure (e.g. introducing drainage, preventing pollution, compaction, improving carbon content)
- protecting soil moisture.

Saving water

Efforts to save water in grain production can take three directions:

- substituting less water indigent crops
- reducing crop water requirements (maximizing yield per unit of water)
- improving irrigation management.

It has been said that irrigated wheat, for instance, requires much less water than wet rice, but how much remains to be verified under proper crop management conditions. Also, roots and tubers (which, in the Chinese terminology, are included among grains) offer the possibility of expanding carbohydrate production with supplementary or no irrigation.

Improved irrigation management, as pioneered by the International Irrigation Management Institute (IIMI), can achieve considerable water savings, for instance by substituting supplementary irrigation in semi-arid regions for perennial irrigation, often with the win-win result of higher yields accompanying the water savings.

Under conditions of water scarcity, optimizing irrigation means striking a balance between attempting to maximize grain yield per land or water unit, measured alternatively in tons per hectare or in tons per cubic meter of water. Any attempt at this kind of optimization requires realistic water prices as well as realistic grain prices.

In the rain-fed areas, introduction of faster maturing, drought tolerant varieties in combination with water harvesting techniques and better soil moisture management promises higher and, more importantly, relatively stable yields.

High precision agriculture based on the application of information technology can furthermore help achieve savings of water and other inputs to the tune of 15-20% over previous best practices.

Raising fertilizer use efficiency

China's agriculture is the world's largest. However, its lead in chemical fertilizer consumption appears more a sign of crisis than of achievement. According to official data, China consumes 309 kg of fertilizers per hectare, as compared to 103 kg/ha in the USA and 155 kg/ha in Europe (1994/95). This comparison is somewhat misleading since in China half the arable land is irrigated, allowing a higher average cropping intensity than can be achieved in Europe or the USA. Calculating fertilizer input per hectare of cultivated land lowers China's consumption to 197 kg/ha (1995) which still appears high. However, the picture becomes somewhat less preoccupying when taking account of the assumption that official data on the size of China's cropland underestimate the actual area planted by between 25 and 44 percent.

If widespread under-reporting by farmers of the planted cropland is the case, actual fertilizer consumption per hectare would also be some 25-44 percent lower than official data suggest. This would bring the Chinese average down to around 130 kg/ha and much better in tune with the European and U.S. averages.

Also, because of the poor quality of some domestic fertilizer products, the actual nutrient content of the material could be lower than indicated on the label. Any calculation of fertilizer consumption based on label information is therefore likely to overestimate fertilizer use in terms of pure nutrient equivalent[42].

While Europe's and America's fertilizer use can be considered an economically admissible aspect of an affluent agriculture sector, high fertilizer expenses are a heavy burden for China's struggling agriculture and economy. S. Fan has calculated that chemical fertilizers are the most important cash input, now accounting for 12-13 percent (1991-95) of total input costs in Chinese agriculture, approaching half the level of labor costs. As the third most expensive input after land (44%) and labor (28%), fertilizer costs exceed the combined costs of the next most important inputs, irrigation (6%), and machinery (5%),[43] Yet, despite the heavy expenditure on fertilizer many farmers do not have regular access to fertilizers, and only part of the country's requirements are met

Pingali and Rosegrant, for instance, mention that two thirds of China's agricultural land is now deficient in phosphorus because of the increased cropping intensity and the predominance of year around irrigated rice production systems.[44] A CAAS study found that in 1995 "nearly 3 million tons of applied nitrogen and just over 2 million tons of applied phosphates were not used by crops and livestock, but potash removal was 6.7 million tons greater than the amount applied."[45] The excess nitrogen and phosphate added to pollution whereas the serious shortfall in potash application diminished the efficiency of nitrogen and phosphate uptake.

Farmers following the "3H," principle (high yield, high quality, high efficiency) tend to apply too much fertilizer with the concomitant danger of leaching of nutrients into aquifers, especially in irrigated farming[46]. Better balancing fertilizer application should increase yields by 16-50 percent, according to the World Bank[47]. Low nutrient content and pollution of locally produced fertilizers with toxic metals are other widespread problems.

In future, plant improvement and crop management research in China can be expected to emphasize the need to slow down and, if possible, reverse the rapid growth trend of fertilizer consumption[48] by

• optimizing fertilizer use through a combination of realistic fertilizer price policies, adequate extension work, and free marketing and trade practices[49]

• improving fertilizer quality

• developing more nutrient efficient crop varieties[50]

• promoting adoption of integrated plant nutrition (IPN) which combines use of natural fertilizers (manure, crop residues, green manuring) with nitrogen fixing crops and judicious use of mineral fertilizers.

Integrating soil, water and nutrient management

Given the fact that land, water and fertilizers are the most conspicuous constraints of China's future agricultural development there seems to be considerable logic in tackling all three issues together, in an integrated fashion. A number of international organizations and CGIAR centers, with inputs from the CGIAR Technical Advisory Committee (TAC), are cooperating in a Soil, Water and Nutrient Management Program (SWNM) which focuses on soil-water relationships as part of a more consistent, systematic and environmentally sensitive integrated natural resources management (INRM) in ecoregional approaches to research. TAC stressed the following linkages between productivity-enhancing and resources conserving research:

• managing water and nutrient supplies for greater efficiency and sustainability; research on the efficiency of water and nutrient use by crops, especially to prevent degradation of irrigated land, considering both economic and biophysical efficiency;

• research on the processes underlying the long term, less obvious forms of soil and water degradation;

• managing soil fertility (organic matter, mineral nutrients, acidity).

Efforts to maximize yield per land, water or fertilizer unit pursue competitive objectives. In the past, Chinese grain policies tended to stress yield per *land unit* rather than per unit of *water or fertilizer*. A policy of joint optimization of the three parameters would give up some of the yield per land unit in order to achieve substantial economies in water and fertilizer consumption, and correspondingly boost the yields per unit of water and fertilizer. The foregone environmental damage would be a positive externality of this policy, as urged by TAC. Eventually, the iterative optimization process should lead to higher total availability of grain because the water and fertilizer quantities saved can be used to reclaim or upgrade other land, thus expanding the productive base.

Seed distribution, grain procurement and marketing

Liberalizing and modernizing both the seed distribution system and the grain procurement, trade and marketing system would help unleash China's current production potential. Research in policies and institutions could help to improve the performance of state agencies, privatize them, and open up China to international seed and grain trade companies, thus allowing the economy to reap the benefits of competition. Farmers need access to timely deliveries of quality seed, quality fertilizer and other inputs; all barriers to internal and external grain trade and transport should be removed; and the system of carryover and food security grain stocks needs fine tuning to avoid the costs of holding excessive stocks as well as the dangers inherent in understocking.[51]

CONCLUSIONS

China's current yield levels in food crop production are probably substantially overstated. Large yield gaps still exist which offer potential for future growth. Wheat and maize appear to be the most promising grain crops because they combine good potential for yield increases with strong market demand.

Much of the current production potential is held back by policy and institutional rigidities and shortcomings in distribution. Relatively simple steps to liberalize the grain economy can be expected to eliminate spatial distortions in production patterns, thus substantially benefiting productivity, while reducing chemical input consumption and dangers of long term environmental damage. Once the rigidities have been eliminated it will be much easier to provide farmers everywhere with seeds of existing improved cultivars and related inputs. The grain area can be expected to shrink, with a concomitant positive effect for average productivity. Trade liberalization will lead to increasing international specialization in high value crops, and agricultural self reliance will replace grain self sufficiency in the sense that additional high value farm exports will permit rising grain imports, with a positive effect for China's agricultural terms of trade.

The above steps should increase grain productivity sufficiently to satisfy China's medium term grain demand increases. The breathing space thus obtained will be needed to bridge the time lag between the initiation of new research ventures and the availability of results in farmers fields. Streamlining and strengthening national agricultural research while enhancing collaboration with the global agricultural research system will help to harness the best of modern science in the service of the Chinese farmer. Three yield parameters need to be jointly optimized land, water, and fertilizer inputs -- for maximum productivity and long term resource conservation, whereas the fourth -- labor/mechanical/animal power -- can still be considered relatively abundant.[52] The excellent past performance of Chinese scientists and research institutions indicates that they, if given adequate conditions, will not fail to master the challenges of the future and help farmers maintain China's agricultural self reliance well into the coming century.

[1] How Efficient Are Modern Cereal Cultivars? CGIAR News 4/4(1997), 2
[2] For instance as a result of tax discrimination of agriculture, abolition of farm or input subsidies, low-price cereal imports, etc.
[3] By no means a new phenomenon: about 2400 DC, wheat yields in irrigated production in Mesopotamia were about 2 tons/ha; by 1700 BC they had declined to one-third of the earlier level, "probably due to salinization of the irrigated soils". Donald L. Plucknett: Saving Lives Through Agricultural Research. CGIAR 1991, 2
[4] Although rice monocropping is a traditional production method, the conversion to intensive double and triple cropping keeps rice paddies flooded for most of the year without adequate drying period, leading to build-up of salinity, micronutrient (mainly zinc) deficiencies and soil toxicities (especially iron toxicity), subsoil compaction, decline in nitrogen-supplying capacity, and increased pest build-up. Result: A "rapid degradation of the paddy environment." Pingali and Rosegrant 1993,9 and Prabhu L. Pingali: Technological Prospects for Reversing the Declining Trend in Asia's Rice Productivity. World Bank Conference on Agricultural Technology 1991, 6
[5] Pingali 1991, 7
[6] From 1990 to 1996, average cereal yields in developing countries rose 1.6 % a year.
[7] There is hence no proof for the popular assumption that the Green Revolution ended sometime in the early 1980s. It would be more accurate to consider the Green Revolution an ongoing process fuelled by the

application of modern science to developing country agriculture.
[8]For a discussion of expected long term growth rates of grain yields see
N. Alexandratos: China's consumption of cereals. Food Policy 22/3, June
1997, 261
[9]USDA World Agricultural Outlook Board: International Agriculture and
Trade (China). (July 7, 1997), 20
[10]Information Office of the State Council of the People's Republic of
China: White paper "'Grain Issue in China", Oct. 1996, 2
[11] See chapter 23: International cooperation.
[12]Information Office 1996: White Paper, 5
[13] Information Office 1996, 3. This figure is considered understated.
World Bank estimates of 120 million and 350 million for two different
poverty lines are mentioned by Alexandratos 1997, 257
[14] USDA 1997, 22
[15] "An overemphasis on grain growing during the 1960s and 1970s le to
elimination of some crops, orchards, and trees; neglect of animal
husbandry; and environmental damage." R. Crowley, 3
[16] " One of the greatest constraints is self-imposed: ideology. For
thousands of years China's officials have equated overflowing grain
stocks and self-sufficiency with successful agricultural policy.. Many of
these ideas are antiquated and the pursuit of such goals is done primarily
for political gain and at a higher economic cost." S. Rozelle and M.W.
Rosegrant: China's past, present and future. Food Policy 22/3, 199.
Incidentally, in 1996, thanks to two consecutive bumper grain harvests,
China was again a net agricultural exporter with a surplus of $0.9
billion. China's custom statistics in USDA 1997, 8
[17] USDA 1997, 27
[18] Shenggen Fan and Mercedita Agcaoili-Sombilla: Why Do Projections
on China's Future Food Supply and Demand Differ? IFPRI 1997, 26
[19]USDA 1997, 23
[20] "A policy of linking the appropriation of land for non-agricultural
construction purposes 10 land.development and reclamation will be
implemented." Information Office, White Paper 19%, 8. WRI says:
"China is developing a comprehensive zoning plan for the whole country.
One goal is to prevent net loss of farmland." World Resources Institute;
Food Production and Agriculture in China. World Resource 1994-95. It
should be noted that the White Paper was compiled as a response to
Lester Brown's (Who Will Food China,1995) wake-up call and might err
in the opposite direction. Its ambitious land reclamation plans are likely
to encounter heavy environmental trade - offs.
[21] Information Office 1996, 5

[22] Concern has been expressed about the extent of grassland destruction through deforestation, expansion of agriculture, and overgrazing. "Damaged converted grasslands are difficult to restore to productive pasture land use." S. Rozelle et al. Poverty, population and environmental degradation, in Food Policy 22,3, 232.

[23] Information Office 1996, 6

[24] T.C. Tso: Population, Food and Energy in China. Unpublished, 1996, 3; also: The American Association for Chinese Studies (AACS) Bulletin 23/1, June 1996, 8. Recently, the Land Administration indicated that China's cultivated land is actually 124 million hectares. Binsheng Ke 1996, in Shenggen Fan and Mercedita Agcaoili-Sombilla: Why Do Projections on China's Future Food Supply and Demand Differ? IFPRI 1997, 24

[25] World Bank 1997: At China's Table; Food Security Options, 4

[26] The question arises whether these slow rates of irrigation expansion result from underreporting of land by farmers, and whether a revision of the arable land data would entail a revision of irrigated land data.

[27] China Council for International Cooperation on Environment and Development (CCICED), 7th report of the WG on Environmental Scientific Research etc. (1996) 12

[28] Information Office 1996 8

[29] "In the few locations where rational irrigation water pricing has been implemented, water demand has declined without affecting yields. World Bank 1997, 26

[30] Less than 0.4% of China's total irrigated area is currently under trickle irrigation; sprinkler systems are use on about 2% of irrigated land. Water use efficiency in general is low. CCICED 1996 22

[31] CCICED 1996 chapter "Water Resources" in World Bank 1997.

[32] Robert Crowley et al., U. of Illinois at Springfield web page: China, economy, 3

[33] USDA 1997, 24 (State Statistics Bureau estimate); FAO reports China's total fertilizer consumption at 29.6 million tons (1 995)

[34] FAO: Fertilizer Yearbook 1995

[35] USDA 1997, 27

[36] Alexandratos 1997, 257

[37] Rozelle and Rosegrant 1997, 198.

[38] China's official average wheat yield is about 1.36 (rice 1.63; maize 1.25) times the world average. Deducting 2544 percent, as suggested by Tso, would mean that only China's rice yield is clearly above world average.

[39] Science 273, 7/96, 13

[40] If the higher arable land figures apply which are mentioned in the preceding section "Land", China's share of the world's arable land would rise to about 8.6 percent.

[41] In 1996, China produced between 21 and 22 percent of the world harvests of cereals and grains (Chinese definition). It is the world's leading producer of rice, wheat, meat and cotton, 8

[42] T.C. Tso: personal communication

[43] S. Fan: Production and productivity growth. Food Policy 22/3, 222

[44] Prabhu L. Pingali and Mark W. Rosegrant: Confronting the Environmental Cotisequences of the Green Revolution in Asia. Unpublished AAEA paper (1993), 12

[45] World Bank 1997, 16

[46] T.C. Tso, personal comm.

[47] World Bank 1997, 17

[48] As manifest in Europe since 1990.

[49] T.C. Tso sees considerable scope for fertilizer savings and reduction of pollution through better timing and methods of fertilizer application. (Tso 1996, 4.) S. Rozelle et al. (1997, 247) observed a steep increase in chemical fertilizer and pesticide use in high population growth counties, accompanied by a decline in organic manure application. In low population growth counties, fertilizer and pesticide use was maintained or decreased whereas manure application mostly increased. The authors subsume that due to rigidities in interregional trade and the existence of production quotas, high population growth has forced farm households to intensify production on a fixed land base, thus resulting in spatial distortion of geographically optimal production patterns. Free agricultural marketing and trade would hence relocate production away from zones of high population density toward agriculturally more favorable areas, thus optimizing production intensities by decreasing chemical input consumption in high population density regions and increasing it in low density regions. The result would be an overall reduction in chemical input consumption and less long term degradation of soil and water resources caused by intensification.

[50] How Efficient Are Modern Cereal Cultivars? CGIAR News 4/4(1997), 2

[51] J.Y. Lin: Agricultural growth in China. Food Policy 22/3, 210

[52] FAO figures indicate that China's 876 million farm population (1996) represents 35% of the agricultural labor in developing countries. Despite rural-urban migration, the Chinese agricultural population continues to increase at 0.74% per year. For the near future, labor shortages are therefore likely to remain localized and seasonal.

CHAPTER 4.23

FOOD DEMAND AND SUPPLY IN CHINA

FOR THE 21ST CENTURY

Shu Geng, Colin Carter, Yin-yen Guo
University of California - Davis, California, USA and

Hiroyushi Kawashima
National Institute of Agro-environmental Sciences, Japan

Since 1950, China has increased its grain production 4.5 fold, from 100 million tons to 450 million t in 1994, reflecting an impressive annual growth rate of 3.4%. During the same period, China's population increased from 400 million to 1,200 million, at a growth rate of about 2.5% per year. The greater production rate compared to the population growth rate, particularly during the 7 years immediately following the cultural revolution, has provided each Chinese on average 1 kg grain per day, a record unmatched in recent Chinese history. Since the amount of cultivated land has steadily decreased for more than four decades, the increased production of grain is due to increased yields (Figure 4.23.1). To put things into perspective, 4.23..2 shows the cereal yield trends for China versus the United States and the world average. China's yield has surpassed the world average since the late 1970s and is rapidly approaching the US level. Cereal yield growth rates for the last 35 years have averaged 4% in China, 2% in the US and 2.3% for the world. In addition to cultivar improvement, China has expanded its irrigation area from 30% of the cultivated land in the1960's to 50% in the 1990s. Fertilizer application increased from 8 million t to 36 million t during the last 30 years to about 375 kg/ha in 1995. Today, China is the largest grain producer in the world and has the most intensive production system. To maintain or improve its high productivity, China's agricultural system requires tremendous external inputs from outside the farming system. The long-term sustainability of such high input system is questionable.

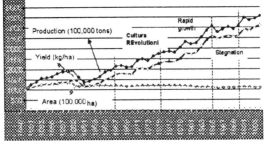

Figure 4.23.1 Chinese grain production   (production million t; yield = x.10 kg. ha^{-1} ; area million ha.

Agriculture is not the only sector that has made significant progress in China. China's overall GNP has grown at a rate over 10% a year for the last 10 years. The per capita GNP increased from 855 yuan in 1985 to 4,754 yuan in 1995 in current price or 1,570 yuan when consumer price index was adjusted. Township enterprises first appeared in China in 1956, but it was not until the late 1970s that their development gained momentum. Today, there are 23 million township enterprises in China employing 135 million people with 1.76 trillion Yuan in output. And township and village based enterprises produce 17% of the country's electronic goods, 26% of the machinery, 40% of the coal and cement, 43% of the processed food and drinks, and 80% of the garments. Overall, this sector accounts for one-third of the China's GDP in 1996.

Future development issues in China are outlined in China's Agenda 21 which was prepared by the Chinese State Planning Commission and the State Science and Technology Commission. The document listed 69 priority programs in 9 chapters. The full text of the programs can be found on the web (http://plue.sedac.ciesin.org/ china/ policy/ acca21/ acca21.html). The agricultural agenda was reported in Chapter 2, which includes the following projects:

1. A strategy for intensive-sustainable agriculture in China and demonstration projects in ten counties;

2. Sustainable agriculture in rural areas of the Huang-Huai-Hai plain;

3. Sustainable agriculture and protection of the environment in the hilly red-yellow soils of southern China;

4. Integrated development and management of the mountain-river-lake region of Jiangxi province;

5. Sustainable management of water resources for agriculture and a demonstration project;

6. Development of biological pesticides and "green" foods.

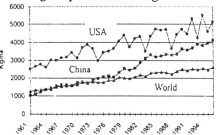

Figure 4.23.2 Average cereal yields (Rice is milled)

Although sustainable agriculture was high-lighted in these priority projects, no formal definition on sustainability was given in the Agenda. It is not clear from the brief descriptions of these projects whether the goals of agricultural production would in concert with the goals of environmental protection. Brown (1996) analyzed some of China's key resource factors and estimated that China would import a huge amount of grain by the year 2030. Though many of the Brown's concerns on land and water usage were listed in the China's Agenda 21, the Chinese government's responses to Brown were largely defensive. China has repeatedly assured the world that it can feed itself and will not import more than 3% of the needed grain for the next 30 to 40 years. Specifically, China issued a White Paper on food demand and agricultural production goals by 2030 (China Daily, October 26, 1996.) Table 4.23.1 shows the China's official estimates of grain demand. According to the paper, the population will increase to 1.6 billion by 2030. At that time, the total demand for grain will be about 640 mmt, half of which will be used for feed.

The Chinese government outlined a series of measures to take and concluded that demand is within its production capability (China Daily, Oct. 25, 1996). However, no consideration was given to the potential

Table 4.23.1. Chinese Official Estimates of Population and Grain Demand

Parameters \ Year	1990	2000	2010	2030
Population (100 million)	11	13	14	16
Demand (Kg per person)	380	385	390	400
Total Demand (100 m t)	4.5	5.0	5.5	6.4
Yield Growth rate (%)		1.76	1	0.76

impact of an additional 200 mmt of grain production on the sustainability of natural resources and the environment. The increased purchasing power and available choices that will arise with continued economic growth were not considered either.

In this paper, we will present an analysis of and a projection for the possible level of grain production in the year 2030. We have no doubt that China is fully capable of producing sufficient food to feed its own people. The increased income, however, will empower its people with a freedom of choice over commercial products in a way that China has never experienced before. The question is will China produce all the needed grain by itself or will China choose to purchase a certain amount of grain in order to conserve its natural resources such as water, to maintain flexibility in land use and to supply a greater diversity in the diet and higher quality food?

THE CHALLENGES

The challenges for China to produce an additional 200 mmt of grain are multiple and difficult to overcome. We identify a few of these below.

Arable Land.
Rroughly 10% of China's land is suitable for crop production. However, there has been considerable debate in recent years on the exact amount of arable land in China. Official statistics indicated that there were 92 million hectares of arable land in 1994. Recent satellite images and surveys, however, suggest the actual arable land may be 40% higher than previously reported. Regardless of the actual amount of arable land, there is strong evidence that the base is steadily declining with an average of 300,000 hectares a year or a 9% loss since 1961 (Figure 4.23.3). At the same time, the multiple cropping index (i.e. the ratio of sown area to arable land) has increased from 1.35 to 1.57: a 16% increase in 35 years.

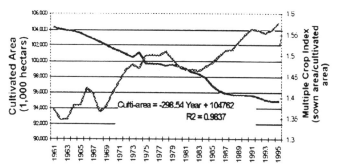

Figure 4.23.3 Cultivated area and multiple crop index in China 1961-1995.

Today, China's agriculture is one of the most intensified production systems on earth, in terms of land use and labor inputs. Further intensification will require both new genetic (high yielding, early maturation and disease resisting cultivars) and managerial technology (inter-cropping, overlay cropping, and multiple cropping) breakthroughs and improved efficiencies in resource use (water and chemical applications). The success of further intensification may not result in a production increase unless the arable land is halted. This will only happen if the government introduces a clearly delineated land use policy which must then be executed and implemented. The first challenge for China is to maintain its arable land for food production despite the unavoidably increasing competition amongst land users generated by economic and population growth. So far China has not demonstrated that this is possible. The same problem has occurred elsewhere in the world when comparable conditions exist.

Grain sown area.
In China's Statistical Yearbook, grain includes cereals, soybean and tuber crops. Unlike arable land, which is mainly affected by factors other than agriculture, the ratio of grain-sown area to the total sown area reflects a decision made by the farmers and agricultural agencies in response to the market values of the grown commodities. Figure 4.23.4 shows the dynamics of the sown area and the ratio of grain area to sown area from 1952 to 1995. Although the sown area fluctuated greatly, it more or less stabilized at 150 million hectares. The ratio of grain-sown area clearly declined at a rate of 3% a year. The relative price among crops may be the direct cause of these trends. The policy of maintaining an artificially low price of grains in China may also have contributed the reduction

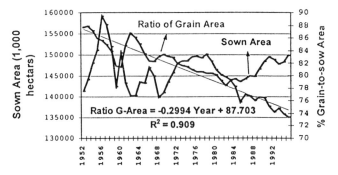

Figure 4.23.4. Total shown area and ratio of grain area in China.

and prompted farmers to grow more profitable alternative crops. Nevertheless, the ratio of grain area relative to other countries in the world is still high: 73% in 1995. As long as the grain quotas are enforced in China, this ratio will remain high. The major portion of the reduction in grain area is due to reduced rice cultivation in south China. Rice grown area has declined about 244,000 hectares per year since 1976. Thus, the second challenge for grain production is whether incentives can be provided to farmers who are willing to grow grain crops even in areas where other cash crops are equally suitable to grow?

Rural and Urban Income Disparities
Both China's gross national product and gross domestic product have increased on average 9.8% per year since 1979. In 1996, the per capita GDP reached 5,568 in current yuan, 13 times higher than in 1979 (Data taken from Asia Development Bank, http://internotes.asiandevbank.org/ and are shown in Figure 4.23.5). The disparity between city and rural income has increased. The rural per capita income was 58% of the urban per capita income in 1985 and was 41% in 1995. Chinese survey data (China's Statistical Yearbook, 1995) also showed the difference between rural and non-rural consumption rates to have widened. For instance, the rural consumption rate was 43% of non-rural consumption rate in 1985, but was reduced to 27% in 1994 (Figure 4.23.6). Even among the rural communities, there existed a wide difference. The per capita rural income ranged from 489 to 2,226 yuan, a 4.6 fold difference between rich and poor regions in 1992 (Figure 4.23.7). The disparities between urban and rural incomes caused rural labors to migrate to cities and in some cases even to abandoned grain fields. The uneven economic growth among

regions and provinces will have a significant impact on grain production.

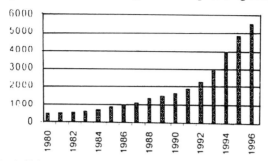

Figure 4.23. 5 Chinese per capita GDP 1980-1996

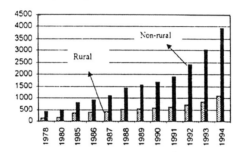

Figure 4.23.6 Per capita consumption (yuan). Rural versus nonrural

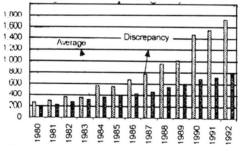

Fgure 4.23.7 National average and regional discrepancies for rural per capita income (Yuan)

Agricultural Policies

Figure 4.23.8 shows the percentage of China's total capital construction investment allocated to agriculture and in government and party agencies. (Data before 1985 are taken from Carter et al, 1996; after 1985 are taken from China's Statistics Yearbook, pages 146-147, 1995.)

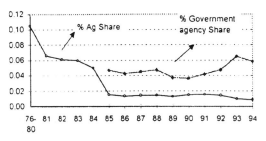

Figure 4.23.8 Chinese Government construction investment 1976-93

The investment in the agricultural sector was less than 2% of total investment from 1985 and was drop below 1% in 1994. Scientific research and polytechnic services sector also received an investment support of less than 1% of total country's investment in 1994. Investment in governmental agencies, on the other hand increased from 4% in 1985 to 6% in 1994. Those sectors receiving the most significant investment were transportation, manufacturing and energy production and supply. The agricultural sector has certainly benefited from those investments as well, but there is no substitution for direct support. For example, the greatest progress in China's agriculture occur during 1976-1984, coincide with the period it received the greatest governmental support. The challenge here is investment priority. The biggest question is whether China will re-instate its investment in agriculture and research and science sectors at a level comparable to years before 1985. The answer to this question will have enormous impact on the likelihood of China being able to feed its population when it reaches 1.6 billion by 2030.

Farm labor
According to the1996 State Statistical Yearbook , there are over 855 million rural people, which means 71% of the total population is classified as rural. The available rural labor force is about 62% of the rural population or 530 million people. Table 4.23.2. shows the estimated agricultural labor force according to the labor days required for each crop (Geng and Zhou, 1990). Furthermore, we assume that 270 work days equals a full time employee. The total number labor days divided by 270 yields a required labor force for cropping needs in China of about 235 million. This estimate does not reflect seasonal peaks in labor demand and, therefore, may under-estimate the total labor requirement. In addition, there are roughly 10 million people engaged in animal husbandry, fisheries and forestry and 135 million employed by rural and

township enterprises in 1996. The estimated total requirement for the rural labor force is therefore about 380 million people. Based on these estimates, there are almost 150 million surplus workers in agriculture.

Table 4.23.2. Farm Labor Demand Estimates

Crops	Harvested area: 1997(million hectares)	Per hectare labor days	Total labor days (billion)	Estimated Labor Requirement (millions)
Rice	31.3	296	9.3	34.3
Wheat	30.0	207	6.2	23.0
Corn	22.5	222	5.0	18.5
Soybean	8.4	143	1.2	4.4
Cotton	12.3	533	6.6	24.3
Rape Seed	6.8	269	1.8	6.8
Vegetables	12.1	1,000	12.1	44.8
Others	26.6	800	21.3	78.8
Total	150.0	3,470	63.4	234.9

This estimate is consistent with the previous estimation provided by Geng and Zhou (1990) when adjusted for population growth. Carter et al. (1996) carried out a detailed calculation based on provincial data and also found the excess labor supply in agriculture to lie between 140 and 170 million (Figure 4.23.9).

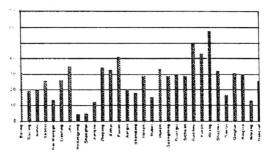

Figure 4.23.9 Estimated rural labor surplus by Province in percentage (From Carter, Zhong and Cai, 1996)

Having 235 million people directly engaged in agricultural production is perhaps still too many and this gives rise to a paradox: Unless the size of the farm labor can be further reduced, improvements in farming efficiency and farmers' incomes will be limited. But if the farming population is reduced (e.g., 50% reduction by 2030), then there could be a serious unemployment problem. In any case, the problem of farm labor employment would pose a major challenge to the government.

Natural Resources and Environment

The continent of Asia holds about 32% of the world's water resources as measured by annual runoff. This is by far the largest, twice as large as North and Central America combined. But in terms of available per capita water, the projection for the year 2000 suggests that Asia has the smallest, 3,300 cubic meters per person per year, while Europe has the second smallest 4,100 cubic meters (FAO statistics). China, however, has only about 2,300 m^3 per capita per year. This number is approaching the biological limit, or 2,000 m^3, which is the minimum amount water that is required to produce an adequate annual per capita diet (FAO). On the supply side, China has an amount of water that ranges between 1,100 billion cubic meters in a wet year to 650 billion cubic meters in a dry year, 70% of which is used for irrigation (Chinese Agricultural Resources and Regional Planning, 1987). With a growing population and a rapidly growing economy, the water demand could exceed 650 billion cubic meters in near future. The chance that agriculture will receive less rather more water allocation is real. An adequate water supply is essential for grain production, particularly in Central Northern and Western China. In addition to a potential water shortage problem, there are serious erosion, desertification, and environmental pollution problems. All these resource and environmental problems pose a real threat to the sustainability of the China's intensive production system.

Water is the most critical resource issues, because future grain production growth depends on an adequate water supply. This issue is particularly exemplified by the case of the Huang-Huai-Hai (HHH) plain, the largest Plain in China. The plain embraces about 20% of China's population and arable land but produces almost 25% of the total grain. The rainfall ranges from 400 to 900 mm per year decreasing from the south to the north and from the east to the west. The average available water (surface and ground water included) in HHH region is 5 m^3 per hectare, merely 20% of the national average. The dense population and the intense cultural practices have been consistently depleting the underground water reserve on the average of 1 metre per year and has greatly exacerbated the problem of water shortage in the region (Sanders, 1994).

The Plain is the cradle of Chinese civilisation. It represents the glorious past of China; it also bears enormous responsibility for feeding the nation's increasing population. The task of further improving the agricultural productivity without rapidly depletion of non-renewable natural resources is one of the greatest challenges that China has ever faced. It is a task that must be successfully accomplished if future generations are not to suffer from depleted environmental quality.

PROJECTION METHODS

We take two approaches to projecting future grain needs and supply. First, we use a simple aggregated approach that only considers cereal yields and harvest areas. In 1995, the national average grain yield was about 4,664 kg/ha. The yield ranged from 3,000 to 15,000 kg/ha among locations and crops. The Chinese government presented the goal of grain yield in 2030, which is 6,000 kg/ha (China Daily, October 25, 1996). This goal can be accomplished with an annual growth rate of 1% from 1996 - 2010 and 0.7% from 2011 - 2030, or an average growth rate of 0.8% for the next 35 years. Under these yield growth scenarios, China would produce more than 640 mmt grain by 2030, assuming the grain-sown area remains at 1995's 110 million hectares. However, the grain sown area has been declining for the last 40 years (Figure 4.23.4 and 4.24.5), and there is no indication that it can be entirely stopped in the near future. Brown pointed out that a declining grain area is the norm as it has been experienced in other East Asia countries such as Japan, South Korea and Taiwan. In recent years, the grain area as fallen by 0.21% per year. We use these yield and area growth rates to calculate potential grain production for the next 35 years up to 2030. This approach is called the Aggregated Top-down approach.

The second approach is based on a regional analysis and projections and we call it a Regional Bottom-up approach. In this approach we partition China into three economic regions according to the level of per capita rural income. Within each of the regions, we analyzed grain yield, grain sown area, and rural economic conditions. Association analyses were performed to identify the relationship between grain production parameters and the rural economic changes. The association analyses include statistical regression and correlation analysis and Geographic Information Systems analysis. Results were used to develop a simulation model, which is run by Stella, a dynamic simulation program. The simulation steps were first, multiply cultivated land area by the multiple cropping index to estimate the total sown area. The sown area was then multiplied by the grain sown area ratio to generate the grain-harvested area. The grain harvested area was then multiplied by the yield to give an estimate of total grain production. The process is expressed by the following equation:

Production = (cultivated land area) x (multiple cropping index) x (Grain-sown area ratio) x (yield).

The growth rates of each of these variables were estimated from historical and regional data sets.

Rates for Regional Simulations.   Initial values and growth rates for simulations were determined from 1984 - 1995 data, as described below.

Population Growth Rate: Based on estimates from 1984-95, the population growth rates are 0.008462, 0.006996, and 0.008117 respectively for low-income, middle-income, and high-income areas. These numbers are used to estimate per capita statistics.

Arable Land and Multiple Cropping Index Annual Growth Rate: The estimates from 1984 - 95 data (shown below) were used for projections. After 10 years, middle income area estimates were substituted for the low-income estimates and the high-income area estimates were substituted for middle income area estimates.

Economic Regions	Cultivated Area growth rate	Multiple Cropping Index growth rate
Low	-0.00119	0.00931
Middle	-0.00276	0.00830
High	-0.00812	0.00036

Grain to Sown Area Ratios.   Figure 4.23.11 shows the relationship between rural per capita income and the grain sown area ratio. Though there are many factors which affect the grain sown area ratio, a simple regression analysis shows that 25% of the variation of the ratio can be explained by the rate of rural income change. The higher the income growth rate is, the smaller the ratio is. This association allows us to estimate the economic impact on grain sown area.   The method of calculating the growth rate of the ratio is described below.

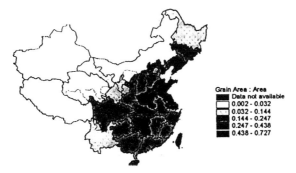

Figure 4.23.10 Intensity of grain cultivation in China

1. The impact of income growth on the grain sown ratio is calculated from the following regression equation,
Growth rate in grain sown area ratio = -0.1662 (income growth rate) + 0.0171; R^2 = 0.25.
2. The initial income growth rates for each region were estimated from 1984-95 data and these rates were gradually decreased to 5% after 20 years of high growth, at which point in time per capita income will approach that of other developed countries.

Year	Low	Middle	High
1995 - 2005	0.11	0.14	0.16
2006 - 2010	0.11	0.10	0.10
2011- 2015	0.11	0.10	0.07
2016 - 2030	0.05	0.05	0.05

3.The growth rate of grain sown area ratio for a certain income growth rate was calculated from the above regression equation plus the initial growth rate estimated from 1984-95 data.
(a) . The initial rates were: -0.01707 for high-income areas, -0.00459 for middle-income areas, and -0.00761 for low-income areas.
(b) Thus the actual growth rates of grain sown area ratio used in the simulation were as follows,

Year	Low	Middle	High
1995 – 2005	-0.0105	-0.0108	-0.0266
2006 – 2010	-0.0105	-0.0041	-0.0166
2011– 2015	-0.0105	-0.0041	-0.0116
2016 - 2030	0.0012	0.0042	-0.0083

Yield growth rate:
1. High-income area - 0.015 for 10 years and then 0.007 for 25 years.
2. Middle income area - 0.015 for 15 years and then 0.007 for 20 years
3. Low income area - 0.01 for 15 years and then 0.007 for 20 years.

RESULTS

Aggregated top-down approach

Results of the top-down projections are presented in Figure 4.23.10. China's official projection line is shown as the top line and Brown's projection as the bottom line. The Chinese official projection of grain production is 640 m t in 2030 compared to Brown's projection of a 20% reduction in production by 2030 from the 1995 production level.

As a result Brown concluded that China will soon have to import a significant amount of grain and would eventually import 260 m t to meet demand by 2030. Our top-down projection line is the middle line in 4.23. 10. Under the top-down scenario that China will achieve a yield level of 6,000 kg/ha but that grain area will continue to fall at 0.21% a year, the net result is that in 2030 China will produce 560 mmt grain. The discrepancy between supply and demand in this case implies that China would import about 70 mmt grain. This amount is not unreasonable given that China started to import grain in 1980 and imported about 30 million tons in 1995.

Nevertheless, this prediction suggests China will import more than 10% of its needs by 2030, double that of the officially projected percentage.

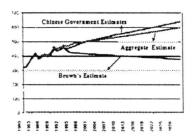

Figure 4.23.11 Projected Chinese grain demands and supply scenarios

Regional bottom-up approach

China's grain production and yields differ widely among provinces and regions. The highest yields occur in Jiangsu and Jilin provinces where average yields are above 5,000 kg/ha. The lowest yields occur in Gansu,

Qinghai, Ningshia, and Shaanxi where the growing season is short and water is limited. When the ratio of grain area is plotted against land area, a concentration pattern is revealed (Figure 4.23.11). Most of the grain crops are planted in the eastern and southern provinces. The vast western part of the country, notably Xinjiang, Tibet, and Nei-mongol, have less than 0.2% of their land planted with grain crops. The rural per capita incomes follow a similar trend. Figure 4.23.12 shows the provinces that have per capita incomes less than 999 yuan, between 1,000 to 1,499, and above 1,500 yuan. Obviously the coastal provinces are economic growth centers. Table 4.23.3 shows the summary statistics on rural incomes and grain production. The per capita income of these regions is positively associated with grain yield and the multiple cropping index, but negatively correlated with the ratio of grain to total sown area. The richest region is also the most intensively cultivated region, which is reflected in the multiple cropping index (1.92).

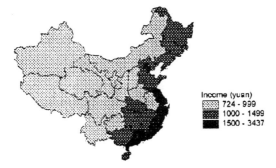

Figure 4.23.12 Rural household per capita income in China (1994)

But it is also the region with the lowest grain to total sown area ratio, namely, 72%. In other words, cash crops account for 28% of the sown area for the high income region, 27% for the middle income region and 25% for the low income region. Though less than 40% of the population lives in the middle income region, this region supplied more than 50% of the national grain output. Simulation results are presented in Table 4.23.4.

Population Projections. The population is expected to increase by 31% to almost 1.6 billion by 2030. However, the distribution of population among regions is not expected to change significantly. It is predicted that the low-income region will have 647 million, the middle income region will contain 600 million and the high-income area will have 333 million people.

Table 4.23.3. Regional grain production statistics (1995).

	Low-income	Middle-income	High-income
Population (million)	481 (40%)	470 (39%)	251 (21%)
Income (yuan)	887	1,247	2,213
Grain Yield (kg/ha)	3,324	4,553	5,305
Grain Production (mmt)	150	241	78
Per capita Grain (kg)	311	512	311
Grain sown area (million hectare)	45 (40%)	53 (47%)	15 (13%)
Total sown area (million hectare)	60 (39%)	72 (47%)	21 (13%)
% grain-sown area	75	73	72
Cultivated area (million hectare)	43 (45%)	41 (44%)	1 (11%)
Cropping index	1.41	1.74	1.92

Projection on Income. .Under the previously defined rate scenarios, the per capita income could increase dramatically in China. So far China has made enormous progress in terms of economic growth. The per capita GDP is estimated at 5,568 yuan in 1996, which is 12 times greater than the income in 1980. Our projected income in 2030 will be 16,199 yuan for the low-income region, 24,680 yuan for the middle income region and 45,844 yuan for the high-income region. By then, China will have a GDP in excess of 5 trillion US dollars, which should be very close to Japan's total GDP and probably second only to the US. The impact of the income increase on agriculture is two fold. The first impact occurs through the competing use of the land, water and other resources by other sectors of the society; agriculture probably will receive less public and governmental support just as its need is increasing. The second impact is the increased purchasing power and options of consumers. As a result, China's agriculture will not only be concerned with grain yields, but also be competitive on international markets in terms of price and quality. At that point in time, the question is no longer whether can China produce sufficient food to feed its people, but whether China's product is the preferred choice by its consumers. Or to what extent can China afford to allocate scarce resources to agriculture for food and feed production purposes? The real question is how much will be produced at what market price.

Grain Area Projections. Although we project a 15% reduction in cultivated land by 2030 (from 95 million hectares to 80 million hectares), the total sown area is not expected to decrease. In fact it is expected to increase slightly from 150 million hectares to 154 million hectares. We chose to use the "Official" statistics on arable land as a starting point for the simulation. We fully recognize that the "un-official" figures could be much higher.

Table 4.23.4. Projected regional grain production statistics (2030).

	Low-income	Middle-income	High-income
Population (million)	647 (41%)	600 (38%)	333 (21%)
Income (yuan)	16,199	24,680	45,844
Grain yield (kg/ha)	4,437	6,545	7,329
Grain production (mmt)	205	276	48
Per capita grain (kg)	317	460	144
Grain sow area (million hectare)	46 (49%)	42 (44%)	6 (7%)
Total Sow area (million hectare)	75 (49%)	63 (41%)	16 (10%)
% grain-sown area	61.64	66.96	41.8
Cultivated area (million hectare)	39 (49%)	33 (41%)	8 (10%)
Cropping index	1.90	1.91	1.94

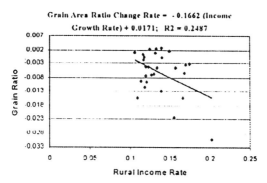

Figure 4.23.13 Rural income growth rate and the rate change of grain shown area ratio.

The more important numbers in our simulation are total sown area and grain sown area. The underestimation error of arable land on grain production is partly adjusted by using a high growth rate scenario for multiple cropping index in the simulation. The multiple cropping index is allowed to increase further to lie above 1.9 in all regions. This high level of intensity in cropping systems will not occur without strong governmental support programs in land and water use and investment of research. The greatest reduction in cultivated land will occur in high-income provinces (-24%) where grain cultivation is most extensive and productive. This region will also lose 55.5% of its grain area. In contrast, the low-income area will actually increase its crop sown area (from 60 to 75 million hectares) and grain harvest area (45 to 46 million hectares).

Yield Projections. The average grain yields will increase in all regions. The yield for the low-income area will increase from 3,325 to 4,437 kg/ha, the middle income area from 4,553 to 6,545 kg/ha and the high-income area from 5,305 to 7,329 kg/ha. These increases depend on the successful development of a number of technological innovations. The high yielding and early maturing cultivars that are suitable for multiply- and inter- cropping systems must be developed. Efficient irrigation and fertilization systems that would increase available water and nutrient to crops without requiring additional water and chemical inputs must be developed and implemented in order to achieve the expected 40% grain yield increase.(Figure 4.23.14)

Figure 4.23.14. Projection of China's grain production.

Poduction Projection. Grain production will increase from 468 mmt in 1995 to about 529 mmt in 2030. The moderate increase (13%) in total grain production reflects the impact of economic growth on grain production. The per capita grain production will be 331 kg in 2030, a reduction from 390 kg per person in 1995. This is by no means a signal of impending disasters. This level of production, however, would not meet the desired level of food diversity and meat consumption as outlined in the China's White Paper (China Daily 10/25/96).

DISCUSSION

The results that we have presented are not intended to be taken literally as point prediction. Instead, they represent a plausible scenario. The question we are addressing is not whether can China feed itself. The answer to that question is trivial: of course, China can feed itself. If the main purpose of China's future development is to produce an amount of grain necessary for self-sufficiency, and if China is willing to invest a large share of its resources in grain production, then China may not only feed itself but could export a significant amount of grain. Indeed, it is not an interesting question. A similarly posed question is: can Japan feed itself? Of course, it can, but instead Japan imports about 70% of its required grain. The result has nothing to do with Japan's production capability but has everything to do with economic and policy choices. Will China produce all its needed grain in the future, even if it is fully capable of doing so? This is an interesting and complex question. The answer hinges on a number of factors: political, economic and technological. according to all reasonable predictions, China's economy will continue to grow at a rapid pace. As a result, new purchasing options and alternatives will open to Chinese people in a way as has never occurred in China before. Whether China will produce all its grain is then a decision of market choice. It is within this context that we examine the likely consequences of economic growth on grain production.

Brown (1995) argued that if countries become densely populated before they industrialize, they inevitably suffer a heavy loss of cropland. Since Japan, South Korea, and Taiwan have all lost their cropland at a rate of more than 1% per year over the past four decades, Brown concluded that China's grain production will have to fall by at least 20% from 1990 level in 2030 due to cropland loss. As a result, he projected that China will have to import 200 to 369 million tons by 2030. Brown's conclusion triggered a series of passionate responses from both Chinese officials (China Daily, Oct. 25, 1996; Mei, 1995) and academics from both in and outside China (Adam, 1995; Huang et al, 1995; Penning de Vries, 1995). Most of these reports challenged Brown's assertion and argued a rosy future. Rosegrant of IFPRI (Science, Random Sample 275(2503):1071) explained that while production of feed grains for meat-producing animals will rise rapidly, it will be offset by a decline in direct grain consumption by humans. With appropriate investment and technological inputs, he predicted that China will again be exporting grain by 2020.

Our results are neither as pessimistic as Brown's nor as optimistic as IFPRI. We showed a gradual increase of imports that could eventually reach 120 mmt by 2030. In our analysis we combined agricultural production factors and economic impacts on grain sown area. Grain-land has been steadily declining in China for several decades and is particularly accelerated in provinces with rapid economic development. Farmers prefer to grow cash crops for economic reasons. We feel the impact of economic growth on the fundamentals of grain production is negative. Arable land will continue to decline and will be subject to the pressures of housing, township enterprises, transportation and other land uses. The potential increase of total sown area through multiple cropping, however, will not necessarily benefit grain production because other cash crops be more profitable.

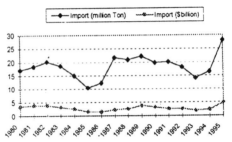

Figure 4.23.15 Chinese cereal imports: Amount and cost 1980-1995

The issue of the actual amount of arable land has become a hotly debated topic in recent years. It has been argued that the actual amount of arable land could be at least 40% bigger than the official numbers, or about 140 million hectares in 1995 (Crook, 1993; Johnson, 1995; Smil, 1996; Mei, 1997). The implications of this extra land are not clear. How much of the increased arable land is devoted to grain? Was the underestimation uniform throughout China? How will the error affect the official statistics on total agricultural sown area? How much will the error affect the grain yield estimates? to which crop and at which location are these yields affected ? The answers of these questions are extremely important in order to predict China's grain production and yield with a reasonable confidence.

As we can see from Figure 4.23.11, grain production is concentrated in the eastern half of China, particularly in the south east coastal provinces (Zhijiang, Jiangsu, Fujian, and Guongton) and the north east (Jilin). These provinces are most fertile and advanced in technology development.the Huang-Huai-Hai plain (HHH) which is one of the 12 agricultural zones in China is well known for its sophisticated inter-cropping and multiply-cropping (wheat-corn-cotton) systems. The HHH region produces more than 50% China's wheat and cotton and is a leading producer of maize, oil seeds, fruits, and meat. Yields in these and other central China principal grain producing regions have been well studied and are reported as similar to the yields of the US (Lin et al, 1996; Stone, 1990; Goldberg, 1990; Geng, 1991, 1993). Thus, the average grain yields for high- and middle- income regions that are reported in China's Statistics Yearbook, are realistic. If this is true, then the error in the amount of arable land may not be all relevant to these regions. The vast western low-income region is most likely responsible for the major errors in arable land reports (Figure 4.23.12). Since we did not have any basis for adjustment, we had to use yield estimates from China's Statistics Yearbook. The yield was 3,324 kg/ha in 1995 for the low-income region, which could be an overestimate if the grain-sown area was underestimated.

In a review article, Smil (1996) criticized Brown's book and suggested that a number of gradual improvements could potentially result in impressive gains in available grain. The improvements he suggested were more efficient uses of water and fertilizer; reductions of the vast amount of post-harvest waste; and improved efficiency in pork, broilers, fish and dairy production. It should be noted that more efficient use of resources will not necessarily lead to increased grain production. The issue of whether China will produce more grain was not addressed by Smil. The problem with Brown's assertion was not his analysis but his projection. He used China's historical data and correctly pointed out some of the problems in grain production. His projections of China's future grain production, however, was solely based on other countries' data (Japan, South Korea, and Taiwan). He gave little consideration to potential technological developments, policy responses , or farm level economic responses to higher prices.

The underlying assumptions of our projections were: China will improve its grain yields through plant breeding, cropping system management, policy, and research and extension efforts. At the same time, there will be a negative impact on grain production due to the rapid development of China's market economy. The population will increase to 1.6 billion by 2030. By then, the per capita income will increase almost 18 fold from 1995 (without adjusting for inflation). With increased disposable income comes increased purchasing choices. By 2030, the eastern part of China will reach a living standard comparable to Taiwan or South Korea today. The food diet will be highly diversified. The quality of food will be of primary concern. The per capita grain consumption in China will increase and about one-half will be used as feed. The competition for land and water use among the sectors of the society will be extremely fierce. Agriculture may no longer dominate resource allocation policy in China. In addition, external factors such as world market prices and trade balance concerns will also favor a policy of grain import to China. It is our belief that China will import a significant amount feed grain. Under normal weather conditions, the import volume could reach 120 mmt by 2030. The alternative to limiting grain imports may adversely undermine China's economic growth. It is a matter of choice for China's policy makers.

REFERENCES

Adam Margit. 1995. Can China Feed its People? The Answer May Be No, Then Yes. CERES: 26-29.

Brown, Lester R. 1994. Who willfeed China? Wake-up call for a small planet. World Watch Institute.

Carter, C.A., F. Zhong, and F. Cai.1996. China's Onging Agricultural Reform. The 1990 Institute, South San Francisco, CA 94080.China's Food Problem. China Daily, Oct., 25, 1996

Crook, F.W. 1993. Underreporting of China's cultivated land area: implications for world a gricultural trade. Special Article, China/Rs-93-4, USDA

Crook, F.W. 1994. An Introduction to China's Rural Grain Supply and Use Tables: China, International Agriculture and Trade Reports, ERS, USDA, WRS-94-4, August 1994.

Geng, S. and Yixing Zhou. 1990. Chinese Agriculture Resources: From the Present to the Year 2000. In Agriculture Reform and Development in China (Ed. T.C. Tso), pp. 199-231.

IDEALS, Inc., Beltville, Maryland.

Geng, S., D. Bessatt, J. Sun. 1991. An Evaluation of the Research Projects Funded by the Second Phase World Bank Loan to the Center of Cotton, Wheat, and Maize Research of the Chinese Academy of Agricultural Sciences. Chinese Ministry of Agriculture and World Bank Review report.

Geng, S. and S. Kaffka. 1993. A Review of China's Dry Land Agriculture Projects. Chinese Ministry of Agriculture and World Bank

Review report

Goldberg J. R. 1990. Grain options for China - 1990-2000. In Agriculture Reform and Development in China (Ed. T.C. Tso), pp. 114-122. IDEALS, Inc., Beltsville, Maryland.

Huang Jikun, S. Rozelle and M. Rosegrant. 1995. China and the Future Global Food Situation. IFPRI 2020 Brief 20. Washington D.C.

Huang, J. and S. Rozelle. 1996. Technological Change: Rediscovering the Engine of Productivity Growth in China's Rural Economy. J. of Development Economics 49: 337-369.

Johnson, G D. 1995. China's Future Food Supply: will China Starve the World?. Seminar Note. State Statistical Bureau. 1995. Statistical Yearbook of China 1995. Beijing: China Statistical Publishing House.

Mei, F. 1995. Sustainable Food Production and Food Security in China. FAO Regional Office for Asia and the Pacific (RAP) Food and Agriculture Organization of the United Nations, 15th World Food Day, Bangkok, RAPPublication 1995-37.

Penning de Vries, H.van Keulen, R. Rabbinge, and J.C. Luyten. 1995. Biophysical Limits to Global Food Production. IFPRI 2020 Brief 18.

Sanders D.W. , R.N. Roy, A. Kandiah and S. Geng. 1994 . Integrated Management of Soil, Water and Plant Nutrition for Sustainable Agriculture and Rural Development -Chinese National Programme for Sustainable Agricultural Development in the North-eastern Plain. UN-FAO Mission Report

Smil Vaclav. 1996. Is There Enough Chinese Food? The New York Review (Feb. 1) PP. 32-34.

Stone,B. 1990. The next stage of Agricultural development: implications for infrastructure, technology and institutional priorities. In Agricultural Reform and Development in China. Edited by T. C. Tso, pp. 47-93. IDEALS, Inc., Beltville, Maryland.

CHAPTER 4.24

Plant Resources of China

Dai, Lun-kai
Institute of Botany, Chinese Academy of Sciences.
Tai William
Editorial Committee Flora of China
University of Maryland, College Park, MD. U.S.A.

Plants are the material basis for human survival. The use of plants is constantly expanding through long time practices. Development in science and technology, has enabled the utilization of plants to reach a new horizon. All plant materials that can be used by man are called plant resources. The plants that are widely cultivated, utilized, and developed commercially, are called economic plants. However, those only used rarely or have been used in the past, are called resource plants.

The complexity of natural conditions and vast land area, have made China a country very rich in plant resources, not only in quantity, but also in kind. The Chinese started early in the use of plant resources, particularly, the use of plants to treat human diseases. Li Shi-zeng recorded the effectiveness for using plants in treating human diseases in his famous book "Compendium of *Materia Medica* (Ben-Cao-Gang-Mu)". Plant fibers were used to make paper and cloth. Plant pigments were used in dye business and food coloring. Lacquer and paints also came from plant materials. The Chinese people have accumulated rich experience in the use of plant resources.

In the late 1950s and early 1960s, for the purpose of improving agricultural production, China conducted a nation-wide general survey of its plant resources. The survey not only covers the plants that had been used traditionally in the history of China, it also investigated the use of wild plant species by people of minority races. There are more than 11,000 species classified as resource plants, only a small portion of which have been developed for commercial use and cultured on a large scale. The rest of them are planted commercially and mainly used as wild plant species. Based on the usage, the plants can be divided into two groups, resource and non-resource. Resource plants can be subdivided into plants for construction, fiber, starch and sugar, oil and fat, tanning, fragrance, resin and rubber, pharmaceutical, pesticides, honey, food, dye, feed, and hosts for economical insects. The non resource plants include plants for maintaining green areas, ornamental, and green manure, indicator and pollution-resistant plants, germplasm and genetic materials. Each category is described in detail as follows:

Plants used as construction materials:

Mainly the woody parts, the trunk and branches, of the plants are used in construction. Lumber has been used extensively in construction industry and is an integral part for national economic development and social welfare.

Based on tree species and woody structure, different lumbers may have different physical and chemical characteristics, including differences in wood grain, hardiness, splitting, rot-resistance, and strength of wood. Based on these properties, some woods can be used in architectural construction, mining construction, transportation, military, sporting goods, and furniture. Others can be used for pulp to make paper, packaging, arts and crafts such as sculptures. China has more than 1000 tree species which can be considered as high quality commercial species. They belong to approximately 340 genera and more than 90 families. More than 300 species have extensive commercial use, many of them are endemic native species of China. Those cultivated commercially include plants of the families of *Pinaceae, Taxodiaceae, Lauraceae, Fagaceae, Betulaceae*, genus *Populus* of the family *Salicaceae* and the bamboo plants of the family *Graminosae*. The lumber with beautiful color and grain include *Dalbergia oderifera, Ormosia microphylla* and the plants of the genera of *Buxus, Eyrus, Cinnamomum*, and *Phoebe*. These are good materials for sculpture and arts and crafts.

Fiber Plants:

The main sources are fibers of roots, leaves, fruits and seeds, and particularly the fibers of the stem. Plant fibers are mainly used in the industries of textile, pulp, ropes, packaging, plastics, and explosives. The stems and leaves of monocotyledenous plants can be used for weaving to make hats, sandals, bedding mats, and kitchen and farm vessels and containers.

The fibers used in the textile industry, except cotton, are usually the phloem tissue of the plants. The xylem (woody) tissues are used in pulping to make paper. Due to differences in physical and chemical properties, e.g. length, strength, and plasticity of the fibers, they are used in different industries.

China has more than 460 species, wild and cultivated, of fiber plants belong to the families of *Urticaceae, Moraceae, Tiliaceae, Malvaceae, Sterculiaceae, Thymelaeaceae,* etc. The bark of many tree species is a good sources for fibers.

Starch Plants:

The sugars (monosaccharides, disaccharides, and polysaccharides) including starch are extensively stored in different plant parts. The monosaccharides include glucose, fructose, rhanmose, mannose, galactose, and arabinose. The disaccharides include sucrose which is condensed from a combination of glucose and fructose. The polysaccharides are made of monosaccharides. Starch is made of glucose, and is stored mainly in the fruits, seeds, roots and tubers. Many of these plants are cultivated agriculturally. There are more than 400 wild starch species in China belonging to the families of *Fagaceae, Convolvulaceae, Polygonaceae, Leguminosae, Menispermaeceae, Hycrocaryaceae, Nymphaeceae, Liliaceae, Araceae, and Dioscoreaceae.* Some of the ferns also have high starch content.

Many species have high sugar content in their stems, e.g. sugar cane, sweet sorghums, etc. More plants store their sugar in the fruits and root tubers. The berries, nuts, and melons have high content of sugar. Many wild plant species have high sugar in their fruits, e.g. plants of the families of *Rosaceae, Vitaceae, Myrtaceae, Actinidiaceae, Rutaceae, Rhamnaceae, Ebenaceae, Elaeagnaceae, Moraceae, Sapindaceae, Ericaceae,* etc.

Plant resins are mainly polysaccharides, mainly made of glucose, galactose, mannose, arabinose, rhamnose, and xylose. There are more than 60 species of useful plant resins, particularly the seeds of *Ficus pumila* of the family *Moraceae, Cassia occidentalis* L., *Sesbania cannabina* (Retz.) Pers., *Gleditsia sinensis* Lam., *Sophora japonica* L. of the family *Leguminosae.* Resin content in the seeds can be as high as 84%. *Amorphophallus rivieri* Durieu, *Bletilla striata* (Thunb.) Rchb. f.

and *Prunus persica* (L.) Batsch. have high quality resins. In addition to being main food sources., starch and sugars are used extensively in light industries (pulping, textile, pharmaceutical, and brewing industries). Starch is also used in heavy industries such as glue and for precipitation in metallurgy, molding and ore screening. Plant resins are used extensively in printing, textile, chemical engineering, cosmetics, pharmaceutical, dying, food processing, mining, and oil industries. The resin of *Sesbania cannabina* is used in machines for oil exploration.

Oil Plants:

Oil and fat are commonly present in plants, mostly stored in the fruits and seeds. Different plants have different oil contents made of different fatty acids. There are two major kinds of fatty acids. The saturated fat in solid state under ambient temperature, and unsaturated fat which is in liquid state under room temperature. The fatty acids together with phospho lipids, glycerol, wax, phenolic compounds, lipoproteins, pigments, and free fatty acids that determine the physical and chemical properties of the oil or fat. Although phopholipids, glycerol, wax, lipoproteins and pigments pose no health problems in foods, it reduces the purity of the cooking oil and thus decrease its quality. Usually, these impurities are removed during processing. The phenols are harmful, e.g. gossipinol in cotton seed oil is poisonous and should be eliminated before the oil can be used in foods. On the other hand, the vitamins, particularly those oil-dissolving ones, e.g. vitamins E, K, and carotenoids are beneficial for human health.

Fats are important resources for daily living and in various industries. Fats are used to make soap, paints, and lubricants. After hydrolysis, the fatty acids and glycerin have wider industrial applications. They are used in the food industry, pharmaceutical, cosmetic, tanning, and metal industries. Glycerin is also needed in military and mining industries. Oil plants are cultivated extensively. There are more than 400 wild oil plant species, particularly in the families of *Pinaceae, Lauraceae, Euphorbiaceae, Theaceae, Leguminosae, Rosaceae, Rutaceae, Caprifoliaceae, Celastraceae, Compositae.* Some of the species may have oil content higher than 8%. *Mesua ferrea* L. of the family *Guttiferae* has 79% seed oil content; *Corylus chinensis* Franch., and *Corylus heterophylla* Fisch. of the family *Betulaceae*, 65%; *Pistacia chinensis* Bge. of *Anacardiaceae*, 56.5%; and *Deseurainia sophia* (L.) of

Cruciferae, 44%. China is rich in oil plants and the future looks bright. Species suitable for planting in mountainous regions can be selected not only for their oil content, but also for greenization and conservation.

Tannin Plants:

Tannin is present in roots, stems, and fruits. Barks and seed coats are particularly rich in tannin. Tannins are poly-phenol compounds. They are used in the leather industry and fish net treatment to provide corrosion resistance. Tannin can also be used as a water softener in boilers, and in the industries of dying and printing pharmaceutical, ink making, and oil exploration.

Surveys conducted in China revealed more than 300 species of plants having high content and high quality tannins including seed plants and ferns. Different species and different organs of the same plant have different tannin contents. In *Pteridophytes, Dryopteris crassirhizoma* Nakai of *Aspidiaceae,* and *Pteridium aguilinum* (L.) Kuhn. of *Pteridaceae;* in the gymnosperms, plants of the families of *Pinaceae, Cupressaceae, Taxaceae,* and *Cephalotaxaceae;* all rich in high quality tannins. In the angiosperms, plants of the families of *Fagaceae, Rosaceae, Polygonaceae, Betulaceae, Juglandaceae, Aceraceae, Anacardiaceae,* and *Rhizophoraceae* have high content of tannins. Many of the herbaceous species, e.g. *Polygonum bistorta* L. and *Rumex_acetosa* L. of *Polygonaceae,* their roots may contain 25% of tannins. *Sanguisorba_offficinalis* L. of *Rosaceae* and *Limonium gemelinii* (Willd.) 0. Ktze. of *Plumbaginaceae* have 15-20% tannin. These herbaceous plants can be cultivated in large acreage. China has large need of tannin for the leather industry.

Essential Oil Plants:

Essential oils are secondary plant products and mostly exist in roots, stems, leaves, fruits and seeds. They are divided into two major categories according to their chemical contents. First, those having mono- and sesquiterpene compounds, they are easy to evaporate and are called evaporative oils. Second, non-terpene compounds which have more complicated chemical compositions, e.g. oils in garlic and onion. Usually, people call those with mono- and sesquiterpene compounds as essential oils. Essential oils are used extensively to make perfumes and in food processing industries (to make beverage, cakes, etc.), cosmetics, cigarettes, soap and toothpaste, office supplies, and pharmaceuticals. China exports essential oils. Natural essential oils having no toxic substance and side effects are more and more preferred by users. Their future development looks very promising.

China has more than 300 species of plants with essential oils. Those cultured extensively include *Cymbopogon citratus* Stapf, *Rosa Chinensis* Jacq., *Jasminum sambac (L.)* Ait., *Michelia alba* DC, etc. Wild species include plants of the families of *Magnoliaceae, Lauraceae, Rutaceae, Oleaceae, Labiatae, Verbenaceae, Umbelliferae, Compositae,* and *Graminosae*. Many of them contain high quality and high quantity of essential oils, e.g. *Chimonanthus praecox* (L.) Link. of *Calycanthaceae, Cyperus rotundus* L. of *Cyperaceae, Zingiber officinale* of *Zingiberaceae, Pinus massoniana* of *Pinaceae*. Some plants may have high quality essential oils, but the quantity is too low for further development.

Resin Plants and Rubber Plants:

Plant resins are mixtures of complicated chemical compounds with high molecular weights. They are divided into: (1) Acid resins: e.g. pine resin; (2) Alcohol resin: resin of *Siyrax tonkinensis* (Pierre) Craib ex Hartw. from Viet Nam. (3) Non soluble alkaloids: plants of the genus *Eulphorbia* of the family *Euphorbiaceae*. Pine resin is used to make rosin and turpentine. Rosin has wide uses in paper, soap, paint, electrical, national defense, and plastic industries. Turpentine is used to make camphor and in other perfume and pharmaceutical industries. The resin from *Rhus verniciflua* Stokes is rot-resistant, and used extensively in paints for construction, furniture, ship building, machinery industries.

Rubber is a high molecular weight, unsaturated carbohydrate, a polymer of isoprene (polyisoprene). Its characteristics are high elasticity, non water- and nonair permeable, durable, and corrosion resistant. No other material has comparable traits. China first imported *Hevea brasiliensis* Muell.-Arg. of *Euphorbiaceae,* and planted *Manihot glaziovii* Muell.-Arg. in Southern China. In the meantime, a wide range exploration was also conducted to search for wild rubber plant species. More than 20 species were found to have industrial potential: *Toenongia tonkinensis* Stapf. and *Ficus pumila* L. of *Moraceae, Chonemorpha erioslylis* Pitard, *Ecdysanthera utilis* Hay. et Kaw. and plants of the genera *Parabarium* and *Parameria* of the family *Moraceae*. Herbaceous rubber plants include *Taraxacum kok-saghyz* Rodin of *Compositae* Hard rubber plants include *Ficus elastica* Roxb., *Artocarpus hypargyraea* Hance of *Moraceae, Jurinea souliei* Franch. of *Compositae, Eucommia* of *Eucommiaceae, Evonymus alata* Regel. E. Bunizeana Maxim, *E. grandiflora, E. Laxiflora* Champ. and *E.Myriantha* Hemsl. of *Celastraceae.*

Chemical content of hard rubber is similar to that of regular rubber but different structurally. Hard rubber has less elasticity, easier to be oxidized in air and becomes brittle.

Rubber has wide industrial uses and is important in economical development. They are used in transportation equipment, industrial and agricultural facilities, construction, national defense, medical equipment, electricity and communication industries, scientific research facilities, cultural and sporting equipment, and equipment used in daily life.

Pharmaceutical Plants:

Plants have been used very extensively in Chinese herb medicine to treat different diseases and to maintain people's health. Herbal medicine has had very long history in China. The first Chinese literature, The Book of Herbal Medicine by Shen-Nong (Shen-Nong Ben Cao Jing), appeared in the year 280 AD. A series of medicinal books published afterwards which recorded the experiences of using herbal medicine of the Chinese people. Li Shi-zheng of Ming Dynasty collected all accomplishments on Chinese herbal medicine before the 16' century and published his classic book Compendium of *Materia Medica* (Ben-Cao-Gang-Mu). He listed 1015 species as having medicinal uses. After the establishment of the People's Republic of China, the government set the top priority on the health of its people. The use of herbal medicine received support for further development. Research organizations on Chinese herbal medicine were established. The studies of herbal medicine attained a new level of height and depth. Surveys on herbal medicine reached every corner of society. Anatomical studies have been conducted on important medicinal plants. In order to raise its effectiveness and to remove toxic components, chemical content of many medicinal plants were analyzed. New pills, liquids, and capsules have been developed.

China is very rich in medicinal plants. The "Dictionary of Chinese

Medicinal Plants" recorded 4773 species of medicinal plants. The people of minority races also use plants for medicinal purposes extensively. Its is estimated that the Chinese uses more than 11,000 plant species in medicine, among which more than 400 species are used regularly. Medicinal plants are planted in large farms. Wild species are also used. Medicinal plants are present in all families, particularly the families of *Ranunculaceae, Papaveraceae, Rosaceae, Leguminosae, Solanaceae, Umbelliferae, Rubiaceae, Loganiaceae, Labiatae, Araliaceae, Apocynaceae, Euphorbiaceae, Berberidaceae, Compositae, Liliaceae, Amaxyllidaceae, Dioscoreaceae, Orchidaceae, Stemonaceae;* among *Gymnosperms, Ephedraceae, Ginkgoaceae;* and among *Pteridophytes, Lycopodium clavatum* L. *Selaginella tamarriscina* (Beauv.) Spr., *Equisetum hiemale* L., *Lygodium barometz* (L.) J. Sm., *Cyrtomium forturiei* J. Sm., *Woodwardia japonica* (L.f.) Sm., *Pyrrosis sheareri* (Bak.) Ching.

All plant parts, roots, stems, leaves, flowers, fruits, and seeds, can be used in medicine. Their chemical structures are complicated including plant alkaloids, saponins, terpenes, ketones, organic acids, etc. China is rich in medicinal plants, however, more research is needed to use these plants more effectively, particularly the studies on useful ingredients. Wild species need to be domesticated for higher yielding capacity. Many important Chinese herbs are exported to foreign countries all over the world.

Pesticide Plants:

To use plant components to kill plant pests has always been an important farm practice. It has the advantage of being safe for human and animals, the pests do not
develop resistance as fast as they do against synthetic pesticides, easy to degenerate and leaves no harmful residues. It is particularly important to use them on vegetables and fruits that are usually eaten without cooking. Some of these pesticide plants can also serve as a green manure to improve crop production without polluting the environment. In recent years, scientists have studied the use of ecdysone to prevent insects from normal metamorphosis.

China has developed more than 200 kinds of pesticides from plant sources. Effective ingredients are well distributed in all plant parts, some parts are richer than others. Chemical structure is also complicated including plant alkaloids, saponin, essential oils, fats, and ketones.

Plants of the following families contain pesticide components: *Ranunculaceae, Leguminosae, '.Rutaceae, Euphorbiaceae, Solanaceae, Compositae, Stemonaceae,* and *Araceae;* include the following species: *Anemone hupeliensis* Lemonine, *Clematis chinensis* Osbeck, *Pulsatilla chinensis* (Bge.) Reg., and *Ranunculus.japonicus* of the family *Ranunculaceae; Derris ferruginea* Benth, *Gleditsia sinensis* Lam., *Millettia pahycarpa* Benth, *Pachyrrhizus erosus* (L.) Urban., *Pueraria peduncularis* Grah., and *Sophora flavescens* Ait. of the family of *Leguminosae.* Those have shown good results and have been used extensively include *Groton tiglium* L. *Euphobia fischeriana* Stud., *Euphorbia helioscopis* L. and *Ricinus communes* L. of *Euphorbiaceae; Tripterygium widfordi* Hook. f., and *Celastrus angulatus* Maxim of *Celastraceae; Sapindus mukorossi* Gaertn. of *Sapindaceae; Camellia oleifera* Abel of *Theaceae; Stellera chamaejasme* L. of *Thymelaeaceae; Eucalyptus robusta* Smith of *Myrtaceae; Nicotiana rustica* L., and *Datura metel* L. of *Solanaceae; Strophanthus divaricatus* (Lour.) Hook. et Am. of *Apocynaceae; Periploca sepium* Bge. of *Asclepiadaceae;* and *Veratrum nigrum* L. of *Liliaceae.* More detailed research is needed to study the effective ingredients, and the toxicological mechanism to warrant future uses of the pesticides from plant resources and for synthetic pesticides.

Plants as Honey Sources:

Bee honey and pollen are foods rich in nutritional value and well liked by people. Plants abundant in flowers, fragrance, and rich in honey are preferred by honey bees. Many plants have honey, particularly in the wild species of *Leguminosae, Rhododendron, Rosaceae, Oleaceae, Labiatae,* and *Compositae.* Plants of the genera *Pinus* and *Salix* rich in flower and pollen, are frequently visited by honey bees.

Food Plants:

Discussed here are mainly fruits, seeds and vegetables of wild species. Fruits of wild species include *Torreya grandis* Fort. ex Lindl., *Myrica rubra* Sieb. et Zucc., *Costanea henryi* (Skan.) Rehd. et Wils; plants of the genus *Castanopsis,; Crataegus -cuneata* Sieb. et Zucc., *Pyrus Pyrifolia* (Burm. f.) Nakai, *Rosa laevigata* Michx. and

Rubus corchorifolius L. f. of *Rosaceae*; *Vaccinium bracteatum* Thunb. of *Rhododendraceae*; and *Zizyphus sativa* Gaertn. var. *spinosa* (Beg.) Schneid., and *Hovenia duleis* Thunb. of *Rhanmaceae*. Wild vegetables include: *Pteridium aquilinum* var. *latiusculum, Portulaca_oleracea* L., *Amaranthus viridis* L., *Capsella bursa-pastoris* (L.) Medic., and *Oenanthe javanica* (BI.) DC.

These wild fruits and vegetables contain no pollutants. They taste good and are well liked by people. Leaves and fruits of some plants can be used to make beverage and herbal teas, e.g. *Gynostemma pentaphyllum* and *Hippophae ssp.*

Pigment Plants:

Natural pigments usually come from plant sources. These pigments are usually used for food coloring or dying. Common food pigments include yellow pigment from the fruits of *Gardenia jasminoides* Ellis, red pigment from *Rosa laevigata* Michx. Other food coloring pigments include those from plants of the genera of *Rubus, Citrus, Vaccinium Lonicera*, and *Rubia*.

Plant pigments are used as dyes. In addition to the plants used in food coloring pigments, they also use *Pterocalya stenoptera* DC., *Mirabilis jalava* L., *Sapium sebiferum* (L.) Roxb., *Rhamus crenata* Sieb. et Zucc., and plant of the genus *Indigofera*. Natural pigments are generally used by the common populace. The textile industry usually uses synthetic dyes. Natural pigments may be better for human health and should be studied in more detail.

Plants as Animal Feed:

Plants are the main source of food of domestic animals, birds, fresh fishes, and economic insects. In addition to forage crops, there are more than 500 species of plants used in animal feed. Some genera of the *Pteridophyta*, e.g. *Azolla* and *Pteridium*, are good food sources for pigs. In *Angiosperins*, plants of families *Leguminosae, Compositae, Graminosae*, and *Cyperaceae* are good to feed cattle, goats, and horses. Some plants of the algae family can be used to feed fishes. Insects of the silk family are fed with tender leaves of *Morus australis, Cudrania tricuspidata*, and plants of the genera of_*Cinnamomum* and *Caotanea*.

Host Plants for Economic Insects:

These are plants used as hosts for *Laccifer lacca, Melaphis chinensis,* and *Ericerus* pela. *L. lacca* usually live on the plants of *Dalbergia hupeana* and *Pterocarva stanoptera* DC. The host plants for *E. pela* are leaves and twigs of *Ligustrum lucidum,* and *Ligustrum sinense* of *Oleaceae* or *Fraxinus insularism.* These insects are useful in the industries for pharmaceutical, chemical, and other light industries.

Plants not as Natural Resources:

Man cannot live without plants. Plants decorate the living environment improve the living conditions, regulate the climate, water and conservation, balance carbon dioxide and oxygen content in the atmosphere, absorb harmful gases and heavy metals and therefore cleanse the air. Plants of many families can be used as indicators to show soil pH and metal content in the soil.

Green manure is very important in any crop rotation system. Not only do they provide organic fertilizers for the farm, they also improve the physical properties of the soil. Plant of the families *Leguminosae* and *Graminosae* grow root nodules and fix the nitrogen from the air.

Plants are used extensively to beautify the human living environment. People always chose ornamental plants, plants with beautiful crown, and lawn to enrich the environment surrounding their living quarters.

Plants resistant to tough environmental conditions, e.g. salt-resistant and draught-resistant, are usually used as pilot plants to improve undesirable ecological conditions. Although these plants do not serve as food or manufacturing resources, they are very important material bases, people cannot live without them.

Plant Germplasms:

Different plant species have different genetic bases. Genetic engineering technology has been used to introduce genes from one plant into a different plant and thus improve the quality and yield of the receiving plant. Genetic engineering has expanded the possibility to combine good characteristics in different plant species. And therefore, the preservation of plant germplasm and the preservation of genetic diversity has become all the more an important task. Particularly important is to preserve those rare and endangered plant species.

China is one of the countries in the world that is rich in plant species, many of them are endemic to China and cannot be found anywhere else in other parts of the world. To protect valuable plant species, is not only important for China, the whole world as a whole can also be benefited. China is very rich in plant resources. Rational development, utilization, and protection of plant resources is a problem that every Chinese has to consider seriously. Only rational development and reasonable use of plant resources, a sustainable supply of valuable resources can be maintained. The only way that plant resources can be developed is to be done on the basis of preservation and protection. Only when plant diversity is protected, can a long term supply of plant resources and a good living environment be maintained. We have to keep a good balance between the use and the protection of plant resources, not to destroy and not to pollute the environment, in the meantime to provide sustainable resources for economic development to warrant a good living environment. A rational plan is needed for the utilization of our plant resources.

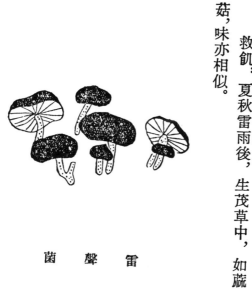

救飢：夏秋雷雨後，生茂草中，如蔬菇，味亦相似。

菌　聲　雷

CHAPTER 4.25

Horticulture, Vegetables and Fruits

Miklos Faust

Beltsville Agricultural Research Center
Beltsville, MD, USA

Scenarios dealing with supply-demand situations in global agricultural systems usually deal with demand for grains, which is roughly one-half of global consumption of food energy. Experience in South Korea shows that rapid urbanization and fast-rising per capita income is accompanied by a choice among food items and the demand for rice decreases in the overall diet while demand for coarse grains as animal feed increases (Havener, 1993). As per capita income increases vegetables and fruits, the so called *horticultural crops,* along with animal protein, also become more important part of the human diet. Horticultural crops consist of hundreds of vegetable species and dozens of fruit species. Types of vegetables and fruits grown in a given area depend on the local climate and the desire of the people who consume these crops. In a large country such as China the number of vegetable and fruit species grown is large. Nevertheless, overall production and consumption data allow us to draw inferences about their use and value in the overall diet. The same data is also useful to project the necessary steps China should make to ensure the supply of fruits and vegetables for its population by 2030.

Vegetables

Since the establishment of the People's Republic, many changes have occurred in vegetable production and marketing in China. In pre-1949 China vegetable producers owned their land and responded to market conditions to supply vegetables to rural and urban consumers. Small farmers also produced vegetables for self-consumption. By 1958 farm families were mobilized into economic collectives. Free market operations were replaced by state-organized specialized vegetable production units around urban areas. The Ministry of Commerce formed state-owned vegetable companies to purchase, transport, store, and retail vegetables to urban residents (Crook, 1996). Farmers still produced vegetables on their own private lots some of which was sold on free markets. During the Cultural Revolution, free markets were closed and state companies were the sole performers to wholesale and retail vegetables in urban areas.

During the early 1980s, local rural free markets were reopened. Production teams of state owned vegetable companies were disbanded after 1984. Farm families and rural economic cooperatives raised vegetables for their own consumption and for the rural local open markets. Direct marketing began as farm families were permitted to bring the produce to urban areas for direct sales. After 1985 the amount of vegetables moving through open market increased. By 1995, about 80% of vegetables were marketed through local open markets and only 20% passed through state-owned channels (Crook, 1996). Between 1979 and 1994, the number of rural open market increased from 36,767 to 66,580. The increase in open market activity in urban areas was even more rapid, rising from 2, 226 markets in 1979 to 17,880 in 1994 (Crook, 1996).

Vegetable output rose dramatically beginning in 1979. Area of vegetable production grew from 3.3 million ha in 1978 to 8.9 million ha in 1994. While vegetable production moving through open markets increased, vegetable production of government-owned vegetable companies decreased. Crook (1995) reports that vegetable sales of the Anhui Vegetable Company fell by two-third from 85,000 t in 1985 to 30,000 t in 1994. In contrast the Dazhongxi open market in Beijing (one of the five major vegetable markets in the city) increased vegetable sales from 13,3000 t in 1986 to 500,000 t in 1994.

Clearly a large portion of vegetables now are marketed through open wholesale and retail markets. Producers responded to market forces and invested in plasticulture and greenhouse technology to produce vegetables earlier. Some entrepreneurs also transport vegetables from southern locations to consuming northern areas. With the increase of vegetable production the export of some specialty vegetables, such as mushrooms, special beans, canned bamboo shoots, asparagus, edible ferns, and fungi, also started. The 1995 level of export was 3.1 million t and aimed mostly toward Japan, Hong Kong, Singapore and the Republic of Korea.

Vegetable consumption in China is high especially among the farmers. about half of the Chinese Provinces vegetable consumption is twice as high, in other Provinces about equal, with that in the United States (Table 4.25.1).

Although vegetable production is high in China, it decreased somewhat during the last 25 years. Rural consumption fell from 142 kg in 1978 to 108 k in 1994, while urban consumption fell from 144 k in 1985 to 121 kg in 1994. Crook (1996) assigns the reason for decrease in vegetable consumption to shifting consumer preference. He theorizes that with the increase in per capita income vegetable consumption shifted from heavy root crops to light leafy vegetables which causes a decrease in weight of consumed products. Urban per capita income rose from 749 RMB in 1985 to 3.500 RMB in 1994. Urban consumer spending on fresh vegetables rose from 41 RMB in 1985 to 152 RMB in 1994. The overall consumption data does not specify the variety of vegetables used in the

Table 4.25.1. Per capita vegetable consumption of farmers in selected Provinces of China, 1993

Province	Kg	Province	Kg
Liaoning	187.9	Jilin	142.9
Heilongjiang	159.9	Jiangsu	142.7
Fujian	108.5	Jianxi	145.9
Shandong	105.3	Hubei	177.1
Hunan	132.7	Guangdong	106.6
Sichuan	118.9	Guizhou	132.8
Yunnan	140.0	Beijing	95.3
Tianjin	90.9	Shanxi	86.0
Inner Mongolia	99.6	Shanghai	86.2
Zheijang	72.5	Anhui	92.6
Henan	63.5	Guanxi	66.8
Hainan	79.0	Tibet	21.5
Shaanxi	62.4	Gansu	29.3
Quinghai	37.2	Ningxia	71.0
Xinjiang	62.2		
United States	66.0	of which 51.2% consumed fresh and 48.8% processed.	

Data from China Agriculture Yearbook, 1994, and Agricultural Statistics, 1993.

diet. This requires that in the upcoming three decades China need to increase production of vegetables by only about 0.5% per year to provide the existing high standard for the increase in population.

Presently 8.9 million ha of vegetables are planted in China. One-half per cent production increase necessitates that vegetable production is enlarged yearly by 44,500 ha. Since vegetable production in China produced a high income per land area, thus, increasing production seems not difficult. As it is clear from Table 4.25.1, there are areas within China where vegetable consumption is still low. These are the areas where either enlarging the production area or increasing production on the existing area is a real possibility.

In the future more vegetables will also be needed in highly urbanized areas. Increasing vegetable production around urbanized areas is difficult. As the city grows transport from the growing area to the city's markets is becoming longer and, therefore, vegetable areas can not be extended to a large distance from the center of the cities. In cooler areas, increasing vegetable production would also necessitate of using heated greenhouses for winter production, and plastic covered houses to extend production to for early production during spring and for late production during fall. The near city production also utilizes prime land which sooner or later needed for housing. Sociologists foreshadow the development of large mega cities around the world. Fortunately, in China, the development of only four mega-cities of 25 million people or more (Beijing, Tianjin, Shanghai and Shenyang) is expected by 2025 (Asian Development Bank Annual Report, 1996), but the other cities are also large which renders the utilization of near-city land to be a particular problem.

The question must be raised whether the production of vegetables around the cities can be increased or alternative methods should be explored. The alternative would mean that vegetables are grown far from the consumption centers and the produce is transported to the consumers. When vegetables are grown far from the consumption centers there are usually two reasons for such production system. One reason is that the climate of the particular area where the vegetable is grown is exceptionally suited for its production and higher yields can be obtained or continuity of production is better assured than in any other areas. The other reason is that winter vegetables can be grown in a warm climate and outdoor production does not require heated houses or plastic overhead structures. However, production far from the consumption centers requires transportation. In the United States and in Europe the transportation cost is less that the benefits obtained from energy saving or income received from higher yields. Thus, the choice is often producing vegetables distant from the production centers.

Transporting vegetables for long distances usually changes the marketing system. There are two major systems utilized in marketing of produce. One predominantly exists in the United States, the other in Europe. In the United States, producers and buyers both are large. This is important especially in the case of buyers. Buyers, usually large supermarket chains, require large shipments of produce that arrive to their distribution centers and shipped to the large supermarket stores usually in large lots. Quality standards of the produce are well known by the producer and the supermarket distribution center and the transaction can be done by telephone. The buyer does not need to see the produce before purchasing it.

Although, there are supermarkets in Europe that operate like those in the United States, a large portion of European stores is small, they consist of individual establishments. These stores pick up their daily need early morning at a central wholesale market where the produce arrives during the night. The wholesale markets are covered large market halls which can handle railroad cars or large trucks bringing in produce as well as small producers bringing in and selling their products. In each wholesale market there are a relatively large number of wholesalers, each usually specializing in a single crop or a combination of related crops. They know the production of the crop, they are people who purchase truckload or carloads of produce they specialize in. The small store owners, in turn, deal with the various wholesalers, finding the produce they need, buying their daily need, often a few boxes of each kind, and transporting the produce themselves from the wholesale market to their store.

Most cities built and operate central wholesale produce markets, often license wholesalers, ensure railroad and road access to the wholesale market and are responsible for the sanitation of the facility. In many Chinese cities a rudimentary system somewhat resembling the European system exists. Often the wholesale market is an open field where farmers bring their products and local merchants come to get their daily need. In most cases store owners purchase the produce directly from the producer, which necessitates that producers spends most of the day on the market selling their produce. In some cases there is a roof above the market but often the wholesale area is muddy when rains. Presently there is no need for this fields to be connected to railroads or large roads because no produce is transported to the wholesale market in railroad car or truck-size lots.

By 2030 production and marketing systems will certainly change. Vegetable productions need to be increased. Where should the increase take place? It is unquestionable that some of the near-city production of vegetables will remain. However, it is also certain that the production in distant areas from the large cities will also gain importance. Therefore a comprehensive plan needed to be developed that takes all phases of production, transportation and marketing into consideration. Planning of locations of wholesale markets near existing railroad lines is of utmost importance. Wholesale markets require a large space and such space has to be set aside now.

For distant production of vegetables, especially for winter production, southern part of China should be considered. Hainan Island, or the entire southern border area of China is highly suitable for vegetable production. Since produce must be transported from here in large lots to the northern cities, production needs to be organized to be a synchronized activity. In general the Chinese farm size is small. This is even more so for vegetable farms. As a result hundreds of farmers need to organize their production activities to produce a uniform product and be able harvest at the same time. They must learn packaging techniques, apply them, and make ready their produce for shipping. This is a formidable organizational and educational task. An extension activity need to be set up and many years of conscious efforts exerted before the system will function as it should. It involves the development of a quality control system, a compensation system for the produce, and the development of a trust between the wholesale buyers and the farmers. Concurrently with the initiation of long distance vegetable production the cities must establish wholesale markets, one, perhaps more than one per city. The system of wholesalers should be established and the strict sanitation system in the market should be put in place. Since vegetable transport from distant production areas is already started the establishment of wholesale markets and the licensing of private wholesalers need to be started. Vegetable production around the cities will remain, but even these growers need synchronize their activities to be able to deal with larger more uniform lots of produce. The central wholesale market will be far for many growers. They need to send their products with the products of others on large trucks to the market. This requires almost the same type of organization described above. Thus in vegetable production I foresee a need for organization and a modest need for increasing production to achieve the goals set for 2030.

Crook (1996) describes the present efforts of various authorities to deal with the increasing problem of vegetable supply of large cities. The Ministry of Internal Trade is vigorously promoting government-owned chain stores as a means to recapture a larger share of food retailing. Others propose that the government exercises effective regulation and control through economic, legal, and necessary administrative means, while state cooperative commercial enterprises play a role as the main channels of the distribution system. Beijing municipality published ceiling prices which were established by collecting wholesale prices from the 6 largest open markets of the city. Then a committee composed of representatives from the Price Bureau, the Industrial and Commercial Management Bureau, the Agricultural Bureau, and the Vegetable Company, Second Division of the Commerce Bureau, meet to set a guiding price (*zhidao jiage*) as the ceiling price. These are all short sited temporary methods and will not work. The Chinese authorities must be more visionary to find solutions which may take several years to develop but will assure a future solution.

The following steps need to be taken to assure a smooth flow of vegetables from the growers to the consumers:
• Determine the best production areas for vegetables before the year of 2000
• Organize growers to produce uniform products in large quantities shippable to distant cities. This is an ongoing activity and need to be started in areas where such grower-association already exists.
• Build large wholesale market in cities connected to railroads and truck lines. Location of such market should be decided upon by the year of 2000. Construction of the first market should be started by 2005.
• Establish a vegetable/fruit wholesale system by 2005
• Operate wholesale markets including sanitation of the market. Sanitation should be taken over by he city immediately.

Fruits

In contrast to vegetable production, the problems of fruit production are very different. China produces a wide variety of temperate-zone fruits such as apples, peaches, pears and grapes, and tropical and sub-tropical fruits such as bananas, citrus, mangoes, papayas, guava. Agriculture reforms instituted in the early 1980s encouraged farmers to expand fruit production. Orchard area rose from 1.8 million ha in 1980 to 8.1 million ha in 1995 (Crook, 1997). Fruit production increased from 6.8 million t in 1980 to 39.9 million t (UN data) or 46.5 million t (Chinese statistic) in 1996. Before 1980, state-owned enterprises handled fruit marketing. Marketing Cooperatives purchased fruit from farmers, stored the product and transported it to state-owned, township, and village-owned processing plants. As economic reforms were implemented in the 1980s, an increasing portion of fruit was marketed through open markets.

Even though China's population is much larger than that of other countries, its fruit production is not much higher than those of India, Brazil, and the United States (Table 4.25.2).

This indicates that the fruit consumption of the Chinese people is much lower than those of most countries. Compared on the basis of population the planted area in China is able to produce the amount of fruit needed, but production of the planted area is far below of the worldwide average production.

Table 4.25.2. Fruit production in selected countries in 1995.

Country	Production 1,000 t	Share in world production %
China	39,957	10.0
India	39,086	9.7
Brazil	32,494	8.1
USA	30,459	7.6
Italy	17,771	4.4

Data are from FAO handbook.

Thus, in fruit production the yield per unit of land must be increased rather than the planting area enlarged. Area and output of several fruits in China and the United States are given in Table 4.25.3.

In addition to the fact that insufficient fruit is available for consumption, the consumption of fruit in China is lopsided. During fall a lot of apples and pears are available, but during the rest of the year these fruits are not obtainable. Some citrus and bananas are occasionally on the market. A comparison of the proportions of fruit produced in China and the US. Illustrates this point.

In future years the situation will be worst. More apples were planted since 1994 Of the present 3.2 million ha only 60% are in production. The rest of it young orchards that will produce apples in the year 2,000 or later. According to Chen Li, Haisheng Fresh Fruit Juice Co., Xian,

Table 4.25.3. Area and output of fruits.

Fruit	China			United States		
	Area 1,000 ha	Total Production 1,000t	Yield t/ha	Area 1,000 ha	Total production 1,000 t	Yield t/ha
Bananas	196	2,700	13.77			
Apples	2,228	9,030	4.05	189	5,306	28.07
Citrus	1,125	6,560	5.83	353	5,871	16.63
Pears	597	3,217	5.38	29	458	15.79
Grapes	141	1,354	9.60	297	6,051	20.37
Pineapple		457			550	
Chinese dates		522				
Persimmon		789				
Others		5,486		183	11,423	
Total orchards	6,434	30,112		1,051	29,659	

Data from China Agricultural Yearbook 1994; and Agricultural Statistics 1993.

the Chinese government solution is the burgeoning apple production is to produce juice concentrate (Apple News, 1997).

A majority of concentrate is produced in Shandong and Shaanxi provinces. China domestic apple-juice market is small, limited only to prosperous coastal cities, with an estimated potential of 120 million consumers the most. Consequently, the juice concentrate is produced for export. In China most fruit is consumed fresh. Juice and canned fruit are too expensive for the average Chinese persons. In other countries, consumption is divided between fresh fruit, juice, canned and/or frozen products, and between subtropical and temperate-zone fruits. In the United States per capita fruit consumption is 63.4 kg of which fresh citrus 8.7; fresh non-citrus 32.1; canned fruits 5.6 ; frozen fruits 1.8; and fruit consumed as juice 15.2 kg. Thus in China not only the productivity of orchards, but also the variety of available fruits need to be increased.

Talking to many horticulturists in China, I gained the impression that the knowledge generated worldwide in fruit production in the 1970s and 1980s is missing from the general knowledge of horticulturists. Basic understanding of photosynthetic processes in a tree, proper nutrition of an orchard, and biochemical processes initiated during ripening are not in he the knowledge bank of university graduates. The management of the orchards is highly unsatisfactory. Misconceptions are abundant among horticulturists. I have heard everywhere that "we have limited land, therefore we have to put trees closer together." Scientists, horticulturists seem to ignore the fact that tree productivity is governed by light rather than by land and crowding the trees blocks light out and decreases rather than increases productivity. Essentially, all the orchards I have seen are oversupplied with nitrogen and deficient in potassium. Fruit number per tree would increase by decreasing nitrogen and fruit size would greatly increase by increasing potassium. Thus, productivity could be doubled by proper nutrition. Nobody seems to comprehend that prescribing nutrition of trees based on leaf analysis can do more for increasing productivity than many heavy labor-requiring processes farmers do today. In my opinion, increasing orchard productivity must start by making sure that the university education produces orchard managers or fruit extension persons who are knowledgeable in the basic details of fruit production, and they know the elements of high productivity.

Apples, apples, and apples are everywhere. Essentially, other fruits do not exist. Peach and apricot, native fruits of China, are produced in so low quantities that there are no statistics kept on them. There is a continuing desire to bring in new cultivars from abroad. This is acceptable in fruit types that are not native of china such as cherries, strawberries, thornless blackberries, or raspberries. However, with imported cultivars `one always faces the unknown: how the cultivar will adapt. When native material is available the progress is much faster by selecting exceptional individuals and propagate them. Such approaches should be followed in apricot, peach, walnut, kiwi, and oriental plum. I could never rationalize why a nation that lifted the oriental pear to a very high level of quality would not use the same method to improve some of the other fruits.

Table 4.25.4 Comparison of distribution of fruit species in the Chinese and U.S. fruit production in 1995 (Data from Crook1997)

Item	China 1,000 t	%	United States 1,000 t	%
Apples	14,011	33.2	4,870	16.7
Citrus	8,222	19.5	14,333	49.1 *
Pears	4,942	11.7	860	2.9
Bananas	3,125	7.4	6	0
Grapes	1,742	4.1	5,358	18.4
Pineapples	539	1.3	313	1.1
Dates	782	1.9	20	0.1
Persimmons	969	2.3	0	0
Other	7,814	18.5	3,404	11.5

* Large portion is consumed as juice.

Propagation of a selected or imported cultivars is a problem in China. There are no large nurseries producing dozens of cultivars of a given species and catalogs to advertise their production. Such nurseries were the leading elements of the North American fruit industry starting with the Price Nursery of Long Island in 1760. China yet to have such a nursery. The United Nations Development Program financed the development of a large nursery that is partially functional in Shaanxi today. The nursery is equipped with virus detection systems, screen houses where virus free mother trees can be kept and the modern requirement of such a plant producing system. This nursery is still not at the stage that it would have a catalog and a standard list of many species produced for sale. Several of such nurseries are needed in China. Good nurseries are the basis of good fruit production and China should establish them if it wants its fruit production function at the average world level.

Storing the fruit within hours after harvest is essential for assuring an extended fruit supply or make sure that the fruit is transported to the city in good condition. This obviously requires sorting machinery and storage. Storage can be placed only in the center of sufficiently large production areas. The problem of organizing small growers to produce uniform product, plant the same cultivar, handle the fruit in the same way, harvest the fruit in a synchronized manner , and transport the harvested fruit into the storage immediately, creates similar difficulties vegetable growers face to produce a uniform product. Thus, fruit production has a large socio-economic component as well.

To successfully bring fruit production to the world level by 2030 the Chinese Government and the fruit growers should:
- Remodel the teaching of fruit production at the Agricultural Universities
- Bring knowledge of fruit growers to the turn of the century level
- Increase the variety of fruit species grown
- Place emphasize on selection of excellent individuals of native fruits in addition to experimental orchard of imported cultivars.
- Produce a few large nurseries which can produce virus free trees in large numbers and in a sufficiently large number of cultivars per species.
- Organize growers into interactive - cooperating units that large quantities of uniform product can be produced in a given area by the large number of growers.

Conclusions

The production of horticultural crops has the potential to keep up with the population increase and supply the Chinese people with vegetables and fruit at the world-wide average level, or even at a higher level. However, a number of organizational improvement will have to be made to assure the year around supply of horticultural product, increase the variety of the produce, especially during the winter months, and smoothen the distribution of such products. It is expected that in 2030 a large portion of vegetables will be produced at a large distance from the cities and for this the transport and distribution system must be improved. The concept of city wholesale markets must be developed. Agriculture Universities must improve the knowledge of the graduates. Growers must be educated how to produce uniform products in large quantities and how to organize in grower associations to achieve such production. Overall quality of the produce must be improved and storage and packing houses at the production centers established. None of these are easy tasks, but without them production of horticultural crops will not enter into the 21st century in China.

Cited literature

Agricultural Statistics 1993.

Apple News. 1997. 28:6. U.S. Apple Association, McLean, Virginia.

Asian Development Bank. 1996. Annual report. Manila

China Agricultural Yearbook. 1994. *Zhongguo nongye nianjian* . Zhongguo Nongye Chubanshe, Beijing.

China Agricultural Yearbook. 1995. *Zhongguo nongye nianjian* . Zhongguo Nongye Chubanshe, Beijing.

China Agricultural Yearbook. 1996. *Zhongguo nongye nianjian* . Zhongguo Nongye Chubanshe, Beijing.

China Agricultural Yearbook. 1995. English Edition. China Agricultural Press, Beijing

Crook, F.W. 1996 (June). The development of China's vegetable markets. China: Agriculture and trade report. Situation and outlook series. U.S. Department of Agriculture, Economic Research Service, WRS-96-2. p.40-44.

Crook, F.W. 1997 (June). Introduction to China's horticultural economy. China: Agriculture and trade report. Situation and outlook series. U.S. Department of Agriculture, Economic Research Service, WRS-97-3. p.47-51.

Havener, R.D. 1993. Environment and agriculture. Winrock International, Morillton Arkansas

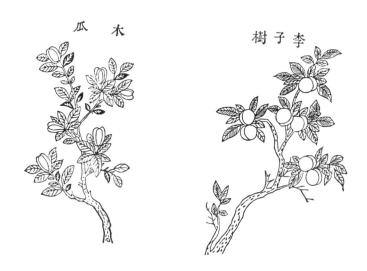

CHAPTER 4.26

Importance of plant protection on the development of agricultural production of China in the coming century.

Wei Fan Chiu(Qiu)
Department of Plant Pathology,
China Agricultural University, Beijing, China 100094

The agricultural production is systemically controlled by a series of factors, such as varieties or cultivars of crop plants, cultivation practice and type of field management including the control of plant diseases and insect pests. All these factors are influenced by the soil and climatic conditions. However, people know how to improve certain soil factors, but up to the present they are unable to control the drastic change of climate in natural conditions, unless under covers of limited areas. Agriculturists nevertheless can create the best conditions for plant growth to obtain a bumper harvest, among which the application of the protection measures has been regarded as practicable. Some rudimentary measures of plant protection were early known by Chinese farmers and had been recorded in documents of ancient China dated back to about 239 BC. The present paper deals only with plant protection problems for China in the coming century.

Present Status of Plant Protection in China

Since 1952, plant protection is under the auspices of the Ministry of Agriculture and carried out by the Plant Protection Stations of various levels and locations throughout China. Recently a Center for Extension of Agricultural Techniques has been established, which includes the extension of plant protection knowledge and methods of controlling plant diseases and insect pests. Such works are carried out by plant protection experts at various ranks of Plant Protection Stations (PPS) established at levels down to counties and villages.

PLANT PROTECTION IN CHINA

To mention some of the most damaging plant diseases and insect pests seems to be necessary. The following list shows the most prevalent and damaging insect pests and diseases so far occurring yearly on the main crop plants, fruit- or forest trees in China and they are arranged in the order of their relative importance to agricultural, horticultural or forestry production.

Paddy rice
Insects: Rice plant hopper, Rice borer, Rice midge; Diseases: Rice blast, Bacterial leaf blight, Rhizoctonia root and sheath rot.

Wheat
Insect pests: Cereal aphids, Wheat midge, Oriental army worm; Diseases: Stripe and stem rusts, Wheat scab, Powdery mildew, and Yellow dwarf virus.

Maize
Insect pests: Asian corn borer; Diseases: Northern leaf blight, *Pytium* and *Fusarium* wilt, and Dwarf Mosaic virus.

Cotton
Insect pests: Cotton bollworm, Cotton aphid, Cotton red spidermite; Diseases: *Fusarium* and *Verticillium* wilt, cotton ball rot, cotton seedling anthracnose.

Rape(for oil seeds)
Insect pests: Diamond back moth, Cabbage aphid, Imported cabbage worm; Diseases: Downy mildew, Turnip mosaic virus, Sclerotinia stem rot.

Soybean
Insect pests: Soybean moth, Greenish hawk moth, Lima bean pod borer.

Cabbage and Chinese Cabbage
Insect pests: Imported cabbage worm, Diamondback moth, Cabbage aphid; Diseases: Essentially similar to those of rape.

Cucumber
Insect pests: Cotton aphid, Greenhouse white fly, vegetable leaf miner; Diseases: Downy mildew, *Phytophthora* blight and qilt, and Cucumber mosaic virus.

Tomato
Insect Pests: Green house white fly, Green Peach aphid; Diseases: Tomato mosaic virus, Powdery Mildew, and Downy Mildew.

Potato
Insect Pests: Potato tuber worm, Potato lady beetle, Eggplant lady beetle; Diseases: Potato degeneration, Potato late blight.

Apple tree
Insect Pests: Apple leaf roller, Apple mite; Diseases: Apple valsa canker, Apple brown rot.

Pine
Insect Pests: Masson pine caterpillar, Long horned beetle; Diseases: Pine nematode wilt, Pine needle cast.

The writer is not prepared to enumerate all of the important plant diseases and insect pests of all crop plants for agriculture, horticulture and forestry in China, however, the above mentioned list shows the most important plant diseases and insect pests which must be controlled in order to obtain a bumper harvest. Proper plant protection practices may result in 10-35% yield increase. In order to achieve this goal the following suggestions are proposed.

Enforcement of an Efficient Extension System

During the past years, PPSs at different levels in China have already done a commendable job, although it can not be considered as perfect. The plant protection experts in the extension station at the lower levels have been discouraged by the low compensation and very inadequate living and working conditions. They can rarely get any opportunity to renew or improve their knowledge of plant protection. Therefore some of them can not stay on the job, but to leave for other professions. The original total number of competent technicians of plant protection assigned to the extension stations at local levels are too few at the beginning and the loss of some of them intensify the seriousness of the problem. The writer is of the opinion that to improve such condition, compensation for the plant protection experts especially working at the lower levels of PPS should be promptly increased and their living and work conditions should be adequately improved. In the same time, the number of competent technicians of plant protection assigned to the stations has to be greatly increased, so that the plant protectors working at lower levels will be encouraged to carry out their works and accomplish their tasks more efficiently, resulting in saving a lot of agricultural products from the attack of insect pests, plant diseases and some other harmful animals.

Training of Teams of Competent Technicians or Plant Protection Experts

In addition to the systematic training from department of plant protection in colleges of agriculture, further training in the form of short courses pertaining to plant protection are to be offered by the Center of Extension for the technicians from the provincial and prefectural stations. Such training is to be held at least once every two years so that the trainees will be able constantly to receive new knowledge and technical innovations of obliged to organize training courses for the technicians from the county and village stations every year at farming leisure. In this time there is an opportunity for training the literate young farmers who engage directly in farming and know what diseases or insect pests need to be given serious attention.

The knowledge of fungicides, pesticides and weed killers are first of all to be put into consideration in the courses of training, because the proper handling of the toxic chemical agents by farmers not only guarantee the full effect of the used chemicals in the control of diseases and insect pests, but also makes possible to protect the life of farmers who might accidentally mishandle such highly toxic chemicals.

Development of Bioplant Protectants

The development of biological control measures for plant diseases and insect pests is practicable since the biological agents do not or rarely cause environmental pollution. This is not something new in China, because the Chinese farmers have long experience to use some predatory insects to control herbivorous insect pests. Recently some endemic or parasitic bacteria and fungi which do not cause any damage to crop plants are used to control insect pests. The same principle is also applicable to control plant diseases, since there exist in the nature certain organisms antagonistic to plant pathogens. Such research projects should be assigned to scientists of senior levels in laboratories of colleges or research institutes to make further development. Furthermore some nontoxic and systemic chemicals which are able to induce the expression of the latent resistance or tolerance of plants to the attack of certain plant pathogens are to be studied and employed as plant bio-protectants. Although such induced tolerance or resistance of plant is not inheritable, it is rather practicable and feasible in field application when no adequate resistant -varieties or cultivars have yet to be developed.

Genetic Engineering in Plant Protection

The routine work in the breeding of resistant plant varieties has contributed a lot to the plant protection. Since the rapid advance of science and technology in the world from the later part of 20th century, genetic engineering has been widely applied to create ideal new resistant species, varieties or cultivars of crop plants which are able to combine the genes of remote origins which are resistant to plant pathogens or insect pests with some genes of superior quality and high yield. Such ideal varieties of crop plants may not be able to achieve by routine breeding works during a comparatively short period. To the present, such research works have been started elsewhere in the world and also in China, such as the transfer of several genes of anti-virus nature into the susceptible tobacco plants causing the later resistant to the infection by related viruses. Likewise, similar works have also been done on wheat to get new varieties of wheat resistant to Yellow dwarf mosaic virus. However such works have just begun in China and will be greatly developed in the coming century. Nevertheless such researches and developing works need a considerable amount of investment which might be obtained either from the Government's special fund or from the sponsorship of enterprises. Strict regulations are need to be observed for field use of such genetically modified organisms (GMO).

Modernization of Plant Protection Apparatus

For either chemical fungicides and pesticides or bioprotectants, the sprayers and atomizers should be designed to suit the large scale applications. The automation of plant protection apparatus with high efficiency should be developed adaptable for various topography in China such as the terrace fields in mountainous regions. The use of airplane for spraying pesticides has already been tried in China especially for the control of forest insect pests, but the structure of airplane for such special purpose should be further improved.

Research must be also carried out by integrating the dripping irrigation system with the application of fungicides and insecticides for control of the soil-borne plant pathogens and soil-inhabiting insect pests.

The genetic engineering in plant protection is not only limited to the creation of new varieties or cultivars resistant to plant pathogens and insect pests but also to be employed in the creation of new bioplant protectants, because the transfer of some anti-fungal or anti-insect genes into some lower organisms is rather easier than into high organisms. Such transgenic organisms with very strong anti-fungus or anti-insect capacity might be used in field control as bioplant-protectants without any side-effect of environmental pollution.

Plant Quarantine be Modernized with Biotechniques

Plant quarantine is one of the effective measures to expel foreign plant diseases, insect pests and some bioagents which are new to China from being imported. The routine procedures are based on the technique employed by classic systematic mycology, bacteriology, nematology, entomology etc. These techniques are not applicable to the inspection and identification of plant viruses. Bio-technique of course are to be employed in the inspection of viruses in seeds and some other plant materials. Modern biotechnology can also be able to aid the more accurate and quick identification of target organisms.

Popularization of Knowledge of Plant Protection

In parallel with the organization of training courses for the different ranks of plant protection experts, another measure to promote the efficiency of plant protection for the increase of agricultural production is the use of the popularized publications, including bulletins, circulars or pamphlets pertaining to the descriptions of plant diseases, insect pests and some important harmful bio-agents. Recommendations of the most feasible and effective measures to combat or control them are nevertheless necessary. Certain circulars are to be specially related to the proper use of various highly toxic chemicals and explanations are made for the properties of each chemical and the care of handle them. Some emergent remedies in case of accidental intoxication before sent to a hospital are to be given. Such publications should be written by specialists and distributed free to different ranks of plant protection experts , especially to literate young farmers. Some countries in the world such as USA have done popularization of agricultural sciences and techniques in this way for many years and the effect has been known as significant.

In the coming century, in order to meet the need of agricultural production by the increasing population of China, the modernization of techniques in plant protection for China is imminent to keep pace with the modernization of other branch of sciences. To realize this task, a very practicable plan for timely action must be taken by the Government with the cooperation of specialists coordinated by the China Association Of Sciences And Technology.   In the same time a proper amount of investment should be assigned to competent researchers in colleges or institutions. No doubt, by the improvement and promotion of the science and technology of plant protection in China, the agricultural production will be able to meet the need of increasing population of China in the coming century.

References (All in Chinese)

Chiu(Qiu), W.F. 1951. Plant Protection in Ancient China.
>Bul.Agri.Sciences 3(5)10-12; 1996 in Some scattered Records of My Agricultural Career pp.1-8, China Agri. Univ. Press, Beijing.

Li, G.B.et al 1990. 'The Integrated Control of Plant Diseases, Insect Pests and Rodents For Wheat Production pp.11-13 China Agr.Sci.& Techn. Press, Beijing.

Du, Z.W. 1991. Strategy and Techniques on The Integrated Control of Rice Diseases and Insect Pests Agri. Publ. House, Beijing.

Zhu, G.R. 1992.Recent Advances in The Techniques of Controlling Diseases and Insect Pests for Vegetable Crops_China Scien.& Techn. Press, Beijing.

Editor Board .1995. A Yearbook of Agriculture of China, 1995. Agri. Publ. House, Beijing.

Guan, Z.H. 1995. Phytomedicine, An Introductory Treatise_China Agri. Univ. Press, Beijing.

Qiu, S.B. 1996. Recent Advances on The Researches for Plant Protection in China China Scien. & Techn. Press, Beijing.

CHAPTER 4.27

China's Livestock Industry: Present and Future

James R. Simpson

Ryukoku University, Japan

China's animal agriculture has undergone dramatic changes since the country was opened two decades ago. The story is dramatic and revealing, not only from a historical viewpoint but also what it portends for the future. Nevertheless, the really exciting and beneficial time spent on analysis of what, where, when, why and how changes will continue to take place. In large part the story rests on technology development and adoption. China is now part of a global society and, in stark contrast to the situation a few decades ago, a considerable portion of the analysis can and should be devoted to analysis of the impact on China from the global livestock industry and related research. That is the approach used in this chapter with the belief that while the reader can easily obtain up-to-date statistics on China, information for a detailed analysis of those statistics is much harder to obtain. Suffice it to say that China has the world's largest inventories of pigs, horses, chickens and ducks. It ranks second in goats, sheep and buffalo, fourth in cattle, 10th in camels and 18th in dairy cattle.

The main objectives of this chapter are to provide a sufficient overview of production systems, techniques and technologies that the reader can make informed decisions on how China's livestock sector might change over the next several decades, demonstrate that great advances can be made in animal productivity, show the enormous impact of productivity improvements on feedstuffs requirements, develop an appreciation of the wide variety of feedstuffs available, and explain how they will be used in Chinese animal production.

Although it may seem strange as a point of departure, it is useful to visualize how far it is to 2010 and 2025, years that are often used as projection targets in long term projections of China's agriculture. For livestock and agricultural projections it is vital to have a "feel" about the length of time involved. The year 2010 is 12 years away from 1998, and 2025 is 27 years away. Consider United States agriculture 12 and 27 years ago. Vast changes have taken place—and the U.S. has been a well developed country during this entire period. Think about the enormous changes which took place in European or Japanese agriculture in the 12 years from 1950 to 1962, and the 27 years from 1950 to 1977. There was a revolution in agricultural structure. Japan, for example, essentially shifted from animal to mechanical power in less than a decade. China has been making breathtaking changes in the past decade, and their agriculture will be completely different in 12 and 27 years from now. The main point is that when thinking about the future, technological and structural changes must be taken into account.

There are a number of misconceptions about China, derived from its chaotic past and the fact that it was only relatively recently opened to the West, which dramatically affect analysis and understanding of the livestock industry. One misconception that greatly affects analysis of the food and feed supply side, and definitely animal agriculture, is that China is a very densely populated country. It is most unfortunate that some well-known public figures, and even academics who specialize in the area, use other heavily populated East Asian economies as points of reference for China's agriculture. I can only deduce that these ethnocentric comparisons have evolved as a result of geographical proximity as well as some similarities in racial phenotype and writing systems.

On a persons per hectare basis, China shares more similarities with the United Kingdom and Germany than with other Far Eastern economies. For example, in terms of persons per hectare of arable land, while China has about 13 persons, Japan has 30; the Republic of Korea has 24; and Taiwan has 23 (Table 4.27.1). On the other hand, the United Kingdom has 10 persons, and Germany has 7. Care must be taken to use comparable data sources, in my case by using the Food and Agricultural Organization (FAO) *Production Yearbook*. Regrettably, some analysts have used Chinese data on cultivated land—which is quite different than arable land—when calculating figures for China as a comparison with other countries.

The amount of permanent meadows and pastures is especially important for cattle and small ruminant production. Clearly, there are differences in use potential between countries. Nevertheless, China has a density of only 3 persons per hectare of permanent meadows and pastures while Japan has 207, and the Republic of Korea has 550 (Table 4.27.1). Data are not available for Taiwan, but it is undoubtedly similar to Japan or

Table 4.27.1 Population density in China compared with selected other countries and regions compared with that projected in China in 2025.

Country or region	Total land area	Arable Land	Perma nent meadows and past ures	Arable Land permanent crops and meadows
	Person per hectare			
World	0.4	4.2	1.7	1.2
China	1.3	13.1	3.0	2.4
Japan	3.3	30.3	207.2	23.9
Korea, Republic	5.0	24.0	499.9	21.0
Taiwan	5.9	23.1	Na	10.9
United Kingdom	2.4	9.9	5.3	3.4
Germany	2.3	6.9	15.5	4.7
China, 2025 proj.	1.7	17.1	3.9	3.2

Korea. In contrast, there are 5 persons per hectare of permanent meadows and pastures in the United Kingdom and 16 in Germany.

It is true that China does have a substantial population density when compared to a country like the United States, where there are just 1.3 persons per hectare of arable land. Nevertheless, density in China is not as great a concern as many investigators would lead us to believe. This is especially true when it is recognized that, in contrast to the United Kingdom and Germany, a substantial portion of China's crop land is situated in warm areas and thus China's crop index, the measure of multiple cropping per year, is much higher than these temperate countries. Apart from feedstuffs production potential, the population density issue is important when studying animal agriculture as it has a great influence on types of production systems and thus projections. In some countries which are often alluded to, such as Japan, most cattle spend virtually their whole life in a shed, completely different from China which has vast rangelands, pastures and croplands on which animals are raised.

Livestock Production Systems and Animal Productivity

Let us turn more specifically to the animal side of the equation. Change in animal production efficiency is a crucial subject which has been overlooked in projections about China's ability to feed itself. Due to space limitations, emphasis in this section is on delineating productivity levels rather than describing systems in keeping with the objective of carefully explaining the importance that technology development and adoption will have on Chinese animal production.

Pigs

Pork production per head of inventory in China was less than half that of France in 1990. Five years later it had grown 33 percent to 88 kg (Table 4.27.2). As economic development takes place in China, pigs will increasingly be raised in commercially oriented units, which will vary from small operations of 20-40 head to 2,000 or more animals, as opposed to the traditional one to two-head backyard operations. Vast improvements are taking place, and will continue to take place in genetics and management, fueled by joint ventures with foreign breeding companies as well as wholly Chinese breeding centers. This technological adoption will be paralleled by phenomenal changes in quality and utilization of feedstuffs. As a result, even though human per capita consumption of pork will still increase, and population will grow, swine inventory will grow proportionally less, and feed per Kg of pork produced will decline.

Slaughter rate (the number of head slaughtered divided by beginning inventory) is another productivity measure. Although the rate for hogs grew 30 percent in China from 1990 to 1995, to 114, it was still very low when compared to a rate of 170 in France and 160 in the United States (Table 4.27.3). The other species shown also reflect very rapid increases in slaughter rates, much higher than my co-authors and I expected when our projections (summarized in a later section) were made using 1989-91 as a base.

Table 4.27.2 production per head of inventory, China, France, and the United States, 1989-91 and 1995

Commodity	China	France	USA
		Kg	
Beef			
1989-91	12	87	109
1995	17	93	112
Percent increase	41.7	6.9	2.8
Pork			
1989-91	66	146	131
1995	88	147	135
Percent increase	33.3	0.7	3.1
Cow milk			
1989-91	1,562	4,797	6,673
1995	1,606	5,437	7,462
Percent increase	2.8	13.3	11.8
Chicken			
1989-91	1.8	8.2	8.1
1995	3.3	9.2	7.6
Percent increase	83.3	12.2	-6.2

Source: Derived from FAO Production Yearbook, 1995

Beef and Draft Cattle

As of 1990, China produced 12 Kg of beef per head of cattle in inventory, in pointed comparison with France and the United States, which produced 87 and 109 Kg, respectively (Table 4.27.2). By 1995 China had increased its output 42 percent, to 17 kg. This level is still comparatively quite low, pointing to the vast increases in productivity still available from structural (mainly systems) changes as well as adopting technology. It is of paramount importance to realize that China has vast grazing areas, a substantial portion of which produce at much less than their potential, partly as a result of management problems. Another reason is lack of incentive caused by remoteness, prohibitive transportation costs, and institutional legacies which still inhibit structural production system adjustments (Li, Ma and Simpson, 1994).

Table 4.27.3. Slaughter rate of selected livestock and poultry in China, France and the United States, 1987-96

Country/year	Hogs	Cattle	Sheep and goats	Poultry
			Percent	
China				
1987	78	7	34	92
1988	84	9	38	105
1989	85	10	40	96
1990	88	11	42	105
1991	91	13	47	113
1992	95	15	50	110
1993	98	18	54	122
1994	107	22	60	150
1995	116	25	69	176
1996	119	Na	Na	Na
1989-1991	88	11	43	105
1994-1996	114	25	69	176
Percent increase 1989-91 to 1994-96	30	127	60	68
France	170	29	77	Na
United States	160	36	48	Na

Slaughter rate is slaughter divided by beginning inventory
Poultry includes chickens, ducks, and geese.
Calculated from FAO *Production Yearbook, 1995.*

Transportation and marketing aspects are crucial not only for beef production, but all of China's agriculture because of its vast size and extreme climatological and geographical differences. Improved transportation will have an important impact on more rational uses of resources in agriculture. For example, as the country develops and becomes mechanized, the grazing lands used for cattle will gradually shift to production of feeder cattle which, rather than being grass fattened, will be followed by minimal feedlot fattening in grain producing areas (Simpson and Li, 1996). In all likelihood, because of economic and institutional considerations, a hybrid of national and foreign fattening systems will evolve.

It is crucial to understand that China has a huge amount of by-product and nonconventional feed resources, such as straw, which are increasingly playing an important role in beef production (e.g. FAO consultancy reports such as Dolberg, 1992). Regardless, production per head of inventory will increase which means that fewer cattle will be required to produce the same, or even expanded, amounts of beef. The lack of attention   given to production complexities has led to misunderstanding about the impact of technology.

Poultry

Poultry for meat are largely raised in confinement, meaning that they must be fed grain or other feedstuffs.   However, as a result of rudimentary procedures, only 1.8 Kg of meat was produced per chicken in inventory in 1990, compared with 8 Kg in France and the United States.   Similar to swine, vast improvements have taken place in productivity so that in the 5 years between 1990 and 1995 meat production per head grew 83 percent. Through commercialization, more efficient use is being made of feedstuffs, resulting in a substantial reduction in feed per Kg of meat produced.   The same situation holds true for eggs.

Dairy Cattle

China's dairy yield actually declined during the 1980s, from about 1,900 kg in the early 1980s, to about 1,500 kg at the turn of the decade because of promoting dual purpose (native beef or draft type crossed with western dairy type) breeds by small producers.   Nevertheless, there are farms with average yields of over 7,000 kg and the industry is modernizing rapidly.   The problem with China's crossbred dairy cattle, and upgraded natives in general, is that they do not have the genetic potential for high yields (Simpson and Wilcox, 1982).   Furthermore, and crucial for a country like China where land is a constraint, is the inordinately large proportions of feed required just for maintenance. That means the total quantity of feed consumed per kg of milk from lower producing crossbred dairy cattle is quite high relative to purebreds or high crosses.

Sheep and Goats

An indication of productivity improvement is gained by observing that the slaughter rate of sheep and goats increased from 34 percent in 1987, to 69 percent in 1996 (Table 4.27.3). Clearly, care must be taken in interpretation of the data, as with all agricultural data, on a year to year basis due to climatic influences as well as the very rapid structural changes taking place. Nevertheless, it is apparent that sheep and goat numbers have increased very quickly during the first half of the 1990s as they are easy for lower income rural residents to raise and they multiply quickly. The number of goats by 2025 will probably be about the same as in 1990, but sheep numbers will be larger since sheep are mainly used for wool.

Work Animals

This category includes asses, camels, cattle, horses, water buffalo and yaks. As the country mechanizes, animals used for transport will gradually be replaced by trucks, and work animals by tractors and mechanized transport. Conservatively, by 2025 the number of asses will likely fall by 30 percent, camels and horses 40 percent, mules about 10 percent, water buffalo about 40 percent and draft cattle at least 60 percent.

Animal Feedstuffs

By the early 1990s there were three major sources of processed feed: the Ministry of Commerce (MOC) which controlled about 70 percent of the market, the Ministry of Agriculture, Animal Husbandry, and Fishery (MAAF or MOA) with about 20 percent, and non-government ones (county run, collective units and private) with 10 percent (Coffing, Colby, Lin and Simpson, 1992). Although great advances have been made in development and manufacture of standardized medium size computerized feed mills appropriate for conditions around the country, low quality of processed feed and improperly formulated rations have been a significant problem largely because most of China's feed mills have operated under extremely rigid and initiative-dampening government controls. If any of their feed grain supplies were received from state grain bureaus, mills were required to sell their resulting products at state-set prices as well as having their profit restricted in the range of between 2 and 7 percent. With a fixed profit margin, there was little incentive to use lowest-cost rations, increase operating efficiency, or insure product quality (problems which have faced all of China's state-owned enterprises) in feed manufacturing. In fact, only a few mills adjusted their rations to changes in ingredient costs, and those that did only as a relatively rare occurrence.

The continual growth in joint-venture or wholly foreign-owned mills is one stimulus which is forcing China's feed mills to improve quality, productivity and efficiency. Growth in the number of private mills will lead to greater mixed feed use than if there were only government owned mills. Despite the influence of various levels of government upon feed mill operations throughout China, it seems that by the end of the 1990s all feed mills will be completely independent in pricing and operation.

China has a long history of adeptly utilizing what are now called nonconventional feed resources (NCFR), crop residues and by-product feeds. As animal feeding has modernized, the use of NCFR, relative to total animal feedstuffs consumed has diminished in commercial pig and poultry operations. However, there will continue to be an important place for NCFR, residues and by-product feeds in China, and they can be expected to be important to the livestock industry, especially for pigs and cattle. Although the definition of NCFR is open to discussion (c.f. Devendra 1992), as used in this chapter means all feedstuffs not commonly found in commercially produced rations used in modern type feeding systems. Thus, minor ingredients such as spent brewers grains and molasses are not considered NCFRs. Rather, they are designated by-products. In addition, ingredients bought and sold internationally such as fish meal are termed by-products

The feedstuffs termed "by-products" are relatively well-known and thus merit little explanation. The main point is that extraction technology, while understood in China, was only becoming reasonably well developed in the early 1990s and then primarily in newer plants. The marketing mechanism is still less than efficiently handling by-products. There are a number of NCFR in China which are either currently important or which have considerable potential as animal feedstuffs. Examples are root and tuber crops, water plants, treated straw, azolla, water hyacinth, leucaena and poultry litter. In our 1994 book Xu, Miyazaki and I estimated that in 1990 about 35 percent of all energy availabilities for which *data are available* or for which *calculations could be made* were from NCFR. We calculated that in 2025, despite a dramatic modernization of the livestock industry, 29 percent would still be contributed by that category. The proportion is actually somewhat higher because there is a substantial amount of NCFR such as water hyacenth and azolla for which calculations cannot be made.

As explained earlier straw, which is widely abundant in most parts of

China, is by far the NCFR with the greatest potential for ruminants. The problem of its low digestiblity and low feeding value has received considerable attention from animal scientists throughout the world (e.g. Doyle, Devendra and Pearce, 1986; Orskov, 1987) and considerable research has been carried out on treatment methods for straw as well as maize and sorghum stovers. There is also a substantial portion of hay produced in China of relatively low quality due to improper curing or to adverse conditions after harvest that could benefit from treatment. The principal straws treated are rice, wheat and barley, all of which are found extensively in China. Sodium hydroxide is one possible treatment method, but is seldom used in most countries. Ammonia has emerged as the preferred on-farm method in most countries where straw, hay and stovers are treated extensively because it is relatively safe and cheap, and also increases energy and protein content (Simpson, Kunkle and Brown, 1993). Although ammonia has been, and continues to be, used in China due to simplicity and widespread availability, urea seems to be more popular. A drawback is that while urea raises protein content, there is a limited effect on palpability. In any event, of all the NCFR, treatment of straw and other low quality forages offers the greatest potential to significantly expand feed resources.

Ironically, China may be in a better position than developed countries to utilize treated straw. The reason is China's communal based production system is ideal for large scale production of calves and fattening cattle in grain growing areas and this system is already providing the impetus for large scale straw treatment. In developed countries, where the production system is based on individual operations, most cow/calf units are too small or producers are simply not motivated to adopt a technology like low quality forage treatment as it requires considerable coordination. In addition, very low priced feed grains often make straw treatment unattractive to cattle feedlots in developed countries. The environmental factor is also important for China. Feeding straw to grazing ruminants is environmentally more friendly than using grain.

It is of critical importance to comprehend the magnitude of error when determining the amount of feedstuffs required to produce a kilogram of meat by simply applying a "standard" feed conversion coefficient, such as eight to one for beef based on production feedlot-type fattening systems, or four to one for pork production, coefficients which are typically used in modeling China. Another serious problem not only for beef production, but for other animals as well, is that modelers typically simply use maize as the energy source and soybeans as the protein source thus completely obfuscating the complexities of a country with many feedstuffs sources

Feed Balance Sheet Modeling: Energy and Protein as a Basis

Review of a linked spreadsheet simulation model which I developed with the cooperation of Xu Cheng and Akira Miyazaki, and explained in detail in our 1994 book *China'a Livestock and related Agriculture: Projections to 2025,* is quite useful to understand how a projection model focusing on technical production aspects can be developed. The purpose of this section is not to stimulate discussion about results, but rather to highlight one method to determine feedstuffs requirements and availabilities, and to show the amount of non-quantifiable feedstuffs used in China. In addition, specification of parameters and variables is a valuable exercise which really forces a researcher to understand the what, where, when, why and how of animal production. This particular model contains two projections. One, referred to as "robust," is based on continued strong economic growth. Slow growth in the economy, coupled with modest growth in per capita consumption of food products, is the basis for the second alternative called "economy sluggish."

Feedstuffs vary greatly in their nutrient composition. Likewise, there is a tremendous difference in nutrient requirements between animals. Cropping patterns and cultivation methods, as well as species composition, also change over time. Because there are so many differences in animals and feedstuffs, the approach used in the projections about China's feed situation was to base calculations on protein and energy requirements, and protein and energy availabilities. A base year, in this case the average of 1989-1991, was selected, the models were validated, and projections were developed for the years 2000, 2010 and 2025. The requirements and availabilities models each have about 300 variables and 1,000 parameters. Prices were ignored as the objective was to determine if, *technically*, China could continue to have a *net* self-sufficiency in animal feedstuffs and animal products

The method used to estimate animal feed requirements was to first project per capita human consumption of animal products, and then multiply the results by population to determine total consumption (Simpson and Ward, 1995). For most livestock commodities, consumption has essentially equaled production in China. Animal inventory was then calculated for food-oriented animals from production requirements. Separate projections were made for animals used for draft or transport. Inventories of all animals, by size and plane of nutrition, were then multiplied by metabolizable energy (ME) and crude protein (CP) daily requirements to obtain nationwide totals. Beef consumption in the economy robust projection was projected to increase from 1.0 kg in 1990 to 3.2 kg in 2025, pork from 20.1 kg to 35.3 kg, poultry meat from 2.9 kg to 18.2 kg, total meat from 25.1 kg to 58.1 kg, and eggs from 7.1 kg to 12.6 kg. The per capita projections may have been somewhat low. Also, the parameters chosen for the models on both the feedstuffs requirements side and the available products side were quite conservative. Nevertheless, the modeling reveals that China's animal inventories will not be as large in 2025 nearly 30 years hence as one might initially believe, despite large increases in human population, due to changes in crop and animal industry structure, management and technology adoption. Although the models were just based on technology known and available in the early 1990s, the benefits from improved efficiency, structural change and mechanization are such that total cattle numbers were calculated to be only about 85 percent higher for the sluggish projection in 2025 compared to 1990, and 60 percent more in the robust alternative. Total pig numbers were just 17 percent and 25 percent greater for the two economic alternatives. Poultry numbers in 2025 were only about double despite poultry meat consumption growing 6 fold and egg consumption nearly doubling.

Review of data for 1994-96 (not shown) indicates that inventories of some animals, such as goats and buffalo, which increase rapidly in response to rapid growth in the early stages of economic development were much higher than projected even for 2000. However, they will begin to decline in a few years as structural changes take place and technology is adopted. The same holds true for pig numbers, which have grown rapidly in response to high economic growth, as technology adoption simply has not been able to keep up with demand for pork in such a short period. Without a doubt the number of pigs required to produce a kg of pork will decline. Poultry numbers follow the same logic.

For the base period 1990 metabolizable energy (ME) requirements by animals were 1.46 times 10 to the 12th power megacalories, i.e., 1.46 trillion megacalories. As a point of reference, one megacalorie is equal to one million calories. Although this chapter is targeted at livestock rather than crop production, it is important to conceptualize how complementary projections can be developed on the availability's side, and to understand complexities related to livestock. Production of each crop for 1990 was multiplied by the amount of ME and CP to determine totals. Many crops have multiple outputs. For example, barley grain is used for human food, animal feed and beer from which the spent grains can be fed to animals after the brewing process. The straw can be used for animal feed with the challenge being to determine the proportion fed, and particularly the amount treated with chemicals to improve its digestibility and protein content.

Feed availability's in terms of ME, which are based on published data, or for sources which could be estimated with some degree of confidence such as pasture lands, were calculated at 0.93 times 10 to the 12th power megacalories for the base year (1989-91). The difference between requirements and availabilities was 0.53 times 10 to the 12th power, or 36 percent. This gap was met by feeding garbage, roadside grazing, and widespread use of nonconventional feed resources such as water plants for which there was no published data. Straw is included in calculations of feedstuffs availabilities.

Results from both the robust and sluggish projections indicate that the calculated gap between ME requirements and availabilities will continue to be 37-38 percent in the years 2000, 2010 and 2025. Because the gap, caused by lack of published data, or data from which feed availability and requirements can be determined, is calculated to remain essentially the same for the next 3 decades, a conclusion can be drawn that *technically* China can *essentially* have a balance between metabolizable energy needs and domestic production basically by just adoption of known technology. The size of the gap is very important as it highlights the importance of feedstuffs which cannot be accounted for in China. Clearly, China will import feedstuffs in some years due to climate induced poor crops.

Technological Change and Productivity

One key to making long term projections about China's livestock sector and the country's ability to feed itself is the realization that, not withstanding remarkable economic growth, it is a developing country with very low income. Consequently, despite remarkable achievements by the rural labor force, much of China's agriculture is still in what is termed the era of horse power (Figure 4.27.1). Hand power is also still prevalent in much of the country. In actuality, substantial portions of China are only now entering the era of mechanical power the period which characterized the United States between World War I and II. The

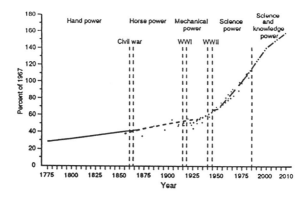

Figure 4.27.1. United States agricultural productivity growth - era of science and knowledge power. Source: Adapted from Lu et al. (1979)

next era referred to in Figure 4.27.1 as the science power era, lasted in the United States for about 40 years. The US and a number of other developed countries are now moving into what I refer to as the science and knowledge power era. China will progress through the eras much more rapidly than the United States and other developed countries did. Barring severe political and/or economic setbacks, Chinese agriculture will be completely different in a few decades than it is now.

Paradoxically, as alluded to with respect to beef production, China's recent history of communism will be an asset in moving the nation quickly toward the era of science and knowledge power. In bright contrast to virtually all other developing countries, China has substantial segments of large-scale crop and livestock agriculture although much of it is still very human labor-intensive. Nevertheless, even the most cursory trip through major grain growing areas quickly convinces one that the economies of scale are large and can easily grow even larger, both from an institutional and land-base viewpoint. China is simply leaping across the smallholder restraint, which is a major problem, if not *the* major problem, of most developing countries. China's historical legacy, coupled with rapid urbanization resulting from a policy of rural industry development which has opened the way for mechanization, provides an optimistic forecast for animal as well as crop agriculture.

Much of animal agriculture is closely tied to crop agriculture and, to a certain extent, the two must be analyzed together. It is important to understand that, because much of China's farm work is still carried out by hand labor and animal power, there is a *de facto* labor shortage during critical farming periods which has become more acute through a vast shift of able bodied people to urban pursuits, and through the development of rural enterprises as a part of the nation's industrialization. However, this labor reallocation is a necessary part of economic development and, as urbanization continues, it will provide positive benefits to agriculture through expanded mechanization and creation of crop and animal producers with higher management skills. Urbanization will also serve to both expand the multiple crop index and to increase yields in both animal and crop agriculture (Simpson,1993). Clearly, discussion about the impact of urbanization and technology adoption in long term projections rests on accurate assessments of how producers will change during the projection period. Many of the technologies—and modern production practices in general—can only be adopted in both China and in economically developed countries by producers with advanced management skills, ones whose continued interest is increased efficiency. I submit that, just as happens all over the world in response to economic development, there will be a shift to urban occupations and that the pejorative term of "peasant," now used to describe China's livestock and crop producers, within a few decades will no longer be an apt designation for those who will be producing the bulk of the nation's agricultural commodities.

It is obvious that adoption of emerging technologies, as well as the multitude of technologies now available but not being used, will vary in China from region to region and even from operation to operation within a district. A salient point is that some producers will leap at opportunities and, through their early adoption, will benefit to a great extent. The slower adopters will eventually drop out. The concept of adoption applies to even the most mundane production techniques as well as sophisticated ones. Furthermore, there are a host of simple, yet unique management practices being developed in a wide variety of countries of great relevance to China.

A virtual revolution has taken place in livestock production and marketing in much of the world over the past few decades. There have been great gains in physical productivity and economic efficiency fueled by changes in industrial structure, vast developments in technology and management practices, and in marketing techniques for the products. The overall goal of this section is to provide a sense of the vast amount of animal technologies being developed worldwide which, in addition to known technologies, will be adopted in China within the next few decades. Many of the techniques are now beginning to be applied in the West and to some extent in China; others are still in their most basic stages. Together they represent a fascinating array of fruits from modern science. Oddly enough, although the methods used in development of the products are often very sophisticated, many of the products, such as vaccines, seeds, and genetic improvement are generally not critically dependent on more than very moderate advances in producer management abilities. Nevertheless, definite synergism's exist between technologies which lead to upward shifts in production functions as well as outward movements along them.

There are numerous genetic oriented technologies which are being developed or have recently been made available in China. Some of these, such as development of lean type pigs through crossbreeding of native and exotic (foreign) breeds, and introduction of exotic breeds of chickens for meat and eggs, are rather obvious. In addition, for large animals in particular, substantial improvements are possible and will dramatically expand productivity. Much of the work in genetic improvement is simply application of known techniques and continued restructuring of the animal industry. However, there are also a whole host of emerging techniques and technologies that will help China speed up the genetic improvement process.

Biotechnology, for example, which refers to a wide array of techniques that use living organisms (or parts of organisms) to make or modify products, to improve plants or animals, or to develop microorganisms for specific uses, will play an important role in China's animal agriculture. A report by the U.S. Congress, Office of Technology Assessment (1992) in the United States provides a good overview of 41 potentially available animal technologies which will become available to China early in the next century through both national and international research and development. The main impact will be on feed efficiency for all animals, reproductive efficiency for beef cattle and swine, milk production for dairy cows, and eggs per layer in poultry. There is a cornucopia of technologies being developed. These are discussed in detail in Simpson, Cheng and Miyazaki (1994). However a few stand out. On the feed side, additives to improve digestibility of feedstuffs are increasingly being used. Genetic manipulation of ruminal bacteria is in its infancy but, as the ruminal fermentation patterns become better understood, it will be possible to boost the rate at which the rumen breaks down cellulose from the current 70 percent to nearly 99 percent. This will be especially important for China which has an abundant supply of low quality roughages such as straw in which treatment techniques such as urea have less than desired impact on digestibility. Microflora developed through genetic engineering will be used to metabolize raw feedstuffs into nutrients more efficiently. Wastes from animals such as poultry litter, hog excrement and cattle manure are increasingly being used as cattle feedstuffs. Recent attention given to ensiling these wastes with other material such as maize, forage or grass hay has yielded promising results. Use of poultry litter is gaining wide acceptance. Rumen-regulating drugs to improve feed digestibility and absorption along with expanded feed intake are being investigated. In most of the developing world, including China, although there continues to be identification of mineral deficiencies, and there are emerging technologies in testing and evaluation procedures, it is clear that management, education and provision of quality products are the limiting factors. In fact, proper mineral and vitamin use, as well as adequate nutrition, are constraints to effective use of many emerging technologies.

Artificial insemination has been used very successfully for the past half century, particularly in the dairy industry. Recent developments in embryo transfer (ET), the process of retrieving one or more embryos from a pregnant donor animal and inserting them in a recipient (surrogate) which then carries the resulting fetus to term, has led to speculation that ET will one day essentially replace artificial insemination (Simpson, et.al. 1996). A principal advantage is that multiple genetically superior animals can be obtained annually from a donor using a relatively low grade, inexpensive recipient. Another advantage is that the genetic makeup of a herd can be expanded much more rapidly than through natural mating or artificial insemination. A virtual revolution in associated techniques such as sexing embryos and sperm, embryo cloning and estrus cycle regulation promise great advances in reproduction and consequently point to greater productivity in the next few decades. China is increasingly engaged in state of the art animal research.

Biotechnology, as well as other research advances, will have a major impact on Chinese as well as world animal productivity in the next several decades. The main recipients will be countries where information flows and industry structural changes are encouraged. China is undergoing a very rapid change in rural policy which is encouraging development and adoption of new production and marketing technologies. Consequently, in this era of globalization, perhaps the best guide to China's future level of productivity can be gained by evaluation of current levels of productivity in other countries, and an assessment of the impact from the technological change process. In contrast to some notable detractors(e.g. Brown, 1995), I believe that (a) productivity growth in developed countries will continue to expand at a remarkable pace following the trend shown in Figure 4.27.1, (b) Chinese agriculture will be a major beneficiary of international work and (c) China will itself become a major player in development of advanced animal as well as agricultural technology.

Although China is pushing hard to obtain, develop and adopt new crop and livestock production technology, researchers and public officials are also quite cognizant of product safety and international controversies surrounding agricultural inputs. International experience demonstrates that as per capita incomes increase, producer, consumer and official awareness of product safety on. the one hand, and efficiency on the other, grow accordingly The pace at which animal production and marketing technologies are being developed worldwide is astounding. Many of these technologies are transferred between scientists at a relatively sophisticated level. Thus, one main criterion for China is to develop a scientific community with the necessary resources to exploit the knowledge being created in economically developed countries. By the early 1990s a major effort to develop the appropriate extension and research infrastructure had been embarked upon. It is ironic, yet significant, that China's vast population is actually an advantage in this endeavor because it enables development of critical masses of scientists for the wide range of technologies being dealt with.

A very interesting question is: what would happen if China really strongly embraced the emerging benefits of research, such as those from biotechnology? The rather conservative projections of even greater technology developed as part of the long term projections discussed earlier indicate that livestock inventory could be 15 to 35 percent lower than in the robust projection. In fact, by 2025, the national inventory of some animal species could be lower than it was in 1990 despite large increases in per capita consumption of livestock products and substantial population growth. Greater production with fewer animals means less feedstuffs requirements.

Much of the technology which has been discussed is sophisticated, yet relatively easy for China to implement. An idea of the impact that much greater productivity could have is obtained by considering pigs. In 1980, China's pigs produced about 36 kilograms of pork per head of inventory. This number is derived by dividing total pork production in the country by the total number of pigs. By 1990, productivity had reached 66 kilograms (Table 4.27.2). By use of considerable new technology and more rapid improvement in management than assumed in the sluggish and robust projections, it is estimated that 128 Kg could be produced per head of inventory in 2025, at a time when most pork will be produced on commercial operations. The level was calculated to be 121 kilograms in 2025 for the robust projection. These numbers are placed in perspective by recognition that the greater-use-of-technology projection of 128 kilograms thirty years hence, is less than the production per pig in the United States back in 1990. In that year the average was 121 kilograms in Europe and 132 kilograms in Japan. In brief, particularly in light of China's performance the past decade, the great technology projections are not out-of-bounds scenario building but rather may be even too conservative, especially when related to recent results for other countries.

Final Observations and Conclusions

This is a decidedly optimistic chapter about the potential to radically reduce the amount of feedstuffs used per kg of animal product produced in China. The story really began when the country's leaders made a radical shift in policy by opening it to the outside world in 1979. Emphasis suddenly shifted to development and adoption of technology. This policy, which was coordinated with price, market and industry policies such as the household responsibility system, development of industry in rural areas, and strong attention to agricultural research and education, has led to great leaps in productivity and efficiency. A major question at this juncture is the extent to which China will continue to absorb macro and micro level management techniques and technologies which have been and are being developed abroad as well as domestically.

China's very low level of animal productivity is one major reason why, on balance, *technically,* it potentially can have a net self-sufficiency in animal feed stuffs and animal commodities for the foreseeable future. On the animal side, as the economy matures there will be a continued shift from backyard to commercial operations, and from draft power to mechanization. As a result of improved management, increased infrastructure like transportation and communication, a wider array of quality inputs, imported and improved breeds, and a host of technologies, the amount of feedstuffs required per kilo of animal product will decrease. The major message in this chapter is the need to be aware of the impact that agricultural technology, technology in general, and management improvement will have on China's livestock industry and thus on analysis of the future. Through careful research mistakes about markets and trade can be avoided, projection results improved and the search for business opportunities made more profitable.

References

Brown, Lester R., (1995). *Who Will Feed China: Wake-up Call for a Small Planet.* New York, W.W. Norton & Co.

Coffing, Art, Hunter, Colby, William Lin and James R. Simpson. (1991). "Chinese Swine Production Makes Big Strides Forward." *Feedstuffs*, Dec. 16, 17-19.

Devendra, C. (1992). *NonConventional Feed Resources in Asia and the Pacific*, 4th ed., Bangkok, FAO.

Dolberg, Frands. (1992). "Beef Production Based on the Use of Crop Residues, Henan and Hebei Provinces, China." Consultancy Report, United Nations Development Programme, FAO, February.

Doyle, P. T., C. Devendra and G. R. Pearce. (1986). *Rice Straw as a Feed for Ruminants.* Canberra: Inter. Dev. Program of Aust. Univ. and Colleges Ltd.

FAO of the United Nations (1995). *Production Yearbook* and *Trade Yearbook*. Rome, Italy.

Li, Ou, Ma Rong and James R. Simpson (1994). Changes in the Nomadic Pattern and its Impact on the Mongolia Steepe Grassland Ecosystem. *Nomadic Peoples.*

Lu, Yao-Chi, Philip Cline, and LeRoy Quance.(1979). *Prospects for Productivity Growth in U.S. Agriculture.* USDA/ESCS Agr. Econ. Rpt. 4-35,

Orskov, E.R. (1987). "Treated Straw for Ruminants." *Research and Development in Agriculture.* 4,2, pp. 65-69.

Simpson, James R. (1993). "Urbanization, Agro-Ecological Zones and Food Production Sustainability," *Outlook on Agriculture*, vol. 22, no 4, pp233-239.

Simpson, James R., Xu Cheng, and Akira Miyazaki. (1994). *China's Livestock and Related Agriculture: Projections to 2025.* Wallingford, UK: CAB International.

Simpson, James R. and Ronald Ward,1995. Analysis Projects Future Livestock Demand in China. *Feedstuffs*, November 13, pp14,16,31.

Simpson, James R. and Ou Li. (1996). "Feasibility Analysis for Development of Northern China's Beef Industry and Grazing Lands" J. *Range Management* 49:560-564, November.

Simpson, James R., William E. Kunkle, and William F. Brown. (1993). *AMMONIA: A Computer Program for Economic Analysis of Forage Ammoniation.* Florida Cooperative Extension Service, University of Florida, SW-086, October

Simpson, James R., et al (1996). *Japan's Beef Industry: Economics and Technology for the Year 2000.* Wallingford, UK, CAB International.

U.S. Congress, Office of Technology Assessment. (1992). *A New Technological Era for American Agriculture.* OTA-F474 (Washington, D.C.: U.S. Government Printing Office, August

車筒轉驢

CHAPTER 4.28

The Potential for Production of Meat from Sheep and Goats in China

Clair E. Terrill*
ARS, U.S.D.A.

The human population of China and the world is continuing to increase at a fairly rapid rate (Figure 4.28.1). This means that food production must also increase rapidly.

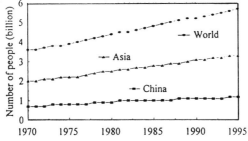

Figure 4.28.1 Trend in population from 1970 - 1996, China, Asia and the World.

However, yields of most food crops have leveled off and some are even falling in per capita production. The loss of family farms in countries like the United States indicates that land in use for food production is also declining. Much of the agricultural land being lost can still be used for grass or forage production. Thus a shift from production of meat from grain to production of meat from forage is beginning.

Sheep and goats are ideal for production of meat from forage because marketable high quality meat can be produced on the poorest of land, on any kind of land, can be started with a very low investment, can be increased in efficiency of production almost every year, for a long time, can be flexible in date of marketing and in variety of products such as meat, milk, cheese, pelts, other byproducts, wool and mohair. The object of this paper is to show the potential for sheep and goats to increase the adequacy of food and fiber production into the next century and possibly until 2030 or beyond. However, the human population is increasing at an average rate of over one percent per year in China and the world.

* Acknowledgment is made to S. Keith Ercanback for calculation of regressions

PRODUCTION OF MEAT FROM SHEEP AND GOAT

The prediction of future trends in food production per capita is extremely difficult and is considered impossible by some but it is worthwhile to consider what can be done to maximize increased food production from existing resources. The best estimate of future trends can be obtained from trends in the recent past. Therefore, trends in production of meat from sheep and goats has been investigated for China, Asia, and the world from data reported in FAO Production yearbooks from the early 1970's through the latest data available for 1995.

Trends in human population increases appear to be slightly slower for China than for Asia but practically identical to the world trend (Figure 4.28.1). Increased production of meat per capita from sheep and goats in China can be compared to production in Asia and the World to determine the progress being made in China.

Trends in numbers of sheep per capita in China increased in the late 1980's and then slowly decreased until the early 1990's while the world showed a definite decline and Asia trended lower in the 1980's and 1990's (Figure 4.28.2).

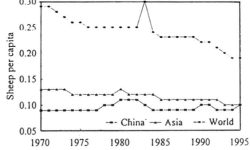

Figure 4.28.2 Trends in sheep per capita from 1970-1995, China, Asia, World

The trend in sheep production to less emphasis on wool production and more emphasis on meat production tended to result in less production of both commodities. China had an increase in goat numbers per capita but Asia and the World showed very little change (Figure 4.28.3). Increase in number of animals tends to increase feed costs but also to increase the human food supply.

Increase in meat produced per animal indicates an increase in efficiency of production which results in increased net returns to both producers and processors. Such increases occurred for sheep for both Asia and the World but increase in China from 1983 to 1993-95 was greater than for Asia and the World (Figure 4.28.4) and was upward for chevron (goat meat) for all three areas (Figure 4.28.5)

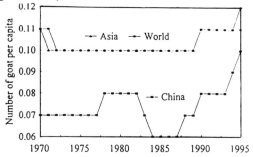

Figure 4.28.3 Number of goats per capita from 1970-1995, China, Asia, World.

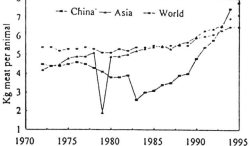

Figure 4.28.4 Meat produced per sheep from 1970-1995, China, Asia and World.

Production of meat by sheep and goats is the most important statistic studied and generally showed an upward trend for both China and Asia (Figure 4.28.6 and 4.28.7). The trend for the World for lamb and mutton was downward but at a much higher level than for China and Asia. The trend for Chevron was upward for all three areas with Asia and China equal in 1995 and the world lowest (Figure 4.28.7). Total meat production per capita showed a greater upward trend for China than for the World and Asia. It appears that the upward trend for China may be responsible for

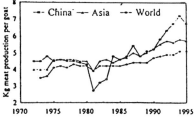

Figure 4.28.5 Meat produced per goat 1970-1995, China, Asia, World.

PRODUCTION OF MEAT FROM SHEEP AND GOAT

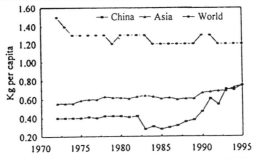

Figure 4.28.6 Meat (sheep) produced per capita 1970-1995, China, Asia, World.

the upward trend for Asia and the world in meat production (Figure 4.28.8). An upward turn in meat production from sheep and goats in China is shown beginning in about 1982 with some acceleration after 1986 and 1989 respectively (Figure 4.28.8). Thus the most recent decade indicates a rapid increase in meat production from sheep and goats in China. Prediction far into the future to the year 2030 cannot be expected to be very dependable even though the expected gain is quite possible.

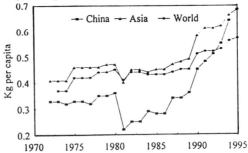

Figure 4.28.7 Production f goat meat per capita from 1972-1995, China, Asia, World.

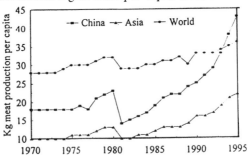

Figure 4.28.8 Trend in total meat production per capita from 1970-1995

The correlation between years and increase in meat production were 0.97 for lamb and mutton and 0.99 for chevron. Regression coefficients were 0.055 + .005 kg per year for lamb and mutton and 0.044 + .003 for chevron. The value for R^2 was 0.94 for lamb and mutton and 0.97 for chevron. The Y intercept values were -4.48 for lamb and mutton and -3.50 for chevron. Application of the regression equations indicate a predicted value of 2.70 kg per capita of lamb and mutton and a value of 2.2 kg of chevron per capita or 4.9 kg of lamb mutton and chevron per capita in 2030 (Figure 4.28.9). Production in China for 1995 is also on an upward trend. These values seem quite conservative but production so far into the future is certain to be questionable. Furthermore we know that selection gains reach plateaus or ceilings sooner or later. These ceilings have generally been reached already for traits depending on weight yields. Practical ceilings for prolificacy or litter size at weaning have yet to be revealed. Artificial rearing makes it much less limiting than was previously true. There are two very important factors which have become effective in recent years which make estimates of meat production from sheep and goats quite optimistic for the future. First is prolific breeds of sheep which were developed in prehistoric times, some of which wean as much or more than three times the level of domestic sheep production (Fahmy, 1996). The second is rapid selection technology for sheep and goats developed by Ercanbrack and Knight in the 1970-80's but which has now been submitted to the Journal of Animal Science for publication. This involves selection for letter weight of the dam and turning ram generations every year. In addition they have developed a more effective method of selecting for feed efficiency and rate of gain (Ercanbrack and Knight 1988). These methods of selection can be applied directly to prolific breeds without the necessity to cross with domestic breeds.

Progress already made in China toward increasing meat production from sheep and goats plus the availability of prolific sheep and the development of new more effective technology for selection for efficiency of meat production in sheep and goats give optimism toward increasing production of meat from sheep and goats for the future in China and the world.

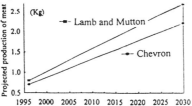

Figure 4.28.9 Projected trends in production lamb, mutton and chevron through 2030.

References

Ercanbrack, S.K. and A. D. Knight. 1988. Selection for efficiency of post weaning gain in lambs. Research Progress Report, U.S. Sheep Experiment Station, Agricultural Research Service, USDA. In cooperation with the Idaho Agricultural Experiment Station, Dubois, Idaho, USA.

Ercanbrack,S.K. andA.D. Knight. 1998. Responses to selection for lamb production. Submitted to Journal of Animal Science.

Fahmy, M.H. 1996. Prolific Sheep. CAB InternationalWallingbord, Oxon

葱　山

CHAPTER 4.29

Hog Cholera (classical swine fever)
Diagnosis and Eradication

I. C. Pan

Tansui Research Institute for Animal Diseases Taiwan
(retired visiting fellow)

China is one of the main driving forces in global agriculture today and will continue to have a tremendous influence on agricultural commodities over the next decade. China has a gigantic population with over 1.2 billion people. The majority are now rising above the subsistence level and consume more sophisticated foods.

In 1984, animal production accounted for only 14.2 % of the total agriculture output value, although much has been achieved over the past 30 years or more in animal husbandry. This is far behind the corresponding figure for the developed countries and falls short of what the National Construction calls for. Therefore, it is an important task to boost animal husbandry so that it will account for a greater percentage of the total agricultural output value and meet the needs of the developing national economy.

According to Forbes magazine, China's population grows by 14 million a year, despite population control. Young people are leaving the countryside, and flocking to the opportunities of the cities. China's urban population, currently over 300 million, should double by 2010.

To fully understand the scale of change that is taking place in China, one can examine the pork market, which supplies China's favorite meat. Over the last five years, Chinese consumption of pork has grown by 11 million metric tons. There are currently 415 million pigs in China

Total meat consumption in China is growing by 10% a year, a staggering four million tons annually, and accordingly, feed consumption is bulging by 15%. Demand for poultry has more than doubled in five years. Beef consumption has nearly tripled during the same period, and the consumption of eggs is skyrocketing; even so, per capita consumption of meat is still only 12% of that of the U.S. or Hong Kong.

Pigs and their farm role have changed dramatically during the past fifty years in the Western world. During the first half of the 20th century, the pigs found on most farms were primarily a waste disposal system that contributed the occasional meal and even less frequent revenue. Scavenger pigs still abound as part of a mixed, subsistence farming system in almost all of the less developed countries. Now, however, these often exist in close association with large, specialized units producing for both local and export markets. As in the rest of China's livestock industry, hog raising has steadily been commercialized. Slow-growing, pot-bellied Chinese pigs are being replaced by lean-meat, fast growing Western breeds raised on scientifically formulated compound feed. The feed conversion ratio per kilogram of pork in China is 8.8 to 11 kilograms of grain-based hog feed. Traditionally, the Chinese fed rice, hulls, potato vines, corn husks and water lilies as well as anything else with nutritional value to their pigs, chicken and carp. Now they feed them corn, soybean and barley.

PORK PRODUCTION SYSTEM

Pigs breed at all times of the year, so pork production is a continuous cycle. Traditionally, sows farrowed periodically. Piglets were retained after weaning at around 8 weeks of age, grew slowly, and were consumed eventually or sold by their original owners. Now pigs may constitute the only type of livestock present on a farm, with the system often further specialized by operators concentrating on specific stages of the production sequence.

Some farms still keep and mate sows, farrow the litters and raise piglets to market weight. These are called "farrow to finish" operations. Other farmers engage in production of weanlings, maintaining breeding herds but selling all their piglets around 8 weeks of age to other farmers. Those who keep no breeding animals but purchase all of their pigs from weanling producers are called "finishing or feeder operators." The pig finishers or feeders then take care of these purchased animals until they reach market weight. Regardless of the type of operation, all hog farmers must arrange for appropriate facilities and care for their stock. In addition to the moral obligation to provide humane care, producers recognize that their livelihoods depend on creating an environment that keeps animals healthy and comfortable so they can achieve their full genetic potential.

Careful and conscientious management is essential for ecologically and environmentally friendly and profitable production; so, it benefits the pigs, the producers and all of society. The over-ambitious, industrial way of pig raising with an over-crowded environment must be avoided, since it is not only inhumane to pigs but also heavily contaminates the environment and brings about many subclinical diseases which do not become apparent under a more relaxed rearing regime. Mixing antibiotics into the feed for pig raising has been falsely believed to accelerate the growth of pigs and also to prevent bacterial diseases. As a result, the practice has produced strains of bacteria resistant to antibiotics, which is the most challenging problem to public health today in certain countries and must be stopped immediately at any cost.

Prophylaxis:
Infectious diseases have always been associated with animal raising. Among many pig diseases, the most feared by farmers has been hog cholera or classical swine fever, because of its rapid spread and high mortality, nearly 100%. Although China has a very effective lapinized virus vaccine against hog cholera (2, 3, 10), the malady still persists in China after implementation of the vaccination program almost 40 years ago (2). Reasons are multiple; however, changes in the practice of pig raising and the modes of vaccination could be considered the most important factors in the continuous existence of hog cholera. Furthermore, the recent outbreaks of hemorrhagic disease of rabbits affect tremendously the production of effective vaccine difficult. Consequently, a vaccine based on cell culture technology must be developed for economic and sanitary reasons. Therefore, porcine cell lines (PK15, MVPK, ST), or the subculturable-rabbit-splenic cells persistently infected with the lapinized vaccine virus, should be used (1). The infected cells should be blindly passaged without any attempts at virus clonings; since there is mounting evidence that both vaccine and virulent virus, respectively, consist of more than a single virus population. Therefore, attempts at cloning the virus for the vaccine production should be avoided, unless a definite plan exists for the systematic research on the antigenicity and virulence of the cloned virus.

Due to ever increasing demands for pork and related products in recent years, the method of pig-raising has been industrialized so that piglets are weaned and vaccinated at 3 weeks of age. The second vaccination is given at 6 weeks of age, and they are sold for fattening at about 8 weeks of age. Vaccination in this manner certainly provides the majority of pigs solid immunity to hog cholera in a clean farm of a "farrow-to-finish" operation where hog cholera has never occurred. However, this vaccination scheme does not ensure that all vaccinated pigs are protected from hog cholera infection in a farm, regardless of the type of operation, where hog cholera has been diagnosed in the past and which is still experiencing sporadic occurrences of hog cholera at present (contaminated farm). A small portion of vaccinated pigs do not "take" and are still susceptible to hog cholera, because a great individual variation exists among the littermates for the amount of intake of the colostrum and the rate of catabolism of the maternal antibody. These are the pigs which grew faster, and susceptible to hog cholera infection, for they were born earlier and, therefore, consumed more colostrum than other littermates. Under the circumstances, more often than not, the quality or efficacy of the vaccine itself is questioned by the farmer and by the responsible veterinarian, even though the vaccine itself may have been quite potent. Thus, hog cholera continues to exist in the contaminated farm as long as susceptible pigs are available for infection, and because of this, a misdiagnosis of some other disease has been quite a common occurrence, due to the high confidence that a veterinarian has in this vaccine.

Thus, the effectiveness of the current program, i.e., vaccinating pigs at 3 and 6 weeks of age, has to be re-evaluated. In this regard, I would personally propose vaccinating pigs for the third time at 8 weeks of age, in addition to the current vaccination practice, right before selling for fattening. Of course the validity of this proposal has to be tested and carefully evaluated on farms contaminated with hog cholera. Alternatively, a single vaccination could be given to each pig at 10 weeks in the clean farms, where the "farrow-to-finish" operation is practiced, because the maternal antibody should be completely catabolized by this time so that all vaccinated pigs should be solidly immunized against hog cholera. If this new scheme works, it will be the first case of achieving elimination of hog cholera by vaccination only without exercising expensive "stamping-out" methods, currently in practice elsewhere in the developed countries. I also anticipate that the proposed new scheme, if it were adopted and exercised, would be able to achieve eradication of hog cholera from China in the very near future.

Diagnosis.

Clinical Diagnosis of Hog Cholera: Since hog cholera is a highly contagious and fatal viral disease of swine, an outbreak of acute, feverish disease with a high mortality (almost 100%) among unvaccinated pig populations is highly suspicious of hog cholera. Affected pigs are huddling together with shivering due to fever (about 40.5 C). They do not respond to treatment with antibiotics and usually die about 4 to 10 days after the onset of fever with a gradual loss of appetite, though still drink water. Coinciding with the onset of fever, leukocyte counts drop abruptly, especially lymphocytes which reach 30 to 50% of normal counts, and the appearance of young forms of neutrophils (metamyelocyte and myelocyte) occur. At the necropsy of carcasses or animals sacrificed at the moribund stage, if anemic and/or hemorrhagic infarcts in the spleen of normal size and texture, peripheral hemorrhages (hemadsorption in lymph sinuses) of lymph nodes, button ulcers in the colon, accompanied by discrete, pin point-sized petechiae scattering in kidneys are found, then a tentative diagnosis of hog cholera can be made with almost certainty. Secondary bacterial infection is a common occurrence, in the form of pneumonia and septicemia, because a state of immunodifficiency occurrs in pigs infected with hog cholera (5). Therefore, the presence of lesions in the internal organs mentioned above is very helpful in the clinical diagnosis of hog cholera.

Definitive Laboratory Diagnosis: In our own experience, the leukocytes, including monocytes, lymphocytes, and neutrophils in a cytospin preparation of buffy coat-cell suspension, prepared from peripheral blood, are seen already infected with the virus by the time of the onset of fever (uncertain for eosinophils and basophils), and the numbers of infected leukocytes increase with the progression of the clinical course, as evidenced by the direct immunofluorescent test, specific for hog cholera. Due to infection of T- (5) and B- lymphocytes (5, 11) by hog cholera virus, a state of immunodeficiency develops, meaning that pigs infected with a virulent hog cholera virus are likely to be incapable of inciting immunune response against any other pathogens, which are unlikely to cause diseases by themselves without the complication of hog cholera. In the neutrophilic series of leukocytes, the young neutrophils (metamyelocytes and myelocytes) are infected first and lobulated neutrophils are seen infected at the terminal stage, which indicates that the infection has taken place within the bone marrow (5). The immunofluorescent test on frozen sections of any organs, except skeletal and smooth muscle (with an exception of arteriolar wall muscle), can be used for an accurate diagnosis. I personally favor examining brain, tongue, skin, and lymphoid organs, i.e., tonsil, lymph node, and spleen. Examination on specimens taken from the ear biopsy can be conducted on a herd basis (6). The hog cholera virus can easily be isolated in PK15, MVPK, or ST cell cultures and identified by immunofluorescent test without any difficulty, because the infected pigs do not produce enough amount of antibody, and the virus titers in organs and blood are quite high . The infected cultures can be stained either by immunofluorescent or antibody-biotin-complex (ABC) methods. The diagnosis of hog cholera by PCR technique requires approximately the same length of time with cell culture methods, and it requires specific reagents, expensive laboratory equipment, and trained personnel; this technique can be used as a research tool in the laboratory, at present (4).

Diagnosis as Hog Cholera Based on the Presence of Encephalitis. Histopathologically, the presence of non-suppurative encephalitis has been widely used as a positive diagnosis of hog cholera; however, there are several other diseases which acquire non-suppurative encephalitis as well, i.e., swine vesicular disease and Aujesky's disease, to name two.
From our own experience, when 35 experimental pigs which were vaccinated either neonatally with a single dose of the vaccine or with two doses at 3 and 6 weeks of age, respectively, and sacrificed at 10 weeks of age, about 60% of them had slight to moderate degrees of vascular cuffing and gliosis observed in their brains though they were clinically healthy at the time of sacrifice (5). Seventy-four percent of nineteen brains of healthy hogs, collected at a slaughter house, had a non-suppurative encephalitis, in which 21% of them were accompanied by a slight degree of non-suppurative meningitis, presumably remnant lesions of the past infection(s) caused by diseases other than hog cholera. Although the number of swine examined were relatively small, they were all healthy at the time of examination. Furthermore, we know from our own experience that the lymphocytes are infected with the virus (immunodeficiency) so that the lymphokines presumably produced in pigs infected with hog cholera virus are either minimum or nil by the

time of the height of the disease; therefore, the encephalitic lesions seen in the paraffin sections are mild, whereas practically all neurons are infected with the virus, as we see in the immunofluorescent test on the frozen sections and/or in the paraffin sections immuno-stained with the ABC technique. Therefore, diagnosis of hog cholera based on the presence of non-suppurative encephalitis alone is unacceptable, in my opinion, without conducting other laboratory tests definitive for hog cholera.

Differential Diagnosis from Other Diseases: Another fatal swine disease that needs to be differentiated from hog cholera is African swine fever (ASF). The necropsy findings in the classical form of ASF, however, are unique so that necropsy findings alone can be used to differentiate it from hog cholera. In ASF, the enlarged, hard and bloody spleen (acute, infectious splenomegally) can be broken easily by the hands. These findings do not happen in hog cholera; instead the spleen is quite normal in size, unless there are other bacterial complications. The swollen and extensive hemorrhages in the regional lymph nodes are quite characteristic of acute ASF. In contrast, hog cholera has slightly swollen lymph nodes with peripheral hemorrhages (hemoadsorption in lymph sinuses) with watery cut surfaces. African swine fever virus grow in the buffy coat-cell culture (7), as hog cholera and swine influenza virus do, but only ASF virus-infected cells (monocytes) show hemadsorption in which infected cells adsorb onto their surface swine erythrocytes which have contaminated the buffy coat-cell culture. Incidentally, swine monocytes infected with swine influenza virus are able to show hemadsorption of chicken erythrocytes, when swine erythrocytes are replaced by chicken erythrocytes (9). Finally, the buffy coat-cell cultures inoculated are fixed and confirmed by immunofluorescent test or ABC test for further confirmation of diagnosis; for, non-hemadsorbing virus does not show hemadsorption in the buffy coat-cell cultures. Although the differentiation of acute ASF from hog cholera by clinical diagnosis is not a difficult task, subacute and chronic forms of ASF require labolatory diagnosis by detection of ASF-specific antibody (8), for there are no definitive lesions that can be differentiated from hog cholera, and there are difficulties in demonstrating ASF virus antigens, perhaps due to the blockage of antigenic determinants by the rapidly rising antibody levels in the infected animals.

Hog cholera also requires differential diagnosis from another acute, contagious, fatal, bacterial disease, swine erysipelas. The onset of swine erysipelas is characterized by a sudden onset of high fever (41-42 C) and abrupt, complete loss of appetite, but still continued water drinking.. Affected pigs generally show high counts of leukocyte and increased monocytes in the blood. The positive response to an antibiotic treatment is dramatic, and generally, fever subsides to a normal level in two to three hours of parenteral administration of a dose of an antibiotic against Gram-positive bacteria, i.e., penicillin and Baytril (Byer) , to name two. If the treatment by antibiotics does not continue for two or more days, however, a subacute form manifested by scattered "diamond skin" lesions appearing everywhere on the skin. Continuous treatment with antibiotics clears the disease within a week or so.

RECOMMENDATIONS AND CONCLUSIONS

1.A scheme for the eradication of hog cholera solely by vaccination is proposed, thus, avoiding expensive "stamping-out" methods currently adopted by developed countries.
2.A thorough field trial should be conducted carefully before implementing the most effective and economical scheme for vaccination.
3.The live vaccine virus grown in cell cultures, persistently infected with the lapinized vaccine virus, should be developed and recommended as a substitute for the original vaccine made entirely of the extract of infected rabbit tissues. The cloning of the virus in cell cultures is not recommended.
4.Methods of hog cholera virus isolation and identification in cell cultures by direct fluorescent test or by ABC stain have been discussed.
5.Careful and conscientious management of a swine farm is essential for keeping balanced ecology and healthy environment. Over-crowded environment must be avoided.
6.Last, but not least, the practice of mixing antibiotics in the feeds should be banned by law to avoid creating antibiotic-resistant strains of bacteria, which is a grave threat to public health.

REFERENCES

1. Fuh, T.F. Personal Communication (1991).
2. Lin, TTC, and R.C.T. Lee: An Overall Report on the Development of High Safe and Potent Lapinized Hog Cholera Virus Strain for Hog Cholera Control in Taiwan. National Science Council Special Publication No.5, (1981).

3. Lin, T.T.C. Pathogenicity of the Hog Cholera Live Vaccine Virus by reverse passages through SPF pigs. Taiwan J. Vet.Med. Anim. Husb. 23: 43, (1973).

4. Liu, S.T., S.N. Li, D.C. Wang, S.F. Chang, S.C. Chiang, W.C. Ho, Y.S. Chang, and S.S. Lai: Rapid detection of hog cholera virus in tissue by the polymerase chain reaction, J. Virol. Methods. 35: 227-236 (1991).

5. Pan, I.C., T.S. Huang, C.H. Pan, S.Y. Chen, Y.L. Lin, an B.Y. Huang: Unpublished data (1993).

6. Pan, I.C., T.S. Huang, C.H. Pan, S.Y. Chen, S.H. Lee, Y.L. Lin, B.Y. Huang, C.C. Lin, N.J. Lin, J.T. Lin, Y.H. Yang, S.Y. Chiu, J.S. Chang, D.K. Hue, H.C. Lee, and C.N. Chang: Skin, tongue, and brain as favorable organs for hog cholera diagnosis by immunofluorescence. Arch. Virol. 131: 475-481 (1993).

7. Pan, I.C. Spontaneously susceptible cells and cell culture methodologies for African swine fever. Y. Becker (ed.) African Swine Fever. Martinus Nijhoff Publishing, Boston (1987).

8. Pan. I.C., Huang, T.S. and Hess, W.R. New method of antibody detection by indirect immunoperoxidase plaque staining for serodiagnosis of African swine fever. J. Clin. Microbiol. 16 (4), 650-655.

9. Pan, I.C. and Butterfield, W. Unpublished data (1980).

10. The Research Group of Hog Cholera Vaccine of the Control Institute of Veterinary Bioproducts Ministry of Agriculture. Studies on the Avirulent Lapinized Hog Cholera Virus: The Susceptibility of Rabbits to the Lapinized Hog Cholera Virus. Monographs of Research Report of the Institute of Veterinary Drug Control, Ministry of Agriculture, China. Vol. 4, 49-63 (1970).

11. Susa, M., Konig, M., Saalmuller, A., Reddehased, M. J., and Tiel, H-J. Pathogenesis of classical swine fever: B-lymphocytes deficiency caused by hog cholera virus. J. Virol. 66: 1171-1175 (1993)

CHAPTER 4.30

Development of Food Industry of China in 1997-2030

Paul C. Ma and Hsien-Jer, Chen
Hangzhou Want Food Ltd.

Impact of Grains Production and Food Demand in China

The population of China will close to 1.3 billion by the end of the Twentieth Century. It occupies 22% of the total population in the world. Furthermore, the Chinese population is still increasing yearly by 14 million, and will reach to 1.6 billion by the Year 2030. China now has 110 million hectars of plowland, which shares only 7% of the total on earth. The total grains production in China has been up to 460 million metric tons in 1995 with an average consumption of 380 kilograms per capita, which is almost tripled when compared to that in 1949. This figure has approached to the medium level of food consumption in the world. Basically, China herself has been able to solve the problem of food self-sufficiency at the present time. Although the grains production may flirther increase to 500 million tons at the year of 2000. But the food consumption may be accordingly up to 400 kilograms per capita as well because of the economic growth. That is to say, China will need 640 million tons of grains at the year of 2030 when her population reaches 1.6 billion. Consequently, as long as China keeps a steady economic growth continuously, the food supply is still a challenge for China in the next century. In order to find out a better solution for food self-sufficiency,

Chinese government has raised a guide and policy as follows briefly:

(1)Exploitation of the Potentiality of Increasing Grains Production.
(2)Protecting land resources and increasing crop productivity on the unit land basis. (3)The total area of plow-land must remain at least 110 million hectares.

Exploiting the non-grain foods production potential.

Exploiting new protein foods by utilizing wasteland or undeveloped plowland on seaside, plains and mountains for cultivating aquaculture and poultry.

Saving food consumption

It is estimated, that China may have lost or wasted at least 10% or equivalent to 45 million tons of the total crops during the stage of farming, harvesting, transportation, storage, processing and consumption. Therefore, a reasonable and efficient control of the food supply is necessary.

Changing Chinese food-intake custom and content.

In the future, the food content for Chinese people will be of "medium calorie, high protein and low fat". On the basis of traditional food intaking custom, the direct-consumed grain foods, like rice and wheat, will decrease, while the indirect protein foods, like seafood, pork, beef, will increase.

From understanding of the background of Chinese agriculture and foods-supply history, apparently Chinese government must put more effective efforts on the control of her population and land resources as well as improvement of agricultural production and technology. Since the food industry is an important extension of agricultural production, it is necessary and helpful for either a value-addition on harvested crops and grains, or a balance tool between food supply and demand. From the viewpoint of industry and technology, two essential aims may be achieved through food technology development:

Firstly, saving the harvested grains and crops from being wasted or damaged. Secondly, a rapid converting the foods forms and platability for human consumption.

This paper proposed some feasible topics concerning the development of the food industry for China in 1998-2030.

Some Suggested Topics for Future Development of Food Technology and Industry in China.

Application of advanced technology for paddy storage and rice grains polishment.

The rice production in China is about 160 million metric tons *(1995)*. It shares one-third of the total grains harvested in China, and is also the main foods for two-thirds of Chinese people. At present, owing to an inadequate paddy drying and storage treatment, the rice grains are usually over-dried and deteriorated. In addition, since most of rice mills in countryside are small in scale and old in facilities, the producing rate of broken rice is generally much higher during polishment, which consequently causes not only an unreasonably lower recovery rate of the white whole rice, but also a significant loss of the rice platabiluty. Providing that the recovery rate of the polished rice is reduced by one percent, it implied a total loss of 1.5 million tons of paddy for whole China. This loss is not so easy to be compensated by the farming technology. Therefore, the development of advanced technology for paddy drying, storage and polishment is of challenge attention in China and must be regarded as important as the agricultural technology development. Some suggested items are as follows:

• Storage technology and facilities of large grains solos.
• Drying and refreshing technology of paddy at higher moisture levels (15%).
• Pretreatment and tempering technology of paddy before polishment.
• Rice polishing and refining technology, including color-sorting, cut-refining and de-odorization.
• Small-package and controlled atmosphere storage of whitened rice.
• Development of the feasibility to store paddy at lake or seas bases.

Developing technology for processing aged or low-grade rice.

The rice grains obtained after a proper drying of paddy may be stored over one or two years without apparent loss of its eating quality. But several problems encountered, like shortage of paddy warehouses, rodent damages, aged-odoring and lower polishing rate, etc. Nevertheless, usually after rice grains being polished, the whitened rice therefrom may not maintain good palatability for more than two months of storage. From a viewpoint of food technology, the rice palatability can be stabilized and well-stored for a much longer time through a series of treatments including mass-cooking, seasoning-addition, fluid-bed drying, compression and puffing, etc. The application of the technology mentioned above may not only save the paddy or grains storage-loss, extend the shelf-life of rice in an alternative precooked and dried form, but increase the added-value of processed rice products.

Development of rice bran relating industry

The brown rice obtained from dehulling of paddy contains about 10% of bran and the latter comprises 15% of high quality oil, 25% of protein and plentiful of carbohydrates as well as vitamins. So the rice bran is a very good resource of edible oil and protein for human consumption. The total rice production in China is 160 million metric tons per year, from which about 14 million tons of brans is obtained. Unfortunately, most of them are merely used as poultry feeds. If all the bran in China can be further processed to extract and/or isolate their components, about 2.2 million tons of rice oil as well as 3.5 million tons of rice protein may be produced. This food source must not be ignored and wasted. During 1970-1980, rice bran oil technology and its relating industry had used to be well developed and installed in Japan, South Korea, Taiwan and America. The only necessary condition for developing rice bran industry is that enough amount of fresh bran (ca 200 tons a day) within about ten hours obtained from rice polishment must be efficiently collected and transported to the oil extraction plant; otherwise, the oil comprised in bran will be rapidly subjected to oxidation, followed by a serious rancidity and deterioration.

Development of the whole corn processing industry

The total corn production in the world is 450 million tons (1995), in which United States, of 200 million tons, is the largest corn producing country, while China, of 85 million tons, is the second one. During the period from 1985 to 1995, China used to export ten million tons of corn every year to Japan and Korea. In many areas of China, corn is not directly consumed as human foods. Instead it is majorly used as animal foods, and minorly goes to the production of corn oil and starch products.

In 1995, only 1.5 million tons of corn grain was used to produce corn starch and fructose/glucose syrup. Therefore the utilization of corn is rather a developing-worthwhile industry in China. For instance, in 1995, the total starch production in China was only 1.26 million tons, among them 81% was from corn, 14% from tapioca, and the rest 5% from potato, wheat and sweet potato.

Nevertheless, corn processing is a capital and technology-intensive industry, which needs a closed incorporation of some other relating industries like foods processing, feeds, fermentation, paper, textile or alcohol. In general, a plant of yearly production capacity at a level above 50 thousand tons of corn is basically reasonable and economical. Particularly in China, corn may be heavily considered to be further processed and derived as a direct human foods.

Promoting development of starch sugars technology and industry

Starch sugar is a general name of the sweeteners derived from a partial or complete hydrolysis of plant starches. Some products like dextrin, maltodextrin, crystallized glucose, fructose or maltose, etc., are solid or powdered form, while glucose syrup, high fructose syrup, or high maltose syrup are the examples in liquid forms. Whatever, starch sugars in many food processing areas can be a valuable alternate of sucrose from sugar cane or beet roots as a functional and industrial sweetener. The starch sugars processing, similar to that of native starches, is a typical localized industry. It is closely relating to the development and progress of local food industry as well as the supply of raw starch materials. In most of the developed countries, the consumption of starch sugars is hardly less than sucrose. Before 1994, the demand of sucrose in China was only 4.5 million tons and reserved a self-sufficiency. But from the Year of 1995 on, along with the market opening and resolution of China, the consumption of sucrose jumped up to 7.5 million tons. As a result, China needs to import about 1.5 to 2.0 million tons of sucrose yearly. In another respects, in 1995, the overall starch sugars production in China, including malto-dextrins, refined glucose solids, maltose syrup and high fructose syrup, was surprisingly only as low as 350 thousand tons. The averaged consumption of sucrose and starch sugars for Chinese were only 6.0 and 0.3 kilograms per capita, respectively. It was far less than the average level of 40 kilograms in the developed countries. For the sake of meeting the urgent requirement of functional sweeteners for food industry uses, a heavy promotion and development of the advanced starch sugars technology and industry must be taken into consideration. In fact, the majority of the sugar consumption increment is going to industrial applications in soft drinks, fruit products, confectionery and baker, etc. In terms of the sweetness and processing functionalities, there are three kinds of starch sugars may be considered to develop to replace the conventional sucrose.

(1)   High Fructose Syrup:

Higher sweetness, convenient for transportation and application in soft drinks or fruit juice processing.

(2)   High Maltose Syrup:

Lesser sweetness, ore physico-chemical stability and functionality, suitable to be used in candy, jam, bean fillings, fruit preserves and bakery.

(3)   High Maltose Solids:

Same as above, but more convenient for transportation, storage and application, even for home use.

Especially, the high maltose syrup or its solid sugar is rather a newly developed starch sugar. It is characterized by its disaccharide content as high as 80% or more when comparing to the conventional maltose syrup which has only 35-40% of disaccharide. Chemically, both maltose and sucrose are the disaccharide and have almost the same physical properties. But maltose itself posses more chemical stability against heat and acids. Furthermore, its sweetness is only 30% of the sucrose; therefore, high maltose syrup or solids can replace sucrose in quantities for most of the food processing. Apparently, China will suffer from a big shortage of natural sweeteners for food industry in the near future. So the development of the specific functional starch sugars is very worthwhile and of economic meanings.

Developing the advanced technology for producing functional soy protein ingredients and texturized soy protein products.

Soybean undoubtedly will become the most important food protein source for humankind. The world soybean production is 110 million tons in

1990. China used to be the world's largest soybean producing country before 1950. But since then, the United States, Brazil and Argentina exceeding over China became the four major soybean growing countries. Their production quantities are United States 52.4, Brazil 20.5, Argentina 10.5, and China 10.8 million tons, respectively. It had been almost increased by 8.9 times from the year of 1935 till 1990, while China only increased doubly during the last 50 years. It is estimated that the soybean production in China will be up to 20 million tons at the Year of 2000.

Soybean contains as high as 40% of superior protein, which is the highest in contrast to any other plant. From the viewpoint of nutrition, soy protein comprises enough essential amino acids except methionine. It can be easily compensated with corn and wheat proteins in combination to become a "complete protein" as good as animal proteins. In another respect, from the viewpoint of land resource, soybean will also become the main stream of food proteins for human consumption in the future. If one acre of land goes to the cultivation of soybean, wheat, corn and rice, the protein produced therefrom can provide one person surviving for 5560, 2217, 1932, and 1930 days, respectively. But in case of cultivating poultry, pig or cow, the surviving days

are only 462, 322, and 192 days, respectively. It is elucidated that the conventional way of utilizing soybeans as animal foods for converting plant proteins into animal proteins is rather low in efficiency. In China, exceeding 80% of the harvested soybean is subjected to oil extraction and its defatted by-products of high protein is majorly used as feeds or fertilizer. This is reasonless and wasteful for the balance of foods supply and demand in China. Supposing that the soy protein thereof can be technically processed and texturized into a form of "meat analogue", it will do much advantages to balance the protein foods demand of China.

Traditionally, in China, many protein foods are prepared from soybeans, and almost daily consumed, like soy milk, tofu, fried tofu, vegetarian meats, etc. But they still cannot replace animal meats in quantities, owing to the limits of production scale, shelf-life, transportation, and texture platability. Since the 1960's, many countries have put much efforts of research application on extraction, functionalities improvement, and texturization of soy protein. Now-a-days, a newly food industry based on soy protein has been well established, in which the functional soy protein isolates and the texturized soy protein or imitation meat products are the two most successful cases. The former is prepared from de-fatted soy flakes and possesses some specialty functions for food processing purposes. It has been widely used in ham, sausage, coffee cream, and bakeries. The texturized soy meat is derived from isolated soy protein through a series process of formulation, cooking extrusion and/or gum reconstitution. Simply saying, soybean can be rapidly modified and converted to a simulated meat product via food technology. In other words, a pork- or beef-analogue can be obtained from soybean protein directly, obviating the breeding pigs or cows. It significantly enhances the protein converting efficiency and land utilization.

CONCLUSIONS

The most important aim of food industry development of China in the next century is to increase the agricultural production and to efficiently utilize the limited food resources. Apparently in the near future, the content of food-intake for Chinese people still may not be the same as that for Europeans or Americans. It should be reevaluated and come back to the traditional Chinese living and food intake. On strategy, the consumption of food calorie and vegetable proteins must be increased. Among them the processed foods from rice and soybean may be the two major grains foods for direct consumption, while corn may be indirectly consumed as foods through a secondary food processing. In order to enhance the utilization and quality of the foodstuffs from crops and grains, an introduction of newly developed grains food technology and the installation of larger-scaled food processing organization must be emphasized.

CHAPTER 4.31

Postharvest Management and Food Processing

Chien Yi Wang and HowardZhang
Beltsville Agricultural Research Center,
ARS, U.S.D.A.

POSTHARVEST MANAGEMENT

Fruits and vegetables are valuable sources of nutrients which are beneficial to human health. In addition to vitamins, carotenoids, and dietary fiber, fruits and vegetables also contain a number of antioxidants, free radical scavengers, and phytochemicals which can trigger increased production of detoxication enzymes to neutralize cancer-causing agents. Virtually all vitamin C in the human diet are obtained from fruits and vegetables. About 50% of vitamin A that our body needs are also from fruits and vegetables. Green and yellow fruits and vegetables also contribute approximately 35% of vitamin B6, 25% of magnesium, 20% iron, and many other nutrients important for good health (Salunkhe et al., 1991). Thus, fruits and vegetables are very important in maintaining our health.

In the past, policy in China emphasized increasing the production of farm products. While it is important to improve the yield of the crops, it is equally important to maintain the quality of these crops after harvest. In particular, fruits and vegetables are highly perishable. Quality declines rapidly after harvest, especially if proper postharvest handling procedures are not followed. Loss of the harvested crops due to spoilage before consumption is practically the same as if these crops were never produced at all. The postharvest losses are particularly painful because these losses occur after farmers have already invested the labor, the time, and the capital. The energy and money spent to grow the food materials would be wasted if these food materials spoil before being consumed. We should realize that the reduction in postharvest losses is essentially equivalent to increasing the production. This increase in production does not have to involve the use of additional farmland. Rather, it occurs as a results of improved postharvest handling systems and refined storage techniques.

Postharvest losses of fruits and vegetables is a serious problem not only in China, but also in every country in the world. Global estimates of postharvest losses in developing countries have been reported by the U. S. National Research Council (Anon 1978), as summarized in Table 4.31.1.

In humid and tropical regions, the combination of high temperature and humidity hastens the microbial growth and physiological deterioration of the harvested crops. Enormous postharvest losses can occur in these circumstances if the food products are not properly handled. Rough handling, lack of cooling and low temperature facilities, lack of sorting to remove damaged and diseased produce before storage, and the use of inadequate packaging materials all contribute to the problem. Even in developed countries,postharvest losses remain to be a serious problem.

Table 4.31.1. Postharvest losses of various crops in developing countries, as estimated by the U. S. National Research Council.

Crop	Postharvest losses of total crops (%)	Crop	Postharvst losses of total crops (%)
Apples	14	Lettuce	62
Avocados	43	Papayas	40-100
Banana	20-80	Plantain	35-100
Cabbage	37	Potatoes	5-40
Carrots	44	Onions	16-35
Cassava	10-25	Raisins	20-95
Cauliflower	49	Stone fruits	28
Citrus	20-95	Sweet potatoes	35-95
Grapes	27	Tomatoes	5-50
		Yams	10-60

It has been estimated that postharvest losses of fruits and vegetables amount to approximately 5 billion dollars per year in the U. S. These losses vary with seasons, locations, market systems, and specific commodities. They can occur at any point during the marketing channel including wholesale, retail, and consumer levels.

Factors Affecting Deterioration of Fruits and Vegetables

A number of factors can cause produce deterioration including (1) metabolic changes associated with respiration, ripening, and aging which affect chemical composition, texture, and color; (2) Moisture loss with resultant wilting or shriveling; (3) bruising and mechanical injury; (4) parasitic diseases; (5) physiological disorders; (6) freezing and chilling injury; (7) flavor changes; and (8) growth changes such as sprouting of potatoes. Rough handling during harvesting or transportation will increase bruising and mechanical damage. The physical injuries predispose produce to decay, increase water loss, hasten respiratory rate, and ethylene production leading to rapid deterioration.

Methods for MaintainingPostharvest Quality of Fruits and Vegetables

Several postharvest techniques are effective in maintaining good quality of fruits and vegetables as described below.

(1) *Refrigeration*: Refrigeration or low temperature storage is the most effective method for retarding spoilage of fresh produce (Hardenburg et al., 1986). Deterioration of fruits and vegetables by their own metablic changes or by microorganisms is reduced as temperature is lowered, so that the marketable life span may be extended. Cold temperature retards: (A) aging due to ripening, softening, and textural and color changes; (B) undesirable metabolic changes and respiratory heat production; (C) moisture loss and resulting in wilting; (D) spoilage due to invasion by bacteria, fungi, and yeasts; and (E) undesirable growth. Rapid removal of field heat and prompt cooling of the harvested fruits and vegetables are essential for quality maintenance. There are three principal methods of precooling: hydrocooling, forced-air cooling, and vacuum cooling. Hydrocooling is the cooling of fruits and vegetables using water to transfer heat from the product to refrigerant coils or to a bed of ice. The heat is transferred by forced convection in a boundary layer film of water at the product surface. Because of its simplicity, economy, and effectiveness, hydrocooling is a popular precooling method. In forced-air cooling, air is forced through the vented container, thus passing by and taking heat from the fruit. Vacuum cooling is the practice of precooling perishable fresh produce by the rapid evaporation of water from the product. Water is vaporized in a flash chamber under low pressure. A regulated flow of makeup water is fed into the flash chamber to replace the water drawn off as vapor. This cooling method is the most effective in cooling vegetables with a high ratio of surface area to volume. Cooling by the forced-air method is usually 4 to 10 times faster than regular room cooling. But hydrocooling or vacuum cooling is usually 2 to 20 times faster than forced-air cooling. It is important to continue to keep the produce at a low temperature after the precooling in order to maintain the best quality. This practice is called "cold chain". However, there are two potential problems which may be associated with refrigeration: chilling injury and freezing injury.

Chilling injury - Certain fruits and vegetables, especially those of subtropical or tropical origin, do not tolerate low storage temperature. Most fruits and vegetables produced in southern China including provinces of Fujian

Guangdong, Guangxi, and Taiwan, are susceptible to chilling injury. When stored at or below a threshold chilling temperature (usually 10°C), tissues of these fruits or vegetables weaken because they are unable to carry on normal metabolic processes. Various physiological and biochemical alterations occur in chilling-sensitive crops in response to chilling stress (Wang, 1990). These alterations lead to the development of a variety of chilling injury symptoms. Products that are chilled often look sound when removed from low temperatures. However, symptoms of chilling such as pitting or other skin blemishes, internal discoloration, or failure to ripen, become evident in a few days at warmer temperatures. Fruits and vegetables that have been chilled are also particularly susceptible to decay. Treatments which have been shown to alleviate chilling injury include temperature preconditioning, intermittent warming, controlled atmosphere storage, pretreatment with calcium or ethylene, hypobaric storage, waxing, film packaging, application with growth regulators or other chemicals, and genetic manipulation.

Freezing injury - Freezing injury occurs when ice crystals form in the tissues of the commodity. Tissues damaged by freezing injury generally appear water soaked. The temperatures usually recommended for storing fresh commodities that are not susceptible to chilling injury are slightly above the freezing point. Different commodities vary widely in their susceptibility to freezing injury. Some may be frozen and thawed a number of times with little or no injury, while others are permanently injured by even slight freezing.

(2) *Controlled atmosphere storage*: Controlled atmosphere (CA) storage is a system for holding produce in an atmosphere that differs substantially from normal air in respect to the proportion of nitrogen, oxygen, or carbon dioxide. For some products, reducing the oxygen level in storage and/or increasing the carbon dioxide level as a supplement to refrigeration can provide extended storage life. Careful control of the concentration of oxygen and carbon dioxide level is essential to obtain the greatest benefit and to avoid the physiological injury. Controlled atmosphere storage was first started in 1920s. It was also called gas storage. These terms are used synonymously to describe the gas tight storage with regulated CO_2 and O_2 concentrations. It is a standard practice to place fruits intended for late season marketing in CA storage. Tremendous progress has been made during the past 60 years on controlled atmosphere storage technology. CA storage has withstood the test of time to prove that it is effective for maintaining quality and extending storage life of a number of horticultural crops. Each type of fruit or vegetable has its own specific requirement and tolerance for atmosphere modification. The optimum CA condition for each commodity is continually being revised and improved. Currently, CA storage

is used successfully for the long term storage of apples and pears. With the increased production of apples and pears in China, long term storage is becoming more and more important. Therefore, CA storage can be very useful in China. The potential benefits are retardation of senescence, suppression of biochemical and physiological changes, reduction of physiological disorders, and inhibition of the growth of certain decay-causing pathogens.

(3) *Hypobaric storage*: hypobaric storage, or storage at reduced atmospheric pressure, is another supplement to refrigeration to maintain quality of fruits and vegetables. In hypobaric storage, the commodity is placed in a vacuum tight and refrigerated chamber and evacuated by a vacuum pump to desired subatmospheric pressure. The absolute pressure may range from near 10 mm-Hg to 80 mm-Hg (versus 760 mm-Hg at normal atmospheric pressure) depending upon the commodity and temperature. When the desired low pressure is obtained, fresh air is admitted to the chamber through a pressure regulator and then through a humidifier to bring the relative humidity to near saturation. The vacuum pump operates continuously, therefore the commodity is constantly ventilated with fresh humid air at reduced pressure at 1 to 4 air changes per hour. This facilitates the removal of carbon dioxide, ethylene, and waste volatile by-products of metabolism. The oxygen partial pressure is directly proportional to the absolute air pressure. Thus, the oxygen content of a hypobaric chamber at 0.1 atmosphere is equivalent to 2.1 % oxygen by volume. Studies have shown that hypobaric storage can extend storage life of many fresh produce. Further research is warranted to determine if this technique is feasible to be used in China for commercial application.

(4) *Packaging*: Proper packaging provides protection against bruising, moisture loss, and the spread of microorganisms. Packaging materials also may serve as carriers for chemicals to control spoilage. Polyethylene sleeves fitted over banana bunches protect the fruit from scars and wounds. Film liners for apple and pear boxes retard moisture loss and shriveling of the fruit during storage. In some cases, the micro-environment created by the sealed package may have a beneficial modified atmosphere effect on the produce. However, many factors, such as types and permeability of the film, temperature, respiration rate, amount of the produce, and time in the package, need to be considered when attempting to use film packages to develop desirable modified atmospheres through respiration of the product. Packaging method is applicable in most of the Chinese marketing systems. With little efforts, Chinese consumers can greatly benefit from adequate use of this technique.

(5) *Waxing and surface coating*: The application of wax or wax emulsion coatings to certain perishable products has been practiced for many years. With some products, an improved glossy appearance is the main advantage. Waxed apples take on a bright lustrous finish which is retained during extended storage. Often, the thickness of wax coatings is critical. Too thin a coating may give little if any protection against moisture loss; too heavy a coating may increase decay and breakdown. It should be emphasized that waxing does not improve the quality of any inferior horticultural product but that it can be a beneficial adjunct to good handling. Wax formulations by themselves do not control decay; but combined with fungicides in a single application, they can be valuable in retarding deterioration. Wax coatings are also sometimes used as carriers for sprout inhibitors, other growth regulators, and preservatives. Application methods include spraying with or dipping into water emulsions, foaming, fogging with solutions made with volatile solvents, dripping emulsions onto rotating brushes, and using brushes that are first brushed against solid cake waxes. After application, the products usually are dried with heated air and polished with rotating brushes. Uniformity of coverage is important for success. The type of applicator used is a very important factor in uniform application of a wax coating. With proper facilities and skill, wax and surface coatings can be applied on many of the horticultural products in China to improve their quality.

(6) *Irradiation*: The use of gamma radiation has been studied for controlling decay, disinfestation, and extending the storage and shelf life of fresh fruits and vegetables. Dosages of 1.5 to 2 kilogray have been effective in controlling decay in several products. However, dosages of 2 kilogray and above can cause discoloration, pitting, softening, abnormal ripening, or flavor loss. Radiation of low doses of 150 to 350 gray (Gy) is effective for ridding papaya, mangos, bananas, pineapples, and grapefruits of insect infestations. Low doses of 80 to 100 Gy will inhibit sprouting of potatoes and onions. Higher doses may increase the susceptibility of potatoes to decay, internal black spot, and after cooking darkening and may cause discoloration of the internal growing points of onions. Commercial application of gamma radiation is limited due to the cost and size of equipment needed for the treatment and to uncertainty about the acceptability of irradiated foods to the consumer. Nevertheless, gamma radiation is an alternative treatment for disinfestation of some fruits and is a technique worth pursuing in China.

(7) *Biotechnology*: Recent developments in biotechnology have increased the possibility for improving postharvest fruit quality through genetic engineering. By manipulating genes which affect ethylene production,

oftening, ripening, or disease and stress resistance, fruits with desirable qualities can be obtained. The U. S. Department of Agriculture and the Food and Drug Administration recently approved the marketing of a genetically altered tomato, 'Flavr Savr' from Calgene, Inc. The tomato contains a marker gene and an antisense polygalacturonase (PG) gene that blocks production of the PG enzyme so as to slow fruit softening. Antisense genes expressed in transgenic plants can be highly effective in suppressing the specific expression of ripening-related genes. Antisense techniques have also been used to identify genes encoding enzymes for carotenoid biosynthesis (phytoene synthase) and ethylene biosynthesis (ACC oxidase). The genetic engineering techniques can also produce crops that are frost-resistant or disease-resistant. These biotechnological methods will have a great impact on the market quality of many fresh produce and is a promising avenue that Chinese scientists need to pursue.

Suggestions for Improving Postharvest Handling Systems in China

Even though there are different marketing systems, culture values, and customs in different countries, the basic principles for maintaining good postharvest quality of fruits and vegetables are the same. Fruits and vegetables are composed of living cells. Therefore, the key to quality maintenance for these commodities is to preserve their normal life processes while retarding deteriorating changes. Factors that need to be considered for reducing postharvest losses and improving keeping quality are: (1) the initial quality of the commodity and harvesting at optimum maturity; (2) rapid removal of field heat and avoiding delay between harvest and storage; (3) providing proper refrigeration throughout marketing period and maintaining the cold chain during transportation, handling, storage, and distribution; (4) providing high relative humidity to minimize moisture loss; (5) careful handling to avoid mechanical injury; (6) using controlled or modified atmospheres during storage and transit; (7) using chemical or heat treatment for the control of decay or physiological disorders; (8) providing protective containers and packaging; (9) Enforcing good sanitation procedures.

The following suggestions for improvement of the postharvest handling system in China may help facilitate achieving the above conditions:
(1) Building precooling facilities, packing houses, and refrigerated storage rooms at or near the growing area of fruits and vegetables. This will reduce the time between harvest and storage, and will be able to cool down the commodities quickly.
(2) Maintaining "cold chain" throughout the marketing period. This includes the use of refrigerated trucks for transportation and refrigerated holding

facilities at the distributing areas.

(3) Establishing a germplasm center to collect, coordinate, and evaluate various cultivars of different fruits and vegetables from domestic and foreign sources. This will provide farmers and growers with the best varieties of fruits and vegetables for planting.

(4) Providing education to farmers and growers and sponsoring shortcourses for solving specific problems. This will minimize postharvest losses due to lack of proper knowledge and skill.

(5) Emphasizing the role of county agents and extension specialists. This will provide farmers and growers with a source of information and a place to obtain advice.

FOOD PROCESSING AND PRESERVATION

Food either is prepared fresh or processed prior to consumption. In the United States, 90% of all the food is processed. The food industry is a multi billion-dollar industry that employs one out of every seven working persons. In order to meet the demand for food, the food processing industry is expanding tremendously.

A Brief History

Dried foods

Dried foods are considered one of the oldest forms of food preservation. In Biblical times, foods were sun dried and stored for later uses. Today, the industry still sun dries foods; however, most of the commercially dried foods are dried by using controlled temperature and relative humidity (commonly called dehydration). Dried or dehydrated foods constitute a major portion of man's diet, including the vast amounts of flour, sugars, salt, spice, cereals, fruits as well as many prepared products, like potatoes and other root crops, and other vegetables. Spray drying and freeze drying have been developed for heat sensitive products. Most dried foods have a shelf life in excess of two years.

Canned Foods

The canning of food started in the mid-1800's and has progressed to the point today where each consumer uses some 500 cans or packages of food per year. Canning food is a process whereby food is heated aseptically to destroy organisms of public health significance. Food is first filled in the container, sealed, and finally cooked to destroy harmful microorganisms. The processed canned products have a shelf life of approximately 30 months for most items (Gould, 1996).

Frozen foods
First commercially processed around 1930 frozen foods are relatively new. Food is prepared as for canning but frozen in the package or individually quick freezing (IQF) of each piece of food and then packaged and held at -18°C until ready for consumption. Much of the frozen food today is accomplished as a secondary procedure, that is, the processor manufactures and packages complete dinner entrees using individual frozen items as portions of the entree. In either case, the product has a shelf life of approximately one year.

Minimally Processed Foods
Minimally processed foods are refrigerated or shelf-stable. Thermal pasteurization or sterilization is performed to provide the needed microbial inactivation. High-temperature-short-time (HTST) process was developed in the 1960's and ultra-high-temperature process was developed in the 1970's through 1980's. These relatively new procedures result in food quality that is significantly better than retort-canned food. Ionizing irradiation, high hydrostatic pressure (HHP), pulsed electric field (PEF), ultrasound, ozonation and bacteriocins are new technologies under research and development (Gould 1995). All these new developments point to a trend in food preservation that eliminates thermal degradation.

Reasons for Processing Foods

The following are some of the many reasons why food processing is important:

(1) Prevent spoilage and extend the period during which a food remains wholesome by preservation techniques which inhibit microbiological or biochemical changes and thus allow time for distribution and home storage.
(2) Increase variety in the diet by providing a range of attractive flavors, colors, aromas and textures in foods. These qualities are referred to as *eating quality, sensory quality* or *organoleptic quality.*
(3) Preserve the nutritional quality of raw materials and provide the nutrients required.
(4) Generate income for the manufacturing company.
(5) Eliminate waste.
(6) Make seasonal fruits and vegetables available year round.
(7) Convenience for the consumer.
(8) Develop new products, that is, cucumbers to pickles, tomatoes to catsup, cabbage to kraut, oats to Cherrios, corn and potatoes to chips.
(9) Safely put foods away for emergencies.

(10) Increase the value of the product.

Recent outbreaks of pathogenic microorganisms in the US and Japan, such as Salmonella and *E. coli O157:H7*, heightened the general public's awareness of the safety of their food supply. One of the major efforts in food processing is to reduce and/or eliminate microbial growth in food. Food processing, to a large extend, is food preservation.

In industrialized countries, the market for processed foods is changing. Consumers no longer require a shelf life of several months at ambient temperature for the majority of their foods. Changes in family lifestyle, and increased ownership of freezers and microwave ovens, are reflected in demands for foods that are convenient to prepare, are suitable for frozen storage or have a moderate shelf life at ambient temperature. There is also an increasing demand by some consumers for foods that have undergone fewer changes during processing that closely resembles the original material and have a 'healthy' image. Each of these is an important influence on changes that are taking place in the food processing industry (Fellows, 1996).

Location and Operation of a Commercial Food Processing Operation

The following are some of the basic factors that must be considered in the establishment of a food processing business (Gould, 1996):

(1) Adequate Water Supply - The water must be potable and low in mineral salts (calcium, magnesium, sulfur and iron). Upwards of 10,000 liters of water are required to process a ton of fruits or vegetables.
(2) Sewage Disposal Facility - Wastes from fruit and vegetable processing facilities are high in organic matter, consequently the BOD is high and this must be lowered before discharging even to municipal systems.
(3). Available Raw Materials - Primary food processing plants are generally located close to where fruit or vegetable crops are produced. In order to attract growers to want produce a crop that meets specific quality standards, the produce must yield a sufficient amount.
(4) Adequate Labor Supply - This is necessary for the temporary processing operations and often satisfied with seasonal labor that moves from one area to another and willingly provide the necessary labor if adequate work and pay are available.
(5) Sufficient Capital - This can be a problem, but all big companies were small once and growth is inevitable if a firm is operating efficiently and

effectively. General rule of thumb has been that $25,000 is required per 1,000 cases of a given fruit or vegetable per day. This may seem high, but the fruit and vegetable industry is almost always a seasonal industry.

(6) Adequate Markets - Today, the food firm looks beyond the borders of his own local area and some even think globally. Marketing requires transportation and this may be a major problem, as most foods in the U.S. today move by truck.

(7) Management and Technical Support - A food process firm must have the following personnel on board: General Manager who is knowledgeable in working with money, people and the given commodity; Superintendent of Production, with ability to work with production problems, that is, manpower, materials, methods and measurement practices; Procurement personnel to deal with agricultural production concerns for each given crop; and Quality Assurance personnel.

Magnitude of US Food Processing Industry

According to Bureau of Standard, food processing industry is divided into six categories:

2032   Canned Specialty Products, such as Baby Foods, Nationality Specialty Foods, Health Foods, and Soups, except Seafood

2033   Canned Fruits and Vegetables, Fruit and Vegetable Juices, Catsup and Similar Tomato Sauces, Preserves, Jams and Jellies

2034   Dehydrated Fruits, Vegetables and Soups (from Dehydrated Ingredients)

2035   Pickles, Sauces and Salad Dressings

2037   Frozen Fruits and Vegetables, including Fruit Juices

2038   Frozen Specialties, that is, Frozen Dinners and Frozen Pizza

There are some 2,049 firms utilizing some 211,900 employees, with the value of shipments more than $45,191,800,000. Approximately 100% of value is added to the raw material by the processing operations. Further statistical details are found in Table 4.31.2.

Major Quality Loss Reactions

Whilst most preservation techniques are primarily based on prevention of the growth of food spoilage and poisoning microorganisms, the preservation of the other attributes of foods is of additional concern. With few exceptions, all foods, following harvest, slaughter or manufacture, lose quality at some rate or other in a manner that is very dependent on food type and composition, formulation, storage conditions and so on (Gould, 1989). Quality loss may be accelerated, or minimized, at any of the stages indicated, and the total

Table 4.31.2. 1992 Census Data for Specific Food Processing Categories

SIC No	Total Number of Establish ments	Total Number All Employees	Production Worker Hours(M	Value Added by Manufacture (M$)	Value of Shipments M$)
2032	220	20,800	35.5	3,224.1	6,300.3
2033	684	63,900	109.8	6,970.2	14,876.4
2034	150	11,400	18.7	1,133.0	2,334.5
2035	375	21,100	32.2	3,640.5	6,244.3
2037	255	48,000	79.0	2,935.9	7,598.0
2038	362	46,700	73.3	7,838.3	4,100.1
TOTALS	2,046	211,900	348.5	22,004.4	45,191.8

preservation of a food is therefore often highly multi component in that it seldom relies on one factor alone. Nevertheless, the principal quality loss reactions, which are therefore the principal targets for preservation, are well known, and they are relatively few. They are listed in Table 4.31.3. Quality degradations are essentially microbiological, chemical, enzymatic and physical.

Table 4.31.3. Major Quality Loss Reactions

Microbiological	Enzymatic	Chemical	Physical
Growth or presence of toxinogenic microorganisms	Hydrolytic reactions catalyzed by lipases proteases etc.	Oxidative rancidity	Mass transfer movement of low MW components
Growth or presence of infective microorganisms	Lipoxygenase	Oxidative and reductive discoloration	Loss of crisp textures
Growth of spoilage microorganisms	Enzymatic browning	Non-enzymic browning Nutrient losses	Loss of flavors Freeze losses induced damage

When preservation fails, the consequences range broadly (Table 4.31.4) from extreme hazards, e.g., if some of the toxinogenic microorganisms are not controlled, to relatively trivial quality loss, e.g., resulting from color loss.

Microbiological Quality Loss Reactions

The major targets for the antimicrobial preservation of foods are listed in Table 4.31.5. They include some that are of great importance including, as mentioned above, the prevention of food poisoning, and also the prevention of many types of food spoilage. Some of these at first may seem to be of less concern in the sense that, if not controlled, they are not life-threatening. However, many of the spoilage reactions listed may nevertheless seriously

Table 4.31.4. Consequences of Loss of Quality

Nature of quality loss	Consequence
Presence of toxins	Hazard to the consumer
Presence of pathogenic microorganisms Microbial spoilage	Loss of food
Potential microbial hazard	Inability to distribute and market product
Unacceptable rate of oxidation	
Development of rancidity	Limitation of shelf life
Unacceptable change in texture	
Color change	Increased packaging and distribution costs
Flavor loss	
Texture change	
Poor keepability, color, flavor, texture	Low quality of marketed food

reduce the shelf life of foods. Some of the less severe quality loss reactions listed maybe insufficiently serious to lead to major losses of foods but may nevertheless lead to low quality in marketed products with consequent consumer dissatisfaction, loss of sales etc., and thus still be of economic importance.

Table 4.31.5. Major Targets for Antimicrobial Food Preservation Techniques

Major targets	Examples
Food Poisoning Microorganisms Presence or multiplication of infective microorganisms	Salmonella, Listeria Campylobacter
Multiplication of toxinogenic microorganisms	Staphylococcus aureus Clostridium botulisum
Food Spoilage Microorganisms Generation of minor metabolic products	Thiols, esters, amines, peroxides generating off-odours, off-flavors, discoloration etc.
Generation of major metabolic products	Lactic acid, acetic acid, arbcon dioxide, hydrogen,causing sounng, blowing etc.
Secretion of enzymes	Lipases, phospholipases, proteases, amylases, cellulases, polygalacturonases, causing flavor and texture changes
Presence of biomass	Visible presence of microorganisms as e.g. slime, haze, mold colonies etc

Factors Affecting Microbial Growth

Since preservation is based first on the delay or prevention of microbial growth, it must operate through those factors that most effectively influence the growth and survival of microorganisms. These factors are listed in Table 4.31.6. Intrinsic factors include those chemical and physical factors that are within the food, and with which a contaminating microorganism is therefore inextricably in contact. Processing factors are deliberately applied to foods in order to achieve improved preservation. Extrinsic factors include those factors that influence microorganisms in foods, but which are applied from outside the food and act during storage. Implicit factors include those factors that are related to the nature of the microorganisms themselves, and to the interactions between them and with the environment with which they are in contact during growth (Gould, 1989) .

Table 4.31.6. Factors Affecting the Microbial Ecology of Foods.

Chemical	Physical Processing	Intrinsic factors	Extrinsic factors	Implicit factors
Nutrients	ERH/a_w	Changes in food composition	ERH/during storage	Microbial growth rates
pH and buffering capacity	Ice and Freeze concentration	Changes in microbial types	Temperature during storage	Synergistic effects
Oxidation reduction potential	Colloid changes	Changes in microbial numbers	Oxygen tension	Antagonistic effects
Antimicrobial substances		Microstructure		

Major Preservation Technologies

The major food preservation techniques that are employed are, therefore, all based on this limited set of factors, so the range of techniques is necessarily limited also. It is summarized in Table 4.31.7 in such a way as to highlight the fact that most of the techniques act through the slowing down or, in some instances, the complete inhibition of microbial growth (Gould, 1989). Few act by direct inactivation of the target bacteria, yeasts or molds. Then, in addition to the main inhibiting and inactivating techniques, there are the procedures that restrict the access of microorganisms to the food product. Some of these more radical approaches are chemically-based, some biological and some physical. They include, for example, a continual widening of the combination procedures that can be effectively and safely used; new applications of modified atmosphere packaging; use of naturally occurring antimicrobials that are animal-derived (e.g. lysozyme, lactoperoxidase, lactoferrin), plant-derived (e.g. herb, spice and other plant extracts) and microorganism-derived (e.g. bacteriocins); new and improved means for the accurate delivery of heat to foods (e.g. by microwaves, by ohmic heating and by induction heating) so as to achieve the minimal processes necessary to ensure stability and safety; the use of high hydrostatic pressures to inactivate microorganisms in foods without the need for substantial heating, and with consequent minimal damage to product quality; the use of high voltage pulsed electric fields for similar purposes; the direct and synergistic application of ultrasonic radiation to pasteurize and sterilize foods with the minimal application of heat; innovative food surface decontamination procedures aimed at greatly improving the safety of some foods of animal origin; radically new approaches to aseptic processing (Gould, 1995).

Table 4.31.7. Major Food Preservation Technologies

Objective	Factor	Mode of achievement
Slowing down or inhibition of microbial growth	Lowered temperature	Chill-storage Freezing and frozen-storage
	Reduced water activity/raised osmolality	Drying and freeze-drying Curing and salting Conserving with added sugars
	Nutrient restriction	Compartmentalization of aqueous phases in water-in-oil emulsions
	Decreased oxygen	Vacuum- and nitrogen packaging
	Increased carbon Carbon dioxide	Carbon dioxide enriched'controlled atmosphere' storage or 'modified atmosphere' packaging
	Acidification	Addition of acids Lactic or acetic fermentation
	Alcoholic fermentation	Brewing, vivification Fortification
	Use of preservatives	Addition of preservatives inorganic (e.g. sulphite,nitrite) organic (e.g. propionate,sorbate, benzoate, parabens antibiotic (e.g. nisin, pimaricin)
Inactivation of microorganisms	Heating (HTST, UHT, Retort ohmic, induction Ionizing irradiation	Pasteurization Sterilization Radurization Radicidation Radappertization
	High hydrostatic pressure	Pasteurization

Continuation of Table 4.31.7

	Pulsed electric field	Pasteurization
	Ultrasonication	Pasteurization
	Ozonation	Oxidation of BOD
	Pulsed light	Surface decontamination
	Addition of enzymes (e.g. lysozyme	Deactivate microorg
Restriction of access of microorganisms to foods	Decontamination	Treatment of ingredients(e.g. with ethylene oxide) Treatment of packaging materials (e.g. with hydrogen peroxide and/or heat, irradiation)
	Aseptic processing	Aseptic thermal processing and packaging without recontamination
	HACCP and GMP	Hazard Analysis and Critical Control Point Good Manufact Practice

New and Emerging Food Preservation Technologies

Hurdle technology

The microbial stability and safety of most traditional and novel foods are based on a combination of several factors (hurdles), which should not be overcome by the microorganisms present. The hurdle effect is of fundamental importance for the preservation of foods, since the hurdles in a stable product control microbial spoilage, food-poisoning and, in some instances, the desired fermentation process(Leistner, 1995).

A typical food may subject to six hurdles, high temperature during processing, low temperature during storage, low water activity, acidity (pH), redox potential of the product, and presence of preservatives. The microorganisms present cannot overcome these hurdles, and thus the food is microbiologically stable and safe.

Natural antimicrobials

Antimicrobial agents exist in nature from origins of bacteria, animals, and plants. Antimicrobials from bacteria are called bacteriocins. Nisin is generally recognized as safe (GRAS) and is effective to Gram positive bacteria. Nisin has been successfully used for more than 40 years in processed cheese and cheese spreads (Hill, 1995) to prevent the out growth of Clostridial spores, such as *C. botulinum.* Antimicrobials exist in animal bodies in the forms of antibodies, metal sequestrants, lysozyme, peroxidases and phenoloxidases. Many spices and herbs and extracts possess antimicrobial activity, almost invariably to the essential oil fraction (Deans and Ritchie, 1987). Recent research, however, is more focused on the antioxidants from plant origin.

Food Irradiation

After five decades of research and development, proof of wholesomeness, political debates and public scrutiny, food irradiation is established as a safe and effective method of food processing and preservation (Loaharanu, 1995). Similar to other food preservation methods, food irradiation offers advantages and disadvantages relative to the types of food to be treated. It is increasingly recognized as an alternative method to fumigation of food to control insect infestation and microbial contamination, as well as facilitating trade in fresh and dried foods. Its role as a method to ensure hygienic quality of the more solid foods, e.g. meat, poultry and seafood, is increasingly advocated and used in some countries. Its introduction into commercial practice has been somewhat slow because many governments required comprehensive data to support the wholesomeness of irradiated food, lengthy regulatory procedures in granting approval for the food or the process, often with the involvement of the public. Irradiation dosage of 10 kGy is generally considered safe and effective.

Fast heating methods

New heating methods include microwave, ohmic, and conduction heating. Microwave heating is based on the dielectrics of water molecules and the loss due to electrolytic carriers in the foods. For all practical purposes industrial applications are carried out at 915 MHZ in the USA, 896 MHZ in the UK and 2450 MHZ worldwide. All commercial and consumer microwave ovens operate at 2450 MHZ (or 2375MHz in the former Soviet block). Microwave heating is fast and convenient. Uniformity of heating depends on the uniformity of water and ions in the foods.

Ohmic heating is also termed as Joules heating, as heat is produced by an electrical current flowing through the food. Ohmic heating is rapid and suitable for continuous heating of a food stream which shortens the come-up time for pasteurization or sterilization processes. Food flows between two non-corrosive electrodes while direct-current (dc), or alternating current (ac) is applied across the electrodes. A higher frequency ac source would reduce the electrolysis reaction and minimize fouling of the electrodes. While heating a particulate food, Ohmic heating can significantly reduce the temperature lag between the liquid phase and the solid phase. By matching the electric conductivities of the liquid and solid phases, one can realize a faster heating rate in the solid particles. Ohmic heated products are commercially available in Germany and Japan.

Induction heating is similar to ohmic heating, the difference being that no metal electrode contacts the food. The technology works as an ordinary electrical transformer. Food flows through a secondary coil, while an electrical field is induced from the primary coil and the magnetic core. The induced the electric field, in turn, produces the heating effect.

Nonthermal Preservation Technologies

Nonthermal preservation techniques have attracted much of the industrial attention in the past five years. Several processes are being developed and validated. High hydrostatic pressure (HHP) treatment at 800 MPa for five minutes is generally recommended for inactivation of vegetative microorganisms (Knorr, 1995). HHP is a batch process and is well suited for pasteurization of foods in flexible packages. Progress has been made in a semi-continuous HHP system. Commercial applications exist in Japan, United States, and Mexico.

Pulsed electric field (PEF) treatment inactivates microorganisms by permeablization of microbial cell membrane. PEF is found to be nonthermal (Zhang et al, 1994). Major efforts have been put in to demonstrate the PEF process in a pilot plant scale (Qiu et all, 1997). PEF pasteurized orange juice retains the same flavor and taste qualities as its fresh counterpart. PEF process is ideal for continuous pasteurization of liquid beverages which are sensitive to thermal degradation.

Pulsed intensive light (PIL) is a process where a very bright, broad spectrum, light is flashed onto the surface of food or packaging materials. The main bactericidal effect is provided by the UV spectrum. PIL process is applied to surface sterilization of packaging materials and drinking water. Experiments were also carried out to decontaminate surfaces of fruits and meat. Due to the low depth of penetration, a smooth product surface is needed.

An expert panel has declared that ozone, one of the most effective disinfectants but one not previously approved for use in food processing in the U.S., is generally recognized as safe (Graham, 1997). Ozone has a very high oxidation power and inactivates microorganisms. Application of ozone is likely to be used in solid food decontamination and treatment of waste water. Ozone is reactive to any biochemical substance, especially food. Treating food with high content of lipids may lead to off-flavor generation.

References

Anon. 1978. Report of the steering committee for study on p o s t harvest food losses in developing countries. National Research Council, National Science Foundation, Washington, DC, 206 pp.

Deans, S. G. and GRitchie. 1987. Antimicrobial properties of plant essential oils. *Int. J. FoodMicrobiol.*, **5**:165-180.

Fellows, P. J. 1996. Food Processing Technology Principals and Practice. Woodhead Publishing, England.

Gould, G. W. 1989. Mechanisms of Action of Food Preservation Procedures. Elsevier Applied Science, London

Gould, G. W. 1995. New Methods of Food Preservation. Blackie Academic & Professional, London.

Gould, W. A. 1996. Unit Operations for the Food Industries. CTI Publications, Baltimore, MD

Graham, D. M. 1997. Use of ozone for food processing. J. Food Technology 51(6):72-74. Anon. 1978. Report of the steering committee for study on postharvest food losses in developing countries. National Research Council, National Science Foundation, Washington, DC.

Hardenburg, R. E., A. E.Watada, and C. Y. Wang. 1986. The commercial storage of fruits, vegetables, and florist and nursery stocks. USDA Agi. Handbk. 66,

Hill, C. 1995. Bacteiocins: natural antimicrobials from microorganisms. In New Methods of ood Preservation. Ed. G.W. Gould. Blackie Academic & Professional, London.

Knorrl, D. 1995. Hydrostatic pressure treatment of food: microbiology. In New Methods of Food Preservation. Ed. G.W. Gould. Blackie Academic & Professional, London.

Leistner, L. 1995. Principles and applications of hurdle technology. In New Methods of Food Preservation. Ed. G.W. Gould. Blackie Academic & Professional, London.

Loaharanu, P. 1995. Food irradiation: current status and future prospects. In New Methods of Food Preservation. Ed. G.W. Gould. Blackie Academic & Professional, London.

Qiu, X., L.Tuhela, and Q. H.Zhang. 1997. Application of High Voltage

Pulsed Electric Field in Non-Thermal Food Processing. ISH'97 Conference Proceedings Volume 6 pp.373-376. August, 1997. Montreal, Canada.

Salunkhe, D. K., H. R. Bolin, and N. R. Reddy, 1991. Storage, processing, and nutritional quality of fruits and vegetables. CRC Press, Boca Raton, FL.

Wang, C. Y. 1990. Chilling injury of horticultural crops. CRC Press, Boca Raton, FL.

Zhang, Q.H., A. Monsalve-González, G.V. Barbosa-Cánovas and B.G. Swanson, 1994. Inactivation of *E. coli* and *S. cerevisiae* by pulsed electric fields under controlled temperature conditions. *ASAE Transaction,* 37(2):581-587.

CHAPTER 4.32

Postharvest management of agricultural products

Jui - Sen Yang

Institute of Marine Biology
National Taiwan Ocean University

Postharvest losses of agricultural products are high around world. However, in China the losses in quality and quantity of agriculture products are extremely serious since the postharvest techniques, facilities, management and policies are not developed very well. Although many techniques and facilities are available in other countries, but postharvest technologies, facilities and handling systems are highly commodity-specific and location-specific. For example, irradiation, a newly - developed technique, is useful for controlling sprouting in garlic, but not so good for fresh ginger. Controlled atmosphere (CA) is very commonly used in apple storage in USA, but not existent in China at present. Therefore whether a particular technique, facility, management or policy is practical or not depends on the kinds and marketing systems, the economic condition, the population education and the social system. A system approach is extremely necessary in the development of posthatvest management for China during the coming 50 years.

The problems in agricultural postharvest handling

Social system

An old traditional social system still exists in China, although it is changing during the 1990s. Farmers are keeping their traditional ways. Extension system didn't function well and farmers do not have the concept how to reduce waste and improve marketability. The society is still rather conservative and the farmers, transport workers and marketers are not very aggressive in improving their technique and providing good quality products to the market. Even so, there is a serious loss during the postharvest period. In the society not many persons pay attention to the postharvest losses, although agricultural institutes have done considerable amount of research in this area. People seem to enable to accept a decayed apple and don't know what is a good apple or tender rice.

Food quality

Agricultural products may be accepted in a relatively low quality if quantity is insufficient and people's income is low. However, the economy in China is growing fast during the last several years. People's income is going up. Therefore, it is time to reduce postharvest losses and to provide high quality agricultural products for the people.

Farmer's need

An extensive review of conventional and newly-developed methods is necessary to provide the fundamentals for practical new methods or technologies for postharvest handling in China. A national or regional program of applied research will be needed. This is the foundation of all future extension and education programs for postharvest-loss- reduction and quality maintenance of agricultural products.

Most farmers in China use the traditional methods handed down from their forefathers to do pre- and post-harvest manipulations in agriculture. The traditional agricultural methods maybe fit for applying to produce food just for a family but not for the markets in modern society. The old methods need to be modified and improved for reducing the quality and quantity losses. Therefore, to import postharvest as well as preharvest handling for producing commercial agricultural products become an important course in China at present.

Poor package and transportation

Packaging and transportation share significant portion of the postharvest handling of agricultural products. In China agricultural commodities are still agricultural products instead of commercial products, and most of farmers produce agricultural commodities for their family consumption and maybe partially for market. Therefore, many farmers are market retailers as well as agricultural producers. Few farmer have thought about commercializing the agricultural commodities. Packaging is considered as a wasteful process in the small traditional farm operations. Generally, traditional packaging of agricultural commodities use waste materials, such as rice hull and newspaper. The packing materials are economic to the farmers in China, but the rice hulls and newspapers provide little protection from pressure and are rough and unclean in appearance. Contamination and infection by mold or other microorganism are commonly found in the commodities with rice-hull or newspaper package. Weight loss during transportation is another serious problem to consider. With heavy loss, poor appearance and bad quality the commodities require improvement in packaging as the first step of progress. Transportation of agricultural commodities require considerable care for preserving good quality, although the transportation techniques are well established in the world. Mechanical damage in the transportation channels is generally reduced by proper mechanical means and often caused but certainly not prevented by using human shoulders, arms or hands. At present, shoulders, animals, animals pulling carts, humans-pulling two-wheel implements are commonly applied for transporting agricultural commodities in China. However, the losses during transportation are not very tremendous in China since most of agricultural commodities are for local market. Recently some big companies manage large parcels of land to commercially produce agricultural products. The companies have their own technician and facilities for pre- and post-harvest handling. Package and transportation should be not a serious problem in the commercial company. Therefore, the traditional farmers need the consultants to help to improve their packing and transporting systems.

Storage

Old and conventional methods, such as common storage, underground storage, and cave storage, are all used today in China for storing agricultural commodities. They are very useful and economical methods in subtropical and temperate countries. Many citrus fruits are stored underground in southern-China and apples in caves in northern China.

However, modern techniques, such as cold storage, modified atmosphere (MA) storage, controlled atmosphere (CA) storage, vacuum storage, low oxygen CA storage, low ethylene CA storage, chemical (such as wax) treatment and irradiation are not yet commonly used in China. To introduce these new techniques to the farmers or producers in China is an important task.

Policy

The traditional farmers are generally small and use manual labor, in comparison with large farms of the world with modern facilities that are automated and working on a large scale. Therefore, how to use modern facilities and techniques in traditional small farmers becomes a serious challenge in postharvest handling in China. Government policy plays an important role in agricultural management. Should we develop some techniques for the traditional small farms or the farmers need to cooperate to become a large operator using modern facilities and techniques? We need to answer this question soon. The government should have a clear policy to guide the agricultural development.

To develop handling and storage techniques for future China

Facilities and techniques for postharvest handling and storage are changing around the world. Studies are necessary in developing methods or facilities for the farmers in difficult country. The facilities and techniques which were developed for American farmers are not necessary suitable for Chinese farmers. Researchers and policy developers should use the information in published the literature to search and develop the suitable methods. For future development in the 21st century China needs to seriously consider and improve postharvest handling and storage for agricultural products.

Grading and packing

Grading is the first and most important work in agricultural commodities after harvesting. Grading can be done during picking, in the field or in the packing house. Each agricultural commodity needs to be studied and a grading and packing method for its storage or marketing established. Research should institute an optimum grading method for each agricultural commodity and extension experts should introduce the method to farmers. Grading may be done at the time of packing. For example, strawberries are generally graded by the pickers during picking and put into containers. The strawberries in the containers will be transported all the way to retail markets to avoid any additional touching by fingers for regrading or repacking. However, substantial washing, sizing, waxing and other treatments are required to perform in large packing house. The fruits such as apple, avocado, banana, citrus fruits, etc require large packing houses. The large company with modern packing room may be more cost efficient than small ones. In the 21st century large cooperation will replace small farmers. Equipment as well as management of a packing house are important in minimizing postharvest losses. It is necessary to evaluate damages caused by each step of operation in a packing house to identify operation points for improvement.

Grading is a very important operation in large companies for shipping to distant markets or for marketing in more complicated marketing system. However, grade standards may not be important for small farmers who sell their agricultural products directly to consumers in farmer's markets since the buyers check every commodity they buy. Anyway, after the year of 2000 large companies and supermarkets will become the major system of agricultural production and business in China. Grading and packing will become more important. Agricultural products with uniform grade and suitable packing shall be welcome and have good price in the future. Of course, good grading and packing will tremendously reduce losses of agricultural products after harvesting.

Washing and precooling

Washing, brushing and polishing should all be avoided in agricultural products if possible. In traditional farming and marketing washing, brushing or polishing was commonly used for improving the appearance of agricultural commodities. Washing the commodities in pails or tubs drastically increase decay. The contaminated water is drawn into the fruit with obviously unfortunate results if the commodities are submerged and if the commodities are warmer than the washing water. However, washing with non-recovery rinses or with fungicides or hydrocholoride will reduce certain types of decay of agricultural commodities. It is necessary to do research case by case in washing techniques.

Precooling can be done by mechanical refrigeration' hydrocooling or air cooling. Mechanical refrigeration is usually beyond the capability of any small grower who field pack. A simple evaporation cooler can be made for the small growers. A practical approach is to put the products, picked during the day, into a cooler overnight. The next morning it is packed without the field heat.

Hydrocooling may be a common and convenient method of precooling for agricultural products in China at present. The products are either showered with or submerged into cold water. Hydrocooling is very useful for peach, banana, cherry, litchi, longon and cucumber. The field heat of the products is effectively and fast removed. The rate of respiration, ripening and deterioration of the products slow down and, then, the postharvest losses are significantly reduced. However, some products, such as strawberries and grapes, cannot tolerate hydrocooling since the process increases decay. Air cooling is another precooling method, and is widely used in all kinds of agricultural products include rice, corn, wheat, strawberry and grape, but not very good for leaf-vegetables. Air cooling can be performed in room cooling and in forced-air cooling. In room cooling the products are kept well-spaced in a cooling or refrigerated room. In forced-air cooling, cold air is forced through the products to contact with the products to speed up the cooling process. Before the year of 2000 the forced air may be commonly used in China. In the 21 st century the refrigerated-room cooling should become the important precooling method. Forward, the forced-air in refrigerated room should be promoted to be the future powerful precooling process.

Cold storage

Although in China common storage, underground storage and cave storage are commonly used at present, cold storage should become popular very soon, just as in Taiwan, since a desirable temperature range can easily be controlled for each of many agricultural products. However, the lowest safe, the highest freezing and the optimum storage temperatures should be examined for each product. In China many agriculture institutes are studying the cold storage. The studies should be done in case of species varieties, growing conditions, harvesting air temperature and humidity and harvesting season of agricultural products. Otherwise, if the temperature of cold storage is not optimum, the agricultural products will appear freezing injury and a tremendous loss in the products will occur. Refrigeration in cold storage requires electrical power and is generally very costly. First of all, a stable electrical power supply is necessary for maintaining the temperature in a very stable situation. Since the quality of agricultural products in cold storage is much better than those in common storage, underground storage or cave storage, cold storage should be commonly applied in China very soon. Therefore, electrical power should be seriously developed and supplied. Before being commonly applied, refrigerated rooms should be studied and designed very carefully for each agricultural product. The temperature variation, the water loss problem and the room size in refrigerated storage should be seriously considered and designed. The refrigerated rooms for storing fruits are different from these for storing vegetable, grains or meats. The rooms for apples are also different from those for bananas. The refrigerated rooms designed for northern China are different from those designed for southern China. Poorly designed refrigeration rooms will cause a huge loss of agricultural products. Therefore, it is necessary to require a close cooperation between engineers and postharvest physiologists and technologists in designing a refrigerated storages for agricultural products.

Controlled atmosphere and modified atmosphere storage

Controlled atmosphere (CA) storage employs reduced temperatures, with reduced oxygen levels to further suppress metabolic activity in stored products. Commercial application of CA storage are mainly restricted to the storage of apples although successful commercial CA cabbage and pear operations are beginning in USA and Canada. Twenty-two other fruits have been reported to have moderate to excellent potential for benefit from CA storage or transportation. These fruits include apricot, avocado, banana, blackberry, blueberry, sweet cherry, cranberry, fig, grape, grapefruit, kiwifruit, lemon, lime, mango, nectarine, papaya, peach, persimmon, pineapple, plum, raspberry and strawberry. Studies of CA storage of vegetables such as celery indicate 3.0% oxygen and less than 1.0% CO_2 are optimum for that crop. The CA storage for apples is generally required in a condition of 1-2% CO_2 and 1-3% O_2 at -3^0C. However, CA storage is very costly and is applied in bulk storage for fruits or crops. The CA storage is commonly used by large farmers or companies as in U.S.A. In China farmers are generally small producers. They do not need and do not have the capacity of using bulk CA storage for their small amount of products. Anyway, in the 21st century if the economy grows fast in China and the farmer associations or companies are organized, the CA storage should be used extensively for some fruits or crops.

The CA storage can be upgraded with the addition of carbon monoxide; and called modified atmosphere (MA) storage. The CA storage with carbon monoxide is useful for some fruits as grapes since the application of CO is effective in inhibiting postharvest decay. However, carbon monoxide is terribly dangerous to human and therefore difficult to handle. Without very careful performance system carbon monoxide will be not suggested to be used in China. A low-ethylene CA storage is highly recommended to storing climacteric crops, such as apples.

Polyethylene chambers are used to store small quantities of crops. The chambers in refrigerated rooms, common storage rooms or underground caves are successful in northern China. The polyethylene chambers are somewhat inferior to modern CA storage, but much better than common storage.

Modified atmosphere (MA) storage is performed with putting certain concentrations of O_2 and/or CO_2 in a sealed polymeric films. Since crops, especially climacteric fruits as banana, papaya etc, respire fast after harvesting, it is difficult to maintain stable O_2 and CO_2 concentrations and to remove condensed water from sealed bags. This creates un-optimum concentrations of O_2 and/or CO_2 and condensed water in the bags and induces the crops to decay. A desirable film for maintaining a suitable balance among O_2, CO_2, ethylene, N_2, etc. is necessarily developed. Otherwise, MA storage is not recommended to apply at present or in near future.

Food irradiation

In China, agricultural output rose steadily from 1978. Postharvest losses of crops are serious. The losses are due to the absence of adequate storage capacity and transport. Facility. Although research in agricultural and plant sciences in China was following conventional patterns to improve disease resistance and productivity in early 20th century, new techniques were studied and developed for improving agricultural production from late 20th century. The agricultural producers increase. Anyway, individual farmers and their families have been made responsible for the safe storage of their seeds and crops. Using food irradiation to preserve the agricultural products becomes difficult. Food irradiation is not a panacea for eliminating waste of food or agricultural products. However, irradiation was used commercially to sterilize herbs, to prevent potato tubers from sprouting, and to disinfest insects in crops. Other suggested uses of irradiation include retarding the ripening of fruits; preventing spoilage of fruits and vegetables; inhibiting growth of the buds of asparagus or the shoots of onions; overcoming quarantine restrictions for certain fruits; removing *Salmonella* from stored meat; inactivating the pork parasite *Trichinella spiralis*; and completely sterilizing herbs and spices by killing disease bacteria and viruses. China should use food irradiation to eliminate waste of crops before agricultural industries as refrigeration, shipping transportation system, canning, drying, frozen, and other processes are developed. Food irradiation is an energy-saving technique of food processes since in many cases refrigeration or/and freezing are not necessary in irradiated crops. Irradiated garlic potatoes or onions are stored at room temperature, and cool storage is not necessary. The second reason for using food irradiation is that the process has no chemical residue. After irradiation crops are not need to be treated with chemicals and irradiation do not produce any toxic chemical residue. The third reason is that irradiation is easy to treat crops after packing. Therefore, the products are kept very clean in packages after irradiation. At the same time the processed amount of crops is huge during irradiation treatment. From 1998 to 2030 China should use gamma irradiation in the agricultural products for reducing the losses after harvesting. However, irradiation should not be used on all agricultural products. The benefit of food irradiation should be evaluated item by item before the technique is applied. Anyway, the research of food irradiation was done very well in China during 1980-1990, but the process system from farmers to consumers should be built up and the promotion should be more aggressive.

Education and extension

Although the development of agriculture seems to be the responsibility of scientists in research institutes, its success depends on well coordinated programs among research workers, policy makers, funding agencies, managers, administrators, farmers, transport workers, retailers and consumers. Research programs should be coordinated with extension and education programs since the application of new techniques, facilities or systems of harvesting will involve everyone from growers to consumers. Researchers need extension and education workers to transfer the new techniques, facilities and information to the people involved in the pre- and post-harvest systems and to collect the feedback information. The workers in the production system should have a perfect education or extension course to gain new knowledge or information and to know how to run a new technique, facility or system.

The education of everyone involved in growing, picking, handling and marketing of agricultural products is the essential requirement to reduce postharvest losses. Everyone involved in the production system has to understand that most of agricultural products are alive and every effort should be made to plan a healthy, attractive and living product in the hands of the ultimate consumer. Therefore, the human factor becomes the most important thing in the postharvest stage.

References

Bramlage, W.J. 1982. Chilling injury of temperate origin. HortScience 17 *:165-168*

Burditt, A.K., Jr. 1982. Food irradiation as a quarantine treatment of fruits. Food Technol., November 1982, p.51-62

Couey, H. M. 1982. Chilling injury of tropical and subtropical origin. HortScience 17:162-165

Grierson, W., J. Sould and K. Kawade. 1982. Beneficial aspects of physiological stress. Hort. Reviews

Grierson,W. and W.F. Wardowski. 1978. Relative humidity effects on the postharvest life of fruits and vegetables. HortScience 13: 570-574.

Grierson, W. 1987 Postharvest Handling Manual US/AID Product 505-0008, Belize, C.A. From : Chemonics International, 2000 M ST, Washington, D.C. 20036, U.S.A.

Lai, C. L., J.S. Yang and M.S. Liu. 1994 Effects of gamma irradiation on the flavor of shlitake (Lentinus edodes Sing). J. Sci. Food. Agric. 64(1): 19-22

Liu,   F.W. 1970. Storage of banana in polyethylene bage with an ethylene absorbent. HortScience 5: 25-27.

Liu, F.W. 1979. Interaction of daminozide, harvesting date, and ethylene in CA storage on 'McIntosh 'apple quality.J. Amer Soc. Hort. Sci. 104: 599-601

Liu, F.W. 1970. Storage of banana in polythylene bags with an ethylene absorbent. HortScience 5: 25-27

Liu, F. W. 1987 Developing practical methods and facilities for handling fruits in order to maintain quality and reduce losses. In : International Seminar on Postharvest Handling of Tropical and Subtropical Fruit Crops, Food and Fertilizer Technology Center and Taiwan Agricultural Research Institute, Taichung, Taiwan.

Liu, F.W., J.R. Thrk, D. Samelson and D.J. Kenyon. 1986. Low-ethylene CA storage of 'McIntosh' apples in a semi-commercial size room. HortScience 21 : 480-484.

Maxie, E.C., N.F. Sommer and F.G. Mitchell. 1971. Infeasibility of irradiation fresh fruits and vegetables. HortScience 6: 202-204

Mukai, M. K. 1987. Postharvest research in a developing country: A view from Brazil. HortScience 22, 1: 7-9.

Murray, D. R. (Au. & Ed.) 1990 Biology of Food Irradiation. Pp.255, Research Studies Press LTD

Purvis, A.C., K. Kawada and W. Grierson. 1979. Relationship between seasonal resistance to chilling injury and reducing sugar level in grapefruit peel. HortScience 14,3:227-229

Rivero, L.G., W. Grierson and J. Soule. 1979. Resistance of 'Marsh' grapfruit to deformation as affected by picking and handling methods. J. Amer. Soc. Hort. Sci. 104,: 551-554.

Wilson, L. G. 1976 Handling of postharvest tropical fruits. HortScience 11(2): 120-121.

Wu, J.J. and J.S. Yang 1994. Effects of irradiation on the volatile compounds of ginger rhizom (*Zingiber officiale* Roscoe). J.Agric. Food Chem. 42(11): 2574-2577.

Wu,J.J., J.S. Yang and M.S. Liu. 1996. Effects of irradiation on the voltile compounds of garlic (*Allium sativum* L.). J. Sci. Food Arric. 70: 506-508

Yahia, E.M., K.E. Nelson and A.A. Kader. 1983. Postharvest quality and storage life of grapes as influenced by adding carbon monoxide to air or controlled atmospheres. J. Amer. Soc. Hort. Sci. 108:1067-1071.

APPENDIX

AUTHOR INDEX

TITLE INDEX